Digital Services in Crisis, Disaster, and Emergency Situations

Lídia Oliveira
University of Aveiro, Portugal

Federico Tajariol
University Bourgogne Franche-Comté, France

Liliana Baptista Gonçalves
University of Aveiro, Portugal

A volume in the Advances in Human Services and Public Health (AHSPH) Book Series

IGI Global
PUBLISHER of TIMELY KNOWLEDGE

Published in the United States of America by
IGI Global
Information Science Reference (an imprint of IGI Global)
701 E. Chocolate Avenue
Hershey PA, USA 17033
Tel: 717-533-8845
Fax: 717-533-8661
E-mail: cust@igi-global.com
Web site: http://www.igi-global.com

Library of Congress Cataloging-in-Publication Data

Names: Oliveira, Lidia, 1968- editor. | Tajariol, Federico, 1972- editor. | Goncalves, Liliana, 1984- editor.
Title: Digital services in crisis, disaster, and emergency situations / [edited by] Lidia Oliveira, Federico Tajariol, Liliana Goncalves.
Description: Hershey : Information Science Reference, 2020. | Includes bibliographical references and index. | Summary: "This book presents recent studies on crisis, disaster and emergency situations in which digital technologies are considered as a key communication element"-- Provided by publisher.
Identifiers: LCCN 2020028050 (print) | LCCN 2020028051 (ebook) | ISBN 9781799867074 (ebook)
Subjects: LCSH: Communication in crisis management. | Crisis management. | Information technology--Management.
Classification: LCC HD49.3 .H36 2020 (print) | LCC HD49.3 (ebook) | DDC 384.3/3--dc23
LC record available at https://lccn.loc.gov/2020028050
LC ebook record available at https://lccn.loc.gov/2020028051

This book is published in the IGI Global book series Advances in Human Services and Public Health (AHSPH) (ISSN: 2475-6571; eISSN: 2475-658X)

British Cataloguing in Publication Data
A Cataloguing in Publication record for this book is available from the British Library.

For electronic access to this publication, please contact: eresources@igi-global.com.

Advances in Human Services and Public Health (AHSPH) Book Series

Jennifer Martin
RMIT University, Australia

ISSN:2475-6571
EISSN:2475-658X

Mission

The well-being of the general public should be a primary concern for any modern civilization. Ongoing research in the field of human services and public healthcare is necessary to evaluate, manage, and respond to the health and social needs of the global population.

The **Advances in Human Services and Public Health (AHSPH)** book series aims to publish high-quality reference publications focused on the latest methodologies, tools, issues, and strategies for managing the health and social welfare of the public. The AHSPH book series will be especially relevant for healthcare professionals, policy makers, government officials, and students seeking the latest research in this field.

Coverage

- Public Welfare
- Access to Healthcare Services
- Medicare and Medicaid
- Social Work
- Domestic Violence
- Assistance Programs
- Youth Development
- Social Welfare Policy
- Public Funding
- Poverty

IGI Global is currently accepting manuscripts for publication within this series. To submit a proposal for a volume in this series, please contact our Acquisition Editors at Acquisitions@igi-global.com or visit: http://www.igi-global.com/publish/.

The Advances in Human Services and Public Health (AHSPH) Book Series (ISSN 2475-6571) is published by IGI Global, 701 E. Chocolate Avenue, Hershey, PA 17033-1240, USA, www.igi-global.com. This series is composed of titles available for purchase individually; each title is edited to be contextually exclusive from any other title within the series. For pricing and ordering information please visit http://www.igi-global.com/book-series/advances-human-services-public-health/102256. Postmaster: Send all address changes to above address. ©

Titles in this Series

For a list of additional titles in this series, please visit: http://www.igi-global.com/book-series/advances-human-services-public-health/102256

Epidemiological Research Applications for Public Health Measurement and Intervention
Simon George Taukeni (University of Namibia, Namibia)
Medical Information Science Reference • © 2021 • 315pp • H/C (ISBN: 9781799844143) • US $295.00

Global Issues and Innovative Solutions in Healthcare, Culture, and the Environment
Mika Merviö (Kibi International University, Okayama, Japan)
Information Science Reference • © 2020 • 366pp • H/C (ISBN: 9781799835769) • US $195.00

Handbook of Research on Leadership and Advocacy for Children and Families in Rural Poverty
H. Carol Greene (East Carolina University, USA) Bryan S. Zugelder (James Madison University, USA) and Jane C. Manner (East Carolina University, USA)
Information Science Reference • © 2020 • 525pp • H/C (ISBN: 9781799827870) • US $265.00

Emerging Trends in Indigenous Language Media, Communication, Gender, and Health
Kehinde Opeyemi Oyesomi (Covenant University, Nigeria) and Abiodun Salawu (North-West University, South Africa)
Medical Information Science Reference • © 2020 • 320pp • H/C (ISBN: 9781799820918) • US $285.00

Impact of Textile Dyes on Public Health and the Environment
Khursheed Ahmad Wani (Government Degree College, Bijbehara, India) Nirmala Kumari Jangid (Banasthali Vidyapith, India) and Ajmal Rashid Bhat (Government Degree College, Bijbehara, India)
Information Science Reference • © 2020 • 335pp • H/C (ISBN: 9781799803119) • US $245.00

Handbook of Research on Health Systems and Organizations for an Aging Society
César Fonseca (Universidade de Évora, Portugal) Manuel José Lopes (Universidade de Évora, Portugal) David Mendes (Universidade de Évora, Portugal) Felismina Mendes (Universidade de Évora, Portugal) and José García-Alonso (Universidad de Extremadura, Spain)
Medical Information Science Reference • © 2020 • 323pp • H/C (ISBN: 9781522598183) • US $365.00

Cases on Cross-Cultural Counseling Strategies
Bonnie C. King (Midwestern State University, USA) and Tiffany A. Stewart (Midwestern State University, USA)

IGI Global
PUBLISHER of TIMELY KNOWLEDGE

701 East Chocolate Avenue, Hershey, PA 17033, USA
Tel: 717-533-8845 x100 • Fax: 717-533-8661
E-Mail: cust@igi-global.com • www.igi-global.com

Editorial Advisory Board

List of Reviewers

Daniel Ikesinachi Nwogwugwu, *Bowen University, Nigeria*
Pablo Parra, *Complutense University of Madrid, Spain*
Filipa Rodrigues Pereira, *Polytechnic Institute of Viseu, Portugal*
Alcina Prata, *Polytechnic Institute of Setúbal, Portugal*
Paola Primo, *Federal University of Espírito Santo, Brazil*
Fernanda Queiroz, *Federal University of Rio Grande do Norte, Brazil*
Xabier Rolan, *University of Vigo, Spain*
Luísa Sales, *University of Coimbra, Portugal*
Arnaldo Santos, *University of Aveiro, Portugal*
Carlos Santos, *University of Aveiro, Portugal*
Rita Santos, *University of Aveiro, Portugal*
Antonin Segault, *University Paris Nanterre, France*
Lim Choun Siam, *University of Kebangsaan, Malaysia*
Telmo Silva, *University of Aveiro, Portugal*
Marica Spalletta, *Link Campus University, Italy*
Paulo Nuno Vicente, *NOVA University of Lisbon, Portugal*
Yan Wang, *Delft University of Technology, The Netherlands*
Cláudio Xavier, *State University of Bahia (UNEB), Brazil*

Table of Contents

Detailed Table of Contents

This chapter presents a usability study of a prototype tool intended to support information gathering during crisis situations, thus helping professional key communicators monitor social media content from several media outlets in the same workspace. The main aim of the study is to investigate how situation awareness can be optimized for key communicators during emergencies. It does so by taking a mixed method approach on usability testing and by combining emotional responses and cognitive processes to better understand how the user perceives the tool. The study contributes to the field by illustrating the importance of designing for good situation awareness in services/platforms intended for crisis communication. A key conclusion is that the use of human-computer interaction (HCI) and usability studies are central for improving digital services in emergencies. High-stress contexts require rapid decision-making and leave no room for improvisation and perception. Therefore, comprehending digital solutions correctly is crucial for overall situation awareness.

This study presents an information monitoring system: SIGDesastre. It is a method of monitoring the various sources of information available on the internet. The monitoring scenario is the failure of the dam in Mariana, Brazil. This event was considered the biggest socio-environmental disaster in the country. The creation of SIGDesastre involves the identification of the sources to be monitored, an automated search system for keywords in these pre-registered sources and the visualization of the results in a friendly environment. The use of information monitoring on the internet is considered an important tool for the post-disaster risk communication process. The monitoring of information on the internet is believed to be a potential device to support managers in institutional decisions and in the formulation of public policies. Also, for the affected population, they will be able to expand access to information about the actions being taken by the actors involved.

Spain is the second country in the world in numbers of missing people, with 114,226 men and women still in mass graves without having been identified and buried by their relatives. Added to them are the families of stolen babies: 300,000 babies were stolen during and after Franco's dictatorship. Faced with these disappearances, a digital health intervention (DHI) for the accompaniment and monitoring of the process of mourning and trauma recovery could improve the situation of this extensive and diverse population, situated throughout Spain. Limitations in resources for the care and accompaniment of this population are due to various logistical, geographical, financial, stigmatic, and demographic factors—this is an ageing population—preventing them from accessing places of treatment. In this chapter, the author proposes a health app for trauma recovery that can be readily standardized for the wide dissemination of evidence-based care and adapted to the needs of this specific population.

In the context of SARS-CoV-2 health emergency, strongly framed in the normalization of logic of risk, the authors analyze three digital platforms of contagion containment and population tracking in order to investigate, through a comparative-descriptive analysis, the relationship between different socio-political-cultural contexts and the respective responses adopted—the Chinese government tool Health Code, the South Korean app Corona100, and the Italian app, Immuni—to counter a single global emergency. The objective is to investigate the framing operations that introduced on a global scale the use of apps in bio-security and immunity regime for which individual privacy increasingly collides with collective security. The authors consider central the opening of a debate on how the logic of risk and worst case scenario are paradigmatic nowadays in the development of increasingly sophisticated systems, potentially invasive of privacy, even in function of complex threats interconnected on a global scale.

Forest fires are widespread in Portugal, particularly in the summer. Recently, in 2017, Portugal had two great fires. As a result, more than 120 people died, hundreds suffered injuries and registered significant economic and environmental losses. Since then, and due to the evolution and democratization of the internet and technology devices, forest fire content is much more common in cyberspace. Thus, to understand this issue, the authors propose to outline a profile of the digital platforms used in forest fire situations. The goal is to understand the uses and commitment arising from forest fires' issues in digital platforms by presenting a conceptual framework in Portugal's specific case. The authors analyzed webpages, mobile apps, Facebook pages and groups, and YouTube channels, focusing on forest fires contents. By understanding the kind of digital platform, its contents, uses, and interaction, this chapter contributes to understanding digital platforms' role in crisis and disaster scenarios such as wildfires.

An organization's survival during a crisis often depends on its speed of response. The introduction of social media into crisis communication discourse has meant that organizations must revisit their crisis communication strategies. This chapter explores a content analysis of the integration of social media platforms into crisis communication based on a comprehensive review of eight purposively selected crisis studies conducted globally. Findings revealed that Facebook and Twitter are increasingly employed as platforms for crisis communication. It was also discovered that responding to crises promptly, and engaging with the publics before, during, and after crises are crucial to managing organizational reputation. Social media platforms are also capable of spreading mis(information) about crises. Thus, organizations are advised to fully integrate and adopt social media into their crisis communication plans. This chapter extends our understanding of how social media platforms contribute to crisis communication discourse.

This chapter discusses crisis communication management in the Portuguese public security police digital landscape, mainly focused on social media. This is an exploratory investigation it is intended to understand how crisis communication management methodologies in the digital environment can be applied in police intervention, so its legitimacy is reinforced. The study is divided into two parts, the literature review, and the empirical research. The data was gathered together through an interview survey, with the participation of four public security police officers and four investigators. The interviews were transcribed, and their content was analyzed. In an emergency situation, most of the actions are communication activities, so its management in the digital landscape—above all social media—is critical for the police success, as an institution under strong scrutiny and whose actions depend on the citizen cooperation. Communication is essential for building reputation and should be used in threats to legitimacy, and as a tool for reaching out to the community.

Among its main goals, crisis management aims at promoting people awareness in respect of the crisis which they are going to face or in which they are already involved. In order to do that, it uses a wide range of communication tools, among which, over the last decade, social media have proved to be of paramount importance. Based on these premises, the chapter analyses a very meaningful case of crisis communication, which consists of the social media coverage of the early stages of the COVID-19 emergency coming from Italian national and local institutions. The media content analysis carried out on Facebook and Twitter confirms a communication strategy aimed at creating people awareness in respect of the health emergency, suggesting citizens which conducts they need to stop or adopt. However, the analysis also shows that the goal of crisis awareness represents the first step of a wider agenda coming from the Institutions' social posting, which aims at transferring their awareness to citizenship and, as a consequence, inspiring citizens' own responsibility.

COVID-19 is a new ongoing pandemic and an alarming public health issue which urges emergency measures at a national and international level. This chapter uses a collaborative message-centered approach and explores how the Department for Emergency Situation (DES) in Romania tailored its verbal and visual messages to this pandemic and how the Romanian online users assessed the DES involvement in the risk and crisis management. The quantitative and qualitative content analyses showed that DES laid an emphasis on the narratives of informational responsibility and responsibility in action, whereas online users focused on the narratives of multilayered rationalization and a multifaceted incompetence.

Journalists play a key role in digital emergency situations. In the midst of the chaos generated by a pandemic like the COVID-19 crisis, the information transmitted by the journalist is crucial to save lives and minimize damage. This chapter explores journalists' reliance on and trust in official and unofficial digital information sources. Specifically, the authors explore the authorities' official information sources involved in crisis management and the unofficial information sources (i.e., victims and those affected). The research suggests a model for journalists' relationship with their digital information sources in covering crises by exploring dependence, type of contact, trust, and purpose. It also synthesizes the main communication errors of the public authorities that manage the response to the crisis.

Although the literature on crisis communication is quite vast, business communication related to global crises (e.g., natural disasters) is largely unexplored. This chapter aims to fill this gap and shed light on brand communication strategies during a pandemic. A netnographic study was carried out with the purpose of identifying brand positioning and communication strategies during the COVID-19 pandemic outbreak and of understanding the engagement of brands' followers during that period. The study included four brands of large Brazilian companies and comprised the analysis of brands' feed on Instagram during the first five weeks of the outbreak in Brazil. Findings enable to identify two distinct profiles: unprepared brands and leading brands. The chapter provides valuable clues for both managers and researchers dealing with crisis communication.

The purpose of this chapter is to discuss the growth in social services using digital platforms to provide crisis interventions for clients, families, and communities. Digitally provided services afford numerous benefits to clients, such as increased accessibility, timeliness of services, and increase of autonomy. However, there are also new ethical concerns of using digital platforms for social services, as well as dangers to providers—specifically, secondary traumatic stress, vicarious trauma, and manifestations of these phenomena in providers' professional and personal lives. The chapter will discuss these potential concerns and the need for prevention and self-care for providers. Finally, the chapter will review the necessity for ongoing trainings and additional research concerning digitally provided crisis intervention services and occupational hazards for social service providers.

For a plethora of decisions we make on a daily basis, we can rely on timely, tailored information delivered via digital services. The technologies and the knowledge needed for the design, development and delivery of digital services have become increasingly accessible. These developments have also made their way to the crisis response domain and resulted in a variety of digital services to deliver information, supporting emergency responders in their decision-making processes. At the same time, affected communities by disasters are also in need of information during such critical events. Timely, relevant information helps affected citizens to understand the situation, make informed decisions, and gain access to life-saving aid. However, designing, creating, and delivering digital services to deliver vital information to communities warrant specific considerations. This chapter per the authors explores the concept of digital services in regard to empowering communities affected by critical events.

Foreword

Humans in Crisis and the Communication Safety

Life on our planet is an act of emergency that arises from contrast and conflict, in which the conditions for its existence or permanence are a constant struggle against crises and disasters. The human species has evolved in the sense of mastering these conditions, in the search for the guarantee of controlled environments, with low variation of contrasts, attenuating conflicts, diminishing the sense of imminent catastrophe. It was this control and lowering of constant immediate crisis that allowed, as a return, the human being to progress in the scientific knowledge of the world and himself. The maintenance of controlled and predictable environments has thus become a central condition of our contemporaneity.

To sustain this control and predictability, human communication is one of the main pillars. It began by serving in alerts and warnings, among those who had already passed through danger and those who still did not, and evolved to the communication of geographically distant warnings that allowed to gain time and prepare the reduction of impacts and shocks. With the advance of scientific knowledge, communication has incorporated technologies such as meters, sensors, accumulators and comparators of the world and human data that have allowed the development of more agile preventive systems, less dependent on human reaction and therefore faster.

Today we have systems that allow us to anticipate volcanic eruptions, earthquakes, tsunamis, but also storms, cyclones, hurricanes or even traffic jams, excessive concentration of people, displaced railway tracks, proximity to aeroplanes, boats, etc. Digital communication systems evolved human communication processes that were applied directly to the systems and technologies allowing them to become "intelligent" in predicting and anticipating disaster.

However, and despite all this evolution, accidents continue to arise, and life itself on Earth continues to present challenges, as we could see with the emergence of COVID-19 in 2020. In this sense, the need for communication systems to maintain life and well-being during emergencies has not disappeared. We can even say that it has increased, since society has become accustomed to living in great stability, so it is less resilient to crises, which makes the need for human communication even greater.

This is where this new book edited by colleagues Lídia Oliveira, Federico Tajariol, and Liliana Gonçalves comes in, and it launches a set of clues on how we are dealing with crisis and emergencies in the contemporary world. Throughout the various chapters, we can see how two pieces of technology are central to this work today: the internet and the smartphone. The first one for the global dimension that it assumed and that it allows, as never before, to connect any part of the planet in real-time. The second, because it allows connecting any person without the need for continuous connection and battery. They are two connection technologies, and we can say that they are today the main tools of human communication, having largely surpassed any of the previous technologies, owing much of its advance to the introduction of information systems that knew how to abstract human communication to transform it into something readable by systems and machines.

For all this, this book is a central work to understand the scope of communication sciences today, not only by how we deal with the crisis but also by what we can learn about the relationship between communication and the digital world, from the abstract, algorithmic information that allowed human communication to evolve to totally new levels of human interaction.

Nelson Zagalo
University of Aveiro, Portugal
October 2020

Nelson Zagalo *is an associate professor at University of Aveiro, Portugal. He has established a master program on interactive media; created the scientific laboratory EngageLab; and founded the Portuguese Society for Videogames Sciences. He is editor of the blog Virtual Illusion, and has published the books "Interactive Emotions, from Film to Videogames" (2009), "Videogames in Portugal: History, Technology and Art" (2013), and "Creativity in the Digital Age" (2015). His main focus of research is centered upon the design of interactive affective experiences, embracing multidisciplinarly the domains of multimedia, narrative and psychology.*

Preface

Over the past three decades, our society has deeply changed its communication processes due to different factors, such as the evolution of the Internet, the democratization of mobile devices and the increasing availability of digital services.

Smartphones, mobile apps and social media play a key role in this digital arena. The massive use of mobile allows us to consume, produce and share information, through user functions that promote even more interconnectedness and participation among people. The significant development of social media and online social networks in the last few years has completely transformed how society expresses itself, containing extreme power to shape individuals and collective opinions. People and formal organizations stay in touch with each other through the frequent usage of Information and Communication Technologies. Through networked digital technologies, we can talk and share information anytime and anywhere, during our everyday life, or in unique moments that represent some positive or negative experiences, such a crisis. This new digital environment is present in almost every realm of our lives, and it takes massive relevance when it comes to deal with disasters, crises and emergency situations.

Our society witnesses several recurrent crises due to social and industrial disasters or natural catastrophes: wars, tsunamis, earthquakes, wildfires, flooding, cyclones, nuclear disasters. These kinds of circumstances are recurrent, whether they have a local impact or a global effect. As we write these words, we deal with coronavirus pandemic crisis that affects all countries from all continents, and no effective solution has been found for this health emergency. This current situation reveals the impact of living in an increasingly globalized world. It highlighted the disparity in policy responses and health services about the coronavirus pandemic.

However, it also made clear the positive and negative impacts that digital platforms perform on citizens' daily lives: either by adapting to distant sociability, without physical contact during the lockdown or by the digital media adopted by official entities to inform and communicate with citizens, whether due to the potential that digital platforms, particularly social media, operate to shape social beliefs, attitudes, and behaviors. Thus, it is crucial to understand the role of digital platforms/services in any crisis, disaster and emergency situations.

The editors of this book shared a set of research questions across social sciences and information technologies about the design and the use of Information and Communication Technologies before, during and after a disaster or a crisis. There are many challenges in the analysis, prototyping, evaluation, and development of digital platforms and services to provide for human beings involved in these circumstances.

To rise to these challenges, we need both communicational and technological approaches. The first approach is expected to allow scholars to understand people's communication wishes and needs in these risky scenarios. This scientific material is a necessary underpinning for the second approach, focused on digital services and platforms' design to create new forms and devices to connect citizens.

Preface

This book presents a set of recent studies showing that digital technologies are key mediators in crisis, disaster and emergency. Fifteen original chapters highlight both multi- and interdisciplinary research findings.

The first seven chapters are dedicated to issues related to mobile features, apps, usability, information monitoring and app services. From the eighth chapter, social media are highlighted with six chapters dedicated to researching the latest risk and crisis communication trends on social media, from different perspectives: public relations, brand management, journalism, governmental and other official sources. The last chapters present some reflection about the benefits and risks of digital post-disaster aid and communities' empowerment. Along with fifteen chapters, this book covers diverse crises, disasters and emergency situations, as air pollution and radiation, flooding, wildfires, war, earthquakes, but mostly, the current pandemic situation of Covid-19. This health emergency is the leading case study of five chapters, which pinpoints this situation's relevance, but mostly, it reveals this book's timeliness.

The first chapter, "Smartphone Solutions for Citizen-Centered Risk Monitoring in Environmental Disaster Situations," presents three case studies about app sensors to air pollution and radiation. The Pocket Geiger (Pokéga) is a radiation sensor for citizens developed immediately after the Fukushima nuclear accident. More than 100,000 Pokéga units have been produced under an open-source license. (2) The Unreal iSOTOPE is a mobile simulator developed for training Japanese law enforcement agencies during radiation disasters. (3) The Pocket PM2.5 Sensor visualizes the distribution of invisible air pollutants indoors and outdoors. This chapter proposes a new kind of democratic risk communication that can be realized through environmental sensing by citizens with smartphones and considers the challenges involved.

The second chapter, "Emergency Response and Post-Disaster Recovery Using Smartphone-Based Applications," proposes an overview of smartphone features' recent trends as a solution to locate users in rescue situations during natural disasters. One of the major concerns of natural disasters is the destruction of communication links, such as power lines and Internet connections, making it challenging to enable communication among the affected people and the rescue teams. However, the evolution of smart devices equipped with multiple short-range communication technologies, such as Bluetooth and Wi-Fi provides an opportunity to form an ad hoc network with co-located smart mobile device users and communicate their positions and other relevant information to the rescue workers. This chapter provides a detailed description of recent advancement in this area and highlights important aspects that are needed to be considered for practical implementation.

Chapter 3, "Information Overload! Investigating the Usability of an Information Tool for Crisis Situations With Biometric Data," presents a usability study of a prototype tool intended to support information gathering during a crisis, thus helping professional communicators monitor social media content from several media outlets in the same workspace. The use of human-computer interaction (HCI) and usability studies are central for improving digital services in emergencies. High-stress contexts require rapid decision-making and leave no room for improvisation and perception. Therefore, comprehending digital solutions correctly is crucial for situation awareness. The study contributes to the field by illustrating the importance of design for good situation awareness in services/platforms intended for crisis communication.

The chapter "Internet Information Monitoring System: A Digital Tool for Emergencies, Crises, and Disasters" introduces SIGDesastre, an information monitoring system. The monitoring scenario is the failure of the dam in Mariana, Brazil. This event was considered the biggest socio-environmental disaster in the country. The monitoring of information on the Internet is believed to be a potential device to support managers in institutional decisions and public policies. Also, the affected population will be able to expand access to information about the actors' actions.

Chapter 5, "Missing People in Spain, an App for Trauma Recovery: A Digital Health Intervention for Survivors," focuses on the post-disaster phase and proposes a health app for trauma recovery called digital health intervention (DHI). This mobile app would bear the mourning process and trauma recovery of Spanish people affected by their missing relatives and stolen babies during the Spanish Civil War and Franco's dictatorship.

Chapter 6, "The Immune Space: Monitoring Narratives in Contemporary Pandemic," analyses three digital platforms of contagion containment and population tracking. Through a comparative-descriptive analysis between three countries (Italy, Korea and China), authors investigate the relationship between the political and cultural contexts and the respective technological responses adopted to counter the current pandemic situation.

"Forest Fire Scenarios in Digital Platforms: The Case of Portugal" is the seventh chapter. Here the authors propose to outline a profile of the digital platforms used in wildfires, one of the major national disasters yearly. The authors analyze forest fire contents spread through several websites, mobile apps, Facebook pages and groups, and YouTube channels. Their goal is to understand the uses and commitment arising from forest fires' issues in digital platforms by presenting a conceptual framework in Portugal's specific case.

The eighth chapter, "Evaluation of Research Trends in Social Media Crisis Communication," explores social media platforms' role in crisis communication. Based on a comprehensive review of eight selected crisis studies conducted globally, this research found out that Facebook and Twitter are increasingly employed as platforms for crisis communication. Thus, organizations must fully integrate and adopt social media into their crisis communication plans. This chapter extends our understanding of how social media platforms contribute to crisis communication discourse.

Chapter 9, "Communication Crisis Management of the Public Security Policy: The Social Media Landscape of the Police in Portugal," presents a perspective from an official source. This exploratory investigation intended to understand how crisis communication management techniques in the digital environment can be applied in police intervention, so its legitimacy is reinforced. In an emergency, communication is essential for building a reputation and should be used in threats to legitimacy and as a tool for reaching out to the community.

Following chapter, "Setting the Aware Agenda of the COVID-19 Health Emergency: The Italian Pas Social Media Coverage," analyses a significant case of crisis communication in Italy, the first European country to report Covid-19 cases and one of the most affected European countries. Authors studied a relevant social media corpus produced by Italian institutions and studied at the Covid-19 emergency early stages.

From Italy to Romania, Chapter 11 is " 'Keep Calm and Get Informed': Risk Communication and Engagement During the COVID-19 Pandemic in Romania." The authors explore how the Romanian Department for Emergency Situation (DES) tailored its verbal and visual messages to this pandemic and how the Romanian online users assessed the DES involvement in the risk and crisis management.

After the lenses of official sources, Chapter 12, "Social Media and Digital Information Sources in News Coverage Of Crisis, Disaster, and Emergency Situations: A Case Study From Spain," presents a journalistic perspective of the Covid-19 crisis. This chapter explores journalists' trust in digital information sources. Specifically, the authors explore official authorities' information sources and unofficial information sources, i.e. victims and those affected by Covid-19. The research presents a model for journalists' relationship with their digital information sources when they cover crises. It also synthesizes the main communication errors of the authorities that manage the response to the crisis.

Chapter 13, "Communication During a Pandemic: An Analysis Through the Lenses of Brand Management Strategy," aims to shed light on brand communication strategies during a pandemic situation such as Covid-19. The chapter presents a netnographic study to identify brand positioning and communication strategies during the Covid-19 pandemic outbreak and understand its followers' engagement during that period. The study included four brands of large Brazilian companies and analyzed brands' feed on Instagram during the first five weeks of Brazil's outbreak.

Most of the previous chapters draw attention to the critical phase of disasters and crises. The last two chapters of this book offer a post-disaster perspective. In Chapter 14, "When Helping Is Dangerous: Benefits and Risks to Providers Delivering Digital Crisis Intervention," authors discuss the growth in social services using digital platforms to provide crisis interventions for customers, families, and communities. Digitally provided services afford numerous benefits to clients, such as increased accessibility, timeliness of services, and autonomy. However, there are also new ethical concerns of using digital platforms for social services and dangers to providers – specifically, secondary traumatic stress and vicarious trauma.

The last chapter, "Information as Humanitarian Aid: Delivering Digital Services to Empower Disaster-Affected Communities," emphasizes values like solidarity and empowerment through digital services. The authors claim that the development of new technologies arises a variety of digital services to broadcast information, supporting emergency responders in their decision-making processes. They also argue that communities need information during critical events since timely information helps concerned citizens understand the situation, make informed decisions, and gain access to life-saving aid. Based on these premises, the chapter explores the digital services' concept regarding empowering communities affected by critical events.

This book brings different perspectives pointed out by researchers with distinct backgrounds, systematizing the interdisciplinary knowledge about crisis, disaster and emergency themes and merging it with the digital scope. Therefore, it constitutes a major reference to the researchers. However, it will also be a helpful tool for a broad set of stakeholders – governments, local institutions, and public corporations, among others. – who deal with crisis, disaster and emergency scenarios. Moreover, this is particularly relevant since the world is living a current global health emergency.

Across the 15 chapters, the book highlights the relevance of society's digitization, its usefulness, and its contribution to the different phases and types of crisis scenarios. By combining this with a deep reflection about devices, social media, digital services, its trends, features, uses, ethical concerns, and awareness, this book covers our digital society's most significant items and threats.

Lídia Oliveira
University of Aveiro, Portugal

Federico Tajariol
University Bourgogne Franche-Comté, France

Liliana Baptista Gonçalves
University of Aveiro, Portugal

Acknowledgment

The editors would like to acknowledge the help of all the people involved in this project and, more specifically, to the authors and reviewers that took part in the review process. Without their support, this book would not have become a reality.

First, the editors would like to thank each one of the authors for their contributions. Our sincere gratitude goes to the chapter's authors who contributed their time and expertise to this book.

Second, the editors wish to acknowledge the valuable contributions of the reviewers regarding the improvement of quality, coherence, and content presentation of chapters. Most of the authors also served as referees; we highly appreciate their double task.

Lídia Oliveira
University of Aveiro, Portugal

Federico Tajariol
University Bourgogne Franche-Comté, France

Liliana Baptista Gonçalves
University of Aveiro, Portugal

Chapter 1
Smartphone Solutions for Citizen-Centered Risk Monitoring in Environmental Disaster Situations

Yo Ishigaki
University of Electro-Communications, Japan

Kenji Tanaka
University of Electro-Communications, Japan

ABSTRACT

Through an analysis of three case studies, this chapter proposes a new kind of democratic risk communication that can be realized through environmental sensing by citizens with smartphones, and considers the challenges involved. The three case studies, which the authors have implemented in the society, are as follows: (1) The Pocket Geiger (Pokéga) is a radiation sensor for citizens developed immediately after the Fukushima nuclear accident. More than 100,000 Pokéga units have been produced under an open source license. (2) The Unreal iSOTOPE is a mobile simulator developed for training Japanese law enforcement agencies during radiation disasters. (3) The Pocket PM2.5 Sensor visualizes the distribution of invisible air pollutants indoors and outdoors. It is particularly useful for fieldwork in developing countries where environmental assessments are inadequate.

INTRODUCTION

Global citizens today carry a sophisticated tool for environmental monitoring in their pockets: a smartphone. Smartphones did not exist at the time of the Minamata disaster, the Deepwater Horizon oil spill, the Three Mile Island accident, or the Chernobyl accident, but it is the conviction of the authors that they hold immense potential for public health. In fact, they could play an essential role in minimizing the

DOI: 10.4018/978-1-7998-6705-0.ch001

damage caused by such disasters in the future by enabling the public themselves to access live information and determine risk at an early stage through mutual communication with experts.

This chapter discusses three mobile participatory environmental monitoring solutions developed by the authors based on this belief: Pocket Geiger (Pokéga) and Unreal iSOTOPE (USOTOPE) for radiation detection, and the Pocket PM2.5 Sensor for monitoring air pollution caused by PM2.5 (atmospheric particulate matter that has a diameter of less than 2.5 micrometers). Integrating physical and social technologies, these systems use smartphones and social media to empower ordinary citizens to monitor their environment, make sense of the results, and take responsibility for their own health.

The chapter begins with an introduction to Pokéga, a low-cost mobile radiation detector. The discussion concentrates on the socially inflected process by which Pokéga was developed and tested, measurement data was shared, and how effective risk communication was facilitated through social media among citizens, sensor engineers, and radiation experts. With over 100,000 sales to date, it is hoped that the case of Pokéga may provide a helpful model for the social implementation of chemical sensing in the environmental field.

The second solution introduced is USOTOPE, a virtual radiation measurement application for smartphones that uses Bluetooth and Wi-Fi beacons to measure electric field strength and simulate the display of a radiation-measuring instrument. As the discussion shows, USOTOPE provides an innovative solution to the difficulties associated with first-responder training for chemical, biological, radiological, nuclear, and explosive (CBRNE) terrorist attacks. Describing a series of drills carried out using USOTOPE, this section explores other possible uses for the application in disaster drills, including zoning by law enforcement, screening of the sick and injured by medical personnel, and cooperation with the private sector and citizens—all of which have hitherto been difficult to achieve.

The third solution addressed is the Pocket PM2.5 Sensor, a device that can be connected to smartphones to monitor air pollution. As the discussion shows, the device has particular potential in developing countries, where air pollution is responsible for millions of premature deaths annually. This section presents a field study being conducted in Rwanda and explores the potential for gamification to raise the awareness of air pollution in households with children.

BACKGROUND

Environmental disasters that cause severe damage to public health and the natural environment are a constant global threat. In traditional risk assessments, such as those described by the United States Environmental Protection Agency (1992), risk management is determined by expert discussion and top-down decision-making. In contrast, the authors believe that citizen-centered participatory monitoring holds the key to effective disaster prediction, prevention, and response.

Burke et al. (2006) originally presented the concept of participatory sensing and predicted that the data collected from mobile sensors held by citizens could be used for public health, urban planning, natural resource management, and documentary filming. With the spread of smartphones, this concept has become a reality. For example, road congestion, consumption behavior, location information, and travel history are widely collected and utilized through Global Positioning System (GPS) sensors and payment modules pre-built into smartphones. The authors take this concept one step further with the addition to smartphones of environmental sensors for crisis communications.

Malone (2004) proposed that society would move from a centralized management approach based on command and control to a more decentralized model of coordination and development. In the context of disaster management, this would mean respecting and promoting citizen-level disaster information gathering, democratic consensus building, risk-averse behavior, and risk communication. Turoff et al. (2010) highlighted the effectiveness in crisis communication of unofficial, user-generated information (e.g., backchannel information or wisdom of crowds), produced by ordinary people through social media. At the same time, they also indicated the risks of misinformation or overconfidence (or over-diffidence) and the difficulty of sharing information with official bodies. These issues were starkly exposed in the radiation measurements after the Fukushima Daiichi nuclear accident in 2011. The Pokéga section of this chapter shows how trust can be improved by the sharing of government and citizen measurements with each other on social media or by citizens and experts engaging in radiation risk discussions. The International Atomic Energy Agency (IAEA) wrote in a report submitted after the Fukushima Daiichi accident: "For government authorities and agencies, crowdsourcing certainly is the 'genie that will not go back in the bottle.' It is necessary to accept that this technology is here to stay and that empowerment of the public is not necessarily a negative development" (2014, p. 40).

In the near future, big data collected through citizen participation using smartphones will be able to create a new social framework for environmental risk assessment, disaster prevention planning, hazard mapping, and disaster prediction or detection in collaboration with experts, governments, and municipalities. Moreover, participatory monitoring and discussions on social media can improve public understanding of environmental hazards, enabling agile responses to unpredictable environmental hazards and conservation of resources by administrations. To facilitate such a shift, the authors argue that it is necessary to (1) provide citizens affordable and simple methods to monitor the environment, (2) share the measured results among neighbors and experts through social media, and (3) establish a participatory-type mobile system for monitoring the environment that allows objective discussion and verification of scientific information so that appropriate action can be taken to mitigate risk.

MAIN FOCUS OF THE CHAPTER

The three cases presented in this chapter demonstrate that, with appropriate tools, ordinary citizens are able to collect and visualize a large volume of pollution information themselves through environmental sensing. This creates opportunities for community environmental monitoring and voluntary decision-making through sharing and exchange with third parties and experts. If citizens are able to monitor and share environmental risks on a daily basis, mutually evaluate situations with experts and municipalities, and avoid excessive dependency on the government for such management, a disaster-mitigated society based on self-decisions can be realized, as Tanaka and Itoh (2003) and Takayama et al. (2018) predicted. In such a society, members of the public are empowered to respond appropriately (e.g., evacuate) on their own in the event of a disaster, without waiting for instructions to be given. This approach represents a new form of environmental disaster prevention in which the general public, experts, and the government share information to form a consensus through scientific discussion via social media, and using universal technology bases that are affordable and widespread, such as semiconducting sensors and smartphones. The authors hope that the cases introduced in this chapter will provide a model for implementing systems that can contribute to expanding and improving risk communication on environmental disasters.

POCKET GEIGER

Pocket Geiger (http://www.radiation-watch.org), hereafter Pokéga, is a mobile radiation measurement device (Figure 1). The detector is small, lightweight, and inexpensive because it employs a universal photodiode as a gamma-ray sensor, and signals are processed through smartphones. The measurement range is 0.05 μSv/h to 100 μSv/h, which is sufficient for practical use, and radiation levels can be shared and visualized using GPS location information. A total of six models (Type 1 to 6) have been developed to date, with the first Type 1 model having been crowdfunded through Kickstarter. As of 2020, over 100,000 kits have been shipped, and more than one million pieces of data have been recorded and accumulated. In addition, discussions on radiation protection have been actively held on a dedicated social media community page on Facebook.

Figure 1. Exterior view of the Pocket Geiger (Pokéga) Type 4. Pocket Geiger is available in models from Type 1 to Type 6.

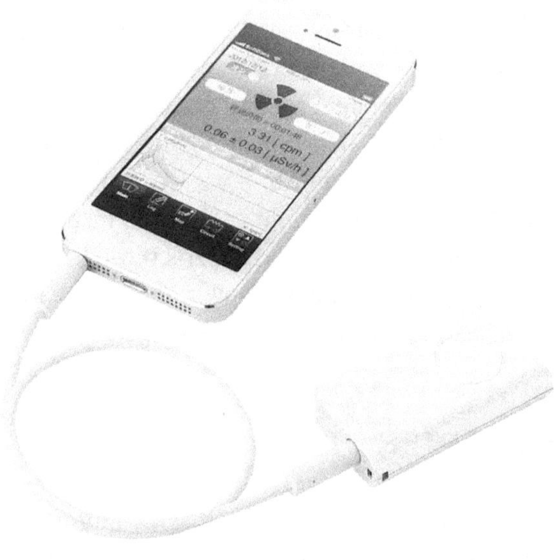

Background

Pokéga was developed by the authors in the aftermath of the Fukushima Daiichi nuclear accident in 2011 to address several perceived risk-monitoring needs. After the accident, radiation doses in the region differed greatly depending on local environmental factors (weather, vegetation, and drainage). For this reason, it became necessary to measure dose information at multiple locations and to share this information correctly among local residents. A need among citizens also emerged for a forum for discussion with experts on correct measurement methods and quantitative risks of radiation. The following discussion explores the responses of industry, experts, government, and citizens after the Fukushima accident, focusing on issues of *measurement*, *sharing*, and *discussion*, which are relevant to the need for, and implementation of, participatory monitoring.

Japanese manufacturers had been selling Geiger–Müller tubes and scintillation radiation detectors before the accident, but they were prohibitively expensive for the general public, ranging from 100,000 to 500,000 yen, and difficult to obtain for several months after the accident due to the rapid increase in demand. As a result, inexpensive, predominantly imported, measuring instruments were introduced into the market. However, a National Consumer Affairs Center of Japan (NCAC) investigation of nine such models revealed that none accurately measured radiation doses due to the low absolute values and high variability of the measurements (NCAC, 2011a). The NCAC received many enquiries related to radiation-measuring instruments – 680 cases from March 11 to the end of November 2011 (NCAC, 2011b) – and there were other cases where incorrect product labeling and usage methods were pointed out (NCAC, 2012), which caused social confusion. It was not until nine months after the accident that a domestic company was able to launch an inexpensive measuring instrument for the general public, but this product did not have a function to share the measured values (S.T. Corporation, 2011).

Compared to these disruptions and delays with commercial products, unpaid volunteer groups of researchers and university officials were quick to respond. For example, five days after the accident, Ryo Ichimiya's Radmonitor311 (https://sites.google.com/site/radmonitor311/) and Safecast (https://safecast.org/) were established as portal sites to centralize radiation information online and publish doses using mobile sensors, respectively, and the geographical distribution of doses became available. However, individual citizens remained unable to measure radiation doses and discuss the risks in their living areas.

By contrast, the response of the government was sluggish. The Ministry of Education, Culture, Sports, Science and Technology (MEXT) established a website (https://radioactivity.nsr.go.jp/en/) to consolidate radiation monitoring information only five months after the accident. Three months after that, in November 2011, it began implementing a real-time dose measurement system in Fukushima Prefecture, and a total of 2,700 monitoring posts were installed in February of the following year (MEXT, 2012). In addition, the Japanese government was slow to release the preliminary results of the Emergency and Rapid Radiation Effects Prediction Network System (SPEEDI) in the immediate aftermath of the accident; it was only due to domestic and international criticism that some results were released two weeks later and the full results two months later (Ikeda & Maeda, 2013, pp. 42–43).

In addition, while the foreign media reported the seriousness of the accident, many experts who appeared in domestic media outlets tended to underestimate it, resulting in a very strong distrust among citizens of the government, mass media, and experts (Masamura, 2013). In order to ensure that accurate information about radiation is disseminated and accepted, it is necessary to develop a scientific discussion based on mutual trust. However, mass communication in the aftermath of the Fukushima nuclear accident resulted in citizens losing faith in experts.

Thus, neither the government nor the private sector has been able to effectively provide citizens with the means to measure, share, and discuss radiation doses in risk communication since the Fukushima nuclear accident. It was in this context in May 2011 that the authors began researching and developing Pokéga as a smartphone-connected mobile dosimeter that would enable all citizens to contribute to measuring radiation. The first version, Type 1, was released in August 2011. More than 15,000 devices were distributed in the first six months because they were the first inexpensive dosimeters for individuals after the earthquake. In addition, Pokéga's use of smartphone technology allows measurement results to be shared and discussed on social media to promote radiation protection.

Design Concept

The design of the Pokéga was determined by the need to quickly develop and disseminate a low-cost device. The authors actively sought to utilize general-purpose parts and off-the-shelf products that were readily available to everyone. Furthermore, releasing the blueprints under an open source license allowed the involvement of many engineers and specialists in the development of the system. The following discussion summarizes characteristic features of the Pokéga design.

General-Purpose Semiconductors

In the past, components such as Geiger–Müller tube and scintillation sensors have been used for radiation measurements. These conventional sensors have the advantage of high sensitivity. However, the sensors and peripheral circuits (photomultiplier tubes, high-voltage circuits, etc.) are expensive and complex and require periodic calibration due to age-related deterioration. Although the principle of radiation measurement using photodiodes is well established (Knoll, 2010, pp. 365–414; Iniewski, 2010, 2011; Spieler, 2005; Dearnaley & Northrop, 1966; Kitaguchi et al., 1996), Pokéga is the first application of a photodiode as a radiation-measuring instrument for the general public.

Smartphones

Connecting the sensor to a smartphone has the following three advantages.

- Low cost, compactness, and light weight are achieved by eliminating parts such as user interfaces and power supplies.
- Easy to improve and add functions through software.
- GPS, communication functions, cameras, etc. make it possible to share information and discuss measurement status.

These features are useful not only for environmental measurements but also for various Internet of Things devices, such as credit card reading and online payment device with smartphones.

DIY Semi-Production

For the initial Type 1 model, a do-it-yourself (DIY) kit method was adopted. Buyers received pre-mounted boards that they then had to place in a FRISK mint candy case, which is readily available throughout Japan (Figure 2). In addition, a 10-yen coin was used as a beta particle shield.

By using these familiar general-purpose materials, it was possible to develop the first Pokéga Type 1 personal dosimeter in three months and distribute it at a price of 1,850 yen. Recently, the appeal of these semi-finished products, such as their low cost and high scalability, has been recognized by society as a new and valuable concept, and it is becoming established as a new style of manufacturing. Examples of this trend include the Maker movement (Anderson, 2014), which involves consumers in product development through DIY, and the "IKEA effect" (Norton et al., 2012), in which customer attachment to a product increases through the assembly of semi-finished products.

Figure 2. Assembling the Pokéga Type 1. The user places the Pokéga board in the mint candy case and uses a 10-yen coin as a shield against beta radiation.

Improvement of a Gamma-Particle Detection Circuit

Over three years and six iterations of the Pokéga (Type 1, August 2011; Type 2, February 2012; Type 3, June 2012; Type 4, August 2012; Type 5, November 2012; and Type 6, December 2014), several improvements were introduced. These include the shift from eight low- to a single high-sensitivity photodiode sensor, which reduced the measurement time from 20 to 2 minutes at a standard air dose of 0.05 μSv/h; energy harvesting technology, which allowed the device to draw power from the smartphone itself; and a built-in anti-vibration circuit and comparator to eliminate noise and control input gain.

Software Development

The software can be downloaded from the Apple App Store or Google Play. Figure 3 shows screen captures of the iOS version. In version 1.0 (left), only counts per minute unit of measurement was supported; the current version 1.4 (center) displays the air dose in μSv/h, along with the standard deviation. The graph also shows the moving average of the dose (solid line) and standard deviation. On the map screen (right), dose values can be plotted using the GPS function to share and visualize geographic trends in dose.

Participatory Development

The development of Pokéga took place with a significant lack of funding and human resources; thus, the authors adopted a participatory development method, as described below.

Crowdfunding

To quickly raise funds for the initial production, the authors relied on Kickstarter, a platform for crowdfunding. The project was successfully funded: the minimum amount needed to produce the initial batch was pledged in four days.

Figure 3. Changes in Pokéga's software: version 1.0 (left), current version 1.4 (center), map screen of version 1.4 (right)

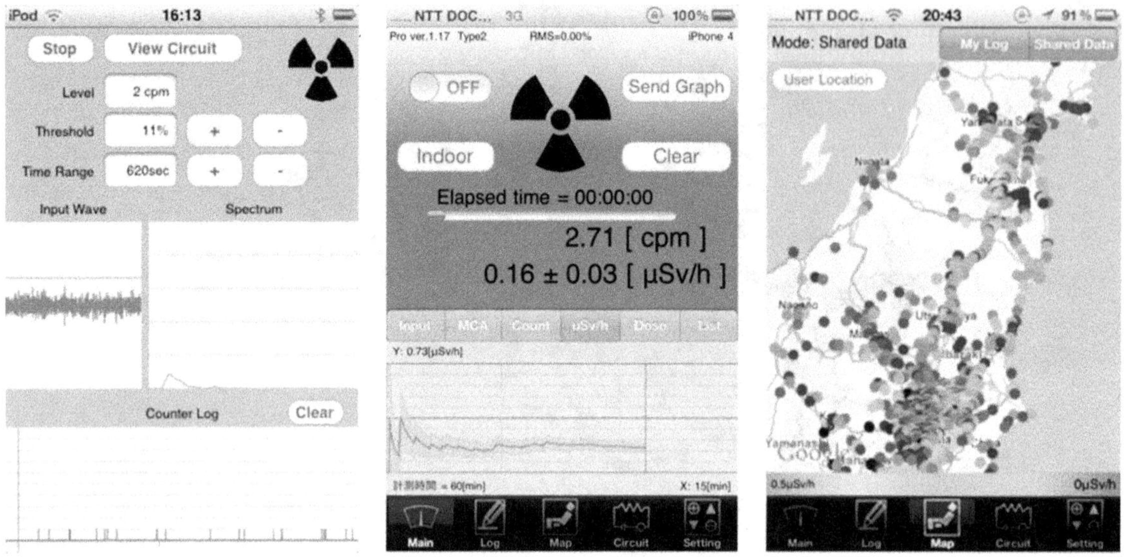

Media Exposure and Publicity

Although there was no advertising budget, the Kickstarter campaign garnered significant media attention, including features and interviews by various media outlets (e.g., Gizmodo, CNET, Make, Gigazine, Le Monde, etc.). As a result, the project gained significant publicity, which attracted further Kickstarter backers and investment.

Open Source Research

Pokéga's experimental data, schematics, and software were released under an open source license from the beginning, and can be used for both commercial and noncommercial purposes. This open source development stance, which does not monopolize or secure the results of research and development but opens and returns them to society as common knowledge (commons), is likely to win the support of many engineers, specialists, and researchers.

In fact, the approach attracted offers of cooperation in the development of Pokéga from experts all over the world. The project has benefited from foreign expertise, calibration tests, field tests, circuit simulations, and technical improvement suggestions. For example, a team from the Dutch Ministry of Defence and the National Metrology Office conducted a classification test and issued a certificate of performance free of charge (Kuipers et al., 2011). According to the interview article with the collaborators (Van de Weijer, 2013), strong identification with the purpose of the project was cited as the motivation for these cooperative actions.

A Social Product

A product that has value in terms not only of its function and price but also the user's sympathy for the project's management attitude and social ripple effect is called a social product, which has its own particular value in the market (Auger et al., 2003). Pokéga is such a product. The production and activity base was located in Ishinomaki City, Miyagi Prefecture, which was the affected area. In fact, once the products were released onto the market, the media began to note favorably that they were made in Ishinomaki.

Risk Communication on Social Media

As Pokéga became more popular, sharing and discussion in a dedicated Facebook group became more active. The content of the 1,549 topics posted by users between July 2011 and July 2012 was classified as follows: dose sharing (54%), feature requests (20%), usage (9%), comparison tests (8%), bug reports (5%), and other discussion (4%). A selection of examples is presented next. These examples provide a glimpse into the information that local residents seek in times of disaster and how social media discussions can help to resolve questions.

An example of dose sharing is shown in Figure 4. In these posts, there was much discussion about the sharing of measurements in living areas (city, park, school, home, etc.) and how to understand these values. There were many comments from experts, resulting in a high level of discussion.

Figure 4. Examples of user reports: decontaminated soil has been piled up and a no-go sign erected in a children's park in Chiba Prefecture (left); residents who have been allowed to go home on furlough are holding Pokéga in and around J-Village in Fukushima Prefecture (right)

Through these social experiments, the authors found that the public and experts require different kinds of metadata (e.g., instrument parameters, error ranges, measurement locations, meteorological conditions, etc.) in radiation measurements. Therefore, in 2016, the authors proposed a guideline on metadata for sharing radiation measurements on social media (Segault et al., 2016).

The comparison tests topic relates to accuracy reports conducted voluntarily by users. Comparisons with commercially available air dosimeters were frequently reported. In addition, there were many posts that compared the indicated values at governmental and municipal monitoring posts with Pokéga's measured values. In almost all cases, the measured values were consistent across devices but differently interpreted. Some users concluded that "Pokéga is inexpensive but accurate," while others said that "I thought that the monitoring posts were deliberately showing low dose values so that the government and local governments could control the situation," and "I have come to trust the monitoring posts after seeing the results of the comparison experiment." This positive change in public trust in government and experts is essential if appropriate risk communication in environmental hazards is to be achieved.

In terms of requests and proposals, many suggestions for hardware and software improvements were made on the Facebook group, mainly by engineers. Based on these suggestions, major upgrades were steadily implemented from Type 1 to 6. In addition, non-engineers contributed to the minor version-up of the system for stable operation by generating bug reports for new features. Other questions about how to use the device were also asked, but most of these were resolved through communication among users.

Analyzing the logs of these social media discussions could help predict what kind of information people want after a disaster. Therefore, the authors prototyped Crowd Talks (Ishigaki et al. 2017), a system that visually groups the content of discussions in a Facebook group by combining feature analysis of documents using latent Dirichlet allocation (LDA) and two-dimensional plots using principal component analysis (PCA). It would be useful for disaster response teams and developers of disaster management systems to know the characteristics and changes in the vast amount of social media discussions without having to read them all.

Discussion

During a nuclear disaster, there is a high level of uncertainty about the situation among people, which requires a high degree of cross-disciplinary response and advanced knowledge of radiation protection, even if it is generated by citizen-level action. In order to understand and respond appropriately to complex situations, there is a need for an increased diversity, or spread of viewpoints on the part of people and organizations, which is referred to as collective requisite variety (Weick, 1987). Pokéga could increase the collection and dissemination of more diverse viewpoints of civil society and improve its ability to respond to disasters.

"Citizen science" refers to scientific investigation and research activities by nonspecialist lay people (Silvertown, 2009). In general, data obtained through citizen science have been considered to be prone to errors, making it difficult to ensure adequate data quality (Kremen et al., 2011). However, if citizens share performance-rating instruments such as Pokéga, based on standardized guidelines, the quality of the data can be improved. Furthermore, expertise in measurement methods and the environment can be acquired by citizens through dialogue with experts. With the use of smartphones and social media, the quality of citizen science will break into new territory, and the barriers between citizens and experts may start to blur.

The timing of Pokéga's launch announcement coincided with the birth of the term "crowdsourcing." Crowdsourcing is now widely recognized as a form of online project in which crowds participate not only for monetary compensation but also for the satisfaction of contributing to society and personal growth (Estellés-Arolas & González-Ladrón-de-Guevara, 2012). In recent years, crowdsourcing has become increasingly important, especially in times of disaster, and several projects were launched in the immediate aftermath of the COVID-19 outbreak (some of which can be found on crowdsourcing mentor Codementor's site; https://www.codementor.io/covid19/developers).

Future Directions

Pokéga has been adopted for Berkeley RadWatch (https://radwatch.berkeley.edu) and is expected to be used as a fixed measurement station, taking advantage of its low power consumption. In addition, the authors have been studying the use of Pokéga for the management of radiation exposure of medical personnel in interventional radiology (IVR) since 2017 (Terasaki et al., 2017; Fujibuchi et al., 2019a; Fujibuchi et al., 2019b). In IVR, high levels of exposure to X-rays are a concern because the surgeon performs the procedure, including catheterization, on the patient near the radiology equipment (Chida et al., 2012). Pokéga can be used as an X-ray measuring device in a medical environment, taking advantage of its compact, lightweight, and inexpensive characteristics.

Pokéga will next be used to monitor radiation from natural resources, that is, naturally occurring radioactive materials (NORM) and technologically enhanced NORM (TENORM), which has led to high radiation zones scattered throughout the world. In particular, field research will focus on zircon sand processing and storage facilities in Bangka, Indonesia. Zircon sand, which is widely used in industrial products, generally contains radioisotopes such as uranium and thorium (Hazin et al., 2008). Some facilities in Bangka measured by the authors using Pokéga in 2020 have air-dose rates of 5–20 μSv/h, which is very high for residential areas (Ishigaki et al., 2020) – equivalent to a radiation-controlled area by the standards of developed countries. It is suggested that intervention be considered according to International Commission on Radiological Protection (2007) recommendations.

Of particular concern is that many sandbag clusters are located right next to ordinary homes. Since there are no entry restrictions or guidance signs, local residents, including children, easily enter these facilities. In addition, workers, most of whom come from agricultural and fishing backgrounds, have limited radiation literacy and are not provided with personal dosimeters and masks. Future research steps will include first visualizing the distribution of radiation doses in the region and then examining reasonably achievable goals for radiation protection and education of residents and workers, including medical personnel.

USOTOPE

From a device for citizen-centered radiation measuring, the discussion now turns to a solution developed by the authors to empower first responders in the event of CBRNE terrorist attacks. Drawing on their experience with Pokéga, the authors developed a radiation simulation application for Android smartphones, called Unreal iSOTOPE (USOTOPE), for use in first-responder training, which has been used in drills with field staff from police and fire departments.

Background

Social measures against CBRNE terrorism are attracting attention globally (National Institute for Defense Studies, Japan, 2015). In the event of a nuclear security incident, a radioisotope handling facility disaster, or a nuclear disaster, the scope of impact is vast. The police, firefighters, local governments, and other government agencies are responsible for the first response. This kind of wide-area counterterrorism and disaster response requires special skills, so training is crucial for first responders to gain skills and experience (Tsuchiya et al., 2018a; Tsuchiya et al., 2018b). Particularly in counterterrorism, where radiation and nuclear materials are handled, practical training by law enforcement agencies in activities such as radiation dosimetry, identification of suspicious objects, zoning (hot, warm, and cold zones), and nuclear forensics is vital. From 1993 to 2015, dozens of cases of illicit possession, loss, and perceived criminal activity of radioactive materials occurred every year (IAEA, 2016), which demonstrates the need for more global vigilance and countermeasure training.

However, in training scenarios, it is dangerous for trainees to handle materials that are *actually* highly radioactive. It is also impractical, even for training purposes, to place or detonate highly radioactive materials in public spaces, such as train stations and parks. For this reason, counterterrorism and disaster preparedness training involving radiation and nuclear materials in public spaces often involve taking only fictitious measurements. For example, a training supervisor may announce a pseudo-radiation dose without taking real measurements, despite having a measuring device. In other cases, handmade, nonfunctional measuring devices are used because it is not possible to provide expensive measuring devices to all members of the unit.

Against this background, since April 2016, the authors have developed USOTOPE as a smartphone-based virtual radiation measurement system that shows a screen very similar to professional radiation detectors used by the police or fire departments but with no actual sources of radiation. Instead, Wi-Fi and Bluetooth (BLE) beacons are used to mimic radiation sources, and the virtually obtained radiation levels based on their electric field intensity are calculated and displayed. The distribution of radiation intensity can be embedded on the in-app map in advance so that the defined radiation levels are shown based on GPS location information.

Implementation

USOTOPE is a general-purpose application for Android and can be used on almost all Android devices with wireless communication capabilities. The system uses a dedicated BLE beacon (Aplix JM1L2S) as an imitation gamma radiation source that transmits signals every 100 ms, but Wi-Fi access points can also be used in the second version of USOTOPE developed in 2018. Figure 5 (left) shows the application in use. With software revisions, USOTOPE is able to emulate various measuring instruments in use around the world: Figure 5 (right) shows two types of radiation detectors (scintillation survey meter and ionization chamber) simulated by USOTOPE.

Figure 5. USOTOPE in use: terminal with the application running (left), screen captures of two types of measuring instruments emulated by the application (right)

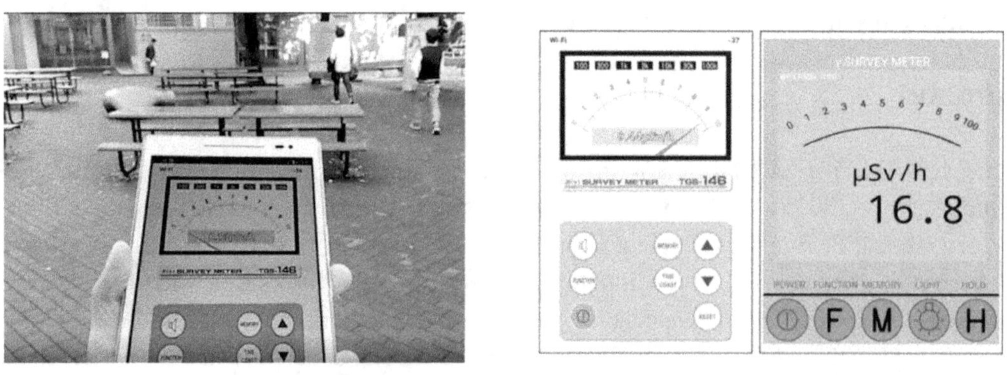

In general, the intensity of radio waves used for Wi-Fi and BLE communications is attenuated according to the inverse square law ($1/r^2$), similar to that of gamma radiation. The gamma radiation dose emulated by USOTOPE was tested and confirmed to obey a nearly inverse square law (Figure 6). That is, a dynamic range of 0.1 to 10 μSv/h of a simulated gamma radiation dose can be obtained within 20 m from the source location, indicating that radio waves for communication suitably mimic gamma radiation.

Figure 6. Simulated dose rate versus distance from radiation source using USOTOPE on smartphones and tablets

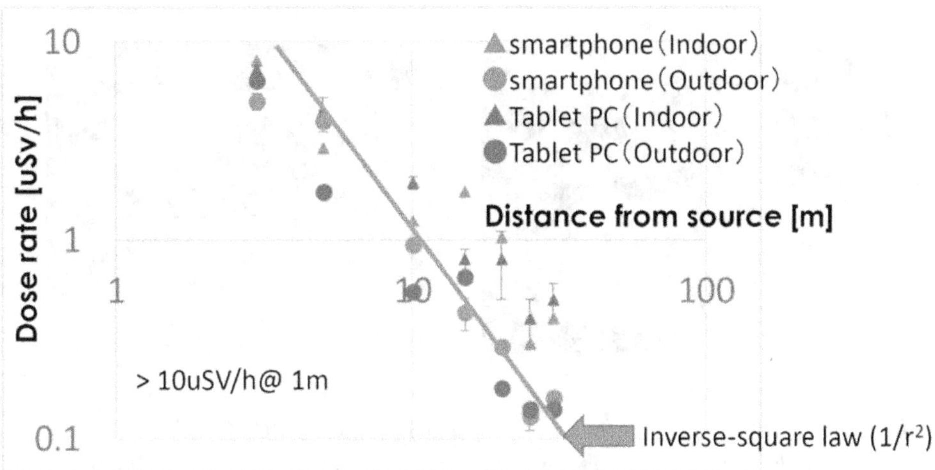

Field Testing

When USOTOPE was first tested in the field in 2016, it was found that users responded so naturally to the application that it was difficult to distinguish it from the actual experience of measuring of radiation. In the test, twelve electronics assemblers with no experience in radiation measurement were given a professional radiation-measuring instrument (Hitachi TCS-172B) and a tablet with USOTOPE installed and set to the task of searching for suspicious objects in three paper bags (one of which concealed both a real gamma radiation source and a Wi-Fi beacon) (Figure 7). The operating principle of USOTOPE was not revealed to the participants until the end of the task, and the order of the instruments used by the participants was changed to reduce sequential bias. The results of a survey on the use of the measurement device, which measured items such as "response and sensitivity" and "ease of measurement" with a seven-point Likert scale, showed no superiority between the real measurement device and the USOTOPE. All the participants believed that USOTOPE could indeed measure radiation. This suggests that USOTOPE is a simple, effective, and realistic tool for education and training.

Through internal testing in the National Research Institute of Police Science in Japan conducted in 2018 (Tsuchiya et al., 2018a; Tsuchiya et al., 2018b), the authors confirmed that USOTOPE is capable of detecting radiological terrorism within a 20-meter radius of a possible threat. After this confirmation, USOTOPE has been used in joint drills between the fire and police departments in Japan since March 2019, as shown in Figure 8, which was taken during a joint exercise involving a railway company at a station in Osaka Prefecture. In this field experiment, the Wi-Fi beacon was also reachable for approximately 20 m. This resulted in the squad members searching for suspicious objects with a sense of reality. In interviews with fire and police crews after the exercise, they rated it favorably because it allowed them to conduct the exercise in a realistic manner.

Figure 7. A participant uses the USOTOPE smartphone app to identify a pseudo-radiation source, but she thinks she is really measuring radiation (left). Participants are looking at a professional scintillation radiometer and the USOTOPE system and answering a questionnaire about the differences between the two (right).

Figure 8. Joint training by police and fire departments using USOTOPE: Radiation source – Wi-Fi beacon hidden in a plastic bag (top left). Surface contamination survey (top right). Team members checking the radiation dose in the air (bottom)

Discussion

Whereas Pokéga contributed to improving risk communication and diversity of thinking in civil society in the post-disaster and recovery process, USOTOPE could be useful in detecting the occurrence of a disaster event itself and quickly identifying its causes. There are three levels of situational awareness: 1) being able to recognize when something happens, 2) being able to identify its cause, and 3) being able to predict the course of events (Endsley, 1995). Level one of situational awareness can be achieved by USOTOPE by detecting high doses of radiation and level two by identifying whether the situation is a nuclear disaster or a nuclear crime and by distinguishing between the types of radiation.

Future Directions

USOTOPE is a unique system that allows people to train for dangerous situations that are difficult to replicate by simulating them on a smartphone. At present, the application can be used for defense training in two of the five possible CBRNE terrorism threats: namely, radiological and nuclear. In the future, USOTOPE could be used for the detection and zoning of chemical, biological, and explosive threats.

In addition, if the detection range of the measurement can be made shorter and more accurate, it may be applied to radiation medicine, such as for training in the screening of disaster victims for surface contamination and preventing exposure to X-rays in hospitals. Conversely, by increasing the detection range of the measurement to an urban scale, it would be possible to simulate a dirty bomb explosion or train people for a nuclear disaster. Figure 9 demonstrates the next-generation USOTOPE prototyped in 2020. The experiment mimics the display of a predetermined amount of radiation at prespecified GPS coordinates. In an actual explosion and diffusion of radioactive materials, wind direction and the presence or absence of shielding would have significant impact; simulations taking these factors into account can be performed by Monte Carlo simulators such as the Particle and Heavy Ion Transport code System (Japan Atomic Energy Agency, 2020). In the future, USOTOPE will be able to integrate Monte Carlo simulator calculation results with its GPS mapping to simulate radiation doses at a large scale even more realistically.

Now that small radiation sensors like Pokéga are widely used, citizens are also important measurement points. In the case of actual wide-area radiological terrorism, coordination between law enforcement agencies and citizens using consumer radiation detectors should be planned. For example, it would be effective for these agencies to compile citizen-measured data shared over social media to serve as reference information for countermeasures. In fact, the 2016 film *Shin Godzilla* includes a scene of precisely such a response. In the future, USOTOPE could be effectively used for joint training between citizens and the government.

Pocket PM2.5 Sensor

The two case studies presented so far have focused on radiation monitoring. The final case study concerns the Pocket PM2.5 Sensor, a solution developed by the authors for participatory monitoring of air pollution, which, like radiation, is largely invisible.

Figure 9. Prototype of GPS-based USOTOPE: assuming that high levels of radioactive material were placed on the campus, radiation levels were visualized based on GPS location

Background

The World Health Organization (2018a, 2018b) announced that seven million people die prematurely each year from lung cancer and respiratory diseases caused by air pollution, such as by microparticulate matter. About 90% of the world's population lives amid polluted air. This pollution is particularly severe in low- and middle-income countries in Asia and Africa, which account for more than 90% of the deaths.

The difficulty for risk communication in these contexts is that PM2.5 (airborne particulate matter with a diameter of 2.5 µm or less) has few acute effects except in certain highly sensitive groups (e.g., the elderly, those with respiratory illnesses, and children). Since most people, except sensitive groups, have no immediate symptoms when they are in a contaminated area, they do not perceive the risks involved. For example, in certain areas in India, the Middle East, and Africa, the authors' field observations show that, even when PM2.5 concentration exceeds 100 µg/m^3 and cityscapes are conspicuously hazy, people seem to remain unperturbed, and few wear masks because they feel nothing is wrong with their bodies. In the short term, PM-related mortality increases by 2.8% for every 10-µg/m^3 increase in PM2.5 exposure (Kloog et al., 2013).

It is in this context that the authors, drawing on their experience with Pokéga, developed the Pocket PM2.5 Sensor (Figure 10) in 2016, a smartphone-connected air pollution sensor that enables anyone to measure the concentration of toxic fine particles in the atmosphere. The Pocket PM2.5 Sensor (hereafter Pocket PM) uses a light-scattering method involving a laser light emitting diode (LED) and photodiode to achieve compactness, low cost, and low power consumption. It can measure the concentration of harmful PM2.5 and PM10 (particles 10 µm or less in diameter) in µg/m^3 within a range of 0–999.9 µg/m^3. Measurement results, along with GPS location information, can be shared in comma separated value (CSV) or Google Earth KML format and visualized on a map, as Figure 11 shows (Ishigaki & Tanaka, 2017; Ishigaki, Matsuno, & Tanaka, 2017).

Figure 10. Pocket PM2.5 Sensor; the first model was connected to a smartphone with a USB cable, but later improvements enabled Bluetooth connection. The next version of Pocket PM2.5 Sensor [PRO] can store a year's worth of records in its built-in memory without a smartphone, as well as location information via GPS.

Figure 11. Data visualization by Pocket PM2.5 Sensor with Google Earth: the concentration is indicated by blue to red markers. An area of high concentration of 160ug/m³ is visualized on the right side of the screen. There is an exhaust vent from a basement restaurant here.

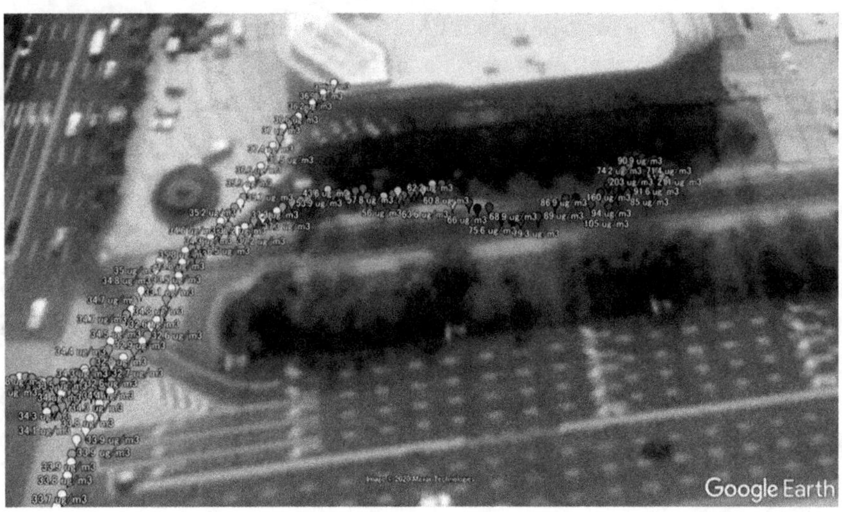

What is more, Pocket PM changes screen color according to the color-coding rules set by the United States Environmental Protection Agency (2012). This allows even the average user to understand the degree of risk associated with the measurements taken.

Air Pollution in Developing Countries

Pocket PM is now being used in field studies in developing countries such as India, Myanmar, Sri Lanka, Indonesia, Rwanda, Uganda, and China. As this fieldwork has progressed, it has become clear that PM2.5 concentration varies greatly locally—a situation unique to developing countries. In these contexts, financial constraints make it difficult to install many instruments for fixed-point observation of PM2.5. The Berkeley Earth site (http://berkeleyearth.org/air-quality-real-time-map/), which collects data from measurement stations around the world, shows that these stations are concentrated in developed countries. Moreover, when the authors actually measured, it was found that PM2.5 concentration varies greatly depending on micro-environmental factors such as road conditions and use areas. This is possibly related to sources of PM2.5 emissions not found in developed countries, such as uninspected cars and motorcycles, open fires, and roadside industries, seen in authors' field studies in Myanmar and China through 2017 to 2018 (Yi et al., 2018; Ishigaki et al., 2017).

At the same time, there is potential for a significant reduction in personal exposure through individual behavioral change, such as reducing these crude emissions, avoiding living near emission sources, and temporarily taking protective measures, such as using masks and air purifiers. This approach to improving the environment through encouraging individual action is called nudging (Hansen et al., 2016), which holds great potential for advancements in the field of public health. From this perspective, in 2019, the authors began a pilot study of 24-h personal exposure assessment study of PM2.5 using Pocket PM combined with a GPS logger in career women and housewives in Myanmar (Win-Shwe et al., 2019).

Household Air Pollution in Low-Income Countries

Household air pollution (HAP), a form of indoor PM2.5 pollution, is also a serious problem in developing countries. It is particularly prevalent in poor countries in sub-Saharan Africa, the Middle East, and South to Southeast Asia, causing 1.6 million premature deaths per year (Ritchie & Roser, 2014). It is argued that poorer households rely more heavily on primitive solid fuels, such as crop residues, dung, firewood, and charcoal (World Health Organization, 2006). When these fuels are burned indoors, large amounts of PM2.5 are emitted into the room.

Since the figure of 1.6 million is epidemiologically calculated, and it is impossible to know the individual details of those who have died, the real risk of air pollution remains a nebulous concept for most people. PM2.5 has limited acute biological effects such as fainting and convulsions, unlike toxic gas; therefore, even if concentrations increase, its adverse effects are imperceptible in the short term. By contrast, if PM2.5 pollution had a similar acute effect as that of carbon monoxide, for example, ventilation structures of houses would certainly have been designed differently. For this reason, a measurement device that complements human perceptual ability may help to perceive, and therefore mitigate, risk.

Fieldwork in Rwanda

In Rwanda, 98% of rural residents still use open fire stoves (so-called three-stone stoves), with wood and charcoal as the main fuels. In a study by the authors, continuous measurements were taken using Pocket PM in five Rwandan households through 2017 to 2019. The results showed that households using solid fuels such as firewood and charcoal were exposed to high concentrations of PM2.5 in excess of 1 mg/m^3 during cooking, as shown in Figure 12 (Ishigaki, Matsuno, & Tanaka, 2017; Yoda et al., 2019). Those exposed were mainly women – highlighting that this public health issue is also a gender issue – and when the nurse on the research team interviewed participants about their respiratory conditions, they reported subjective symptoms such as coughs and tears.

Figure 12. Use of firewood and high PM2.5 concentration in Rwanda. Most of the housewives in these households have no risk perception about PM2.5.

Figure 13 shows the hourly average PM2.5 for the five households where instruments were installed. Peak concentrations appeared to follow the different households' cooking times, but PM2.5 concentration seemed to vary by fuel type. The use of charcoal was less polluting than firewood, while gas contributed the least pollution.

In order to raise awareness in a local Rwandan community about PM2.5, its risks, and realistic protective measures, the authors conducted a trial workshop with Pocket PM in the village of Twishorezo, outside Kigali, in July 2018. The villagers expressed various opinions, such as "I didn't think there was bad smoke in the kitchen," "Sure, there are coughs and tears," and, "But I have to cook."

The availability heuristic is a form of cognitive bias in which greater visibility and communication about a risk results in greater risk awareness, and, conversely, less visibility and communication leads to increased underestimation of a risk (Folkes, 1988). The Rwandan fieldwork revealed a need among local residents for a combination of increased PM2.5 awareness and inexpensive protective measures.

A demonstration experiment was conducted to install solar-powered ventilation fans in households with the highest levels of pollution in January 2019. Figure 14 shows the moment when ventilation fans were introduced for the first time in a low-income village in Kigali. The ventilator significantly reduced PM2.5 during the hours when the kitchen was used, and adverse symptoms appeared to decrease. Unfortunately, the cost of battery maintenance and the installation of the solar panels themselves proved to be too high for widespread adoption of this system.

Figure 13. Hourly average of PM2.5 concentration in five households in Rwanda. Both No. 1 and No. 2 households use both firewood and charcoal, so they show higher peaks in pollution concentration than the other households. No. 3, which uses gas, has the cleanest air, while No. 4 and No. 5, which use only charcoal, have an intermediate level of pollution.

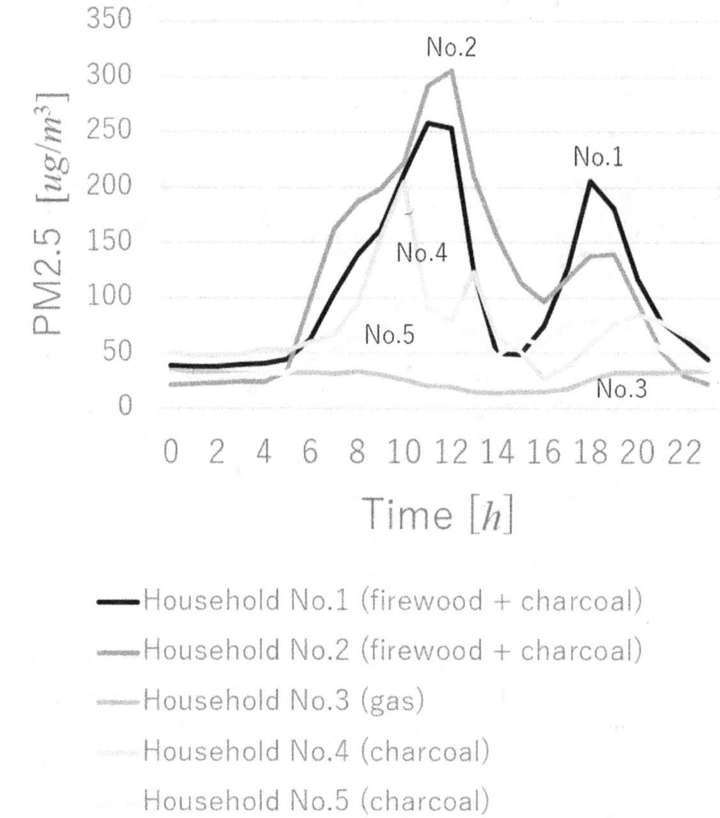

Figure 14. The installation of a solar-powered ventilation system in a house in Kigali, Rwanda. The smoke is discharged from an exhaust vent mounted at the top of the exterior wall. The device is running on solar power and continues to operate at the time of this writing (August 2020).

Discussion

Pocket PM can help residents quantify the current state of the HAP and the effectiveness of measures put in place. Furthermore, continuous data on PM2.5 concentrations collected through the network can be used for epidemiological studies when combined with architecture, aerosol science, or health status.

In addition to technical solutions through improved fuels and equipment, the solution to the problem of HAPs will depend on mechanisms that enable residents to recognize the problem themselves and encourage behavioral change. The process by which individuals gain democratic participation in community life is called empowerment (Zimmerman & Rappaport, 1988). In empowerment for HAP solutions, residents should first have increased awareness and empathy for the problem; second, a sense of self-efficacy that residents themselves can solve problems; and finally, an active participation in effective measures with a sense of ownership. For now, Pocket PM is just a measuring device, but it could be expanded as a tool to achieve empowerment by using smartphones to express information in a variety of ways and by linking with social media, as shown in the case of Pokéga.

An effective solution to HAP in developing countries has yet to be found. Ideas for solutions could perhaps come from a synthesis of diverse disciplines such as technology, community nursing, and psychology. This approach to problem-solving is often defined as design thinking (Cross, 1982). In the next section, some of the new research ideas currently being prepared will be discussed, combining the entertainment sector with measurement technology.

Future Directions

The next trial in 2020, which will be conducted in Rwanda, is an attempt to enhance risk awareness among local children through gamification, a method that introduces an element of play to learning, in order to achieve objectives efficiently and enjoyably. The trial involves a small gaming device the size of a smartwatch that uses Pocket PM. It will run an interactive-type game that integrates location information, similar to Tamagotchi and Pokémon GO, with a character whose behavior changes according to the concentration of PM2.5 in the air (Figure 15). In this context, children are regarded as important players in health education because their literacy rate and education level are higher than that of their parents' or grandparents' generation (Nyirimanzi, 2012). It is hoped that this game terminal will make them aware of the invisible presence of PM2.5 from a young age and raise awareness within their communities in turn. Because many children in Rwanda attend school, the researchers will work with local schools to advance the demonstration experiment by integrating the game into extracurricular activities.

Globally, there is an activity called Improved Cookstoves (ICS) to promote highly efficient cooking equipment in developing countries (Accenture Development Partnerships, 2012). However, the authors found no one using the ICS in local Rwandan villages. According to residents, the metals and ceramics required for ICS are expensive and cannot be bought by the poor. The number one reason for using firewood is "because it's free" (Butare et al., n.d.). Rural communities in Rwanda require solutions based on "frugal engineering" (Le Bas, 2016) that can be achieved with free materials. A complete solution has yet to be found.

Figure 15. Game design for PM2.5 awareness in children. In the left image, the concentration of PM2.5 is so low (13 ug/m³) that the character is living normally, while in the right image, the concentration is so high (300 ug/m³) that the character is crying. The color of the background changes from blue to red depending on the PM2.5 concentration.

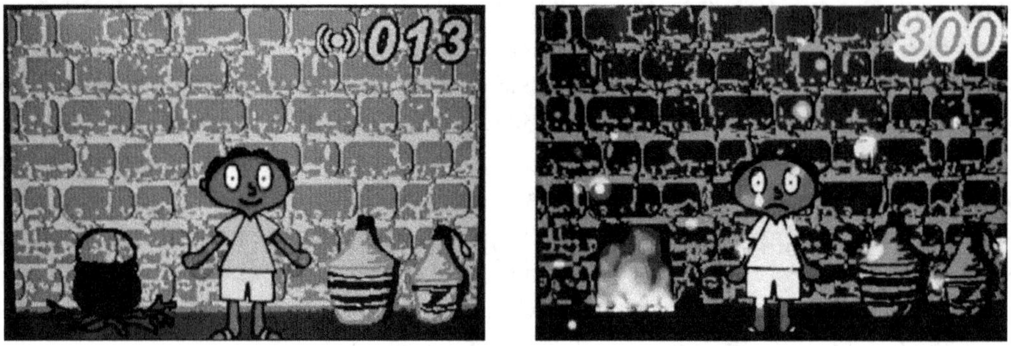

DISCUSSION

The ability to accurately measure and visualize environmental disaster situations with smartphones and discuss them freely through social media and citizen journalism seems to enrich social capital in disaster response and thus increases resilience. This section discusses the challenges therein.

Social capital comprises three elements: trust, potential for cooperation, and networks (Paldam, 2000). In Japanese, the term "kizuna" is used to collectively refer to these three states. A year and a half after the Fukushima nuclear power plant accident, the researchers examined changes in Japanese people's ties and found that people in areas far from the disaster area (e.g., Tokyo and Kyoto) tended to believe that their ties had been strengthened, while people in the affected areas tended to believe that their ties had been weakened. This is due to discrimination against, and rumors about, the affected areas. Therefore, designers of disaster communication systems must give due consideration to fake news and the discrimination the system may bring against infected areas and individuals. In the following section, some specific solutions are considered.

First, appointing a good facilitator helps maintain the neutrality and objectivity of discussions over social media. The authors found that organizing discussions visually through Goal Structuring Notation (GSN) makes it easier for the public to notice scientifically inaccurate opinions and increases the satisfaction and understanding of participants (Matsuno et al., 2016). The visualization of issues and problems through infographics will also aid a smooth discussion. Fortunately, as a result of several individuals, including the authors, acting as volunteer facilitators in the Pokéga's Facebook group, there is no evidence of unscientific or discriminatory discourse. This is partly due to the fact that Facebook is a real, name-based media. If anonymous media were to be used for disaster communications, it would be worthwhile to consider a rating system for the sender, a system of professional checks, and a reporting system.

Privacy and data reliability are also issues. Mapping measurements of radiation and air pollution can lead to estimates of an individual's residence and movement history. Spatial interpolation of the set of measurement points would be one solution. Consideration should also be given to the security of deliberately low or high values recorded as a form of sabotage to the system. Data of questionable spatial statistical reliability should be cross-checked and verified by experts.

CONCLUSION

This chapter provided practical examples of the use of smartphones to measure, share data on, and discuss environmental hazards such as radiation and air pollution in order to manage them. In recent years, new types of environmental accident and disaster risks such as radiation, CBRNE terrorism, and PM2.5 have risen in prominence. These risks share three common issues: (1) special equipment is required for risk assessment, (2) the degree of risk differs greatly depending on the location of only a few meters, and (3) a high level of expertise is required for risk assessment.

Pokéga presents a case of citizen-led radiation monitoring in which a large amount of environmental monitoring information has been collected and visualized by making citizens the main body for measurement; the project has supported not only local environmental monitoring but also independent decision-making through sharing and discussion with experts. Whereas Pokéga measures real radiation, USOTOPE works as a handy simulator. Therefore, USOTOPE will be an effective tool in training and educating not only law enforcement personnel but also citizens during radiation disasters or terrorist incidents. In the emergency management of radiation protection, information sharing should be interactive, not only at the public level of the state and administration but also at the civil level. In particular, when considering social issues such as NORM that involve complex regional risk awareness, legal regulations, and economic promotion, data sharing and social consensus building based on scientific discussions are essential. PM2.5 pollution has similar aspects to radiation in terms of protection: humans cannot perceive it, and the general public is largely ignorant about it. The widespread use of smartphones throughout the world, including in the developing world, could give humanity a new organ of risk perception.

In traditional risk assessment, top-down discussions between the risk assessor and manager determine risk management policy. In these scenarios, data measurement, verification, and monitoring are done as needed, but they are largely optional, for expert decision-making is explicitly regarded as the most important. By contrast, the authors believe that big data collected via citizen participation using smartphones may initiate a new relationship of cooperation between members of the public and experts, government agencies, and local governments that will lead to more efficient, dynamic, and robust disaster prevention planning, hazard mapping, and the detection and prediction of disasters. In this new social framework for environmental risk assessment, citizens will be able to build consensus while maintaining trust in government and experts through open access to and visibility of data from public and professional bodies.

If each individual can measure and share environmental risks on a daily basis, avoid unilateral and excessive expectations of the government, and make decisions through mutual discussions with experts and local governments, the realization of a highly literate and self-determined disaster mitigation society becomes possible, in which citizens themselves are empowered to take appropriate risk-prevention actions in the event of a disaster without waiting for instructions. The key to this new society must be the development of new physical, cyber, and socially connected smart sensing services based on low-cost and widely used general-purpose technology platforms, such as semiconductor sensors and smartphones.

ACKNOWLEDGMENT

This work was partially supported by a JSPS Grant-in-Aid for Scientific Research [15H01788, Pocket Geiger; 19K04980, USOTOPE; 19H03957, Pocket PM2.5 Sensor].

If you are interested in this research, please contact the authors and their research group. We look forward to your participation.

REFERENCES

Accenture Development Partnerships. (2012, April). *Global Alliance for Clean Cookstoves: Rwanda market assessment*. Global Alliance for Clean Cookstoves. doi:10.1163/9789004322714_cclc_2016-0006-001

Anderson, C. (2014). *Makers: The New Industrial Revolution*. Currency.

Auger, P., Burke, P., Devinney, T. M., & Louviere, J. (2003). What will consumers pay for social product features? *Journal of Business Ethics*, *42*(3), 281–304. doi:10.1023/A:1022212816261

Burke, J. A., Estrin, D., Hansen, M., Parker, A., Ramanathan, N., Reddy, S., & Srivastava, M. B. (2006). *Participatory sensing* [Paper presentation]. World Sensor Web Workshop, ACM SenSys 2006, Boulder, CO, United States. https://escholarship.org/uc/item/19h777qd

Butare, A., & Munyampundu, A. (n.d.). *Towards an improved cook stoves program: Market based solutions to eliminate energy poverty. Enquiry on the use of improved cooking stoves in Bugesera, Kirehe and Ngororero Districts*. Draft report. Africa Energy Services Group. (Unknown issue date) https://waterportal.rwfa.rw/sites/default/files/inline-files/SNV_cookstoves%20AESG%20final%20ICS%20draft%20report.pdf

Chida, K., Morishima, Y., Inaba, Y., Taura, M., Ebata, A., Takeda, K., Shimura, H., & Zuguchi, M. (2012). Physician received scatter radiation with angiography systems used for interventional radiology: Comparison among many x-ray systems. *Radiation Protection Dosimetry*, *149*(4), 410–416. doi:10.1093/rpd/ncr312

Corporation, S. T. (2011, July 26). Seikatsusha no fuan wo kaisho- surutame shutodaigaku Tokyo to kyo-udou kaihatsu kateiyou houshasen sokuteiki air-counter wo shinhatsubai [S.T. and Tokyo Metropolitan University launch Air Counter, a home radiation measuring instrument developed in collaboration to alleviate anxiety among consumers]. *S.T. Corporation*. https://www.st-c.co.jp/release/2011/20110726_000266.html

Cross, N. (1982). Designerly ways of knowing. *Design Studies*, *3*(4), 221–227. doi:10.1016/0142-694X(82)90040-0

Dearnaley, G., & Northrop, D. C. (1966). *Semiconductor counters for nuclear radiations* (2nd ed.). Wiley.

Endsley, M. R. (1995). Toward a theory of situation awareness in dynamic systems. *The Journal of the Human Factors and Ergonomics Society*, *37*(1), 32–64. doi:10.1518/001872095779049543

Estellés-Arolas, E., & González-Ladrón-de-Guevara, F. (2012). Towards an integrated crowdsourcing definition. *Journal of Information Science*, *38*(2), 189–200. doi:10.1177/0165551512437638

Folkes, V. S. (1988). The availability heuristic and perceived risk. *The Journal of Consumer Research*, *15*(1), 13–23. doi:10.1086/209141

Fujibuchi, T., Inoue, A., Ishigaki, Y., & Matsumoto, Y. (2019a). Development of a wireless multisensor active personal dosimeter–tablet system. *Progress in Nuclear Science and Technology*, *6*(0), 73–76. doi:10.15669/pnst.6.73

Fujibuchi, T., Nozaki, Y., Ishigaki, Y., & Matsumoto, Y. (2019b). Estimation of the characteristics of gamma-ray dose measurements with an experimental wireless dose monitoring system. *Progress in Nuclear Science and Technology*, *6*(0), 77–80. doi:10.15669/pnst.6.77

Hansen, P. G., Skov, L. R., & Skov, K. L. (2016). Making healthy choices easier: Regulation versus nudging. *Annual Review of Public Health*, *37*(1), 237–251. doi:10.1146/annurev-publhealth-032315-021537

Hazin, C. A. F., Gazineu, M. H. P., & de Farias, E. E. G. (2008). *Uranium and thorium in zircon sands processed in Northeastern Brazil* [Paper presentation]. IRPA 12: 12 International Congress of the International Radiation Protection Association (IRPA): Strengthening radiation protection worldwide, Buenos Aires, Argentina. 10.1148/86.2.354b

Ikeda, S., & Maeda, Y. (Eds.). (2013, March 11). *Emerging issues learned from the 3.11 disaster as multiple events of earthquake, tsunami, and Fukushima nuclear accident*. The Society for Risk Analysis. http://www.sra-japan.jp/cms/uploads/311Booklet.pdf

Iniewski, K. (Ed.). (2010). *Semiconductor radiation detection systems*. CRC Press. doi:10.1201/9781315222608

Iniewski, K. (Ed.). (2011). *Electronics for radiation detection*. CRC Press.

International Atomic Energy Agency. (2014). *IAEA report on international experts meeting radiation protection after the Fukushima Daiichi accident: Promoting confidence and understanding*. International Atomic Energy Agency. https://www-pub.iaea.org/MTCD/Publications/PDF/IEM-6_web.pdf

International Atomic Energy Agency. (2016). *IAEA Incident and Trafficking Database (ITDB): Incidents of nuclear and other radioactive material out of regulatory control – 2016 Fact Sheet*. International Atomic Energy Agency. doi:10.1787/9789264244047-24-en

International Commission on Radiological Protection. (2007). The 2007 recommendations of the International Commission on Radiological Protection (ICRP Publication 103). *Annals of the ICRP*, *37*(2–4). Advance online publication. doi:10.1177/ANIB_37_2-4

Ishigaki, Y., Matsumoto, Y., Pradana, H. A., & Tanaka, K. (2017). *Citizen sensing for environmental risk communication action research on PM2.5 air quality monitoring in East Asia* [Paper presentation]. *The Second International Conference on Cyber-Technologies and Cyber-Systems*, Barcelona, Spain.

Ishigaki, Y., Matsuno, Y., Bando, K., & Tanaka, K. (2017). Wisdom of crowds for reliable discussion and need finding: A case study of information sharing regarding radiation after the Fukushima Nuclear Disaster. *Proc. of the 50th Hawaii International Conference on System Sciences (HICSS)*, 323–331. https://www.researchgate.net/publication/317121029_Wisdom_of_Crowds_for_Reliable_Discussion_and_Need_Finding_A_Case_Study_of_Information_Sharing_Regarding_Radiation_after_the_Fukushima_Nuclear_Disaster

Ishigaki, Y., Matsuno, Y., & Tanaka, K. (2017). *Agile way of risk awareness by smartphone-connected environmental sensors* [Paper presentation]. *14th International Conference on Information Systems for Crisis Response and Management*, Albi, France.

Ishigaki, Y., Pradana, H. A., Permana, S., & Tanaka, K. (in press). Dose-rate mapping using smartphone for risk awareness in local residents and workers at Zircon Sand Facility in Bangka Island in Indonesia (Extended abstract). *Proc. of 15th International Congress of the International Radiation Protection Association (IRPA15).*

Ishigaki, Y., & Tanaka, K. (2017). *Pollution mapping by smartphone becoming real possibility* [Paper presentation]. *International Conference on Civil, Disaster Management and Environmental Sciences*, Bali, Indonesia.

Japan Atomic Energy Agency. (2020). *PHITS: Particle and Heavy Ion Transport Code System.* https://phits.jaea.go.jp/index.html

Kitaguchi, H., Miyai, H., Izumi, S., & Kaihara, A. (1996). Silicon semiconductor detectors for various nuclear radiations. *IEEE Transactions on Nuclear Science*, *43*(3), 1846–1850. doi:10.1109/23.507234

Kloog, I., Ridgway, B., Koutrakis, P., Coull, B. A., & Schwartz, J. D. (2013). Long- and short-term exposure to PM2.5 and mortality: Using novel exposure models. *Epidemiology (Cambridge, Mass.)*, *24*(4), 555–561. doi:10.1097/EDE.0b013e318294beaa

Knoll, G. F. (2010). *Radiation detection and measurement* (4th ed.). Wiley.

Kremen, C., Ullman, K. S., & Thorp, R. W. (2011). Evaluating the quality of citizen scientist data on pollinator communities. *Conservation Biology*, *25*(3), 607–617. doi:10.1111/j.1523-1739.2011.01657.x

Kuipers, T., Van't Wout, C., & Bader, F. (2011). iPhone als stralingsdetector [iPhone as radiation detector]. *Nederlands Tijdschrift voor Stralingsbescherming*, *2*(2), 31–34.

Le Bas, C. (2016). Frugal innovation, sustainable innovation, reverse innovation: Why do they look alike? Why are they different? *Journal of Innovation Economics & Management*, *3*(21), 9–26. doi:10.3917/jie.021.0009

Malone, T. W. (2004). *The future of work.* Harvard Business School Press., doi:10.1007/0-387-28918-6_2

Masamura, T. (2013). Why was the trust for the science and the mass media lost? *Trends in the Sciences*, *18*(1), 42–45. doi:10.5363/tits.18.1_42

Matsuno, Y., Ishigaki, Y., Bando, K., & Kido, K. (2016). *Developing SNS tool for consensus building on environmental safety using assurance cases* [Paper presentation]. International Conference on Computer Safety, Reliability, and Security. https://doi:10.1007/978-3-319-45480-1_5

MEXT. (2012, February 21). *Real-time sen-ryosokutei system no Fukushima-ken naino Kuukansenryo-ritsu no real-time sokuteikekka no kouhyou ni tsuite* [Release of the results of real-time air dose rate measurements in Fukushima prefecture using the real-time dosimetry system]. Ministry of Education, Culture, Sports, Science, and Technology.

National Consumer Affairs Center of Japan. (2011a, September 8). Hikakuteki anka na houshasen sokuteiki no seinou [Performance of relatively inexpensive radiation measuring instruments]. *National Consumer Affairs Center of Japan*. http://www.kokusen.go.jp/news/data/n-20110908_1.html

National Consumer Affairs Center of Japan. (2011b, December 22). Hikakuteki anka na houshasen sokuteiki no seinou dai 2 dan [Performance of relatively inexpensive radiation detectors – Part 2. *National Consumer Affairs Center of Japan*. http://www.kokusen.go.jp/news/data/n-20111222_1.html

National Consumer Affairs Center of Japan. (2012, May 24). Digital shiki kojin senryo-kei no test kekka [Test results of digital personal dosimeter]. *National Consumer Affairs Center of Japan*. http://www.kokusen.go.jp/pdf/n-20120524_1.pdf

National Institute for Defense Studies. Japan. (2015). CBRN defense: Responding to growing threats. In East Asian Strategic Review 2015 (pp. 308–310). doi:10.17070/aeaas.2015.12.40

Nihei, Y. (2014). Fukushima Dai-ichi Genshiryoku-hatsudensho Jikogo no Housyasei-busshitsu kakusan ni-yoru "Fuuhyou-higai" ha Shinriteki "Hanka-higai" Dearu: Shakai-kankei-shihon toshiteno Chiiki-kan Sougo Shinrai no Soushitsu heno Eikyou [Financial damage caused by radioactive fallout from the Fukushima No.1 Nuclear Power Plant is not the result of "damage from rumors", but of psychological generalization of fear: Changes in inter-regional trust as social capital in Japan]. *Hakuoh Journal of the Faculty of Education, 8*(1), 131–162, 2014–05. https://ci.nii.ac.jp/naid/110009814955

Norton, M. I., Mochon, D., & Ariely, D. (2012). The IKEA effect: When labor leads to love. *Journal of Consumer Psychology, 22*(3), 453–460. doi:10.1016/j.jcps.2011.08.002

Nyirimanzi, C. (2012). Rwanda's literacy rate rises. *National Institute of Statistics of Rwanda*. http://statistics.gov.rw/node/1086

Paldam, M. (2000). Social capital: One or many? Definition and measurement. *Journal of Economic Surveys, 14*(5), 629–653. doi:10.1111/1467-6419.00127

Ritchie, H., & Roser, M. (2014). Indoor air pollution. *Published online at OurWorldInData.org*. https://ourworldindata.org/indoor-air-pollution

Segault, A., Tajariol, F., Ishigaki, Y., & Roxin, I. (2016). Sharing radiation measurements through social media: A methodological user-oriented proposal set of guidelines. *International Journal of Information Systems for Crisis Response and Management, 8*(2), 17–30. doi:10.4018/IJISCRAM.2016040102

Spieler, H. (2005). *Semiconductor detector systems*. Oxford University Press. doi:10.1093/acprof:oso/9780198527848.001.0001

Takayama, T., Katoh, K., Imamura, F., Kawata, Y., Murata, S., & Takahashi, S. (2018). Chapter 4: Prevention and Mitigation of Tsunami Disasters. In Tsunami: To Survive from Tsunami 2nd edition, Advanced Series on Ocean Engineering Book 46, English Edition (p. 341). World Scientific Publishing Company.

Tanaka, K., & Itoh, M. (2003). Saigaiji ni tekikaku na kikenkaihikoudou wo michibikutameno jouhou communication [Communication and information inducing suitable danger-avoidance actions from disaster]. *Journal of Disaster Information Studies, 1*, 61–69.

Terasaki, K., Fujibuchi, T., Murazaki, H., Kuramoto, T., Umezu, Y., Ishigaki, Y., & Matsumoto, Y. (2017). Evaluation of basic characteristics of a semiconductor detector for personal radiation dose monitoring. *Radiological Physics and Technology*, *10*(2), 189–194. doi:10.100712194-016-0384-z

Tsuchiya, K., Moritake, T., Ishigaki, Y., & Kosukegawa, N. (2018a). *Development of a radiation-detection simulator with smartphones and beacons for first responders against radiological threat* [Paper presentation]. 4th International CBRNe Workshop, Rome, Italy.

Tsuchiya, K., Moritake, T., Ishigaki, Y., Kosukegawa, N., Uehara, K., Matsumoto, Y., Tanabe, K., Kurosawa, K., Akiba, N., Kakuda, H., Imoto, D., Hirabayashi, M., Hawai, Y., & Kuroki, K. (2018b). *Development of effective training methods for first responders against radiological threat* [Paper presentation]. 8th European Academy of Forensic Science Conference, Lyon, France. 10.3408/jafst.759

Turoff, M., van de Walle, B., & Hiltz, S. R. (Eds.). (2010). Emergency response information systems: Past, present, and future. In M. Turoff, B. van de Walle, & S. R. Hiltz (Eds.), Information systems for emergency management (pp. 369–388). Routledge. doi:10.4018/978-1-60960-609-1.ch002

United States Environmental Protection Agency (US-EPA). (1992). *Framework for ecological risk assessment* (Report no. EPA/630/R-92/001). Risk Assessment Forum, United States Environmental Protection Agency. doi:10.2458/azu_geo_arizona_epasuperfund_1997

United States Environmental Protection Agency (US-EPA). (2012). The National Ambient Air Quality Standards for Particle Pollution. United States Environmental Protection Agency. doi:10.2458/azu_geo_arizona_epasuperfund_1997

Van de Weijer, B. (2013, July 13). Een onderzoeker in ieders broekzak: Smartphones in het regenwoud [A researcher in everyone's pocket: Smartphones in the rainforest]. *de Volkskrant*, 32–33.

Weick, K. E. (1987). Organizational culture as a source of high reliability. *California Management Review*, *29*(2), 112–127. doi:10.2307/41165243

Win-Shwe, T. T., Thein, Z. L., Aunf, W. Y., Yi, E.E.P.N, Maung, C., Nway, N. C., Thant, Z., Suzuki, T., Mar, O., Ishigaki, Y., & Nakajima, D. (in press). Improvement of GPS-attached Pocket PM2.5 measuring device for personal exposure assessment. *Journal of University of Occupational and Environmental Health*.

World Health Organization. (2006). *Fuel for life: Household energy and health*. World Health Organization. https://apps.who.int/iris/handle/10665/43421

World Health Organization. (2018a, May 2). *9 out of 10 people worldwide breathe polluted air, but more countries are taking action*. doi:10.1163/2210-7975_hrd-9841-20180002

World Health Organization. (2018b). *WHO Global Ambient Air Quality Database (update 2018)*. World Health Organization. https://www.who.int/airpollution/data/cities/en/

Yi, E. E. P. N., Nway, N. C., Aung, W. Y., Thant, Z., Wai, T. H., Hlaing, K. K., Maung, C., Yagishita, M., Ishigaki, Y., Win-Shwe, T. T., Nakajima, D., & Mar, O. (2018). Preliminary monitoring of concentration of particulate matter (PM2.5) in seven townships of Yangon City, Myanmar. *Environmental Health and Preventive Medicine*, *23*(1), 53. doi:10.118612199-018-0741-0

Yoda, T., Elisephane, I., Ishigaki, Y., Kim, M., Matsukawa, M., Matsumoto, Y., Munyegera, G., Yamaguchi, K., & Kondo, H. (2019). *Seven months observations of indoor air pollution in Kigali Rwanda* [Paper presentation]. *15th International Conference on Atmospheric Sciences and Applications to Air Quality*, Kuala Lumpur, Malaysia.

Zimmerman, M. A., & Rappaport, J. (1988). Citizen participation, perceived control, and psychological empowerment. *American Journal of Community Psychology*, *16*(5), 725–750. doi:10.1007/BF00930023

ADDITIONAL READING

Baumont, G. (2018). Nuclear crisis preparedness lessons learned from Fukushima Daiichi. In M. Bourrier & C. Bieder (Eds.), *Risk communication for the future: Towards smart risk governance and safety management* (pp. 45–60). Springer. doi:10.1007/978-3-319-74098-0_4

Bijak, M. (Ed.). (2018). *CBRN. Security manager handbook*. Łódź University Press. https://wydawnictwo.uni.lodz.pl/wp-content/uploads/2018/10/CBRN-1-1.pdf

Fassert, C., & Hasegawa, R. (2019). *Shinrai research project: The 3/11 accident and its social consequences – Case studies from Fukushima prefecture* (Report no. IRSN/2019/00178). Institut de Radioprotection et de Sûreté Nucléaire. https://www.irsn.fr/FR/connaissances/Installations_nucleaires/Les-accidents-nucleaires/accident-fukushima-2011/fukushima-2019/Documents/IRSN-Report-2019-00178_Shinrai-Research-Project_032019.pdf

Fukushima Booklet Committee. (2015). *10 lessons from Fukushima*. http://fukushimalessons.jp/en-booklet.html

Iwarsson, E. (2018). Tin mining conflicts in Bangka-Belitung, Indonesia. *Environmental Justice Atlas*. https://ejatlas.org/conflict/bangka-belitung-mining

Ritchie, H., & Roser, M. (2014). Indoor air pollution. *Published online at OurWorldInData.org*. https://ourworldindata.org/indoor-air-pollution

Temper, L., del Bene, D., & Martinez-Alier, J. (2015). *Environmental Justice Atlas*. https://ejatlas.org/

United States Joint Chiefs of Staff. (2018). *Operations in Chemical, Biological, Radiological, and Nuclear Environments* (Joint Publication 3-11). Joint Chiefs of Staff. https://www.jcs.mil/Portals/36/Documents/Doctrine/pubs/jp3_11.pdf?ver=2018-12-07-091639-697

Wahana Lingkungan Hidup Indonesia (Friends of the Earth Indonesia). *Grim portraits of Bangka Belitung tin mining*. Milieudefensie (Friends of the Earth Netherlands). https://issuu.com/walhi/docs/milieudefensie-rapport-walhi-j-lr

KEY TERMS AND DEFINITIONS

Crowdfunding: A method of fundraising via online platforms that allows details of any project to be published on the web and investment to be collected from individuals who are interested in investing in it. If put to good use, it can yield not only funding but also good public relations and networks.

Nudging: A method that encourages people to voluntarily choose the desired action rather than forcing them. For example, providing incentives, gamifying and entertaining, and engaging the unconscious through affordance in the design of products, architecture, graphics, and information.

Open Source Research: Research conducted in collaboration with researchers and engineers from around the world through the publication of research results and resources under an open source license. It has been adopted by many in the software sector, but there are some examples in the hardware and healthcare sectors.

Participatory Development: A method of rapidly developing products, goods, and services to meet social needs by involving a range of people (e.g., users, engineers, and researchers) right from the development stage, rather than confining development within a single domain organization, such as a manufacturer or service provider that provides value to consumers in a single direction. Examples include crowdfunding, open source research, DIY use, and social media use.

Participatory Monitoring: A situation where the public takes the initiative in sensing, voluntarily sharing measurement data, and engaging in discussions on social media. It is expected to speed up environmental risk-awareness, improve risk literacy, and create voluntary risk-avoidance behaviors.

PM2.5: Airborne particulate matter having a diameter of 2.5 μm or less. Because it is so small, it can pass through the cells of the embryo and enter the circulatory system, causing death from respiratory and cardiopulmonary diseases.

Social Product: A product that is valued because of the social context in which it was made and its ethical significance. In addition to the traditional values of brand, price, and function, it is considered a fourth value in consumer purchasing behavior.

Chapter 2
Emergency Response and Post-Disaster Recovery Using Smartphone-Based Applications

Shahriar Kaisar
RMIT University, Australia

ABSTRACT

The number of natural disasters, such as tsunamis, earthquakes, flooding, cyclone, and bushfires, is rapidly increasing globally, and they are claiming thousands of lives while destroying numerous properties. One of the major concerns of these natural disasters is the destruction of communication links, such as powerline and Internet connections, which make it difficult to enable communication among the affected people and the rescue teams. However, the evolution of smart devices equipped with multiple short-range communication technologies, such as Bluetooth and Wi-Fi provides an opportunity to form an ad-hoc network with co-located smart mobile device users and communicate their positions and other relevant information to the rescue workers. This chapter provides a detailed description of recent advancement in this area and highlights important aspects that are needed to be considered for practical implementation.

INTRODUCTION

The number of natural disasters, such as Tsunami, earthquake, flooding, bushfire, etc. and the destruction of lives and properties is increasing globally. The international disaster database suggests that in 2018, 315 natural disasters occurred across the world that resulted in 11,804 deaths while affecting a total of 68 million people and causing economic damage of US$131.7 billion (Natural disaster 2018, 2019). During the recent bushfire incident in Australia (2019-20), around 17 million hectares have burned, which is more than the size of South Korea and 33 people lost their lives including nine firefighters (Richards, Brew, & Smith, 2020). Approximately, 3000 homes have also been destroyed (Richards, Brew, & Smith, 2020). During a natural disaster, people often get trapped into damaged properties and need to be rescued within the first 72 hours, which is known as the golden relief time (Mezghani, Kortoc, Mitton, &

DOI: 10.4018/978-1-7998-6705-0.ch002

Francesco, 2019). However, the destruction of electric lines and cellular networks during such natural disasters make it difficult to establish communication with the affected people, locate their positions, and rescue them (Lu, Cao, & La Porta, 2017; Mezghani, Kortoc, Mitton, & Francesco, 2019; Sciullo, Trotta, & Di Felice, 2020). The rescue workers need to know the exact position of the affected people, their current conditions, and the severity of the damage before making a well-informed rescue decision and conducting a rescue operation. On the other hand, emergency information diffusion is also important in these scenarios to inform the affected people about the upcoming warnings or provide them with rescue instructions. The destruction of powerline hampers the traditional medium of communication, such as TV and radios. The destruction of cellular towers can also impact the availability of the cellular network, which ultimately makes it difficult to broadcast information through mobile messaging or phone calls. The availability also suffers due to high demand during a disaster. In these scenarios, making contact between rescue workers and the survivors become challenging and crucial.

The advent of smart mobile devices equipped with short-range communication technologies, such as Wi-Fi and Bluetooth, has opened new possibilities to address the above-mentioned issues. Smart mobile devices, such as smartphones, tablets, and PDAs have become extremely popular among users worldwide. The number of smartphone users around the world is currently 3.5 billion, which is approximately 45% of the world's population and it is expected that by 2021, the number of smartphone users will reach 3.8 billion (O'Dea, 2020). A similar trend is also observed for other smart mobile device usages. The short-range communication capability of these devices allows them to form a peer-to-peer type ad-hoc network among co-located devices to exchange messages, images, videos, and/or location information. The maximum distance for this type of communication between two devices can be 10m ~200m based on the underlying technology. In this case, Bluetooth offers a shorter range and consumes less energy while Wi-Fi does the opposite. However, it is possible to extend this range using multi-hop communication. Figure 1 shows the use of multi-hop communication in forming an ad-hoc network to share messages or contents including pictures, videos, or texts. In this case, the application installed in a source node (i.e., smart device user) is trying to send a message (can contain text, videos, or images) to a receiver node. Although an end-to-end communication path is shown in the figure for ease of presentation, such a persistent communication path is usually not available in an ad-hoc network due to node movement and intermitted connectivity. Therefore, the actual message delivery happens through multiple message-forwarding events that may occur at different timesteps rather than instantly. In such a network, a few devices play the role of a forwarder node who keeps carrying the message until they reach the destination or another suitable forwarder. This kind of message forwarding technique is called store-carry-and-forward message forwarding. Further details about this are discussed later in the routing module section. This type of communication is also referred to as opportunistic communication, and the path used for this communication is termed as an opportunistic communication path as demonstrated in Figure 1.

Multi-hop communication technique presented above can be employed in an emergency post-disaster response scenario to establish bi-directional communication among the affected people and the rescue workers. Such bi-directional communication will enable rescue workers to obtain necessary information from the survivors and spreading warnings, relief, and rescue information. In this case, a smartphone-based emergency application can use the short-range communication capability of smart mobile devices to form the network and exchange messages. The use of such a smartphone-based application will be highly beneficial in a disaster scenario as the infrastructure-based communication is reliant on the availability of cellular networks and its availability may be compromised due to the disaster. Researchers

Figure 1. Formation of an Ad-hoc network using Multi-hop communication among smart mobile devices. Although an end-to-end path is shown in the figure for ease of presentation, such persistent communication path may not be available in an ad-hoc network. Devices keep carrying the message until they meet another suitable forwarder, or the destination and the message forwarding may occur at different timesteps rather than instantly.

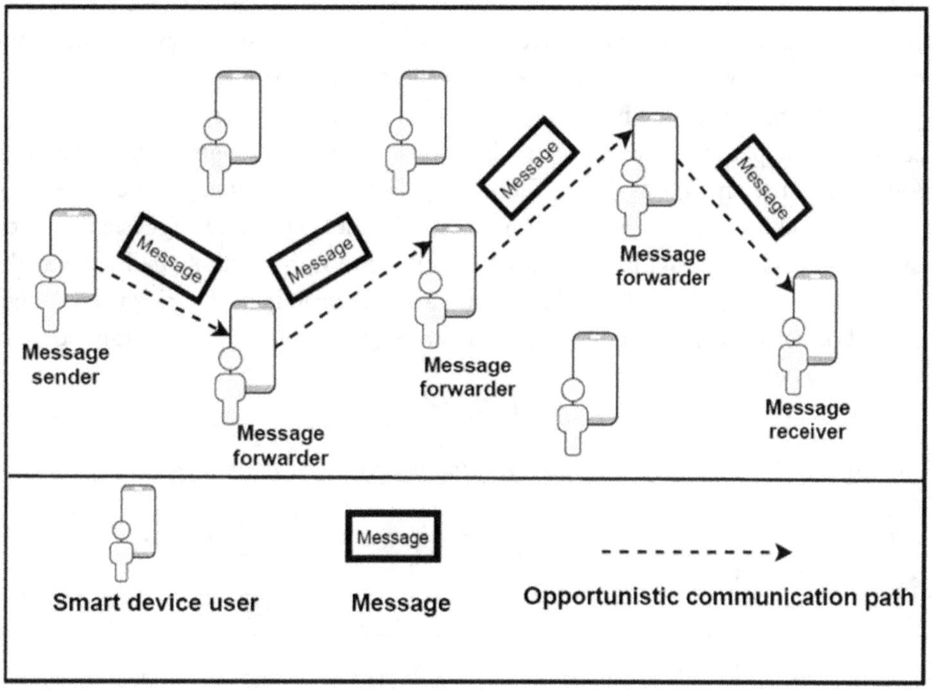

have recently proposed several techniques (Sciullo, Trotta, & Felice, 2020; Hossain & Ray, 2018; Yatbaz, et al., 2018; Pu & Zhou, 2019) to investigate the use of smartphones for emergency response and post-disaster recovery applications. The development of such a smartphone-based tool needs to address issues related to the underlying communication technology, message routing from survivors to rescuers and vice-versa, and the development of an application to transfer different types of data, victims localization, privacy and security issues. This chapter will systematically summarize and critically analyze the existing smartphone-based emergency response and post-disaster recovery methods and highlight the scope for future contributions.

BACKGROUND

A multi-hop communication network among nearby smart mobile devices provides a way for the nodes to communicate and exchange messages or contents. However, there are various challenges associated with the implementation of this type of network. Node mobility is one of the major challenges as nodes are dynamic in such a network and can move from one place to another to create disconnection. Other issues associated with this type of network include energy consumption, limited connectivity, frequent disconnection, high delivery delay, low delivery success rate, security, and privacy. Since smart mobile devices are used to form the network energy consumption is a critical issue as a power-hungry or poorly designed content sharing solution would unnecessarily drain the battery. Limited connectivity remains a major concern for this type of network due to the limited communication range of devices, a limited number of available devices, and problems with identifying or pairing with a nearby device. Frequent disconnection also occurs in this type of network due to node mobility and the sleep cycle of network interfaces used for conserving energy in these mobile devices. Multi-hop ad-hoc networks also suffer from high delivery delay and low delivery success rates. The delivery success rate is measured in terms of the number of messages successfully delivered to the destination to the number of sent messages while the delivery delay indicates the time it takes for a message to reach the destination from a source node. Since a fixed end-to-end path is unavailable in this type of network messages may not reach the destination or can observe a high amount of delay. Table 1 highlight the challenges associated with multi-hop ad-hoc network and show a rating scale to highlight their impact. The table suggests that node mobility, energy consumption, and frequent disconnection remain the most important challenges with multi-hop ad-hoc networks. Delivery success rate and delay are also the prominent metrics used for performance evaluation of ad-hoc networks and hence a proposed strategy needs to achieve a higher delivery success rate and low delivery delay.

Table 1. Challenges associated with multi-hop ad-hoc network and their severity

Challenges	Severity and impact of the challenge
Node mobility	★★★★★
Energy consumption	★★★★
Limited connectivity	★★★
Limited bandwidth	★★
Storage limitation	★★★
Frequent disconnection	★★★★★
High delivery delay	★★★★★
Low delivery success rate	★★★★★
High message overhead	★★★

Although several challenges are associated with multi-hop ad-hoc networks as highlighted above, they are useful in situations where infrastructure-based communication services are partially or fully unavailable. In this regard, a disaster-response and post-disaster recovery scenarios are ideal situations where internet connectivity through cellular networks may be partially or fully unavailable. In this case, smart device-based multi-hop communication can be employed. The number of smart device owners has grown significantly around the globe. Therefore, it creates an opportunity for rescue workers to communicate with the affected people through smartphone-based emergency-response applications. Several works (Sciullo, Trotta, & Felice, 2020; Hossain & Ray, 2018; Yatbaz, et al., 2018; Pu & Zhou, 2019) in literature have addressed this issue. Lu *et al.* (Lu, Cao, & La Porta, 2017) proposed a platform called TeamPhone to enable communication in disaster recovery scenarios, where nearby smart mobile devices can collaborate to provide data communication facilities. They used Wi-Fi and Cellular communication capability of smartphones for seamless connection among rescue workers and survivors in infrastructure-constrained or infrastructure-less scenarios. Along a similar line, works in (Meurisch, *et al.*, 2017) proposed the use of ad-hoc communication capability of smart devices to provide three types of services in a disaster scenario, which includes social networking (i.e., message and microblog exchange) among nearby nodes, diffusion of rescue instruction (e.g., evacuation guidance), and a self-rescue system to notify the location and condition of affected people.

Victim localization is an important aspect of the rescue operation. A few smartphone-based applications (Rosyidi *et al.*, 2018; Yatbaz *et al.*, 2018) are proposed in the literature in this regard. Rosyidi *et al.* (Rosyidi *et al.*, 2018) proposed the use of an Internet-of-things (IoT) device, which can detect Wi-Fi signals coming out from the affected person's smartphone to locate them. In contrast, Yatbaz *et al.* (Yatbaz *et al.*, 2018) considered using both a pro-active and reactive approach for victim localization. In the pro-active approach, the proposed smartphone-based application periodically upload the user's location to a central server before the disaster. The reactive approach occurs just after the disaster and it tries to locate the user based on its last known location and the location of its nearby devices. An interesting work proposed by Mezghani *et al.* (Mezghani, Kortoc, Mitton, & Francesco, 2019) considered the use of the unmanned aerial vehicle (UAV) to enable communication in a disaster scenario in addition to smart mobile devices. UAVs can certainly help in reaching survivors and collect information from them as they can safely go to places where the human being cannot reach after a disaster. These UAVs also allow a faster exchange of communication. Overall, the challenges associated with establishing a multi-hop network as highlighted at the beginning of this section has been addressed in many ways by researchers. This chapter highlights the proposed approaches and analyzes potential opportunities and requirements for future research.

Figure 2. Ad-hoc communication network for smartphone-based emergency response application. Here, the central command center represents the government agency responsible for managing the situation and the local operation base indicates their base station for current situation. The ellipse represents the disaster zone. Please note that infrastructure-based cellular communication is unavailable only in the disaster zone and hence ad-hoc communication is used.

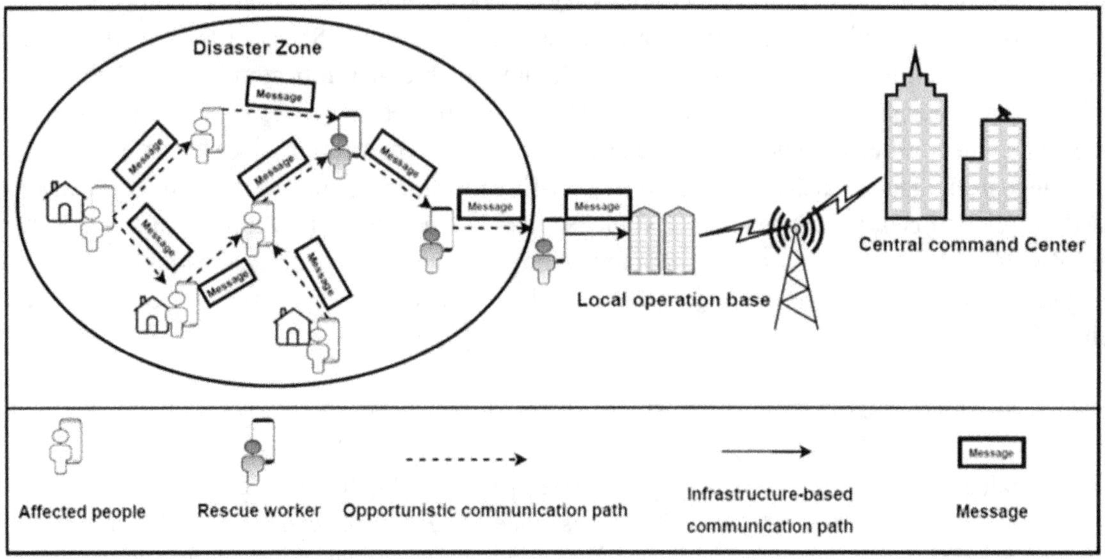

SMARTPHONE-BASED EMERGENCY RESPONSE APPLICATION

A schematic representation of a smartphone-based emergency response application is presented in Figure 2. In this figure, the central command center represents the government agency responsible for managing the situation and the local operation base indicates their operation center for handling a specific event. The ellipse represents the disaster zone where infrastructure-based cellular communication is unavailable. Therefore, opportunistic communication is used within the disaster zone to enable communication. The figure shows that a few members of the emergency response team may enter the disaster zone with their smart mobile devices to communicate with the survivors. Please note that infrastructure-based communication through the cellular network is available outside the disaster zone where rescue workers can utilize this for communicating with the central command. The rescuer team within the disaster zone maintains communication among themselves and with the survivors using multi-hop ad-hoc network communication technology. The figure also shows that a message has been sent by an affected person, which reaches a rescue worker through multiple forwarder nodes. Table 2 presents recent advancements in smartphone-based emergency response application development and highlights their contributions.

The emergency response application uses different modules, such as communication, routing, and application module. The communication module works at the physical layer and is responsible for maintaining communication among end-users. The routing module is responsible for carrying a message from one device to another device or the central command center. Finally, the application module enables end-users to exchange data in the form of messages, voice, video, or images. The details of these modules are discussed below.

Communication Module

The communication module is responsible for enabling communication in a smartphone-based emergency network. Due to the natural disaster, the cellular network may be partially or fully unavailable in this scenario. The objective of the communication module is to provide an opportunity for the rescue workers and the survivors to communicate with each other and with the central command center. If the cellular connectivity is fully available within the disaster zone, it makes it easier to communicate with the survivors using existing communication channels, such as through SMS alerts or voice call alerts. However, if the cellular network is partially or fully unavailable then it becomes more challenging to establish communication. In this case, temporary opportunistic networks among co-located devices are established where cellular connectivity is unavailable. The communication module can utilize various short-range technologies, such as Bluetooth and Wi-Fi. In addition, a few recent studies have also proposed the utilization of Unmanned Aerial Vehicle (UAV), drone, and Long Range Wide Area Network (LoRaWAN) technologies to facilitate communication (Jagannath, Furman, Jagannath, & Drozd, 2019; Mezghani, Kortoc, Mitton, & Francesco, 2019; Sciullo, Trotta, & Di Felice, 2020). The details of these technologies are discussed below.

Bluetooth-Based Communication

Smart mobile devices nowadays are equipped with Bluetooth technology, which enables them to form an ad-hoc network with nearby devices. In this case, one of the devices can act as a master node while other devices may act as a slave node. A Bluetooth network of such interconnected devices is called a

Table 2. Recent Advancement in Smartphone-based Emergency Response Application

Article	Year	Communication Technology	Routing Protocol	Application	Platform	Performance evaluation	Security
TeamPhone (Lu, Cao, & La Porta, 2017)	2017	Wi-Fi Ad-hoc mode	AODV for sending to the base station, Direct routing, and epidemic routing for ad-hoc communication	Enable emergency message exchange including location and details of survivors	Android	Testbed evaluation with four devices	Not addressed
Nicer911 (Meurisch, et al., 2017)	2017	Wi-Fi Ad-hoc mode	Serval routing developed for delay tolerant network	Enables three types of communication between survivors and rescuers	Android	Usability and user experience survey with 12 participants	Public key encryption
SEDCOS (Kohnhauser, et al., 2017)	2017	Wi-Fi and Bluetooth	Epidemic	Message exchange among survivors and rescuers	Not addressed	Simulation with 1000 nodes	Public key infrastructure with certificate revocation
COPE (Mezghani & Mitton, 2017)	2017	Wi-Fi, Bluetooth, and cellular network	Aggregated data diffusion to the cluster head	Alert message diffusion	Not addressed	Simulation-based evaluation with 35 survivors	Not addressed
Android P2P (Yatbaz, et al., 2018)	2018	Wi-Fi Direct	Not mentioned	Collects location and medical information to share with the server	Android	Not addressed	Encryption, password protection, and firewall
IoT-based Searching (Rosyidi, Puspita, Kashihara, Fall, & Ikeda, 2018)	2018	Wi-Fi	Direct communication	IoT-device capture Wi-Fi probe request for localization	Raspberry pi and iOS	Tested with one iPhone and Raspberry pi	Not addressed
StayTuned (Zhang, Pal, Kant, & Vucetic, 2018)	2018	Cellular network	Direct communication	Filters out emergency messages from twitter to disseminate in affected areas	Not addressed	Measured suitability of filtering through simulation with collected data from Twitter.	Not addressed
Drone-assisted recovery (Mezghani, Kortoc, Mitton, & Francesco, 2019)	2019	Wi-Fi, Bluetooth, and cellular network	Aggregated data collection at the cluster head	Data exchange from cluster head to base stations	Not addressed	Simulation-based evaluation by varying number of survivors in an area of 10 X 5 km	Not addressed
PDCRPT (Hazra, et al., 2019)	2019	Bluetooth or Wi-Fi direct	Direct layered communication while using optimization to allocate network resources	Message exchange in the form of texts, images, audio, and videos	Not addressed	Simulation in an area of 3 X 3 km	Not addressed
HELPER (Jagannath, Furman, Jagannath, & Drozd, 2019)	2019	Wi-Fi and LoRa	Cross-layer position based	Exchange message between end-users and with emergency personnel	Web-based app	Not addressed	Not addressed
RescueMe (Pu & Zhou, 2019)	2019	Wi-Fi and Bluetooth	Cluster-based routing	Exchange message	Not addressed	Simulation with 1-9 nodes	Not addressed
LOCATE (Sciullo, Trotta, & Felice, 2020)	2020	Bluetooth and LoRa	Routing protocol optimizes emergency resolution time	Message exchange to send location and emergency requests	Android	Proof-of-concept applications are developed, and simulation-based evaluation is done using an area of 5X5 km	Not addressed
E-Darwin (Pal, Raj, Kant, & Das, 2020)	2020	Wi-Fi tethering	Energy-delay aware forwarding and Hot-potato based forwarding	Message exchange	Android	Proof-of-concept applications are developed, and simulation-based evaluation is done using an area of 200X200 m	Not addressed
Context-aware self-adaptive routing (Rosas, Garay, & Hidalgo, 2020)	2020	Bluetooth	Context-aware routing protocol selection based on delivery rate and overhead	Message exchange	Not addressed	Simulation with varying number of nodes within 80-500	Not addressed
EAR (Chao, et al., 2020)	2020	Not addressed	Forwarding decision based on remaining energy and routing risk	Message exchange	Not addressed	Mathematical analysis and simulation	Not addressed

piconet. The number of devices connected to a master in a piconet is limited to seven, however, the role of the master and slave can change anytime. Multiple piconets can be connected to form a scatternet. Figure 3 shows the formation of such a scatternet. To enable communication between Bluetooth devices, the pairing process occurs. In this case, one of the devices enter the inquiry mode and look for available devices within its range while the other devices must be in the discoverable mode to be found. When the devices find each other, they exchange authentication information to form the initial network. This phase may also require user input to allow a connection. Typical smartphones are equipped with class 2 Bluetooth technology, which operates at 2.4GHz and has a communication range around 10 m (De Dominicis, et al., 2012) while the new version of Bluetooth (i.e., Bluetooth 5.0) looks very promising and expected to provide a communication range up to 200m in an outdoor environment with higher throughput (Yaakop *et al.*, 2017). Bluetooth-based communication technology consumes less amount of energy and hence it has been used for forming device-to-device communication networks (Kohnhauser, *et al.*, 2017) and delay-tolerant networks (Rosas, Garay, & Hidalgo, 2020) to enable communication among smart mobile devices in disaster scenarios.

Figure 3. Formation of Scatternet among co-located Bluetooth devices

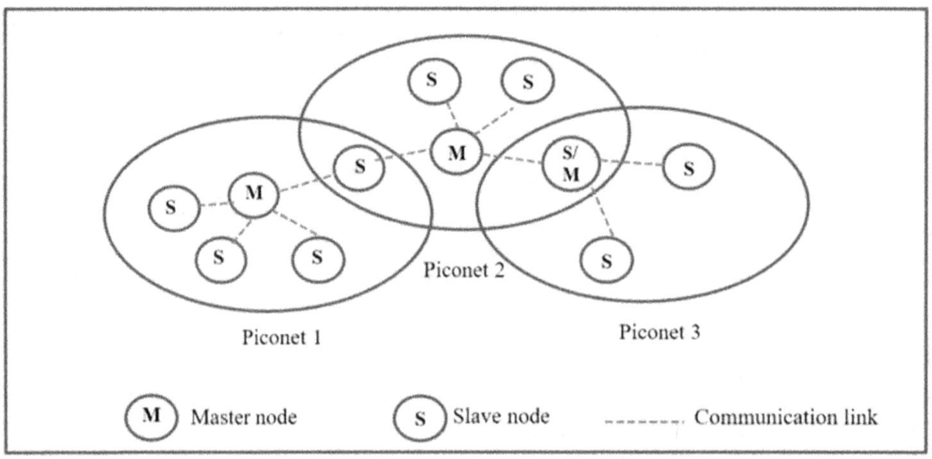

Wi-Fi-Based Communication

Wi-Fi-based communication allows smart mobile devices to extend the range of the network. Practical experiments using popular smart mobile devices suggested that the range of Wi-Fi communication in smartphones varies within 30 to 200 meters (Su *et al.*, 2012; Yoo *et al.*, 2014)). Wi-Fi communication can use multiple modes, namely infrastructure, ad-hoc, Wi-Fi direct, and Wi-Fi tethering modes (Xing *et al.*, 2009; Pal, Raj, Kant, & Das, 2020). In infrastructure mode, communication requires the establishment of an access point and all the messages must travel through the access point. For example, if three devices A, B, and C form a Wi-Fi network and device C is set up as the access point then device A and B will not be able to communicate directly. If A and B want to communicate, they will have to pass the message to device C who will ultimately forward it to the receiver. Infrastructure mode is generally used by smart mobile devices for connecting to the Internet through a wireless access point. In the case of the ad-hoc mode, devices can create a peer-to-peer network among themselves to exchange messages or share content. Although it is more useful for our considered scenario since devices will need to communicate among themselves, it is not widely supported and more often require root access (Pal, Raj, Kant, & Das, 2020).

Wi-Fi direct (also known as Wi-Fi P2P) provides a promising solution for creating an opportunistic network among co-located devices in a disaster scenario. This technology allows one-to-one and one-to-many connections where one of the devices work as the group owner while others play the role of group members. This type of communication is used in (Yatbaz, et al., 2018). However, this is ideally suited for a one-hop network where devices are located within one-hop distance. Another issue with this technology is that it is not available for all smart mobile devices especially the older devices may not have it readily available as part of the operating systems. Another technology widely used for enabling communication among smart devices is Wi-Fi tethering. This technology was originally developed for allowing mobile devices to share their Internet connection with nearby devices. It allows a mobile device to turn into a mobile hotspot where other devices can connect and share its Internet connection. The maximum number of devices that can simultaneously share the connection is limited to 5 to 10 depending on the underlying hardware and the manufacturer. This technique is used for communication in (Pal, Raj, Kant, & Das, 2020).

Overall, Bluetooth based communication offers a shorter range but consumes less energy while Wi-Fi-based communication provides a higher communication range but consumes more energy. Given the nature of the application, in some scenarios, lower energy consumption is preferred over the communication range while in many cases a higher energy consumption is acceptable considering the area the network can cover. A few approaches have also considered other alternative forms of communication technology, such as LoRa, UAV, and drone assisted communication. They are discussed below.

Other Approaches

Recently, the evolution of Long-Range (LoRa) communication technology, specially developed for Low-power Wide-Area Network (LPWAN) to enable long-range low-power communication among Internet-of-Things (IoT) devices, has provided a unique opportunity to enable long-range communication in disaster scenarios. The typical communication range of LoRa is 2-5 km in urban areas and nearly 15 km in suburban areas and it supports a star topology where the modules can be connected to the LoRa gateway which is ultimately connected to the network server. Jagannath *et al.* (Jagannath, Furman, Jagannath, & Drozd, 2019) developed a LoRa based module called HELPER to enable communication in disaster scenarios. It also includes a Wi-Fi module to enable heterogeneous connectivity. The LoRa module was connected to a Raspberry Pi in their prototype design. However, they did not provide any performance evaluation of their prototype. Another interesting work proposed by Sciullo *et al.* (Sciullo, Trotta, & Di Felice, 2020) used a LoRa transceiver module to enable long-range communication and connected it with their smartphone app through Bluetooth interface. Their work focused on efficiently delivering an emergency request from the affected user to a rescue worker and receiving an acknowledgment. Their simulation results demonstrated that the emergency resolution ratio (i.e., the number of emergency requests for which a response is received) reached above 90% when the number of nodes in the network exceeded 20 while the response time was nearly 50 minutes. Considering the nature of the network with partial connection, these results are promising for future development.

Mezghani *et al.* (Mezghani, Kortoc, Mitton, & Francesco, 2019) proposed a multi-tier communication scheme to enable communication among users in a disaster scenario. They considered that smart mobile devices are equipped with multiple network interfaces, such as Bluetooth, Wi-Fi, and cellular. Smart mobile devices can form multiple clusters using one of these network interfaces and produce a network of three tiers (*n1*, *n2,* and *n3* where *n1* is the highest tier). In each tier, devices use the same network interface and in a cluster one of the nodes is selected as the cluster-head. Devices within a cluster only communicate with the cluster-head while the cluster-head communicates with another cluster head in the upper tier to exchange information. They have also used drone assisted communication where drones fly around and communicate with the cluster-heads in *n1*-tier. Although the architecture looks interesting, they have not developed any proof-of-concept application for performance evaluation. Another interesting work proposed by Hazra *et al.* (Hazra, et al., 2019) considered a resource-constrained post-disaster scenario where the network resources, such as communication towers or data mules (emergency vehicles carrying a smart device equipped with a wireless networking interface) are limited. They considered that affected people may take shelter in shelter homes, which are equipped with an information dropbox that can store and transmit information using wireless network interfaces. They also considered volunteer users who carry smart mobile devices and can communicate with the information dropbox as well as the data mule. The data mules drive around to collect information and deliver it to a central command center. This work proposed an optical resource allocation technique to minimize end-to-end delay for message exchange and identified an optimal trajectory for the data mules to collect information. In summary, the proposed approaches look promising and expected to provide acceptable communication in post-disaster scenarios, but most of them lack a proof-of-concept implementation and real-life evaluation with actual users and devices.

Routing Module

The objective of the routing module is to forward messages from source to destination. In this case, both the affected people and the command centers can play the roles of the source and destination. Note that an end-to-end path is usually not available in this kind of network. This type of network is also called a delay tolerant network (DTN) as an end-to-end communication path is unavailable and a tolerable amount of delay is expected for forwarding a message. In this case, a store-carry-forward mechanism is adopted where a device carries a message until it can forward it to the next hop.

Epidemic routing (Vahdat & Becker, 2000) is one of the prominent protocols used in this type of scenario. In the case of epidemic routing, each device copies its message to every other device it meets. Although this method achieves high successful deliveries, it also consumes a lot of energy. Such high energy consumption can be problematic for the disaster scenario as the devices may have limited remaining energy and limited opportunity for charging. Another popular routing protocol is the spray-and-wait (Spyropoulos, Psounis, & Raghavendra, 2005) protocol. This method spreads a specific number of copies of a message in the network and waits until one of them reaches the destination. Figure 4 shows a schematic representation of the epidemic and Spray-and-wait routing protocol in a disaster scenario. Please note that in this figure the ellipses represent two disaster zone where infrastructure-based communication is unavailable and opportunistic communication is used. However, infrastructure-based communication is available outside the disaster zone and can be utilized by the rescue worker to communicate with the central command (a government agency responsible for handling the situation). The figure shows that for epidemic routing (left side) every node forwards a message to every other node for sending it to the command center. On the other hand, for the Spray-and-wait routing, the source node spreads four copies of the message and distributes it among its neighbors who keep a copy for themselves to distribute and forward the other two copies to their neighbors. Finally, one of the nodes successfully forwards the message to a rescue worker who delivers it to the command center. The figure clearly shows that the number of messages required for Spray-and-wait is less than that for epidemic routing and hence it consumes less energy. However, reducing the number of messages also creates the possibility of unsuccessful delivery in this network.

Figure 4. Epidemic (left side) and Spray-and-wait (right side) routing protocol in a disaster scenario

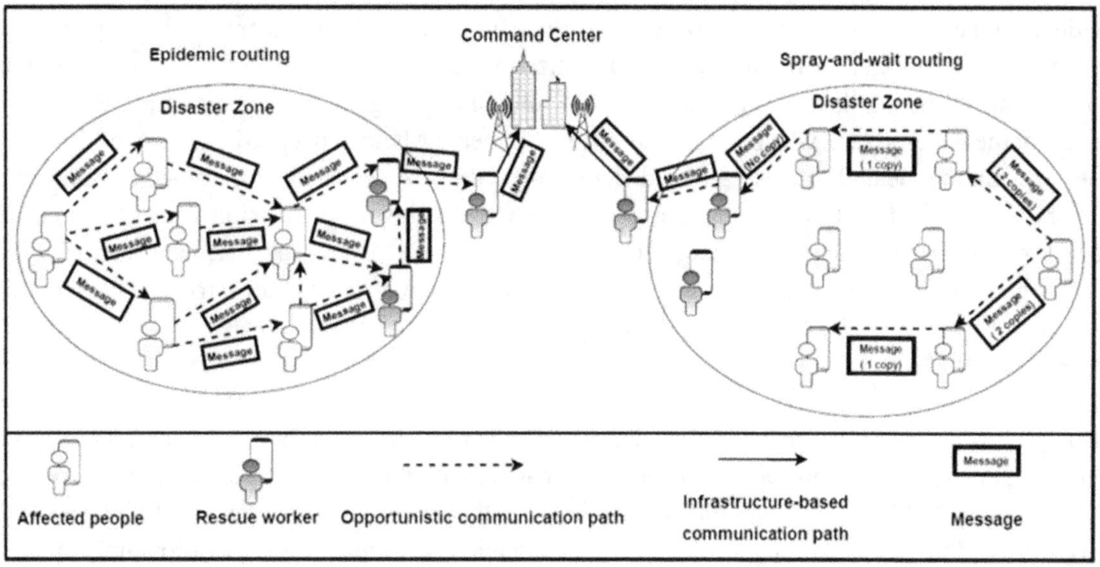

Since energy consumption and successful deliveries are two major concerns for disaster scenarios and as alluded in the previous paragraph that more often they are contradicting goals, an interesting approach is proposed by Rosa *et al.* (Rosas, Garay, & Hidalgo, 2020). They suggested that nodes (i.e., a mobile device) can switch among DTN routing protocols based on their current contexts (calculated in terms of neighbor density and average contact time) and an objective function (calculated in terms of message overhead and delivery success rates). In this case, they considered five state-of-the-art DTN routing protocols including epidemic and Spray-and-wait routing. The simulation results demonstrated that although the proposed approach achieved slightly lower successful deliveries compared to contemporary approaches, it was successful in minimizing message overhead and energy consumption in different scenarios.

Mezghani *et al.* (Mezghani & Mitton, 2018) proposed a cooperative opportunistic alert diffusion (COPE) method suggesting that smart devices are equipped with multiple networking interfaces and have different levels of energy, which should be considered for alert message diffusion. In the COPE approach, nodes with a higher level of energy participate more in message forwarding by frequently waking up their interface and sending communication messages while nodes with lower levels of energy have longer sleeping cycles with no communication to conserve energy. Chao *et al.* (Chao, et al., 2020) have recently proposed an energy-aware approach which modeled the mobility within a disaster scenario as a stochastic multigraph with multiple agents, such as survivors, rescue workers, supply vehicles, and ambulances. They calculated the risk associated with selecting a routing path (obtaining longer delay if a path is selected) using the mean-risk technique and used the max-probability model to determine the maximum probability of delivering a packet to its destination before the deadline. Finally, they used a "λ-optimal" algorithm to find a suitable routing path. Their simulation results demonstrated that their method achieved 8x higher energy efficiency with a 4% lower packet delivery ratio compared to state-of-the-art routing protocols.

In summary, several useful routing methods have been proposed in the existing literature. Some of them achieve higher successful delivery and lower latency while others are better in terms of energy efficiency. Since disaster scenarios require both successful deliveries of a message with a lower delivery delay and conservation of energy to achieve a longer network lifetime, there is still scope for future improvement in this area. A multi-objective optimization technique can be useful in this regard, which can aim at maximizing the delivery success rate while minimizing energy consumption. Based on the context of the scenario (i.e., whether the users require immediate help or need to be connected for a longer time), the routing method can also utilize different algorithms. A scenario with immediate information delivery requirements will need an aggressive delivery method while a situation where the lifetime of the network needs to be longer, an energy conservative approach can be used. The next section will discuss the application module that enables users to communicate with each other through a user interface.

Application Module

The application module sits on top of the other underlying modules and provides a user interface for the users to interact with the application and communicate among themselves. Since users are expected to only interact with this module without the knowledge of the underlying modules, the design and usability of this module remain an important issue. This module is also responsible for providing a platform to the users for interacting with the system for exchanging messages, location information, image files, and videos. The location information can be obtained from GPS or cellular network data. The images, video, or audio files can be captured by the survivors or rescuers. The rescue workers can also send text messages to provide rescue instructions, relief information, or propagate disaster warnings. On the other hand, the survivors may request assistance and highlight their current contexts using text messages. The following section discusses the application modules developed in the existing literature.

TeamPhone (Lu, Cao, & La Porta, 2017) is a prominent work in this area that was developed using the Android platform. The application provides a user interface to configure the system and exchange messages with the rescue workers. It works in two different modes including message exchange and self-rescue modes. The message exchange allows communication between survivors and rescue workers while the self-rescue mode enables the rescue workers to locate the survivor. In this case, the location of the survivor is determined using GPS or if it is unavailable the last known location of the survivor is identified from the operating system and communicated with nearby devices using a one-hop network. Nodes within the vicinity form a self-rescue group and communicate with each other for location information exchange. The waking-up schedule of nearby nodes is also tuned to save energy. For routing, TeamPhone uses the AODV communication protocol when the destination is reachable and DTN-based message forwarding protocol otherwise. The application has been deployed in Samsung Galaxy S3 mobile for testbed evaluation. However, a large-scale user study with actual participants is to measure the usability of the system is not conducted. Pal *et al.* (Pal, Raj, Kant, & Das, 2020) also suggested that they implemented their method on Samsung Galaxy S3 devices to observe power consumption and tune simulation parameters, but they did not provide any other implementation details.

Yatbaz *et al.* (Yatbaz, et al., 2018) developed an application using the Android platform and considered that mobile devices can collect their location information and send it to a centralized server using the cellular network. If a cellular network connection is unavailable, devices can communicate with their neighbors using Wi-Fi direct and request to send this information on their behalf. They suggested that the users will need to go through a registration process before using the mobile app, which will collect their profile information including e-mail addresses, blood types, chronic disease history, and drug allergies. During or after a natural disaster, a device can determine its location using information from the network provider or GPS and forward it to the central server. The central server provides an administration panel to show the location of the survivors on a map and their relevant profile information. However, the authors did not mention any performance evaluation of this system.

Sciullo *et al.* (Sciullo, Trotta, & Di Felice, 2020) recently developed an emergency management application for the Android platform which can be used to send emergency help requests from affected people to rescue workers. In this case, the location of the affected person and the type of emergency (e.g., rescue help or medical help) can be included in the message request. The request is carried and forwarded by nearby users until it reaches a rescuer node, which replies with an acknowledgment message. The app is also connected to a LoRa module through Bluetooth, which extends the range of communication. The users are required to register their information before using this app during an actual emergency. The application provides three functionalities, including (i) creating and forwarding emergency requests, (ii) checking the progress of an existing emergency request, and (iii) configuring the LoRa module setup. The app also allows multi-hop communication where the users can select whether they want to participate in emergency message request forwarding or not.

Although a few proof-of-concept applications have been developed as discussed above, they lack a large-scale user study and performance evaluation. Ad-hoc networks have been used in other areas, such as wildlife monitoring (Juang, et al., 2002; Wietrzyk, Radenkovic, & Kostadinov, 2008), co-location detection and prediction (Haus, Ding, & Ott, 2020), opportunistic communication (Bigwood, Rehunathan, Bateman, Henderson, & Bhatti, 2008), and opportunistic sensing (Sofia, Firdose, Lopes, Moreira, & Mendes, 2016). However, the contexts of these studies were completely different. They were mainly focused on monitoring wildlife animals (Juang, et al., 2002; Wietrzyk, Radenkovic, & Kostadinov, 2008), and monitoring human movement and their interaction patterns during regular life with a small number of devices (Sofia, Firdose, Lopes, Moreira, & Mendes, 2016 used four devices) or with a substantial user group, but in an academic environment (Bigwood, Rehunathan, Bateman, Henderson, & Bhatti, 2008; Haus, Ding, & Ott, 2020). The application scenario under consideration (i.e., disaster-affected area) is significantly different than regular life settings as the situation is very different (users under stress) and the user group is expected to be diverse with a likelihood of having technically challenged people for whom the usability of the application is more important than merely their co-location with other users.

A few works (Cabrero, García, Pañeda, & Melendi, 2015; Álvarez, et al., 2018) have collected their dataset considering the disaster scenario. Carero *et al.* (Cabrero, García, Pañeda, & Melendi, 2015) extracted GPS traces from the Geographical Information system (GIS) of a regional firefighter department in Spain. Their study was conducted for a year and they collected these traces from devices embedded in cars, trucks, personal radios, and helicopters. The focus of this study was to identify communication opportunities in disaster scenarios among users. However, the trace only contained information on rescue workers. In contrast, an interesting study conducted by Álvarez, *et al.* (Álvarez, et al., 2018) in Germany explored the user behavior and interaction by mimicking a disaster scenario in three villages with 125 participants. They introduced two fictitious events during the disaster to emulate user experience under stress. They collected information related to communication opportunities, user movement, and their interaction with the application. This work is very interesting and can be used as a baseline for future studies. However, future user studies should also consider the usability of the apps in terms of their ease of use, response time, and error rate, which is not highlighted in that study. The study also considered the exchange of non-multimedia messages, which can be relaxed in future works as the users may find it more convenient to share multimedia messages to exchange useful information with the authority. Finally, during the study, participants were provided with an extra battery pack to last the whole session, which may overlook the energy consumption problem for smartphone-based application in real-life settings. In addition, Android provides an open-source operating system, which makes it easier to configure and change system settings if needed. Therefore, it has been widely used for emergency response application development in various studies (see Table 2). However, a platform-independent application would be more suitable to reach a wider population as the affected people may have different devices with other operating systems.

In summary, a real-life empirical study demonstrating the user behavior and interaction pattern similar to the one presented by Álvarez, *et al.* (Álvarez, et al., 2018) with a medium to large user-population (50 to 100 users or more) is likely to provide useful insight about the usability of smartphone-based app in a disaster scenario. Future user study should also include participants from different demographics too. Faulkner (Faulkner, 2003) suggested that around 50 users were needed to identify 98% of the usability related problems. Since the emergency response application is considered to work in a critical situation, it will be extremely important to identify usability issues beforehand, and hence future research should consider this for conducting their empirical studies. Future user study should also consider the exchange of multimedia contents in the disaster scenario and user devices with different levels of remaining energy to mimic real-life events. The evaluation metrics should include information such as message delivery success rate, delivery delay, network lifetime, and energy consumption.

DISCUSSION

One major concern for the practical implementation of a smartphone-based emergency and post-disaster recovery application is privacy and security, as the communication utilizes wireless medium and ad-hoc networks without infrastructure support or Internet connection. Although a few works (kohnhauser *et al.*, 2017; Yatbaz *et al.*, 2018) have addressed this issue by employing key-based encryption technique and the use of a firewall, further studies in this area are needed to deal with security attacks, such as denial-of-service (DoS), man-in-the-middle and passive data collection or overhearing. Another important aspect is low energy consumption while achieving satisfactory quality-of-service (QoS). In this regard, solutions such as low power Bluetooth and LoRa may be useful, but more in-depth experimentation in real-life scenarios is needed before practical adoption. In addition, large-scale user studies with a substantially large population in different geographic settings will help in establishing the usability of the system and identify potential issues for improvement. Overall, the practical adoption of such an emergency response application is quite feasible if the security and privacy issues along with the usability of the system are tested with a diverse user population in real-life settings with a medium to large (e.g., 50 to 100 users or more) population.

CONCLUSION

Smart mobile devices have enabled ubiquitous communication in different scenarios including the ones with disaster-affected areas where infrastructure-based cellular networks may be partially or fully unavailable. Emergency communication between rescue workers and survivors is needed in this situation to spread useful information related to the rescue plan, relief work, victim localization, and disaster warning. Several studies have been conducted in the existing literature to address these issues and facilitate communication in emergency response and post-disaster recovery phase. These works addressed issues related to the selection of appropriate communication technology, message forwarding, and routing techniques, and the development of smartphone applications. The proposed approaches are very promising and interesting. However, further research in this area is needed for widescale adoption and practical implementation. This chapter has systematically analyzed existing approaches and highlighted potential areas that require further attention and future development. Overall, the practical adoption of such an emergency response application will be quite feasible if the above issues are addressed properly

REFERENCES

Álvarez, F., Almon, L., Lieser, P., Meuser, T., Dylla, Y., Richerzhagen, B., & Steinmetz, R. (2018). Conducting a large-scale field test of a smartphone-based communication network for emergency response. *Proceedings of the 13th Workshop on Challenged Networks*, 3-10. 10.1145/3264844.3264845

Bigwood, G., Rehunathan, D., Bateman, M., Henderson, T., & Bhatti, S. (2008). Exploiting self-reported social networks for routing in ubiquitous computing environments. In *IEEE International Conference on Wireless and Mobile Computing, Networking and Communications* (pp. 484-489). IEEE. 10.1109/WiMob.2008.86

Cabrero, S., García, R., Pañeda, X. G., & Melendi, D. (2015). Understanding opportunistic networking for emergency services: Analysis of one year of GPS traces. *Proceedings of the 10th ACM MobiCom Workshop on Challenged Networks*, 31-36. 10.1145/2799371.2799381

Chao, M., Chenji, H., Yang, C., Stoleru, R., Nikolova, E., & Altaweel, A. (2020). EAR: Energy-aware risk-averse routing for disaster response networks. *Ad Hoc Networks*, *103*, 1–15. doi:10.1016/j.adhoc.2020.102167

De Dominicis, C., Mazzotti, D., Piccinelli, M., Rinaldi, S., Vezzoli, A., & Depari, A. (2012). Evaluation of Bluetooth Hands-Free profile for sensors applications in smartphone platforms. In *Sensors Applications Symposium Proceedings* (pp. 1-6). IEEE. 10.1109/SAS.2012.6166305

Faulkner, L. (2003). Beyond the five-user assumption: Benefits of increased sample sizes in usability testing. *Behavior Research Methods, Instruments, & Computers*, *35*(3), 379–383. doi:10.3758/BF03195514

Haus, M., Ding, A., & Ott, J. (2020). Multimodal Co-Presence Detection with Varying Spatio-Temporal Granularity. *IEEE International Conference on Pervasive Computing and Communications Workshops (PerCom Workshops)*, 1-7. 10.1109/PerComWorkshops48775.2020.9156105

Hazra, K., Shah, V., Bilal, M., Silvestri, S., Das, S., Nandi, S., & Saha, S. (2019). A Novel Network Architecture for Resource-constrained Post-disaster Environments. In *11th International Conference on Communication Systems & Networks (COMSNETS)* (pp. 328-335). IEEE. 10.1109/COMSNETS.2019.8711166

Hossain, M., & Ray, S. (2018). A smartphone-assisted device-to-device communication for post-disaster recovery. *Springer International Conference on Smart Grid Inspired Future Technologies*, 11-20. 10.1007/978-3-319-94965-9_2

Jagannath, J., Furman, S., Jagannath, A., & Drozd, A. (2019). *Energy Efficient Ad Hoc Networking Devices for Off-the-Grid Public Safety Networks. In 16th IEEE Annual Consumer Communications \& Networking Conference (CCNC)*. IEEE.

Juang, P., Oki, H., Wang, Y., Martonosi, M., Peh, L. S., & Rubenstein, D. (2002). Energy-efficient computing for wildlife tracking: Design tradeoffs and early experiences with ZebraNet. *10th international conference on Architectural support for programming languages and operating systems*, 96-107.

Kohnhauser, F., Stute, M., Baumgartner, L., Almon, L., Katzenbeisser, S., Hollick, M., & Freisleben, B. (2017). SEDCOS: A secure device-to-device communication system for disaster scenarios. In *42nd Conference on Local Computer Networks (LCN)* (pp. 195--198). IEEE. 10.1109/LCN.2017.47

Lu, Z., Cao, G., & La Porta, T. (2017). Teamphone: Networking smartphones for disaster recovery. *IEEE Transactions on Mobile Computing*, *16*(12), 3554–3567. doi:10.1109/TMC.2017.2695452

Meurisch, C., Nguyen, T., Wullkotte, S., Niemczyk, S., Kohnhausar, F., & Muhlhausar, M. (2017). NICER911: Ad-hoc Communication and Emergency Services Using Networking Smartphones and Wireless Home Routers. In *Proceedings of the 18th ACM International Symposium on Mobile Ad Hoc Networking and Computing* (pp. 1-2). ACM. 10.1145/3084041.3084075

Mezghani, F., Kortoc, P., Mitton, N., & Francesco, M. D. (2019). A Multi-tier Communication Scheme for Drone-assisted Disaster Recovery Scenarios. In *30th Annual International Symposium on Personal, Indoor and Mobile Radio Communications (PIMRC)* (pp. 1-7). IEEE. 10.1109/PIMRC.2019.8904140

Mezghani, F., & Mitton, N. (2017). Opportunistic alert diffusion in disaster scenario—Stay alive longer! *28th Annual International Symposium on Personal, Indoor, and Mobile Radio Communications (PIMRC)*, 1, 1-5. 10.1109/PIMRC.2017.8292230

Natural disaster 2018. (2019). Retrieved from https://www.emdat.be/publications

O'Dea, S. (2020, August 20). *Number of smartphone users worldwide from 2016 to 2021*. Retrieved August 25, 2020, from Statista: https://www.statista.com/statistics/330695/number-of-smartphone-users-worldwide/

Pal, A., Raj, M., Kant, K., & Das, S. (2020). A Smartphone-based Network Architecture for Post-disaster Operations Using WiFi Tethering. *ACM Transactions on Internet Technology*, *20*(1), 1–27. doi:10.1145/3372145

Pu, C., & Zhou, X. (2019). RescueMe: Smartphone-based self rescue system for disaster rescue. In *9th Annual Computing and Communication Workshop and Conference (CCWC)* (pp. 832-837). IEEE. 10.1109/CCWC.2019.8666565

Richards, L., Brew, N., & Smith, L. (2020, March 12). *Australian bushfires—frequently asked questions: a quick guide*. Retrieved August 25, 2020, from Parliament of Australia: https://www.aph.gov.au/About_Parliament/Parliamentary_Departments/Parliamentary_Library/pubs/rp/rp1920/Quick_Guides/AustralianBushfires

Rosas, E., Garay, F., & Hidalgo, N. (2020). Context-aware self-adaptive routing for delay tolerant network in disaster scenarios. *Ad Hoc Networks*, *102*, 1–17. doi:10.1016/j.adhoc.2020.102095

Rosyidi, M., Puspita, R., Kashihara, S., Fall, D., & Ikeda, K. (2018). A Design of IoT-Based Searching System for Displaying Victim's Presence Area. In *42nd Annual Computer Software and Applications Conference (COMPSAC)* (pp. 8--13). IEEE. 10.1109/COMPSAC.2018.10195

Sciullo, L., Trotta, A., & Di Felice, M. (2020). Design and performance evaluation of a LoRa-based mobile emergency management system (LOCATE). *Ad Hoc Networks*, *96*, 1–17. doi:10.1016/j.adhoc.2019.101993

Sofia, R., Firdose, S., Lopes, L. A., Moreira, W., & Mendes, P. (2016). NSense: A people-centric, non-intrusive opportunistic sensing tool for contextualizing nearness. In *18th International Conference on e-Health Networking, Applications and Services (Healthcom)* (pp. 1-6). IEEE. 10.1109/HealthCom.2016.7749490

Spyropoulos, T., Psounis, K., & Raghavendra, C. (2005). *Spray and wait: an efficient routing scheme for intermittently connected mobile networks. In ACM SIGCOMM workshop on Delay-tolerant networking*. ACM. doi:10.1145/1080139.1080143

Su, K.-C., Wu, H.-M., Chang, W.-L., & Chou, Y.-H. (2012). Vehicle-to-vehicle communication system through wi-fi network using android smartphone. *International conference on connected vehicles and expo (ICCVE)*, 191-196. 10.1109/ICCVE.2012.42

Vahdat, A., & Becker, D. (2000). Epidemic routing for partially connected ad hoc networks. *Technical Report CS-200006*, 1-14.

Wietrzyk, B., Radenkovic, M., & Kostadinov, I. (2008). Practical MANETs for pervasive cattle monitoring. *Seventh International Conference on Networking*, 14-23. 10.1109/ICN.2008.78

Xing, B., Seada, K., & Venkatasubramanian, N. (2009). *An experimental study on wi-fi ad-hoc mode for mobile device-to-device video delivery. IEEE INFOCOM Workshops*. doi:10.1109/INFCOMW.2009.5072111

Yaakop, M. B., Abd Malik, I. A., bin Suboh, Z., Ramli, A. F., & Abu, M. A. (2017). Bluetooth 5.0 throughput comparison for internet of thing usability a survey. In *International Conference on Engineering Technology and Technopreneurship (ICE2T)* (pp. 1-6). IEEE. 10.1109/ICE2T.2017.8215995

Yatbaz, H., Cinar, B., Gokdemir, A., Ever, E., Al-Turjman, F., Nguyen, H., & Yazici, A. (2018). Hybrid approach for disaster recovery using P2P communications in android. *43rd Conference on Local Computer Networks Workshops (LCN Workshops)*, 46-52.

Yoo, S., Shin, Y., Kim, S., & Choi, S. (2014). Toward realistic WiFi simulation with smartphone "Physics". In *Proceeding of IEEE International Symposium on a World of Wireless, Mobile and Multimedia Networks* (pp. 1-6). IEEE.

Zhang, S., Pal, A., Kant, K., & Vucetic, S. (2018). Enhancing disaster situational awareness via automated summary dissemination of social media content. In *IEEE Global Communications Conference (GLOBECOM)* (pp. 1-7). IEEE. 10.1109/GLOCOM.2018.8647340

ADDITIONAL READING

Lu, Z., Cao, G., & La Porta, T. (2017). Teamphone: Networking smartphones for disaster recovery. *IEEE Transactions on Mobile Computing*, *16*(12), 3554–3567. doi:10.1109/TMC.2017.2695452

Mezghani, F., & Mitton, N. (2018). Opportunistic disaster recovery. *Internet Technology Letters*, *1*(2), 1–6. doi:10.1002/itl2.29

Pal, A., Raj, M., Kant, K., & Das, S. (2020). A Smartphone-based Network Architecture for Post-disaster Operations Using WiFi Tethering. *ACM Transactions on Internet Technology*, *20*(1), 1–27. doi:10.1145/3372145

Rosas, E., Garay, F., & Hidalgo, N. (2020). Context-aware self-adaptive routing for delay tolerant network in disaster scenarios. *Ad Hoc Networks*, *102*, 1–17. doi:10.1016/j.adhoc.2020.102095

Rosyidi, M., Puspita, R., Kashihara, S., Fall, D., & Ikeda, K. (2018). A Design of IoT-Based Searching System for Displaying Victim's Presence Area. *42nd Annual Computer Software and Applications Conference (COMPSAC)* (pp. 8--13). IEEE. 10.1109/COMPSAC.2018.10195

Sciullo, L., Trotta, A., & Di Felice, M. (2020). Design and performance evaluation of a LoRa-based mobile emergency management system (LOCATE). *Ad Hoc Networks*, *96*, 1–17. doi:10.1016/j.adhoc.2019.101993

Yatbaz, H., Cinar, B., Gokdemir, A., Ever, E., Al-Turjman, F., Nguyen, H., & Yazici, A. (2018). Hybrid approach for disaster recovery using P2P communications in android. *43rd Conference on Local Computer Networks Workshops (LCN Workshops)*, (pp. 46--52).

Zhang, S., Pal, A., Kant, K., & Vucetic, S. (2018). Enhancing disaster situational awareness via automated summary dissemination of social media content. *IEEE Global Communications Conference (GLOBECOM)* (pp. 1--7). IEEE. 10.1109/GLOCOM.2018.8647340

KEY TERMS AND DEFINITIONS

Ad-Hoc Network: Networks that can be informed instantaneously without requiring pre-existing infrastructure support.

Communication Protocol: Rules that allows two or more entities to follow a specific procedure for exchanging messages or interacting with each other.

Delay-Tolerant Network: Networks where end-to-end communication paths are usually unavailable, and a longer delay is acceptable for sending data from source to destination.

Device-to-Device Communication: A communication technology that allows user devices to communicate and collaborate without the involvement of the base station.

Internet-of-Things: Refers to a system where things or objects are equipped with sensors and communication capabilities and can be connected.

Opportunistic Network: Networks that are formed temporarily through wireless connection among nearby devices.

Peer-to-Peer Network: Peer-to-peer networks are formed among devices without requiring an external server where the participating devices have an equal amount of rights, permission, and responsibilities.

Chapter 3
Information Overload!
Investigating the Usability of an Information Tool for Crisis Situations With Biometric Data

Jenny Lindholm
Åbo Akademi University, Finland

Klas Backholm
Åbo Akademi University, Finland

Joachim Högväg
Åbo Akademi University, Finland

ABSTRACT

This chapter presents a usability study of a prototype tool intended to support information gathering during crisis situations, thus helping professional key communicators monitor social media content from several media outlets in the same workspace. The main aim of the study is to investigate how situation awareness can be optimized for key communicators during emergencies. It does so by taking a mixed method approach on usability testing and by combining emotional responses and cognitive processes to better understand how the user perceives the tool. The study contributes to the field by illustrating the importance of designing for good situation awareness in services/platforms intended for crisis communication. A key conclusion is that the use of human-computer interaction (HCI) and usability studies are central for improving digital services in emergencies. High-stress contexts require rapid decision-making and leave no room for improvisation and perception. Therefore, comprehending digital solutions correctly is crucial for overall situation awareness.

DOI: 10.4018/978-1-7998-6705-0.ch003

INTRODUCTION

During a crisis, information and communications are fundamental aspects of crisis management. Technical advantages in the form of social media platforms, such as Twitter and Facebook, are continuously transforming crisis management and communication (Veil et al., 2011). Likewise, smartphones and widespread internet access are reshaping the way humans connect. Social media offers an inexpensive platform with a global outreach (Wright & Hinson, 2009), and on-site and real time crisis response activities have become an important part of crisis communication (Palen et al., 2007). Social media usage during crisis situations encompasses, among other areas, an assessment of the situation, sense-making, and validation of the information (Palen & Anderson, 2016). Hence, the design of digital platforms for use during crisis situations is an important area of research.

This chapter presents a usability study of a prototype tool intended to support information gathering in crisis situations. The tool is intended to be used during emergencies, helping key communicators monitor and handle social media content from several media outlets in the same workspace. The main aim of the study is to investigate how situation awareness can be optimized for key communicators during emergencies. It does so by taking a mixed method approach on usability testing and by combining emotional responses and cognitive processes (Picard, 2003) to better understand how the user perceives the tool. This study contributes to the field by illustrating the importance of designing for good situation awareness in services/platforms intended for crisis communication.

Background

A crisis can be defined as a situation where societal structures and values are threatened (Boin et al., 2005). In this article, the concept is used in a broad sense, representing crisis situations caused by everything from natural disasters to terrorist attacks. During a crisis, the public needs to know what is happening, and journalists and other key communicators play an important role in getting the message across (Iannella & Henricksen, 2007). Moreover, the use of social media increases during crisis situations (Sweetser & Metzgar, 2007). Social media can be defined as digital tools that facilitate content exchange and interaction among and between publics and organizations (Wright & Hinson, 2009). The social media platforms vary both in design and features, but the communication process is characterized by higher interactivity with a broader reach than traditional media (Xu, 2020).

The research field of crisis informatics (Palen et al., 2009) looks at social media and technology in crisis management, ranging from the role of official responders to that of the public. The field is multidisciplinary and combines, for instance, social science with computing knowledge (Palen & Anderson, 2016). In a literature review of crisis informatics research (Reuter, Hughes et al., 2018), four broad fields of research were identified. The first field of research concerns empirical investigations of social media use (e.g. Wiegand & Middleton, 2016). The second field concerns the collection and processing of social media data generated during crisis events (e.g. making sense of large amounts of data, Castillo, 2016). The third field concerns the design, construction and evaluation of systems and technical solutions (e.g. by directly designing systems, Hughes & Shah, 2016, or by building models for understanding social media crisis communications, Liu, 2014; White & Plotnick, 2010). The final field concerns cumulative and longitudinal research with new theoretical and practical perspectives (e.g. Veil et al., 2011). This article contributes to the third field of research, which covers system design and the evaluation of technical solutions.

One early identified challenge for crisis informatics was how to gather and analyze the vast amount of data available during a crisis. Palen and colleagues concluded that "We need tools that enable the integration of multimodal data sources—including on-site eyewitness video, interactivity on multiple social media sites, conventional and alternative news updates, and official notifications — into a coherent, time-based, and replay-able information visualization environment" (Palen et al., 2009, p. 478). This challenge persists ten years later, as concluded by Reuter, Hughes, & Kaufhold (2018). One of the main difficulties in a constantly changing social media landscape is designing for the interaction between different actors, such as authorities, citizens, and key communicators. The use of human-computer interaction (HCI) and usability studies are two ways of improving the design of different systems, and especially systems intended for use during crisis situations (Reuter & Mentler, 2018).

Social Media, Crisis Communication and Situation Awareness

Regarding the use of social media as a form of crisis communication, best practices and theoretical research are still taking form (Lin et al., 2016; Rasmussen & Ihlen, 2017). Nevertheless, crisis situations are by nature unpredictable, sudden and novel events - characterized by uncertainty and chaos. This means that communication plays a central role in crisis response, which becomes more challenging when citizens actively participate in the discussion. Social media is making whole communities part of crisis communication.

One central challenge is source credibility (Hovland et al., 1953; Spence et al., 2013). Especially on social media platforms, credible and trustworthy information can be difficult to locate (Edwards et al., 2013). Eyewitness reports from people on the scene spread among networks, mixing with official statements, speculations, and rumors. This means that information is re-posted, and new content is added, making the trustworthiness more difficult to verify. Misinformation can be posted by anyone with a social media account, either in the form of intentional or unintentional rumors (Lin et al., 2016). For key communicators, such as journalists working with crisis coverage, another challenge is the threat of information overload (Gryszkiewicz & Chen, 2012). Research shows that when using information handling systems, individuals search for the least cognitively challenging inputs or information that is most easy to understand (Metzger et al., 2010). In a crisis, this could contribute to the spreading of rumors.

One solution to both these challenges is technical innovations that enhance and promote sense-making and fact checking (Hiltz et al., 2020). To improve sense-making during a crisis, the role of cognition is a relevant link in the process. Having a clear mental model of how a system works helps the workflow (Comfort, 2007). This means having the capacity to combine prior knowledge, incoming information, and system performance into a valid assessment of the situation. Several important features for social media tools that help emergency responders have been identified, such as monitoring multiple platforms at the same time, better algorithms for geographic location identification and flexibility in categorization (Hiltz et al., 2020). However, the technical software development is still a challenge, as well as how to incorporate the users' mental models into the design process (Berndt & Herczeg, 2019).

A situation awareness model (Endsley, 1995) emphasizes how comprehension of a certain situation involves understanding feedback from the environment and forecasting what to do next. Three levels of situation awareness can be identified, starting from perceiving the situation, to comprehending sensory cues, and finally, projecting a likely outcome. In a crisis, situation awareness can be about finding perception cues in the environment, trying to understand the cues through reasoning and hypothesizing, and reaching a forecast on what the probable outcome is. Thus, "maintaining situation awareness of an incident allows watch officers to effectively plan and implement responses, anticipate and manage requests for information about incidents" (Cameron et al., 2012).

Social media provides perception cues that can assist in enhancing situation awareness in a crisis (Cameron et al., 2012). Information can be spread rapidly and benefit the public at large, victims, or emergency managers trying to restore safety. A study of how Twitter communication can contribute to situation awareness in crisis situations identified situational updates about evacuations, sheltering and animal management, for example, as important information (Vieweg et al., 2010). Nevertheless, a big challenge in establishing an overview and comprehension of the event are the rich varieties of information sources and the large streams of data (Cameron et al., 2012). A system design that handles the challenges of large datasets needs, for instance, to take the online clustering of topics, geotagging, and visualization into account.

Usability to Improve Situation Awareness

When designing technical solutions for crisis communicators, it is essential to understand the user's needs as well as design a good and useful product. The international standard, ISO 9241-11 (ISO 9241, 2018; Wallach & Scholz, 2012), defines usability as: the extent to which a product can be used by specified users to achieve specified goals with effectiveness, efficiency and satisfaction in a specified context of use.

Traditionally, usability studies investigate users' experiences by asking respondents to express their opinions, or by observing the person. However, trying to draw conclusions on the user's processing of tasks and the affective responses poses several challenges (Kula et al., 2018). Firstly, asking for self-reports requires an awareness by the respondent as well as a willingness to express an accurate affect. Secondly, drawing from observations relies on the researcher's understanding of hidden psychological states from the respondent's external and visible reactions. The recommendation for gaining a better understanding of the person's experience is to use multiple measurements, both subjective (e.g. interviews, questionnaires) and objective (e.g. skin conductance) (Kukula et al., 2010). Hence, this study looks at task performance as the main predictor of the experience and situation awareness, and incorporates the emotional experience into the analysis.

The emotional system in humans is twofold. On the one hand we have conscious reactions, involving the combination of cognitive and emotional responses, and on the other hand, we have unconscious reactions triggered by autonomic processes in the brain (Poels & Dewitte, 2006). Affect refers to basic psychological states, often described by two properties; hedonic valence (pleasure/displeasure) and arousal (activation/sleep). Affect influences both sensory and cognitive processes (Duncan & Barrett, 2007).

When conducting usability studies, one goal is to reach an understanding of which emotions a person experiences during the use of a certain product. The use of surveys and questionnaires provides us with important insights into some of the user's conscious emotional and cognitive responses. However, the data might be affected by memory bias or other factors (Poels & Dewitte, 2006). Also, self-reported questionnaires involve elements of social control, i.e. the participant may alter their response according to what is expected or socially acceptable. Hence, research on emotions in HCI is turning to the possibilities of unconscious measurements in the form of biometric and psychophysiological data. Such data is often subtle and imperceptible to the individual but can provide invaluable information on how the individual responds to stimuli. The ability to use skin response (e.g. hand sweat) or eye movements (e.g. what the user looks at), for example, are valuable for identifying stressful situations that might be caused or worsened by a system's interface. Especially in a crisis situation, it is important not to cause any extra negative stress or extraneous cognitive load. Understanding the experience and what causes negative feedback provides us with an opportunity to address the challenges.

Specified Aim and Research Questions

This study aims to improve a system prototype designed for use during emergencies. The tested tool was developed within the [removed during peer review process]. The main goals of the project are to study how key communicators use, evaluate and react to social media messages in emergencies, and how their awareness and response can be optimized. The tool is developed following a three-step, user-centered design process comprising the following: a mapping of user needs, development of a prototype based on this research, and two usability tests of the prototype. The feedback from the first usability test was used to develop the prototype to a close-to-working tool [removed during peer review process]. A usability test of this second prototype, conducted in a laboratory setting with one intended user group, news journalists (N = 15), is described and analyzed in this study.

Three biometric measurements are used to enhance the understanding of the user's experience: eye tracking, skin conductance and facial responses. In addition to the biometric measurements, pre- and post-interviews were done with the participants, and subjective measures (AttrakDiff; Hassenzahl, 2006) on the tool functionality were collected. The combination of self-reported experiences and biometric data are used to analyze different types of usability problems as well as the user experience. The usability problems analyzed range from high-level problems (understanding the bigger picture and concepts used in the tool), to low-level problems (problems related to design elements such as visual hierarchy or contrasts). This leads us to three general research questions that will guide us through the remainder of the chapter:

RQ1: How do users' high-level usability problems, concerning the overall understanding of the prototype, affect overall situation awareness in emergencies?

RQ2: How do users' low-level usability problems, concerning specific prototype features and elements, affect overall situation awareness in emergencies?

RQ3: How should an information-handling tool intended for monitoring, gathering, and handling social media and web content in emergencies be designed to contribute to users' overall situation awareness?

Nature of the Close-to Working Tool Prototype

The main idea behind the tool is to use one workspace, instead of several social media platforms and websites to work with events, such as an unfolding crisis. The content for a certain event is gathered from different social media sources, for example Twitter and Instagram, and other internet feeds such as RSS or specific websites using search criteria (e.g. hashtags or keywords). This data is visualized in a feed, similar to the news feed feature in Facebook, and serves as a first overview of the event. From this feed, content of a particular interest can be saved for further analysis, visualization and evaluation. This feature would help key communicators sort through the vast array of social media content during an emergency and save content that is especially relevant, e.g. tweets from eyewitnesses at the scene.

The saved content can be rated on trustworthiness and importance, given a geographic location, and commented on for internal communication. For instance, a co-worker might know the eyewitness, and can post a comment that they will get in contact with the person. To summarize, the strength of the tool

Figure 1.

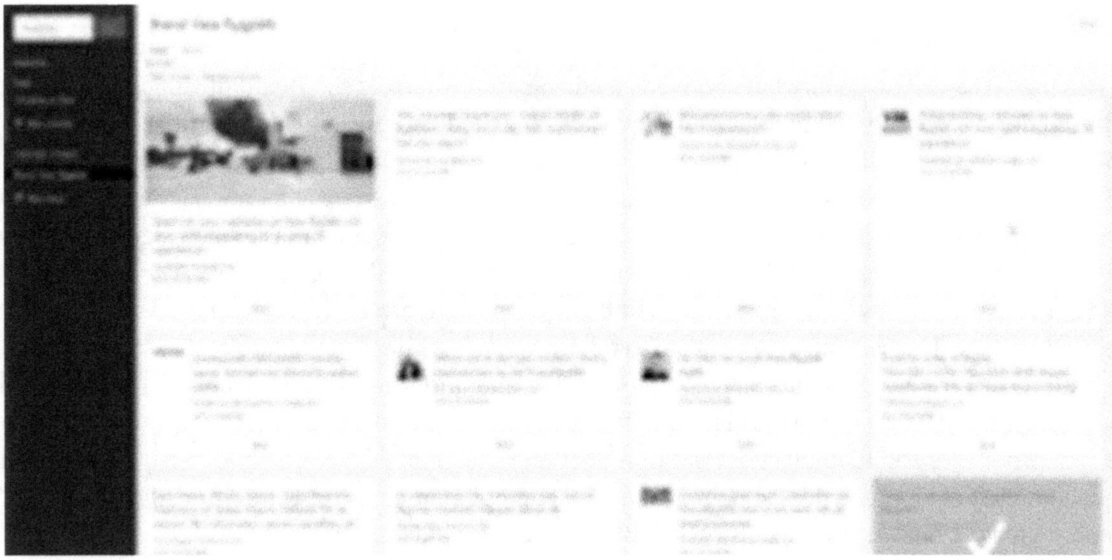

is that it takes social media and web content from a wide range of outlets and pages and places it in one workspace, which allows users to conduct extensive content searches and evaluate the data.

The close-to working prototype used in this usability test had a reactive interface with live data (Twitter and RSS). Thus, the test person could set up a new event and modify the sources displayed in the feed. The feed is divided into a monitor function, displaying a feed from searches on social media, and an ongoing stories function, where content can be added directly from the monitor view or from a search. Moreover, the central features of the tool were to save important content from the monitor view into a more detailed feed, as well as to manually evaluate the importance and trustworthiness of content. The saved content could be visualized on a geographic map or in a fourfold graph (trust/importance).

METHODS

Sample

For this study, 17 news journalists were approached via email or phone, and 15 chose to participate (9 females, 60%; age 28–61, $M = 40$). The sample had a work experience of between 4 and 36 years ($M = 14$), as well as personal and work experience of the most common social media platforms, such as Facebook, Instagram and Twitter. The majority of the participants produced journalistic material for the web (87%); moreover, they worked with newspapers (60%), radio (33%) and TV (20%). All the participants had previously taken part in an earlier usability test of a rudimentary version of the prototype tool [removed during peer-review process]. Hence, participants had a basic understanding of the tool before the test.

Test Procedure and Subjective Measures

Data was collected during the last week of January and the first week of February 2017 in a laboratory setting. The test procedure has received an ethical approval from the Board of Ethics [information removed during peer review process]. All data was collected, stored and analyzed in accordance with guidelines provided by the [information removed during peer review process].

The test was conducted in a room furnished as an office, using the tool at a stationary computer. Data from one participant was collected during each test session. Each session lasted between one and two hours, and the session was recorded. Two researchers were present during each test. A test session consisted of three parts. Firstly, the personnel and the test site were presented when the participant arrived at the laboratory. The participant was asked to relax on a sofa while the structure of the study, as well as the participant's rights, were presented. Also, possible alterations to the demographic information that was collected in the first test, were clarified. The participant was asked how much they remember from the first prototype, and then given a brief introduction to the aim of the prototype. This was to ensure that all the participants had the same information before completing the test. Additionally, the researchers presented the new measurements, which had not been included in the first test.

Secondly, the participant and one of the researchers went to the test room, and the participant was seated in front of a 24-inch computer screen while the researcher introduced the test procedure. The second researcher observed the test from the control room, with a live view stream and audio from the test room. The participant was asked to complete 36 tasks in a predefined order, e.g. "Please add a Twitter hashtag to the search" (see Appendix for information on all tasks, and [information removed during peer-review process] for a more detailed analysis on tasks not included in this chapter). Six of these tasks, regarding the most important features of the tool (monitoring an event; locating sources and adding new sources; saving content; and evaluating the importance and trustworthiness of content), are included in the analyses presented below. In these tasks, the test person was quiet during task completion. This was to ensure that the biometric reactions would not be affected by their voice.

A simulation of a crisis (a fire at the regional airport) was used to test the tasks in a situation more resembling a real-life crisis. The simulation was based on a pre-made dataset of Instagram images, tweets, and Facebook posts, appearing at certain time intervals throughout the test. Hence, the test situation and data displayed could be repeated for all the participants.

After completing the tasks, the participants could provide instant feedback or comments. The observing researcher in the control room took notes. Both researchers observed the tasks and graded each one as a "pass" (no problems), or "struggle/fail" (either completed or did not complete the task after a considerable struggle). The criteria for reaching each level had been defined prior to data collection and tested in a pilot study.

Thirdly, a semi-structured interview was conducted in front of the computer screen after the participant had finished all the tasks. The prototype was shown on the screen and the participant was asked to click around while answering questions. This was to allow the participant to show any problem areas or give suggestions and improvements. The interview included seven questions covering positive and negative feedback and the participant's overall impression of the second prototype and the development in comparison to the first version.

After the interview, the participant completed an AttrakDiff questionnaire (Hassenzahl, 2006). In the AttrakDiff, participants are asked to rate a product according to opposite word pairs, on a 7-point scale. The AttrakDiff consists of four scales, pragmatic aspects PQ (how well the product works), hedonic identity HQ-I (how much the user identifies with the product), hedonic stimulation HQ-S (how stimulating the product feels), and attractiveness ATT (how good the product looks), with a total of 28-word pairs.

After the test session, the participant could ask questions and discuss more general items. Directly after the participant left the laboratory, the researchers went through the observation ratings and discussed any unclarities and discrepancies. Most discrepancies consisted of human error; therefore, rating consensus could be reached.

Biometric Measurements

In this study, we combined observations, interviews, self-reports, and biometric data to identify usability problems and a potential negative user experience.

Three biometric measurements were used: eye tracking, skin conductance and facial responses. Eye tracking is a technique used to follow a user's gaze and eye movements. Information about the visual attention of a person provides objective cues on cognitive processing and distractions (Cowley et al., 2015). When analyzing computer interfaces and task performance, eye tracking provides valuable information (Goldberg et al., 2002). One benefit of eye tracking is that it can help us to understand and pinpoint mismatches regarding where the user thinks or assumes that information should be placed (where they first look) in comparison to the actual design of the interface.

Eye tracking data was collected using the Tobii Pro X3 120 binocular eye tracker with a sampling rate of 120 Hz. The software used was Tobii Studio 3.4.8. Participants were seated in front of a 24-inch full HD widescreen monitor, at a viewing distance of 60cm from the screen. When seated, the calibration procedure was initialized, measuring the characteristics of the user's eyes. This calibration process was successfully completed by all the 15 participants with a gaze sample percentage of between 41% and 96% ($M = 78\%$) during data collection. The lower gaze percentage is due to some of the test persons turning their heads away from the screen to talk to the test leader between tasks.

The eye-tracking metrics used in this study are Fixations before, Time to first fixation, Total visit duration and Fixation count. Areas of Interest (AOI) were created for the metrics, leaving approximately 1cm between different AOIs according to the "1 degree" guideline of the manufacturer. The Tobii I-VT fixation filter was used with a minimum threshold for fixations of 60ms.

The fixations before count measures the number of times the participant fixates on the media before fixating on an AOI or for the first time. This measurement reflects how fast a user finds a specific object on the screen (Roth et al., 2013).

Time to first fixation measures how long it takes before a test participant fixates on an AOI for the first time. This measurement represents the visual saliency of an object or stimulus element (Calvo et al., 2013) and the ability of a stimulus element to attract attention (Oliveira et al., 2016).

Total visit duration measures the duration of all visits within an AOI. This measurement represents the visual attention (Mackert et al., 2013), as well as the depth of cognitive processing of a stimulus element (Josephson, 2005).

Fixation count measures the number of times the participant fixates on an AOI. This measurement represents attention to a stimulus or stimulus element (O'Malley & Latimer-Cheung, 2013) and interest in a stimulus element (Mackert et al., 2013).

Skin conductance was measured in the study, as it is seen in HCI research as a reliable indicator of stress (Healey & Picard 2005; Lunn & Harper 2010). This level of sweating on the skin surface reflects a subject's emotional arousal when interacting with the environment (Stern et al., 2001). Electrodermal activity or galvanic skin response (GSR) is the measurement of the conductance of a very small electrical current applied to the skin. GSR can be described using the skin conductance level (SCL), which is the basal activity (tonic), or the skin conductance response (SCR), a phasic component of the response to a stimulus (Kreibig, 2010). In this study, we report SCL since it provides a slowly habituating measurement of arousal. A Shimmer3 GSR+ unit that records SCL via two finger electrodes attached to the participant's wrist was used. Skin conductance data was collected 128 times per second with a collection range of 0.2μS to 100μS (10kΩ to 4.7MΩ), with an auto-ranging feature that selects the most appropriate signal path in hardware. Data was collected using Consensys GSR software.

Facial expressions are an indicator of emotional expression (e.g Ekman, 1994). Facial expressions can be detected from a user camera and is an unobtrusive measurement that can be used to understand the relation between a user's experience and perceived usability problems (Brancoet al., 2005). In this study, facial expressions were collected using Affectiva's Emotion AI (affectiva.com), which measures different emotion metrics such as anger, fear, and joy, as well as facial expression metrics. The Affectiva SDK measures facial muscle activation and the output is a range of values from 0 to 100. Moreover, valence is a measurement of the positive or negative nature of the experience, ranging from -100 to 100 (McDuff et al., 2013).

RESULTS

In this study, six of the total 36 test tasks are analyzed in more detail (Figure 1; Appendix 1). These specific tasks were chosen as they reflect the most important features in the tool. Furthermore, they represent two different levels of cognitive understanding of the prototype tool and are thus divided into two subgroups. Subgroup one includes tasks related to investigating how well participants grasp what the tool is intended to do (high-level problems) and consists of tasks 1–2. In subgroup two, problems related to more detailed prototype design elements such as visual hierarchy or contrasts are investigated (low-level problems). Tasks 3–6 are included in this group.

Figure 2. Task completion for the six analyzed tasks. The majority of test participants passed the tasks without difficulty.

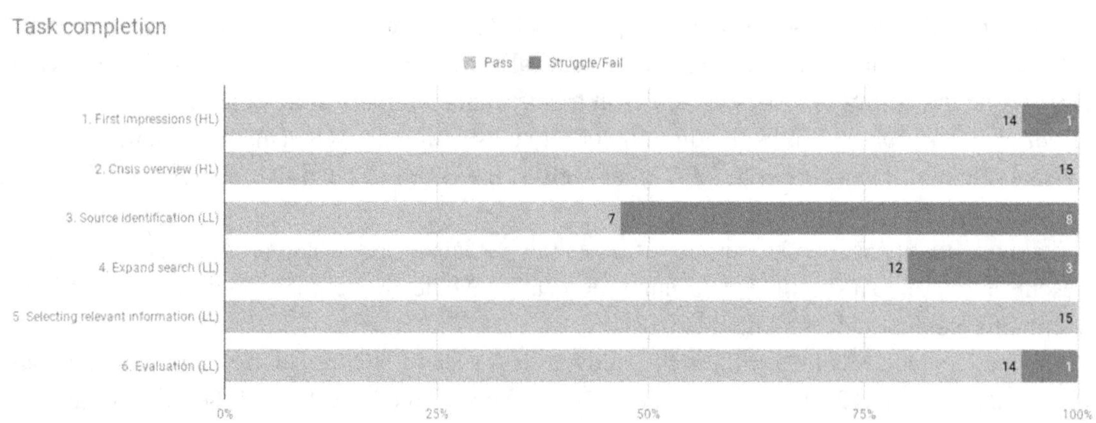

Below, we first present the results related to each task in the two subgroups. The groups are analyzed separately, beginning with the high-level problems. This is followed by results related to the participants' views of the whole prototype. Thus, we begin the results section with more specific features, and then focus on more overarching results. In the final discussion, the results are positioned within the fields of crisis management and communication.

Tasks Related to High-level Problems

High-level problems include tasks related to investigating how well participants grasp more overarching tool features, such as what the tool is intended to do and how well this fit in with the user's existing strategies for how to carry out a similar task. Results reported below are mainly based on subjective measurements, as it is difficult to draw relevant conclusions in relation to overarching features from biometric data. Biometrics are however used for illustrative purposes in the section below. Analyzed tasks related to high-level problems include tasks 1–2 (Appendix 1).

Task 1: First Impressions

This is the first time the test person encounters the prototype tool, and the task is to scroll through the feed, look at the tool and try to make sense of its purpose (without clicking or speaking). During a crisis situation, it is vital that the tool is intuitive and clear. The main objective of the task was to get a basic idea of what the system is about and grasp the main concept of the tool.

Figure 3. Heatmap of the first impressions task. Eye movements showed that the participants tried to comprehend the tool rather than read the included social media posts.

Figure 3 shows the eye-tracking heatmaps of the first impressions of the tool. The area that most participants focused on was the menu located on the left side, as well as the buttons on the top. Hence, the participants were trying to understand the main features of the tool, which was consistent with comprehending the situation. The presented feed consisted of actual Twitter posts from a pre-defined search about the local area, but participants did not focus that much on the content at this stage of the test.

The participants could comment on the features after each task, and most participants understood that this first view was a monitor feed from social media content. Several people said that this is a good feature, not only for crisis work, as it provides a way to keep track of what is going on in a geographic area and what people are talking about. Hence, the monitor view would save time and energy in the daily journalistic work of finding newsworthy stories.

Some participants had troubles understanding where the content came from, especially if they were not that familiar with social media from before the test. One participant suggested that the ongoing stories were automatically created by the tool.

Participants also wanted to broaden the tool and be able to handle their own social media accounts as well as write news stories in the same tool. Using the tool daily, and for several purposes, would increase the effectiveness of use during a crisis.

Task 2: Crisis Overview

In the second task, the test person is introduced to the simulation event (a fire at the regional airport) and told that the first alarm came in about 15 minutes ago. The task is to use the prototype to gain a first impression of what happened by scrolling through the provided social media content feed (created by the researchers for this test). The threat of information overload in a rapidly developing crisis demands an easily understandable task structure. The main objective for this task was to examine if the tool structure facilitated the understanding of a crisis.

The heatmap (Figure 4) shows that in this task, the participants focused exclusively on the content feed, in comparison to the first task where most participants looked at the menus and buttons. At this point in the test, participants already had a good idea of how the tool works and could focus on comprehending the unfolding situation. Not surprisingly, posts containing pictures received more attention from the participants than text-only posts.

All the participants managed to gain a good first overview of the unfolding situation after scrolling through the feed for a couple of minutes. The most common participant response about the task was that a lot of the content shown in the tool was from ordinary people. Some participants saw this as a positive feature, making it easier to find eyewitnesses on the scene. Others thought that this created too much disturbing noise and wanted updates only from trusted sources like news agencies and rescue services. One participant said that it is unlikely that there is time to save or comment on particularly interesting content for further analysis when a crisis is unfolding. However, this would be a great way to get an overview of what happened when some time had passed. These responses reflect two urgent needs for journalists working with an unfolding crisis: finding eyewitnesses who can describe the scene, and first responders/authorities who can verify how the crisis is evolving.

Figure 4. Heatmap of the Crisis Overview task. Participants mainly focused on what was presented about the unfolding crisis in the content feed provided.

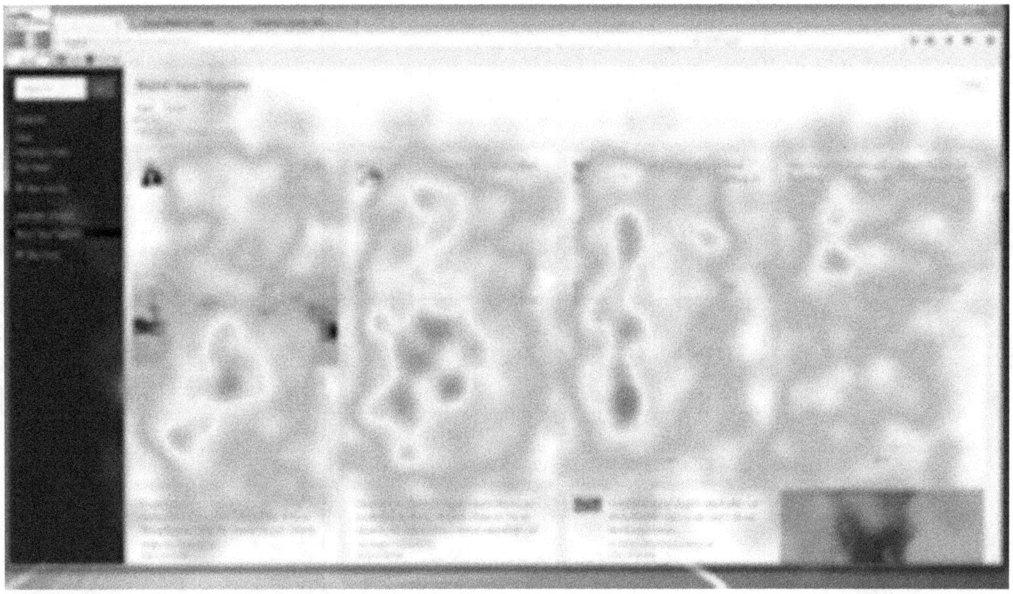

Tasks Related to Low-level Problems

Low-level problems consist of challenges related to detailed prototype design elements. Such elements may include how visual hierarchy or contrasts contribute to whether a participant makes correct decisions in the tool or misunderstands the meaning of a specific feature. Tasks 3–6 (Appendix 1) are included in this group. As low-level problems include detailed features for which specific prototype functions and areas of interest are easily defined, both biometric and subjective results are analyzed and reported below.

Task 3: Source Identification

In this task, the test person is asked to identify which sources are used in the displayed feed. Sources may include Twitter and Instagram hashtags or profiles. This is the first task where the test person can click on features to find information. When evaluating the credibility of information, it is important to know from where the content is derived. The concept of social media/web source identification is a well-known task for journalists. Hence, the main objective of this task is to understand how this feature is implemented in the tool.

Demographic data regarding the biometric measurements for the whole group is presented in Table 1. In general, it took participants around a minute to complete the task ($M = 58$ seconds). Approximately half of the sample completed the task without problem ($n = 7$, 47%; see Table 1), while the rest either found a solution to the task after a considerable struggle or did not complete it at all. Eye tracking heatmaps for the whole sample as well as for those who passed and those who struggled/failed are presented separately in Figure 5. Participants who struggled with the task tried to locate the sources manually, by clicking on the posts, which redirected them to the original content. The "correct" solution was to identify the relevant information directly on the prototype screen in front of them.

Table 1. Demographic data for tasks with low-level problems in total sample

Task	Passed task	Time to conduct task (s)	Fixations before first AOI	Time to first AOI fixation (s)	Total AOI visit duration (s)	Total AOI fixation count	SCL	Joy	Anger	Engage-ment	Smile	Valence
3	*n* = 7 (47%)	*M* = 58 (*SD* = 53)	*M* = 27 (*SD* = 25)[1]	*M* = 25 (*SD* = 29)[1]	*M* = 2 (*SD* = 3)[1]	*M* = 6 (*SD* = 7)[1]	*M* = 2 (*SD* = 2)	*M* = 5 (*SD* = 13)	*M* = 3 (*SD* = 7)	*M* = 27 (*SD* = 26)	*M* = 6 (*SD* = 14)	*M* = -2 (*SD* = 17)
4	*n* = 12 (80%)	*M* = 36 (*SD* = 26)	*M* = 23 (*SD* = 22)	-[2]	-[2]	-[2]	*M* = 3 (*SD* = 2)	*M* = 4 (*SD* = 9)	*M* = 2 (*SD* = 4)	*M* = 23 (*SD* = 19)	*M* = 5 (*SD* = 10)	*M* = -3 (*SD* = 10)
5	*n* = 15 (100%)	*M* = 84 (*SD* = 43)	*M* = 0,1(*SD* = 0,5)	*M* = 0,3 (*SD* = 0,6)	*M* = 10 (*SD* = 7)	*M* = 33 (SD = 24)	*M* = 3 (*SD* = 2)	*M* = 4 (*SD* = 6)	*M* = 2 (*SD* = 3)	*M* = 22 (*SD* = 16)	*M* = 5 (*SD* = 7)	*M* = -5 (*SD* = 9)
6	*n* = 14 (93%)	*M* = 24 (*SD* = 52)	*M* = 2 (*SD* = 6)	*M* = 2 (*SD* = 2)	*M* = 3 (*SD* = 3)	*M* = 7 (*SD* = 8)	*M* = 3 (*SD* = 3)	*M* = 2 (*SD* = 6)	*M* = 8 (SD = 22)	*M* = 26 (*SD* = 23)	*M* = 2 (*SD* = 6)	*M* = -7 (*SD* = 20)

[1] Task 3 could be solved by using either one of two buttons on the screen. Two separate AOIs were thus identified for analyzing eye tracking data. Number and time before first AOI fixation were defined as the time before a participant found the first of these AOIs, and total AOI visit duration and fixation count as the sum of the two AOIs.
[2] Eye tracking data missing from more than 66% of participants.

A series of t-tests were conducted to investigate group differences between the pass and struggle/fail groups in more detail. Potential mean differences on the demographic variables listed in Table 1 as well as on mean age and the four AttrakDiff subscales about the prototype as a whole were compared between the groups.

Regarding eye tracking measurements, it took longer for the struggle/fail (M = 43 seconds, SD = 33) than for the pass group (M = 7, SD = 7; t[10] = 2.62, p = .026) to fixate on one of the two AOI tasks for the first time. The struggle/fail group had a lower number of eye fixations before the first AOI fixation (M = 9 fixations, SD = 13 vs. M = 42, SD = 22; t[9] = 2.95, p = .016). Hence, the struggle/fail participants did see the buttons but did not understand what to use them for.

The struggle/fail group also needed significantly more time to conduct the task (M = 90 seconds, SD = 55 vs. M = 21, SD = 10; t[8] = 3.49, p = .009). Analyses of remaining variables showed no significant differences between groups.

Heatmaps from the two groups showed different gaze patterns when conducting the task. The participants who passed the task (Figure 5) searched for the information on the left side menu and on the top sidebar (where the two buttons were located). In comparison, the group that struggled or failed searched for the information in the content feed.

Figure 5. Heatmaps of the Source Identification task for the groups that passed (left) and struggled/failed the task (right). Those who passed identified the correct menu options in the upper left corner of the workspace.

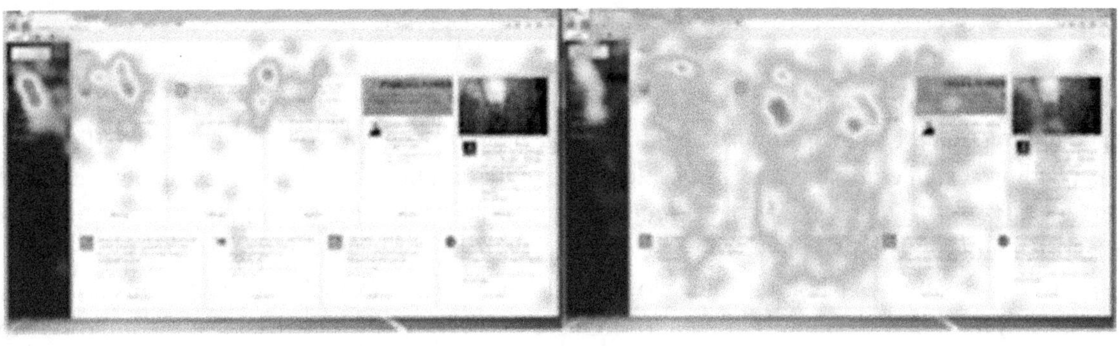

Correlation analyses were conducted with the whole sample to investigate how biometric measurements specifically related to Task 3 were related to age, task completion time and AttrakDiff indicators about the prototype. Significant results are summarized in Table 2 below.

The most significant correlations were found between eye tracking variables and the AttrakDiff. More time or more eye fixations before the first AOI fixation was correlated with not identifying with the prototype while using it (HQ-I) and with finding it less stimulating (HQ-S). The total AOI fixation count was related to more stimulation while using the product (HQ-S) and a higher rating of the general attractiveness of the product (ATT).

Furthermore, an older age was significantly correlated with lower stress as indicated by skin conductance level. Older participants may thus have experienced the task as less stressful. However, this result should be interpreted carefully, as lower skin conductance levels generally have been linked to increasing age.

Table 2. Significant correlations between low-level task biometric variables and age, task completion time, and AttrakDiff subscales in total sample

	Participant age	Time to conduct task (s)	AttrakDiff PQ	AttrakDiff HQI	AttrakDiff HQS	AttrakDiff ATT
Fixations before first AOI fixation				Task 3:-.62* Task 4:-.71**	Task 3:-.62* Task 4:-.81*	
Time to first AOI fixation (s)[1]		Task 3:.84**	Task 3:-.58*			Task 3:-.68*
Total AOI visit duration (s)[1]						
Total AOI fixation count[1]					Task 3: .64*	Task 3: .60*
Skin conductance level	Task 3:-.62* Task 4:-.57*					
Facial expressions: joy		Task 5:.72**				
Facial expressions: anger	Task 4: .70**	Task 4:.60*				
Facial expressions: engagement		Task 5:.52*				
Facial expressions: smile		Task 5:.68**				
Facial expressions: valence				Task 4: .62*		Task 4: .52*
[1]Eye tracking data missing from more than 66% of participants						

Task 4: Expand Search

In this task, the test person is asked to add a Twitter hashtag to a web/social media content search. In the beginning of a crisis, there are usually no clear hashtags or keywords related to the unfolding event, so expanding search criteria as a crisis develops, e.g. with hashtags, could be necessary. Similar to the previous task, this feature contains the execution of a task that is familiar to journalists who use the web for gathering information.

Biometric data for the whole sample is found in Table 1. Eye tracking heatmaps for this task are not presented as most of this data was missing from more than 66% of the participants (see Table 1). Most participants completed the task without problem (n = 12, 80%), and completing the task took M = 36 seconds for the sample.

The two participants who had problems with the task used the wrong prototype features and did not understand the wording used in several of the features (e.g. what is a URL or a source). Those who passed the task still had some questions about the Boolean search alternatives. As the majority passed the task, no group comparison analyses between the pass and fail subgroups were conducted for this task.

Correlation analyses were conducted with available biometric measurements, age, task completion time and AttrakDiff indicators. The results are summarized in Table 2. As in the previous task, more eye fixations before the first AOI fixation was correlated with finding the prototype less relatable to the user (HQ-I) and less stimulating (HQ-S). Regarding facial expressions, more time required to conduct the task was correlated with more anger, and more valence (i.e. expressing positive feelings) with identifying with the tool (HQ-I) and finding it more attractive (ATT). Higher age was significantly correlated with a lower stress level (SCL) in this task, as in task 3 above.

Task 5: Selecting Relevant Information

After looking through the social media and web posts about the simulated crisis in the prototype, the test person in this task is asked to select one relevant social media post and save it. Since the flow of information in social media during an emergency can be very fast-moving, quickly selecting and saving relevant content is essential. This is the first time in the test that the participant tries out the prototype's save feature.

See Table 1 for biometric data for this task. All the participants passed the task in approximately a minute and a half ($M = 84$ seconds), and thus no group comparisons between pass/fail subgroups were conducted. In think-aloud comments about the task, participants described that saving a feature required too many steps, since it also included an evaluation of the content. Some participants were not sure if they were saving information just for themselves or for everybody working on the story.

Correlation analyses showed significant results mainly related to facial expressions. Positive facial reactions (joy, engagement, smile) were positively correlated with task completion time. As all participants passed the task, this indicated that even those who required more time to do the task, enjoyed doing it.

Task 6: Evaluation

When saving social media/web content, the prototype provides an opportunity to subjectively evaluate the trustworthiness and importance of the post in order to facilitate further exploration at a later stage. In this task, the evaluation subpage is opened to display a chart showing the saved posts according to their rating in terms of importance and trustworthiness. The test person is asked to locate the post on the page that has the highest importance rating. The task tests if displaying the ratings in a visual chart is helpful and facilitates comprehension. The task concerns visual implementation of tool features.

Basic biometric data about task completion is found in Table 1. One person did not pass the task, and thus no group comparisons between pass and fail subgroups were conducted. Eye-tracking heat maps from the prototype evaluation subpage are illustrated in Figure 6. The gaze patterns showed that the participants understood the fourfold visualization right away, focusing on the quadrant containing information on the posts that had been rated as important and trustworthy.

The sample completed the task in around 30 seconds ($M = 24$). The participants described the visualization as easy to understand but pointed out that the graph will be more difficult to read if it contains information on too many posts. In this task, none of the correlation analyses between task-specific biometric data, participant age, task completion time, and AttrakDiff subscales were significant.

Views of the Whole Prototype

After task completion, participants took part in a semi-structured interview and completed the AttrakDiff questionnaire (Hassenzahl, 2006) about their impressions of the whole prototype. Nine participants (60%) had a mainly positive first impression of the prototype (mainly negative $n = 2$, 13%; cannot say $n = 4$, 27%). The majority ($n = 11$, 73%) thought that it would fit in well with how they currently conduct journalistic work tasks. Two key features that influenced a positive first impression were that the tool helped the user gain a good overview of an emergency, and that it allowed a user to add information about how trustworthy the found content is. Those who reported a negative or neutral first impression referred to minor issues, such as the wording of some of the features as well as the placement of specific buttons.

Figure 6. Heatmap of the evaluation feature. Most participants could locate the content rating feature right away.

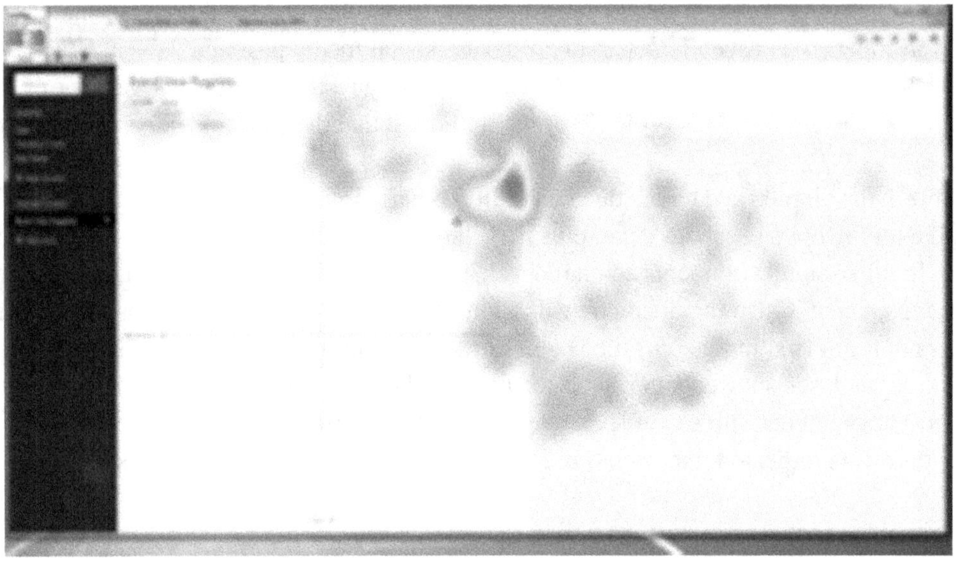

The AttrakDiff subscale means for the total sample were pragmatic qualities PQ $M = 5,2$ ($SD = 0,9$), hedonic qualities-identity HQ-I $M = 5,1$ ($SD = 0,6$), hedonic qualities-stimulation HQ-S $M = 5,0$ ($SD = 0,5$), and attractiveness ATT PQ $M = 5,1$ ($SD = 0,6$). This indicates that the participants rated the prototype design and usefulness as both desired and task oriented (Figure 7). Participants mentioned that the tool is simple to use and does not contain too many options. A correlation analysis between age and the AttrakDiff subscales showed that an older age correlated with lower PQ scores ($r(13) = -.69$, $p = .004$), indicating that older participants found it more difficult to operate the prototype.

SOLUTIONS AND RECOMMENDATIONS

The main focus of the study presented in this chapter was to identify usability problems in a prototype tool intended for use during times of crisis. The study combines biometric measurements with traditional usability surveys (AttrakDiff) and interviews.

Using biometric data to understand the user's experience is helpful. For instance, in this study the participants who struggled or failed to pass a task nevertheless fixated their eyes on the correct button before the group that passed the task. This sensory clue did not help them to find the correct function, as the wording used in the prototype did not fit their mental models (see Berndt & Herczeg, 2019). Moreover, the eye tracking metrics of AOI fixations count and fixations before, correlated with the AttrakDiff ratings. Fixation count represents attention to a stimulus (see O'Malley & Latimer-Cheung, 2013) and interest in a stimulus element (see Mackert et al., 2013). In this study, focusing more on the features also increased the evaluation of the general attractiveness of the tool. The facial expressions did not differ significantly between tasks; instead facial expressions were correlated with time to conduct the task, and a positive score on valence also correlated with higher scores on attractiveness on the self-

Figure 7. Word pairs in the AttrakDiff questionnaire. The sample rated the prototype as relatively useful and pleasant.

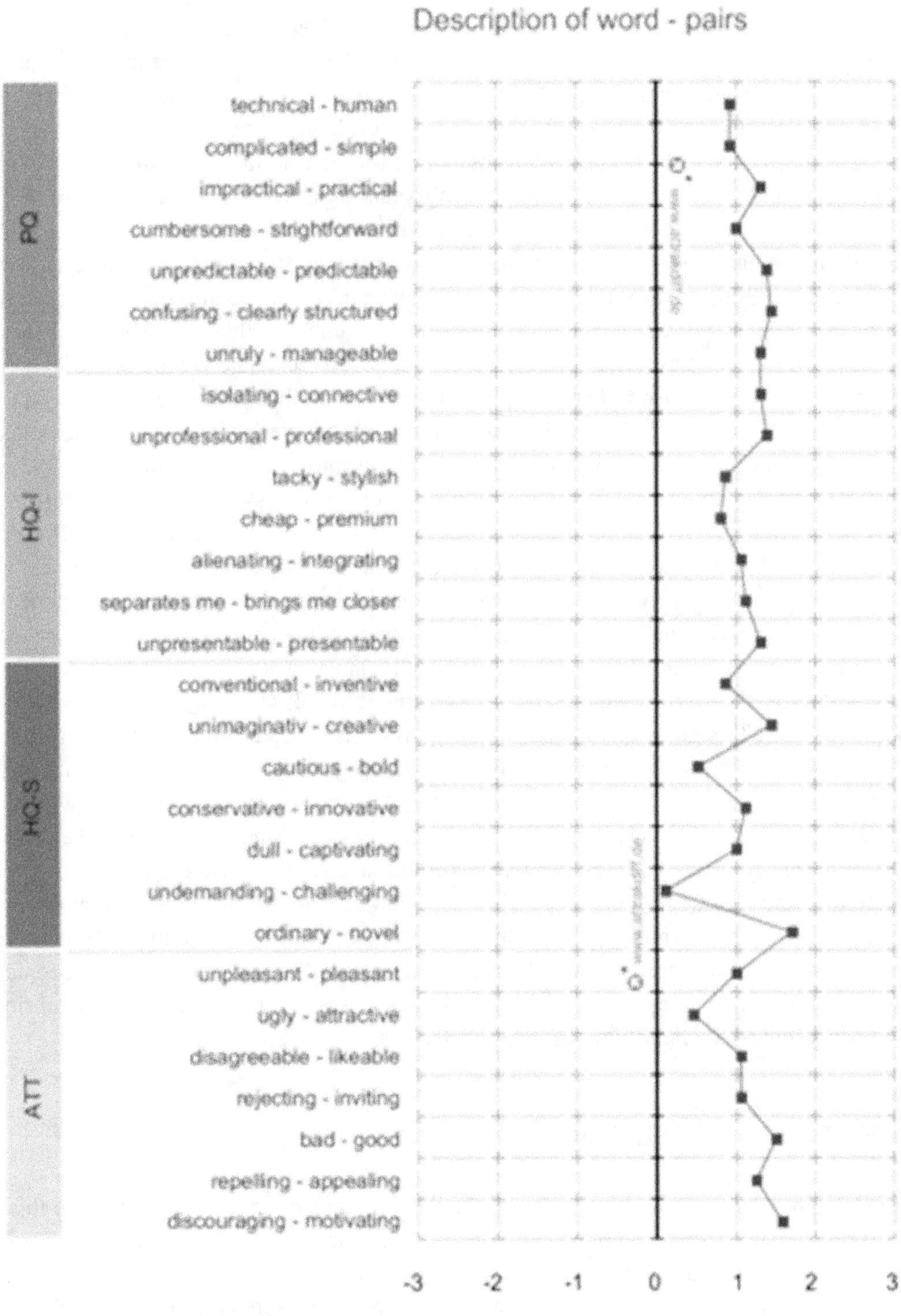

evaluation form. Hence, biometric measurement is an easy and unobtrusive method to use in order to understand parts of the participants' experience and provides valuable additional information to insight gathered through subjective measures.

According to the first impressions of the users, the tool received some very positive feedback, indicating that it is easy to use and clearly structured. Some of the features that were mentioned as helping to create good situation awareness were information gathering from several different platforms into one tool, the save content for further analysis function and the evaluation of content.

In the first research question, we asked how users' high-level usability problems, i.e. concerning the overall understanding of the prototype, affect overall situation awareness in emergencies? The study showed that most of the participants did not struggle with the overall understanding of the tool. Users first focused on getting an overview of the different features, and when a (simulated) crisis occurred, they could focus on getting an overview of the situation at hand. A respondent's mental models of how to approach a new situation will differ, so the tool needs to offer functions that can be personalized according to the needs of the user. One example is how older age groups correlated with lower PQ scores (how well a product works) on the AttrakDiff. People who are less familiar with social media and the logic of such platforms might prefer different visualization options than people who use social media daily.

The second research question focused on how low-level usability problems, concerning specific prototype features and elements, may affect the overall situation awareness of users in emergencies. Some of the low-level problems, related to system presentation and usability, were more challenging. The biggest challenge in the analyzed tasks was source identification. Understanding where the information comes from is vital, so too is being able to change the sources displayed in the feed. To include the right wording in a prototype is crucial and can be a challenge in today's social media landscape, where many terms are in English and might be hard to translate to other languages.

Saving posts from the overall feed to a specific personal feed helps the user create an overview of the situation. In this test, although the feature did not cause any big issues, participants still provided feedback that the wording and colors used in the prototype were challenging. Also, more filtering and sorting options as well as different visualization modes would help the user to gain a good overview of the feeds.

The evaluation feature did cause problems for the participants. Since the tool may be used collaboratively, for instance at a news agency, participants emphasized that content evaluations from different people may vary and that they would need a clear overview of who had done the manual evaluation. A recommendation would be to provide the evaluation history as well as time stamps. Moreover, participants found that it was unclear to know if the presented score was a mean from several people having evaluated a post or just the latest score. Avoiding too many clicks is crucial during stressful situations, and participants suggested simplifying the procedure as well as the scales for evaluation.

FUTURE RESEARCH DIRECTIONS

We end this chapter by tying up the results in relation to future research directions. The third, more summative research question asked how an information-handling tool intended for monitoring, gathering and handling social media and web content in emergencies should be designed to contribute to the overall situation awareness of the user. Crisis situations are by nature sudden and unpredictable, and so is social media (Palen & Anderson, 2016; Reuter, Hughes, et al., 2018). It is difficult to predict the leading platforms in a few years to come and designing for a constantly changing environment is a challenge.

A key challenge for system design is the fact that a usability process is an iterative process, especially in crisis informatics (Hiltz et al., 2020). Testing a tool in a laboratory setting is the first step toward reaching a better product (ISO 9241, 2018). The use of biometric measurements is a suitable method to use for gathering information on an experience without interrupting the participants. However, the usability of digital services should also be tested in real life scenarios, where other types of challenges may arise. Moreover, another challenge concerns the differences between designing tools intended exclusively for crisis situations or for more general use (see e.g. Reuter & Mentler, 2018). Naturally, a "live" crisis creates both stress and situation awareness that are difficult to predict in usability testing with a constructed scenario. A combination of such testing and the subsequent post-crisis feedback of an implemented tool is recommended for improving products designed for emergencies. In this study, participants emphasized the need to use the tool daily in order to have a good understanding of all the functions and to be able to just keep working when a crisis hits. Another limitation with this study is the focus on one single user group, i.e. journalists. Future research should include several different user groups.

CONCLUSION

Although previous research has shown that digital communication during times of crisis poses several challenges, such as information overload or source credibility (Gryszkiewicz & Chen, 2012; Spence et al., 2013), the design and usability of digital services is an important step toward helping key communicators maintain situation awareness (Palen et al., 2009). This study contributes to the field of crisis informatics by taking a human-centered approach to understanding the needs and challenges journalists face when using a digital tool for handling social media.

Situation awareness, in the form of perceiving and comprehending, are important first steps on the way to projecting a likely outcome (Endsley, 1995). In this chapter, low-level problems related to perceiving, and high-level problems related to comprehending were tested in a prototype that gathers data from several different platforms into one feed. Comprehending the functions of the tool and understanding the bigger picture did not cause any problems for the participants. Instead, their wishes and needs were met in this prototype and the majority would start using it straight away. However, some of the low-level problems, e.g. perceiving where to find a function or information, did cause problems. This also means that not reaching the first step of situation awareness, i.e. perceiving, would constrain the next step of awareness, i.e. comprehending. Especially during a crisis, this would cause extra stress and further damage the awareness of the situation (projecting likely outcomes). Hence, using HCI to improve the usability of digital platforms is a key step toward better sentience, and when conducting usability tests, biometric measurements in combination with self-reports and interviews help the researcher to understand the usability problems from several different angles.

REFERENCES

Berndt, H., & Herczeg, M. (2019). The Role of Mental Models and Situation Awareness for Computer System Support in Mass Casualty Incident Management. In *Proceedings of the 31st European Conference on Cognitive Ergonomics.* Association for Computing Machinery.

Boin, A., 't Hart, P., Stern, E., & Sundelius, B. (2005). *The Politics of Crisis Management. Public 212 Leadership under Pressure*. New York: Cambridge University Press.

Branco, P., Firth, P., Encarnação, L. M., & Bonato, P. (2005, April). Faces of emotion in human-computer interaction. In *CHI'05 Extended Abstracts on Human factors in computing systems*. ACM. doi:10.1145/1056808.1056885

Calvo, M. G., Gutiérrez-García, A., Avero, P., & Lundqvist, D. (2013). Attentional mechanisms in judging genuine and fake smiles: Eye-movement patterns. *Emotion (Washington, D.C.)*, *13*(4), 792–802. doi:10.1037/a0032317 PMID:23627721

Cameron, M. A., Power, R., Robinson, B., & Yin, J. (2012). Emergency situation awareness from twitter for crisis management. In *Proceedings of the 21st International Conference on World Wide Web*. ACM. 10.1145/2187980.2188183

Castillo, C. (2016). *Big crisis data - Social media in disasters and time-critical situations*. Cambridge University Press. doi:10.1017/CBO9781316476840

Comfort, L. K. (2007). Crisis management in hindsight: Cognition, communication, coordination, and control. *Public Administration Review*, *67*(s1), 189–197. doi:10.1111/j.1540-6210.2007.00827.x

Cowley, B., Filetti, M., Lukander, K., Torniainen, J., Henelius, A., Ahonen, L. ... Ravaja, N. (2016). The psychophysiology primer: A guide to methods and a broad review with a focus on human-computer interaction. *Foundations and Trends® in HumanComputer Interaction*, *9*(34), 151–308.

Duncan, S., & Barrett, L. F. (2007). Affect is a form of cognition: A neurobiological analysis. *Cognition and Emotion*, *21*(6), 1184–1211. doi:10.1080/02699930701437931 PMID:18509504

Edwards, C., Spence, P. R., Gentile, C. J., Edwards, A., & Edwards, A. (2013). How much Klout do you have… A test of system generated cues on source credibility. *Computers in Human Behavior*, *29*(5), A12–A16. doi:10.1016/j.chb.2012.12.034

Ekman, P. (1994). Strong evidence for universals in facial expressions: A reply to Russell's mistaken critique. *Psychological Bulletin*, *115*(2), 268–287. doi:10.1037/0033-2909.115.2.268 PMID:8165272

Endsley, M. R. (1995). Toward a theory of situation awareness in dynamic systems: Situation awareness. *Human Factors*, *37*(1), 32–64. doi:10.1518/001872095779049543

Goldberg, J. H., Stimson, M. J., Lewenstein, M., Scott, N., & Wichansky, A. M. (2002). Eye tracking in web search tasks: design implications. In *Proceedings of the 2002 symposium on Eye tracking research & applications*. ACM.

Gryszkiewicz, A., & Chen, F. (2012). Temporal aspects in crisis management and its implications on interface design for situation awareness. *Cognition Technology and Work*, *14*(2), 169–182. doi:10.100710111-011-0199-y PMID:32214898

Hassenzahl, M. (2006). Hedonic, emotional and experiential perspectives on product quality. In C. Ghaoui (Ed.), *Encyclopedia of human computer interaction* (pp. 266–272). IGI Global. doi:10.4018/978-1-59140-562-7.ch042

Healey, J. A., & Picard, R. W. (2005). Detecting stress during real-world driving tasks using physiological sensors. *IEEE Transactions on Intelligent Transportation Systems*, *6*(2), 156–166. doi:10.1109/TITS.2005.848368

Hiltz, S. R., Hughes, A. L., Imran, M., Plotnick, L., Power, R., & Turoff, M. (2020). Exploring the usefulness and feasibility of software requirements for social media use in emergency management. *International Journal of Disaster Risk Reduction*, *42*, 101367. doi:10.1016/j.ijdrr.2019.101367

Hovland, C. I., Janis, I. L., & Kelley, H. H. (1953). *Communication and persuasion: Psychological studies of opinion change*. Yale University Press.

Hughes, A. L., & Shah, R. (2016). Designing an application for social media needs in emergency public information work. In *Proceedings of the 19th International Conference on Supporting Group Work*. ACM Press. 10.1145/2957276.2957307

Iannella, R., & Henricksen, K. (2007). Managing information in the disaster coordination centre: Lessons and opportunities. *4th International Conference on Information Systems for Crisis Response and Management*.

ISO 9241. (2018). *Ergonomics of human-system interaction—Part 210: Human-centered design for interactive systems*. Geneva: International Standard Organization.

Josephson, S. (2005). Eye tracking methodology and the Internet. In K. L. Smith, S. Moriarty, G. Barbatsis, & K. Kenney (Eds.), *Handbook of visual communication: Theory, methods, and media* (pp. 63–80). Erlbaum.

Kreibig, S. D. (2010). Autonomic nervous system activity in emotion: A review. *Biological Psychology*, *84*(3), 394–421. doi:10.1016/j.biopsycho.2010.03.010 PMID:20371374

Kukula, E. P., Sutton, M. J., & Elliott, S. J. (2010). The human–biometric-sensor interaction evaluation method: Biometric performance and usability measurements. *IEEE Transactions on Instrumentation and Measurement*, *59*(4), 784–791. doi:10.1109/TIM.2009.2037878

Kula, I., Branaghan, R. J., Atkinson, R. K., & Roscoe, R. D. (2018). Assessing User Experience via Biometric Sensor Affect Detection. In R. Roscoe, S. Craig, & I. Douglas (Eds.), *End-User Considerations in Educational Technology Design* (pp. 123–139). IGI Global. doi:10.4018/978-1-5225-2639-1.ch006

Lin, X., Spence, P. R., Sellnow, T. L., & Lachlan, K. A. (2016). Crisis communication, learning and responding: Best practices in social media. *Computers in Human Behavior*, *65*, 601–605. doi:10.1016/j.chb.2016.05.080

Liu, S. B. (2014). Crisis crowdsourcing framework: Designing strategic configurations of crowdsourcing for the emergency management domain. *Computer Supported Cooperative Work*, *23*(4–6), 389–443. doi:10.100710606-014-9204-3

Lunn, D., & Harper, S. (2010). Using Galvanic Skin Response Measures to Identify Areas of Frustration for Older Web 2.0 Users. In *Proceedings of the 2010 International Cross Disciplinary Conference on Web Accessibility*. ACM. 10.1145/1805986.1806032

Mackert, M., Champlin, S. E., Pasch, K. E., & Weiss, B. D. (2013). Understanding health literacy measurement through eye tracking. *Journal of Health Communication*, *18*(1), 185–196. doi:10.1080/10810730.2013.825666 PMID:24093355

McDuff, D., El Kaliouby, R., Senechal, T., Amr, M., Cohn, J. F., & Picard, R. (2013). Affectiva-MIT Facial Expression Dataset (AM-FED): Naturalistic and Spontaneous Facial Expressions Collected" In-the-Wild". In *2013 IEEE Conference on Computer Vision and Pattern Recognition Workshops*. IEEE. 10.1109/CVPRW.2013.130

Metzger, M. J., Flanagin, A. J., & Medders, R. B. (2010). Social and heuristic approaches to credibility evaluation online. *Journal of Communication*, *60*(3), 413–439. doi:10.1111/j.1460-2466.2010.01488.x

O'Malley, D. A., & Latimer-Cheung, A. E. (2013). Gaining perspective: The effects of message frame on viewer attention to and recall of osteoporosis prevention print advertisements. *Journal of Health Psychology*, *18*(11), 1400–1410. doi:10.1177/1359105312456323 PMID:23188916

Oliveira, D., Machin, L., Deliza, R., Rosenthal, A., Walter, E. H., Gimenez, A., & Ares, G. (2016). Consumers' attention to functional food labels: Insights from eye-tracking and change detection in a case study with probiotic milk. *Lebensmittel-Wissenschaft + Technologie*, *68*, 160–167. doi:10.1016/j.lwt.2015.11.066

Palen, L., & Anderson, K. M. (2016). Crisis informatics: New data for extraordinary times. *Science*, *353*(6296), 224–225. doi:10.1126cience.aag2579 PMID:27418492

Palen, L., Vieweg, S., Liu, S. B., & Hughes, A. L. (2009). Crisis in a networked world: Features of computer-mediated communication in the April 16, 2007, Virginia Tech event. *Social Science Computer Review*, *27*(4), 467–480. doi:10.1177/0894439309332302

Palen, L., Vieweg, S., Sutton, J., Liu, S. B., & Hughes, A. (2007). Crisis informatics: Studying crisis in a networked world. *Proceedings of the Third International Conference on E-Social Science*.

Picard, R. W. (2003). Affective computing: Challenges. *International Journal of Human-Computer Studies*, *59*(1–2), 55–64. doi:10.1016/S1071-5819(03)00052-1

Poels, K., & Dewitte, S. (2006). How to capture the heart? Reviewing 20 years of emotion measurement in advertising. *Journal of Advertising Research*, *46*(1), 18–37. doi:10.2501/S0021849906060041

Rasmussen, J., & Ihlen, Ø. (2017). Risk, Crisis, and Social Media. A systematic review of seven years' research. *Nordicom Review*, *38*(2), 1–17. doi:10.1515/nor-2017-0393

Reuter, C., Hughes, A. L., & Kaufhold, M. A. (2018). Social media in crisis management: An evaluation and analysis of crisis informatics research. *International Journal of Human-Computer Interaction*, *34*(4), 280–294. doi:10.1080/10447318.2018.1427832

Reuter, C., & Mentler, T. (2018). Human–computer interaction and social media in safety-critical systems. *Journal of Contingencies and Crisis Management*, *26*(1), 2–3. doi:10.1111/1468-5973.12192

Roth, S. P., Tuch, A. N., Mekler, E. D., Bargas-Avila, J. A., & Opwis, K. (2013). Location matters, especially for non-salient features–An eye-tracking study on the effects of web object placement on different types of websites. *International Journal of Human-Computer Studies, 71*(3), 228–235. doi:10.1016/j.ijhcs.2012.09.001

Spence, P. R., Lachlan, K. A., Westerman, D., & Spates, S. A. (2013). Where the gates matter less: Ethnicity and perceived source credibility in social media health messages. *The Howard Journal of Communications, 24*(1), 1–16. doi:10.1080/10646175.2013.748593

Stern, R. M., Ray, W. J., & Quigley, K. S. (2001). *Psychophysiological Recording*. Oxford University Press.

Sweetser, K. D., & Metzgar, E. (2007). Communicating during crisis: Use of blogs as a relationship management tool. *Public Relations Review, 33*(3), 340–342. doi:10.1016/j.pubrev.2007.05.016

Veil, S. R., Buehner, T., & Palenchar, M. J. (2011). A work-in-process literature review: Incorporating social media in risk and crisis communication. *Journal of Contingencies and Crisis Management, 19*(2), 110–122. doi:10.1111/j.1468-5973.2011.00639.x

Vieweg, S., Hughes, A. L., Starbird, K., & Palen, L. (2010). Microblogging during two natural hazards events: what twitter may contribute to situational awareness. In *Proceedings of the SIGCHI conference on human factors in computing systems*. ACM. 10.1145/1753326.1753486

Wallach, D., & Scholz, S. C. (2012). User-centered design: Why and how to put users first in software development. In A. Maedche, A. Botzenhardt, & L. Neer (Eds.), *Software for people* (pp. 11–38). Springer. doi:10.1007/978-3-642-31371-4_2

White, C., & Plotnick, L. (2010). A framework to identify best practices: Social media and web 2.0 technologies in the emergency domain. *International Journal of Information Systems for Crisis Response and Management, 2*(3), 25–35. doi:10.4018/jiscrm.2010070102

Wiegand, S., & Middleton, S. E. (2016). Veracity and velocity of social media content during breaking news: Analysis of November 2015 Paris shootings. In *Proceedings of the 25th international conference companion on world wide web*. International World Wide Web Conferences Steering Committee 10.1145/2872518.2890095

Wright, D. K., & Hinson, M. D. (2009). An updated look at the impact of social media on public relations practice. *The Public Relations Journal, 3*(2), 1–27.

Xu, J. (2020). Does the medium matter? A meta-analysis on using social media vs. traditional media in crisis communication. *Public Relations Review, 46*(4), 10947. doi:10.1016/j.pubrev.2020.101947

KEY TERMS AND DEFINITIONS

Biometric: Measurements of human body characteristics such as psychophysiology.
Crisis: A situation where societal structures and values are threatened.
Crisis Informatics: A research field that looks at social media and technology in crisis management.
Eye Tracking: A technique used to follow a user's gaze and eye movements.

Situation Awareness: A theory emphasizing how the comprehension of a certain situation involves understanding feedback from the environment and forecasting what to do next.

Skin Conductance: Level of sweating on a person's skin surface.

Usability: The extent to which a product can be used by specified users to achieve specified goals with effectiveness, efficiency, and satisfaction in a specified context of use.

APPENDIX

Tasks included in the usability test (explanations concerning the feature in question is given in the brackets):

Monitor View

Without clicking, what can you do here? (Interface)
Without clicking, what is in the balk? (Menu)
What does "monitor" mean?
Where is the information displayed in the feed originating from?

Sources

If you want to know what sources are used, how do you do that?
Please add the twitter hashtag #visitvaasa.
Do you know what happens if you search for "#trump OR #hillary"?
If you want to look at a post, how do you do that?

Ongoing Story

How do you find the feed for the ongoing story "Brand Vasa Flygfält"?
Please read through the ongoing story posts.
At the top of the page there is a red balk, what do you think it means (new posts)?
Click on "show new items", which are the new items?

Saved Posts

Some of the posts are red and have a checkmark, what does that mean? (They are saved)
Save a post you find interesting.
What does "importance" mean? (The option to rate importance when opening a post)
What does "trust" mean? (The option to rate trust when opening a post)
What does "location" mean? (Information about the geolocation of the post)
What does "comment" mean? (The option to write a comment)
Click on "save". What does it mean that you save a post?
Click on "feed", what can you do here? (View saved posts)
You want to add a comment to the post made by Mira, how do you do that?
You want to change the post's trust evaluation to a 9, how do you do that?
One of the saved posts are no longer relevant, how do you delete it?

Location

Does any of the posts have a geolocation? Where can you find this information?
Are there posts from people on the scene?
What does it mean that a post has a geolocation?
Where does the importance and trust ratings for this post come from?
How many posts with geographic information is there?

Evaluation

Where can you find more information on trust and importance ratings?
Which post has the highest importance rating?
Please lower the importance rating on this post.
Can you add a comment to the post with information that you changed the rating?
What does a red two in the ring mean? (That two posts have the same rating)

New Story

Can you create a new story about Donald Trump? How would you do that?
Add a twitter search to your story.
You want to add this URL link to the story, how would you do that?

Chapter 4
Internet Information Monitoring System:
A Digital Tool for Emergencies, Crises, and Disasters

Paola Pinheiro Bernardi Primo
Universidade Federal do Espírito Santo, Brazil

Michele Nacif Antunes
Universidade Federal do Espírito Santo, Brazil

Mariela Pitanga Ramos
Secretaria Estadual de Saúde do Espírito Santo, Brazil

Carlos Eduardo Gomes Siqueira
https://orcid.org/0000-0001-8993-3031
University of Massachusetts, Boston, USA

Adauto Emmerich Oliveira
Universidade Federal do Espírito Santo, Brazil

José Manuel Mendes
https://orcid.org/0000-0003-3602-9756
Centre for Social Studies, Faculty of Economics, University of Coimbra, Portugal

ABSTRACT

This study presents an information monitoring system: SIGDesastre. It is a method of monitoring the various sources of information available on the internet. The monitoring scenario is the failure of the dam in Mariana, Brazil. This event was considered the biggest socio-environmental disaster in the country. The creation of SIGDesastre involves the identification of the sources to be monitored, an automated search system for keywords in these pre-registered sources and the visualization of the results in a friendly environment. The use of information monitoring on the internet is considered an important tool for the post-disaster risk communication process. The monitoring of information on the internet is believed to be a potential device to support managers in institutional decisions and in the formulation of public policies. Also, for the affected population, they will be able to expand access to information about the actions being taken by the actors involved.

DOI: 10.4018/978-1-7998-6705-0.ch004

INTRODUCTION

In recent decades records of disasters, crises and emergencies have risen worldwide. In 2009, the influenza A (H1N1) pandemic infected more than 600,000 people and was the first declaration of this type of emergency by the World Health Organization in the 21st century (WHO, 2009). In the early months of 2020, we saw a new worldwide public health emergency due to the new Coronavirus, the SARS-CoV-2, turn into a pandemic

Major socio-environmental disasters have also made history around the world. One important example is dam disasters. According to the database of the World Information Service on Energy (WISE), there are records of about 90 serious dam disasters in the world in the last 50 years, which makes them common and recurrent disasters (WISE, 2020).

In 2015, the biggest socio-environmental disaster in the world took place in Brazil. On November 5, 2015, Fundão dam, located in Mariana, state of Minas Gerais (MG), ruptured generating a "wave" of mine tailings and destruction. More than 50 million cubic meters of tailings were dumped in the Doce River basin. The mining complex where the dam was located belongs to "Mineradora Samarco," a company controlled by the multinationals Vale S.A. and BHP Billiton, each with 50% of the company's shares. About four years after the Fundão dam burst, on January 25th, 2019, the dam of "Córrego do Feijão" mine ruptured, in the city of Brumadinho, also in Minas Gerais. Until September 8, 2019, there were 259 deaths and 11 people missing, according to data reported by government agencies (CIVIL DEFENSE-MG, 2020). This disaster was considered the largest occupational accident ever recorded in Brazil by magnitude of fatalities.

One of the important issues in a catastrophic dam rupture, such as the Fundão's, is risk communication and access to information by the population affected before, during, and after the disaster. Those issues are often neglected in Brazil. We think that by guaranteeing access to quality, transparent and updated risk information for exposed populations, it is possible to reduce social vulnerability. Access to information and communication should thus be considered an important dimension of social vulnerability.

With the advent of the Internet, the volume of information is continuously growing and there is a wide dispersion of information. Integrated, contextualized and networked communication can be used to understand certain problems, including public health ones, either more quickly in identifying outbreaks or forecasting scenarios.

Data collected via the Internet and social media sites have generally been used as a complementary source of epidemiological surveillance data during epidemics and to supplement data for existing outpatient, hospital and laboratory systems (Hempel, 2014; Abdullah; Wu, 2011). Nevertheless, it is still little used for monitoring disasters. Therefore, information monitoring is an important tool to guide the decision making of managers in different areas and, mainly, in the health area. This improvement in knowledge about a given scenario improves the quality of management and the strategy to develop and consolidate public policies in contexts of epidemics, crises and disasters (Cruz and Reis, 2011; Moraes and Vasconcellos, 2006).

One of the strengths of monitoring is to collect a large amount of organized information from different sources (Antunes *et al.*, 2014). In cases where governments need to act quickly, such as in disasters and emergencies, monitoring can be used to automatically track information from the web to support decision making.

This study draws attention to the role of information monitoring in the Internet as a tool for emergencies, crises and disasters, by describing the development of SIGDesastre – an information monitoring system about disasters. SIGDesastre is being developed since 2018, primarily related to the rupture of Fundão dam in the city of Mariana (MG), and it is also expected to be extended to other kinds of disasters.

In this chapter, first, a brief historical context on information monitoring and risk communication will be presented. Then, the proposed monitoring system will be contextualized, as well as the social actors that comprise it. Finally, the SIGDesastre's development will be addressed.

THE INFORMATION SOCIETY IN CONTEXTS OF CRISES, EPIDEMICS AND DISASTERS: ASPECTS OF INFORMATION MONITORING AND RISK COMMUNICATION

Since the last century, information has gained relevance on the world stage, especially after the advent of the Internet. The field of Information Science recognizes the importance of the use and organization of information as a strategic component

Choo (2003) defines information as "an intrinsic component of almost everything the organization does" (Choo, 2003, p.27), and describes the cycle of managing this information within organizations, which includes: information needs, acquisition of information, treatment, organization, storage, dissemination and adaptive reuse. He also emphasizes that this process generates meaning, new knowledge and helps managers to make decisions.

Auster and Choo (1993), Nonaka and Takeushi (1997) think that knowledge management is an effective practice to overcome the challenges and threats imposed by an environmental context. They also emphasize that monitoring is fundamental in the search for information in an organization to obtain new knowledge that will lead to action.

It is necessary, then, to think about information articulated with the technologies that support it, because, despite having their epistemological specificities, they tend to converge to confluent practices and knowledge (Moraes, 2014). In the public health arena, a large amount of information and communication technologies considerably impact health surveillance.

The concept of surveillance has changed over the years, moving from a sector that provided patient care to the systematic monitoring of adverse health events in the community to improve control measures. It is in this sense, moving from the individual to the collective, that it starts to be described as the monitoring of health information (Antunes et al, 2014, p.3).

The WHO (2009) defines public health surveillance as "ongoing, systematic collection, analysis and interpretation of health-related data essential to the planning, implementation, and evaluation of public health practice." Surveillance is conducted to inform disease prevention and control measures.

Data collected via the Internet and social media sites have usually been used as a complementary source for public health surveillance. This complement can be extremely useful, as pointed out by Aiello, Renson and Zivich (2020) (figure 1). The authors underscore some of the main developments related to digital surveillance in public health from 1990 to 2015, including the introduction of Google Trends for the flu epidemic. Two initiatives developed in Brazil were also based in the same framework: the e-Monitor (2006) and the one presented in this chapter, the SIGDesastre (2020).

According to Aiello, Renson, and Zivich (2020), the data for each entry in the timeline were obtained from the following sources: ProMED-mail (Madoff, Woodall, 2005) de Grote Griepmeting (Koppeschaar *et al*, 2017), webpage views used for flu surveillance (Johnson *et al*,2004), Google ads click rate used for surveillance (Eysenbach, 2006), HealthMap (Brownstein, Freifeld and Mandl, 2008), H1N1 missed by Google Flu Trends (Cook *et al*, 2011), Flu Near You (Smolinski *et al*,2015), GFT overestimates (Olson *et al*, 2013), FluSight (CDC, 2013). GFT and GDT are no longer available (Flu Trends Team,2015).

Figure 1. Major events in digital public health surveillance
Source: adapted from Aiello, Renson, and Zivich (2020).

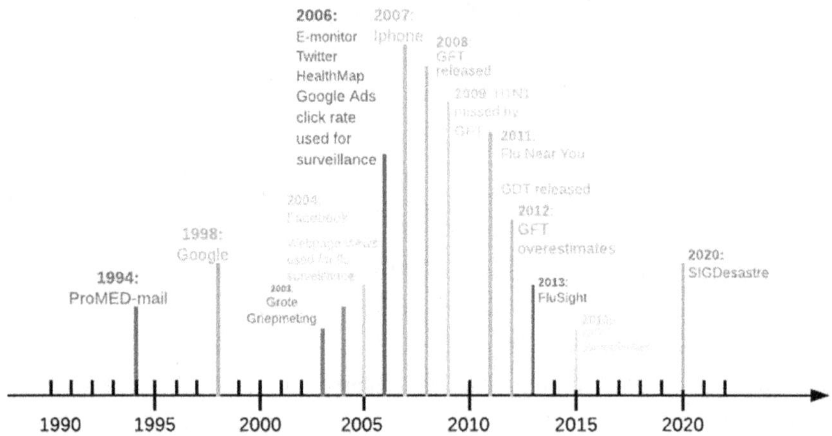

Digital public health surveillance is the inclusion of digital data, particularly from social media or other Internet-based sources, for this same purpose. Digital health surveillance data is often linked to a non-health data source. Twitter and Facebook are used to extract data from posts that may include mention of health-related information (Salathé, 2018).

While there are a variety of new applications in digital public health surveillance, the most utilized have been in the area of influenza surveillance and tracking. Aiello, Renon and Zivich (2020) argue that since the 2000s digital surveillance efforts have largely encompassed three major types of web-based activity: (a) aggregate trends derived from searches (e.g., Google search trends, Wikipedia page views), (b) social media postings (e.g., Facebook posts, tweets), and (c) participatory surveillance efforts such as Flu Near You and Influenzanet. However, all of those efforts have weaknesses

Digital surveillance data based on social media are increasingly used to track diseases, epidemics and pandemics. Hempel (2014) observed in his study of scientific articles on social media and public health that 66% of them found a favorable relationship between data from the traditional epidemiological surveillance system and monitoring results.

Twitter is one of the most used platforms for policy data, allowing public access to a 1% random sample of raw tweets. Another advantage is the availability of geolocated tweets, which can be used to model the spread of a disease as a result of spatial human mobility, potentially offering greater accuracy (Jurdak *et al*., 2015). The most typical use of Twitter involves identifying content through keyword search or natural language processing to identify tweets about health conditions, such as the flu. Epidemic levels are modeled depending on the frequency of such tweets (Rodriguez-Martínez & Garzon-Alfonso, 2018).

Social media is also part of the context of disasters. Houston *et al* (2014) note the potential of social media to facilitate communication and operations before, during, and after a disaster. During and after an event, social media is powerful for both citizens and organizations to document the impact of a disaster. Citizens and organizations can seek information on the progress of actions in disaster response. Therefore, social media is also an essential tool for digital surveillance during disasters, as the data collected can provide a stream of information quickly and consistently about what is taking place and what is needed.

Zhang *et al* (2019) argue that social media has three important functions: (1) to acquire efficiently and effectively information about the awareness of the situation in a disaster, (2) to support self-organized point-to-point assistance, and (3) to allow disaster management to listen to the public.

We can now frame the SIGDesastre, which was developed for monitoring information in the Internet about a disaster in Brazil and its health impacts on the affected population. It can be considered as a system of hybrid digital surveillance for disasters (Brownstein, Freifeld and Madoff, 2009), as it is integrated to more traditional data sources that come from health databases, such as epidemiological data. Furthermore, the data are combined with monitoring news sources, social networks, and major scientific journals, among other sources.

Actually, the SIGDesastre was based on the Influenza e-Monitor system (Silva *et al.*, 2007, Guimarães *et al.* 2008). This digital information monitoring system was developed in 2006 in Brazil to respond to health threats such as a potential flu pandemic. The project intended to be a reliable source of information about the status of avian influenza. A methodology was developed to monitor different sources of information, such as news media, government institutions, social media, and scientific journals. Thus, when forecasting future scenarios, it would assist in planning interventions to prevent risks.

In addition to the principles of information flow and monitoring, SIGDesastre is also based on risk communication. Since risk communication has developed for decades, there is a vast theoretical and conceptual literature in environmental risk studies, disaster management and media studies (Antunes, Oliveira and Rebouças, 2018). Some authors consider risk communication as a dialogical process, built by taking into account the risk factors to which a specific group of individuals is exposed. According to the National Research Council of the United States (NRC), the definition of risk communication reflects an interactive process of exchange of information and considerations between different groups, such as government agencies, institutions, civil groups, and populations about a certain potential risk for the environment and human health (Lundgren and Mcmakin, 2004; Bennett, *et al*, 2010).

Murdoch and collaborators (2003) affirm that risk communication occurs in a complex environment, where a variety of players/actors compete for public attention. In this arena, political institutions and risk managers are caught up in a permanent dispute with other major actors, including scientific community experts, social movements, opposition parties and corporations, not to mention the media and the public. They claim that risk communication requires an interactive model in which key actors continually launch initiatives and respond to each other's movements.

In the development of SIGDesastre, the scenario in which the communication of risk occurs is also a guide for monitoring digital information about the rupture of the Mariana dam. In this arena of dispute, there is also a diversity of social actors involved that play different roles, acting both in the creation as well as the treatment and diffusion of information. In the next section, this scenario will be detailed in order to facilitate the understanding of the proposed system.

FUNDÃO DAM FAILURE: THE MONITORING SCENARIO AND ITS SOCIAL ACTORS

The rupture of the Fundão dam caused great socio-environmental damage. The tailing dumped throughout the basin carried chemical elements and heavy metals that contaminated water and soil. It radically changed the lives of riverine communities. The dam failure released into the environment an estimated volume of over 50 million cubic meters of iron ore waste that destroyed the Bento Rodrigues village in Mariana county and caused damage and casualties throughout the Doce River basin.

Figure 2. The trajectory of ore tailings after the dam burst in Mariana (MG)
Source: Espindola & Guerra (2018)

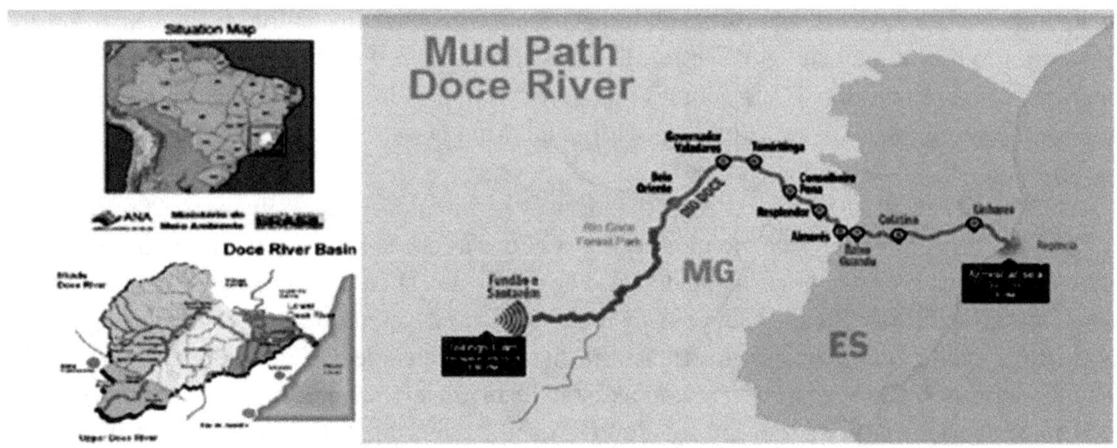

As shown in figure 2, the tailings flowed down the river for more than 600 km, buried aquatic and terrestrial vegetation, killed animals, destroyed natural habitats by modifying and silting river beds, and impacted about forty counties in the states of Minas Gerais and Espírito Santo (ES) (Zhouri *et al.*, 2018). The ore tailings reached the mouth of the Doce River at Regência (ES) in the Atlantic Ocean on November 21, 2015 (Pinto-Coelho, 2015).

The disaster that occurred in Mariana generated a socio-environmental conflict that involved diverse social actors (figure 3), each with a different role (Dias *et al*, 2018).

The recognition of the stakeholders in this disaster is essential to understand the features of the scenario that will be monitored and the complexity of their relationships in events of such magnitude. Understanding the conflicts that already existed and how they grew after the rupture of the dam can assist the populations affected in the implementation of more effective actions and in the management of post-disaster risks, thus preventing new incidents. Table 1 provides a description of the actors.

We must acknowledge that Samarco S/A, the owner of the complex where the dam was located, is directly responsible for this environmental disaster. Reports from residents of the areas affected by the breach make clear that the emergency warning system did not work. Reports issued by environmental agencies indicate that there was no 24-hour monitoring plan for the dams the company manages, and that two other dams also had problems already reported in other documents (Dias *et al*, 2018).

Figure 3. Monitoring scenario and main actors involved

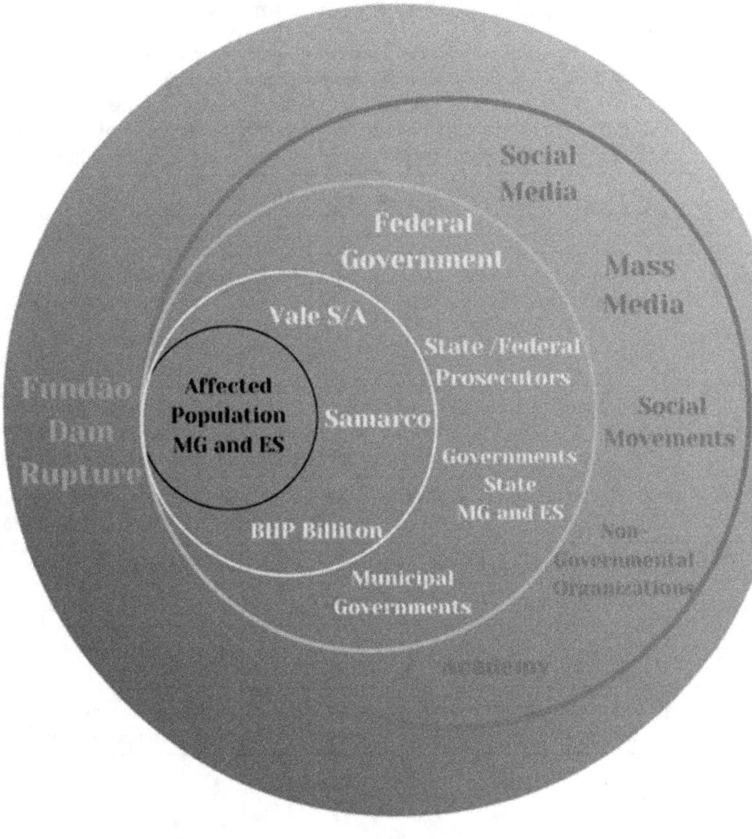

The Emergency Action Plan (EAP), which is mandatory in the Brazilian environmental laws that regulate companies that own high-risk dams, did not include risk prevention and communication procedures. In the 2014 report, Samarco stated that simulation activities were not carried out with the population surrounding the dams. This negligence demonstrates that despite the company claim that it had an early warning system, it did not actually exist (Global Justice, 2016). In addition to actions that were not implemented before and during the emergency, in the post-disaster phase the company made it difficult for the affected population to access information about mitigation actions and reparation of the social and environmental damage (MPF, 2018).

The multinationals Vale S / A and BHP Billiton are co-responsible for the disaster, taking into account that they own the majority of shares of the company. It is worth noting that four years after Fundão dam burst, the company Vale S/A was involved in another major tragic event. On January 25, 2019, a county also in Minas Gerais, another dam ruptured killing hundreds while some bodies are still missing.

Another important body in defending the rights of those affected is the Federal Public Ministry (MPF). In March 2016, a joint initiative by several government agencies and organizations, together with the responsible parties, arrived at an extra-judicial agreement led by the Public Ministry, which established parameters about many topics related to this disaster. The Consent Decree (Termo de Ajuste de Conduta in Portuguese), signed between states, companies and the Public Ministry established programs and

Table 1. Main social actors involved in the Mariana disaster

SOCIAL ACTORS	RULES
Local populations in MG and ES	Affected by the disaster and very important in the environmental recovery process of the affected areas. This is why they need to be heard in decision-making and mobilized to demand respect to their rights. It is important to consider the differences between the needs of the population of the districts of Bento Rodrigues and Barra Longa (MG), which were almost utterly destroyed, and those affected in Espírito Santo who had problems such as water quality, loss of wages, etc.
Samarco S/A	Owner of the dam and directly responsible for the disaster.
BHP and VALE	Partners of Samarco and co-responsible for the environmental disaster
Federal Government	Responsible for environmental legislation, issuing permits and applying fines to responsible companies. Reports point to negligence and failures of government action in these aspects.
Government of Minas Gerais and Espírito Santo and local ones	States whose populations and the environments were affected by the disaster.
Public Ministry (Federal and State)	Judicial branch in charge of investigating the causes and responsible parties for the environmental disaster, and litigation; b) Initially, it proposed a consent decree that mandated the company to pay R$ 1 billion as compensation for social and environmental damage.
Renova Foundation	Non-profit organization created to repair the damage caused by the disaster and manage the funds made available by the company in reparation projects (FUNDAÇÃO RENOVA, 2019)
Academy	Important mainly in post-disaster studies and consulted for research in the most diverse areas.
Non-governmental Organizations and Social Movements	Essential for equalizing the voices between the company and those affected. Mobilize and organize populations in the territories.
Media	Responsible for publicizing the disaster and the actions taken in the post-disaster.
Social networks	Great contacts in the population and the ability to quickly spread news and facts about the disaster. It has an important role in the dissemination of events that escape the mainstream media.

Source: adapted from Dias *et al*, 2018

projects to be implemented by the company, in addition to parameters detailing which parties would be considered affected parties. This same agreement created the Renova Foundation, a non-profit organization responsible for managing recovery and reconstruction funds for the affected areas and reimbursement to the displaced families (Fundação Renova, 2020).

Yet, in 2018, the Public Ministry itself denounced that the populations affected were not properly informed of developments and programs implemented by the states and responsible parties to repair damages (MPF, 2018). The Ministry stressed that information about those programs was not transparent and was disseminated in a dispersed and unclear manner. As a result, a new Consent Decree, known as TAC Governance, included representatives of the affected population in the organizational decision-making structures of the reparatory and compensatory programs proposed by the Renova Foundation (Fundação Renova, 2020). Only after 2018 those affected were able to give their opinion about the programs that would directly interfere in their lives from then on.

Much is still unknown regarding the health of populations that live alongside the river. Reports and studies were disseminated and antagonistic positions emerged. There is scientific evidence that indicates health impacts on the population, as shown by Vormittag, Oliveira and Gleriano (2018), who found psychological, cardiovascular, and respiratory adverse effects in the population of Barra Longa. Another study by Rocha et al. (2016) reported an increase in the incidence of diarrhea and skin disorders in the riverine population of Colatina (ES), associated with changes in the environment and water contamination.

On the other hand, the Renova Foundation website includes reports and assessments from companies claiming that "there is no toxicological health risk due to metals in the studied areas resulting from the rupture of the Fundão Dam" (Ambios, 2019; Fundação Renova; 2020).

This conflicting and opposing information demonstrates that the context of the disaster is also a contested terrain, where information and its sources are paramount in the relationships After established, that includes the population, public agencies, private companies, and the media. Bueno (2018) states that the mass media can be essential for the dissemination of information about disasters, and should be seen as partners, even though sometimes some positions of certain outlets may be at odds with those involved in the disaster.

After the identification of the main actors and how they interrelate, it is possible to establish the potential that post-disaster risk communication has to influence not only individual-level decision-making for risks created by past events, but also the institutional decisions and policy formulation by various stakeholders (Lin *et al*, 2020).

When sharing these listening spaces, including the results of research related to critical analysis of Brazilian media, it seems that the demand for reliable and updated information on post-disaster actions in the affected territories is an emerging issue for both the affected population and those responsible for managing post-disaster actions (Primo *et al*, 2018; Bueno, 2018).

In the post-disaster phase, traditionally the focus has been on the physical restoration of a city or town. The emphasis tends to be on structural recovery, such as reconstruction of buildings, reconnection of communication infrastructures, and restoration of roads. Although these activities are essential, the social components are often overlooked (Dugdale, Negre, Turoff, 2020). In the case of the Mariana dam, the social element has always been neglected, mainly with regard to post-disaster risk communication and access to information on mitigation actions.

It was in this scenario that SIGDesastre emerged. Specifically, when the team of the research group in Communication and Health of the Postgraduate Program in Collective Health of the Federal University of Espírito Santo (Ufes), Brazil, was invited to visit the affected territories in ES and MG and they participated of some meetings related to the collapse, in interaction spaces with managers, community leaders and community members.

For the post-disaster Risk Communication process, and in the context of the information flow, it is believed that the monitoring of information on the internet is a potential device to support managers in institutional decisions and in the formulation of public policies. Also, for the affected population, they will be able to expand access to information about the actions of the actors involved.

SIGDESASTRE: DIGITAL TOOL FOR EMERGENCIES, CRISES AND DISASTERS

As already mentioned, SIGDesastre is being developed to be a tool for monitoring information on the Internet about disasters. In this first phase of development it will bring news specifically about the rupture of the Mariana dam (MG), and it is directed to the managers of the Health Department of the State of Espírito Santo, as an aid in the decision-making process and construction of public health policies aimed at the affected population. Consequently, it is expected that it will be a facilitating tool in the access to information for the population affected in the post-disaster.

The construction process of SIGDesastre is anchored in principles established by theories of co-production of knowledge (Jasanoff, 2004). Hence, it is essential to engage and dialogue with targeted users at all stages of production so that the system can meet their needs. Initially, as it will be explored later, the system is geared to the managers of the Department at the State of Espírito Santo, Brazil. To involve those managers in the design of the system, a survey was performed to gather their expectations about potential responses obtained by the monitoring. The following refer to some examples of questions that managers hope to answer from the information monitoring:

1) Does the population report changes in water quality?
2) Are there reports of changes in food production and impacts on soil quality and agriculture?
3) Are there reports of reduction in population's income in the municipalities affected?
4) What are the complaints of the population and fishermen regarding the skin diseases?
5) Are there indications about the impacts on the mental health of the population and of suicide cases?

The monitoring made by SIGDesastre can help managers find answers to such questions, starting in the moment it integrates data sources obtained through social networks to official health data. Thus, through social networks, managers will be able to listen to the main complaints and reports from the population about its health status in the post-disaster phase and associate it with data from information systems on health.

The expectation is that monitoring will assist managers in actions to prevent and intervene in risky situations after the failure of Fundão dam. Furthermore, it can also help officials in the elaboration of risk, disaster, and emergency communication strategies.

The development of the system involves four phases and the first three stages took place simultaneously: 1) Definition of the keywords to be monitored; 2) Mapping of the online information sources to be monitored and their respective electronic addresses; 3) Technical development of the monitoring system; 4) Virtual space interface design. They will be described below:

1) Definition of Keywords to be Monitored

With the scenario approach, the recognition of the actors involved and the first contact with the managers, the next step was to define the keywords that would guide the developed crawlers required the prior reading of documents related to the monitoring context. The websites of the social actors involved were also consulted, scientific articles published on the topic were read. After this stage, the combination of the following descriptors in Portuguese was defined: 'disaster and Doce River", "disaster and Samarco", "Doce River and Samarco", 'environmental disaster and Vale", "Environmental disaster and Doce River". After defining the search terms, tests were performed for validation and necessary adjustments.

2) Mapping of the Online Information Sources to be Monitored and Their Respective Electronic Addresses

This step is considered of a fundamental importance, as it is related to the monitoring scenario. Initially, it mapped the web addresses of the main social actors involved. In the broader context, using the descriptors defined in the previous step, search engines available on the Internet were used to find other actors who were involved in the disaster, such as news sources, sources related to social movements, among others.

After an exhaustive search, the sources were organized and divided according to their typologies. Initially, five types of sources were considered: governmental, news, social movements, scientific publications and social networks. After the mapping, it was noticed the existence of a high number of web addresses corresponding to companies from the private sector, mainly those related to the mining industry. Thus, the private initiative typology was included, totaling then six types of online information sources, according demonstrated to table 2.

Table 2. Types of online information sources

TIPOLOGY	DESCRIPITION
Private companies	Websites of companies and institutions of private initiative. They usually have in their URL the domain ending in *.com.br*
Governmental	Institutions that exercise state power, whether Federal, State or Municipal. They usually have in their URL the domain ending in *.gov.br*
News	Large news portals and newspaper websites, mainly those located in the municipalities of the Doce River basin.
Social Moviments	Websites that belong to non-profit organizations, whose end of their email address is *.org.br*
Scientific Publications	Websites linked to scientific production, whether periodicals, higher education institutions, institutes and research groups.
Social Medias	Pages and profiles on social networks, such as Twitter and Facebook

This stage generated the Map of Information Sources online, with 1930 sources surveyed and organized according to their typology, as shown in Table 3.It is important to highlight that this information map is constantly updated from the feedback of the system.

Table 3. Number of typologies of sources found

TIPOLOGY	QUANTITATIVE	TOTAL PERCENTAGE
Official	153	7,94%
Scientifical	192	9,96%
News	839	43,48%
Social Moviments	231	11.93%
Private Companies	15	0,77%
Social Medias	500	25,91%
Total	1930	100%

At this moment of the process, the need for the validation of the source map by the managers involved was established. For this, an online electronic form was developed in which all mapped resources were inserted, separated by typology. For this, the managers received the form via email. The sources were presented by typology and the option was made to validate the listed sources and also to point out the insertion of new sources not yet mapped.

The form was initially available for two weeks, with a 40% response rate. There was no insertion of new sources by the respondents. It is worth mentioning that this initial validation does not prevent other managers from being able to do it at other times, nor did it prevent the development of the system, considering that new sources can be inserted at any time.

Additionally, SIGDesastre will also make the epidemiological data of Brazil from the information system of the Unified Health System (SUS) available. This system collects, processes and spreads health information in Brazil, and also contains statistical data on a number of indicators in the country. Then, SIGDesastre will function as a hybrid system, by providing the association of data collected from digital sources with traditional surveillance data.

3) Technical Development of the Monitoring System

The monitoring system is responsible for monitoring the electronic addresses of the sites indicated in steps one and two. All the technology that is being developed was structured using free software, tools for data visualization and analysis, promoting the interface between theoretical and practical aspects of Data Science and big data, data mining and visual analysis of large amounts of data.

When the user requests the information for the system, the API developed uses the database, which contains the addresses of the websites pre-registered in the mapping phase. This bank, in turn, was fed with news captured by crawlers that it was developed to capture the information published on the websites from words and pre-defined e-mail addresses. Then, the crawler cycles through the registered sites daily and when it finds the defined keyword, it includes a new record in the database. This information is organized by each typology, as pointed out in phase 2. Upon finding the information, it returns and is demonstrated to the user through a friendly interface. The development methodology is shown in figure 4.

In the course of developing the system, the team faced some difficulties during the construction process. The main difficulty is related to the retrospective monitoring of the sources defined in step two. As it was necessary to do a retrospective search of those sources and the diversity of the sites given, it was identified that it would be necessary to perform a unitary construction of each crawler, that is, for each source monitored a specific crawler was advanced. In view of the time required in the development schedule, the team chose to categorize the sources by monitoring priority. The sources with the highest priority had their crawlers developed in the first version of the monitoring system. The other sources will be added to the system later.

For prospective monitoring, sites with Really Simple Syndication (RSS) were adopted as criteria. RSS is a format for distributing information in real time over the internet, in which a subset of XML "dialects" that serve to aggregate content or "Web syndication", can be accessed through aggregator programs or websites. It is mainly used on news sites and blogs. RSS technology allows internet users to subscribe to sites that provide RSS feeds. These are typically sites that change or update their content regularly. Then, RSS feeds are utilized to receive these updates, so that the user can stay informed of several updates on different sites without having to visit them one by one.

It is Important to observe that the automation of the system permits it to be feedback at all times and its content is constantly updated. It is essential, therefore, that there is information architecture that allows the insertion of new elements, taking into account user interactions (Silva *et al*, 2007).

Figure 4. Monitoring system development methodology

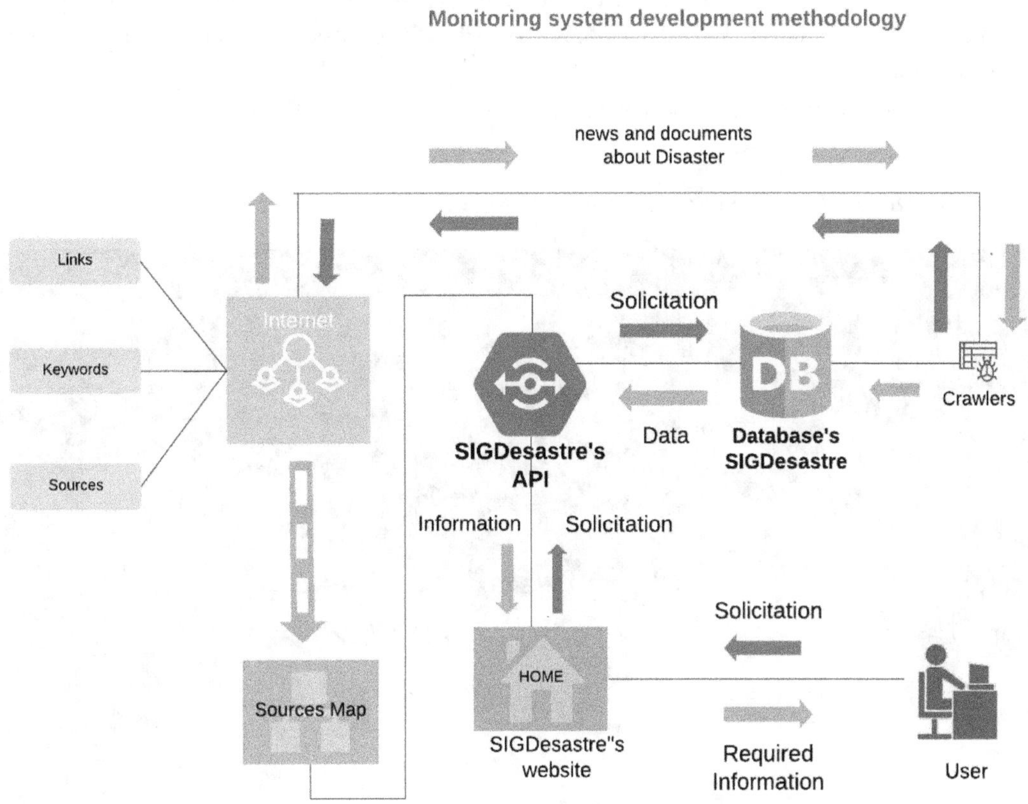

4) Virtual Space Interface Design

This step involves the creation of the virtual space interface, in which the results of the monitoring are made available to the end user. Obeying the principles of Information Architecture, the methodology developed by J.J. Garret (2003) was used. This method focuses on user-centric web interface design. It outlines five steps for producing a website: strategy, scope, structure, skeleton and surface. These surfaces are overlapping, starting from abstract concepts until reaching concrete definitions (Jardim, Santos, Oliveira 2015).

In the strategy stage, the objectives and requirements expected by the proponents and those expected by the user were defined. This stage was divided into two stages, considering that the system is designed to initially serve the managers of the Department of Health, and will later be made available to those affected. Firstly, the needs of managers were raised. Each manager was asked to identify three questions that were expected to be answered through monitoring.

After this point, the scope was started, where the functionalities proposed for the application were divided taking into account the public to be reached. With this scope, which is formed by the characteristics of the system, the structure was built, focusing on the hierarchy to be used, according to the conceptual map prepared with the project team. This concept map, based on authors as Sherratt and Schlabach (1990), "involves the identification of concepts or ideas belonging to a subject, and the description of the relationships between these ideas in the form of a schematic drawing".

The skeleton is then assembled and the structural design of the screens, called wireframe, is developed. This feature allows you to communicate the content and functions of each screen or page of the website, in a simplified version. This step makes it possible to discuss the functionalities and guidance for implementation.

The result of the construction of the visual interface can be seen in figures 5, 6 and 7.

Figure 5. Initial Homepage and system screen showing data found

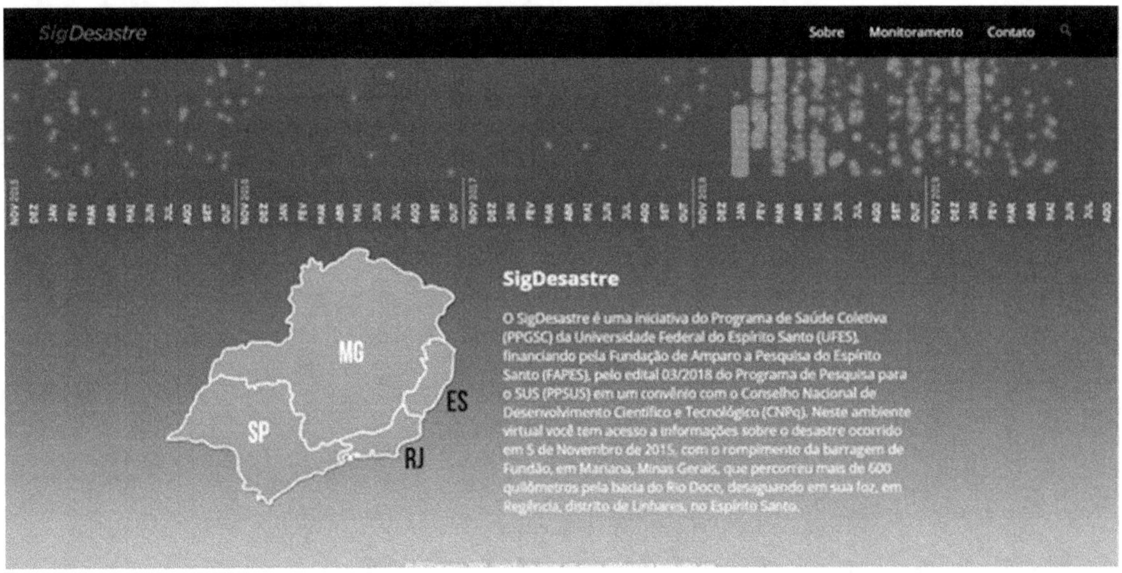

Figure 6. Initial Homepage and system screen showing data found

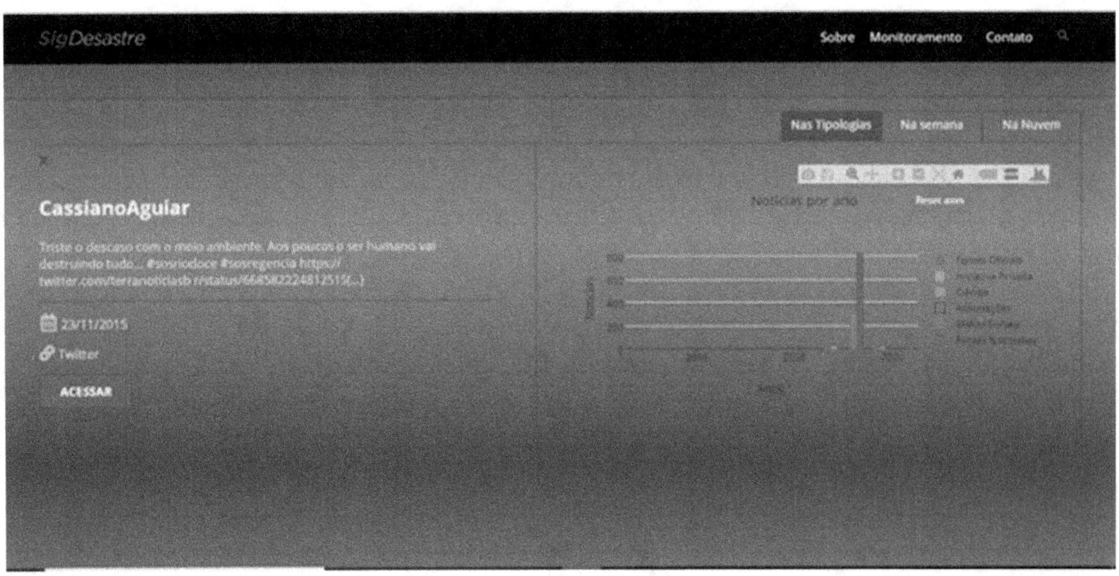

Figure 7. SIGDesastre main page

As mentioned, this system is still under development. The figures shown indicate the position until the completion of this chapter After these steps have been completed, the first version of the SIGDesastre will be made available to targeted users and the usability assessment phase will begin. In this phase, requirements such as user satisfaction, system handling and task resolution are verified, based on the objectives for which the system was created. As a way to guarantee the usability of a website, one must prioritize its non-functional requirements, as a way to ensure that the information given to the user is of quality (Nielsen, 2000; Nielsen, 2002; Pearrow,2000)

It is expected that the tool can be extended to other types of disasters, epidemics and crises, proving to be an important decision-making device for managers, a reliable source of information for scientists and a place of virtual good practices for society in general.

CONCLUSION

This chapter demonstrated the use of information monitoring on the internet as an important tool in risk communication in episodes of disasters, crises and emergencies. The situation described reinforced important issues as the difficulty in accessing information by those affected and the delay in remedial actions by companies. The lack of participation and listening to those affected in the reconstruction and disaster mitigation works were also highlighted in the present study.

The creation of the information monitoring system on the internet, such as SIGDesastre, described in this chapter involves the identification of the sources to be monitored, an automated search system for keywords in these pre-registered sources and visualization of the results in a user-friendly environment.

The SIGDesastre system must act in order to approximate the data, in this case the pieces of information about the breach of Fundão dam, of good social practices (Freitas & Mendes, 2019). With that, both the initial users, managers of the health scope, not to mention the population of the affected territories are considered important and participatory parts in the development of technology.

Plus, risk communication is an important means of deepening and understanding the impacts and processes of post-disaster recovery. It needs to promote the active participation of those involved, especially the most vulnerable ones, consolidating their rights and contributing to the formulation of effective and efficient public policies (Kurosaki 2017; Lin *et al*, 2020)

Although the topic has received more attention in recent decades, the challenges are still huge. In this time of disasters, emergencies and increasingly constant crises, where there is no space to resolve the consequences of an event before another happens, the permanence of a subject on the agenda of these managers can be a hindrance in the maintenance and evolution of the system. The lack of investments and dependence on political positions for continuity of actions can also hinder the continuity of initiatives like these.

It is essential, therefore, that risk communication is seen as a fundamental part of the entire process of coping with crises, emergencies, epidemics and disasters and that the participation of the population involved is increasingly considered.

ACKNOWLEDGMENT

This project was contemplated by the public notice "Projects for the SUS" (PPSUS), number 03/2018, of the Fundação de Amparo à Pesquisa do Espírito Santo (FAPES) in agreement with the National Center for Research and Quality (CNPq), which finances the creation and development of the Monitoring System And support – Institutional Internationalization Program (PRINT) and Coordination for the Improvement of Higher Education Personnel (CAPES). Public Notice CAPES/PRINT number 41/2017.

REFERENCES

Abdullah, S., & Wu, X. (2011). An Epidemic Model for News Spreading on Twitter. *International Conference on Tools with Artificial Intelligence*, 163-169. 10.1109/ICTAI.2011.33

Aiello, A. E., Renson, A., & Zivich, P. N. (2020). Social media – and Internet-Based Disease Surveillance for Public Health. *Annual Review of Public Health*, *41*(1), 101–118. doi:10.1146/annurev-publhealth-040119-094402 PMID:31905322

Ambios. (2019). *Estudo de avaliação de risco à saúde humana em localidades atingidas pelo rompimento da barragem do Fundão – MG*. Retrieved from: https://apublica.org/wp-content/uploads/2019/11/ambios-arsh-mariana-e-barra-linga-final-20190417.pdf

Antunes, M. N., Oliveira, A. E., & Rebouças, E. (2018). Zika and press releases: reflections on the communication of risk and emergency under the perspective of culture and media industries. *Revista Brasileira de Pesquisa em Saúde*, *20*, 110-120. http://periodicos.ufes.br/rbps/article/view/21236

Antunes, M. N., Silva, C. H., Guimarães, M. C. S., & Rabaço, M. H. L. (2014). Monitoramento de informação em mídias sociais: O e-Monitor Dengue. *Transinformação*, *26*(1), 9–18. doi:10.1590/S0103-37862014000100002

Bennett, P., Calman, K., Curtis, S., & Smith, D. F. (2010). *Risk Communication and Public Health*. Oxford University Press., doi:10.1093/acprof:oso/9780199562848.001.0001

Brownstein, J. S., Freifeld, C. C., & Madoff, L. C. (2009). Digital disease detection—Harnessing the Web for public health surveillance. *The New England Journal of Medicine*, *360*(21), 2153–2157. doi:10.1056/NEJMp0900702 PMID:19423867

Brownstein, J. S., Freifeld, C. C., Reis, B. Y., & Mandl, K. D. (2008). Surveillance sans frontières: Internet-based emerging infectious disease intelligence and the HealthMap project. *PLoS Medicine*, *5*(7), 151. doi:10.1371/journal.pmed.0050151 PMID:18613747

Bueno, W. C. (2018). A cobertura jornalística de catástrofes ambientais: Entre a vigilância e aespetacularização da notícia. *Comunicação e Sociedade*, *39*(1), 21–41. doi:10.15603/2175-7755/cs.v39n1p21-41

Centers for Disease Control and Prevention. (2013). *CDC competition encourages use of social media to predict flu.* Press Release. 25. https://www.cdc.gov/flu/news/predict-flu-challenge.htm

Choo, C.W. (2003) *A organização do conhecimento: como as organizações usam a informação para criar significado, construir conhecimento e tomar decisões*. tradution Eliana Rocha. São Paulo: Editora Senac São Paulo.

Choo, C. W., & Auster, E. (1993). Environmental scanning: Acquisition and use of information by managers. *Annual Review of Information Science & Technology*, *28*, 279–281.

Civil Defense of Minhas Gerais. (2019). *Disaster Information Tailings Dam in Brumadinho.* Retrieved from: http://www.defesacivil.mg.gov.br/index.php/component/gmg/page/787-informacoes-do-desastre-barragem-de-rejeitos-em-brumadinho-28-12-19

Cook, S., Conrad, C., Fowlkes, A. L., & Mohebbi, M. H. (2011). Assessing Google flu trends performance in the United States during the 2009 influenza virus A (H1N1) pandemic. *PLoS One*, *6*(8), 23610. doi:10.1371/journal.pone.0023610 PMID:21886802

Cruz, M. M., & Reis, A. C. (2011). Monitoramento & Avaliação como uma das funções gestoras do Sistema Único de Saúde. In Qualificação e Gestores do SUS. Fiocruz.

Dias, A. O., Luz, G. S., Assunção, V. K., & Gonçalves, T. M. (2018). Mariana, o maior desastre ambiental do Brasil: uma análise do conflito socioambiental. In N. I. Ladwig & H. Schwalm (Eds.), *Planejamento e gestão territorial: a sustentabilidade dos ecossistemas urbanos* (pp. 455–476). Ediunesc. doi:10.18616/pgt20

Dugdale, J., Negre, E., & Turoff, M. (2020). Introduction to the Minitrack on Information and Communication Technologies for Crisis and Emergency Management. *Proceedings of the 53rd Hawaii International Conference on System Sciences*, 573-575. 10.24251/HICSS.2020.070

Espindola, H. S., Nodari, E. S., & Santos, M. (2019). Rio Doce: Risks and Uncertainties of the Mariana Disaster (MG). *Revista Brasileira de História*, *39*(81), 141–162. doi:10.1590/1806-93472019v39n81-07

Eysenbach, G. (2006). Infodemiology: Tracking flu-related searches on the web for syndromic surveillance. *AMIA ... Annual Symposium Proceedings - AMIA Symposium. AMIA Symposium*, *2006*, 244–248. PMID:17238340

Flu Trends Team. (2015). The next chapter for flu trends. *Google AI Blog*. https://ai.googleblog.com/2015/08/the-next-chapter-for-flu-trends.html

Freitas, F., & Mendes, J. M. (2019). Disaster, reconstruction, and data for social good: The case of wildfires in Portugal. *International Journal of Disaster Resilience in the Built Environment, 10*(4), 239–247. doi:10.1108/IJDRBE-06-2019-0035

Fundação Renova. (2020). *A Fundação*. Retrieved from https://www.fundacaorenova.org/

Garret, J. J. (2003). *The Elements of User Experience*. American Institute of Graphic Arts.

Global Justice. (2016). *Vale da Lama: Relatório de inspeção em Mariana após o rompimento da barragem de rejeitos do Fundão*. Retrieved from: http://www.global.org.br/wp-content/uploads/2016/03/Vale-de-Lama-Justi--a-Global.pdf

Guimarães, M. C. S., Silva, C. H., & Antunes, M. N. (2008). Monitoramento de informação como estratégia de e-health: Um estudo prospectivo. *Textos de la Cibersociedad, 16*, 216.

Hempel, M. (2014). *The use of social media in environmental health research and communication: an evidence review*. Environmental Public Health.

Houston, J. B., Hawthorne, J., Perreault, M. F., Park, E. H., Goldstein Hode, M., Halliwell, M. R., Turner McGowen, S. E., Davis, R., Vaid, S., McElderry, J. A., & Griffith, S. A. (2014). Social media and disasters: A functional framework for social media use in disaster planning, response, and research. *Disasters, 39*(1), 1–22. doi:10.1111/disa.12092 PMID:25243593

Jasanoff, S. (2004). The Idiom of co-production. In S. Jasanoff (Ed.), *States of knowledge: the co-production of science and social order* (pp. 1–12). International Library of Sociology. doi:10.4324/9780203413845

Johnson, H. A., Wagner, M. M., Hogan, W., Chapman, W., & Olszewski, R. T. (2004). Analysis of Web access logs for surveillance of influenza. *Studies in Health Technology and Informatics, 107*, 1202–1206. PMID:15361003

Jurdak, R., Zhao, K., Liu, J., AbouJaoude, M., Cameron, M., & Newth, D. (2015). Understanding Human Mobility from Twitter. *PLoS One, 10*(7), e0131469. Advance online publication. doi:10.1371/journal.pone.0131469 PMID:26154597

Koppeschaar, C. E., Colizza, V., Guerrisi, C., Turbelin, C., Duggan, J., Edmunds, W. J., Kjelsø, C., Mexia, R., Moreno, Y., Meloni, S., Paolotti, D., Perrotta, D., van Straten, E., & Franco, A. O. (2017). Influenzanet: Citizens among 10 countries collaborating to monitor influenza in Europe. *JMIR Public Health and Surveillance, 3*(3), e66. Advance online publication. doi:10.2196/publichealth.7429 PMID:28928112

Kurosaki, T. (2017). Household-Level Recovery after Floods in a Tribal and ConflictRidden Society. *World Development, 94*(Supplement C), 51–63. doi:10.1016/j.worlddev.2016.12.039

Lin, K. H. E., Khan, S., Acosta, L. A., Alaniz, R., & Olanya, D. R. (2020). The dynamism of post disaster risk communication: A cross-country synthesis. *International Journal of Disaster Risk Reduction, 48*, 101556. doi:10.1016/j.ijdrr.2020.101556

Lundgren, R., & McMakin, A. (2004). *Risk Communication: A Handbook for Communicating Environmental, Safety, and Health Risks*. Battelle Press.

Madoff, L. C., & Woodall, J. P. (2005). The Internet and the global monitoring of emerging diseases: Lessons from the first 10 years of ProMED-mail. *Archives of Medical Research*, *36*(6), 724–730. doi:10.1016/j.arcmed.2005.06.005 PMID:16216654

Ministério Público Federal. (2018). *Official document*. Retrieved from: http://www.mpf.mp.br/mg/sala-de-imprensa/docs/termo-de-compromisso-prescricao

Moraes, I. H. S. (2014). Governança e modelo de gestão da informação e inovação em sistemas e serviços de atenção à saúde. In *Conhecimento, inovação e comunicação em serviços de saúde* (pp. 19–41). Editora FIOCRUZ.

Moraes, I. H. S., & Vasconcellos, M. M. (2006). Gestão da informação e comunicação em saúde: Desenho e implementação de uma proposta de ensino-aprendizagem. In Cenários possíveis: experiências e desafios do mestrado profissional na saúde coletiva. Rio de Janeiro: Editora FIOCRUZ.

Murdock, G., Petts, J., & Horlick, J. T. (2003). After amplification: rethinking the role of media in risk communication. In N. Pidgeon, R. Kasperson, & P. Slovic (Eds.), *The social amplification of risk*. Cambridge University Press. doi:10.1017/CBO9780511550461.008

Nielsen, J. (1999). *Designing Web usability: The practice of simplicity*. New Riders Publishing.

Nielsen, J., & Tahir, M. (2001). *Homepage Usability: - 50 Websites Deconstructed*. New Riders Publishing.

Nonaka, I., & Takeuch, I. H. (1997). *Criação do Conhecimento na Empresa: como as empresas geram a dinâmica da inovação*. Campus.

Olson, D. R., Konty, K. J., Paladini, M., Viboud, C., & Simonsen, L. (2013). Reassessing Google Flu Trends data for detection of seasonal and pandemic influenza: A comparative epidemiological study at three geographic scales. *PLoS Computational Biology*, *9*(10), e1003256. Advance online publication. doi:10.1371/journal.pcbi.1003256 PMID:24146603

Pearrow, M. (2000). *Web site usability handbook*. Charles River Media.

Pinto-Coelho, R. M. (2015). Is there water governance in Brazil? The sudy case: The Fundão Dam Brech, Mariana (MG). In Arquivos do Museu de História Natural e Jardim Botânico. Universidade Federal de Minas Gerais.

Primo, P. P. B., Antunes, M. N., Ramos, M. P., & Emmerich, A. (2018). Diante da dor dos outros: Desastres e a Violação de Direitos Humanos. In Questões sobre Direitos Humanos: Justiça, Saúde e Meio Ambiente, 3, 169 - 192. doi:10.30712/isbn9788565276474.169-192

Rocha, E. M., Moraes, L. G. M., Almeida, L. V., Dalvi, L. R., Andreato, L. C., Bergamaschi, L. K., Bernardina, L. S. D., Pereira, W. B., Gimenez, V. G., Neto, O. C., & Almeida, H. S. (2016). Impacto do Rompimento da barragem em Mariana –MG na saúde da população ribeirinha da cidade de Colatina – ES. *Tempus.*, *10*(3), 31–45. doi:10.18569/tempus.v10i3.1902

Rodríguez-Martínez, M., & Garzón-Alfonso, C. C. (2018). Twitter Health Surveillance (THS) system. *Proc. IEEE Int. Conf. Big Data 2018*, 1647–54.

Salathé, M. (2018). Digital epidemiology: What is it, and where is it going? *Life Sciences, Society and Policy*, *14*(1), 1. doi:10.118640504-017-0065-7 PMID:29302758

Sherratt, C. S., & Schlabach, M. L. (1990). The application of concept mapping in reference and information services. *RQ*, *30*, 60-69. Retrieved from www.jstor.org/stable/25828679

Silva, C. H., Guimarães, M. C. S., Esteves, M. A., Rabaço, M. H. L., Antunes, M. N., Abdala, R., & Marcondes, C. H. (2007). Monitoramento da informação na sociedade de risco: O caso da pandemia de gripe aviária. *Informacao e Sociedade-Estudos*, *17*, 1–10.

Smolinski, M. S., Crawley, A. W., Baltrusaitis, K., Chunara, R., Olsen, J. M., Wójcik, O., Santillana, M., Nguyen, A., & Brownstein, J. S. (2015). Flu Near You: Crowdsourced symptom reporting spanning 2 influenza seasons. *American Journal of Public Health*, *105*(10), 2124–2130. doi:10.2105/AJPH.2015.302696 PMID:26270299

Vormittag, E. M. P. A. A., Oliveira, M. P., & Gleriano, J. S. (2018). Avaliação de Saúde da População de Barra Longa afetada pelo Desastre de Mariana, Brasil. *Ambiente & Sociedade*, *21*, 1–22. doi:10.1590/1809-4422asoc0122r2vu18l1ao

World Health Organization. (2009). *Pandemic (H1N1) 2009 - update 76*. Retrieved from: www.who.int/csr/don/2009_11_27a/en/index.html

World Health Organization. (2009). *Public Heath Surveillance*. Retrieved from: https://www.who.int/immunization/monitoring_surveillance/burden/vpd/en/

World Information Service On Energy. (2020). *Chronology of major tailings dam failures*. Retrieved from: https://www.wise-uranium.org/mdaf.html

Zhang, C., Fan, C., Yao, W., Hu, X., & Mostafavi, A. (2019). Social media for intelligent public information and warning in disasters: An interdisciplinary review. *International Journal of Information Management*, *49*, 190–207. doi:10.1016/j.ijinfomgt.2019.04.004

Zhouri, A., Oliveira, R., Zucarelli, M., & Vasconcelos, M. (2018). O desastre do Rio Doce: Entre as políticas de reparação e a gestão das afetações. In *Mineração, violências e resistências: um campo aberto à produção do conhecimento no Brasil*. Editorial Iguana.

KEY TERMS AND DEFINITIONS

Disaster Management: Management of actions, ideas, policies, and initiatives generated in disaster prevention, mitigation, or responses.

Information Monitoring System on the Internet: Digital technology that monitors, captures, and makes information available on a given topic on the internet.

Monitoring Scenario: The universe in which the disaster is inserted with the identification of social actors and the predominant themes on the subject.

Public Health Surveillance: It consists of the collection, consolidation, analysis, and dissemination of data related to the population's health. It is a systematic and continuous process and aims to plan and implement policies of public health.

Risk Communication: Disaster-focused communication subarea, emergencies, and crises. It can act in all phases of the disaster prevention, preparedness, response or mitigation and it must always be discussed with the affected population.

Social Actors: Individuals, groups of people, organizations, or institutions with representativeness in a certain territory or society in general.

Socio-Environmental Disasters: Disasters of great magnitude that bring in addition to environmental devastation, economic, social, and political problems in an affected area.

Tailings Dam: Constructions made by mining companies. In those places, the waste from the process of processing the mining activity.

Chapter 5
Missing People in Spain:
An App for Trauma Recovery - A Digital Health Intervention for Survivors

Carolina Escudero
University of Missouri, USA

ABSTRACT

Spain is the second country in the world in numbers of missing people, with 114,226 men and women still in mass graves without having been identified and buried by their relatives. Added to them are the families of stolen babies: 300,000 babies were stolen during and after Franco's dictatorship. Faced with these disappearances, a digital health intervention (DHI) for the accompaniment and monitoring of the process of mourning and trauma recovery could improve the situation of this extensive and diverse population, situated throughout Spain. Limitations in resources for the care and accompaniment of this population are due to various logistical, geographical, financial, stigmatic, and demographic factors—this is an ageing population—preventing them from accessing places of treatment. In this chapter, the author proposes a health app for trauma recovery that can be readily standardized for the wide dissemination of evidence-based care and adapted to the needs of this specific population.

INTRODUCTION

This chapter focuses on the challenging issue of a new digital health intervention (DHI) in Spain intended for the families of stolen babies or of missing persons. Basing this work on previous media psychology studies can help to ensure that the app meets the needs of this specific, diverse and large population, and also works towards engendering trust and participation. To this end, the PTSD Coach app and its versions in different countries are taken as a reference, as they correspond to the needs of the population being studied. It also focuses on the relationship between engagement and intervention effectiveness. The author intends to highlight the challenges presented by research into new procedures for adapting the DHI app, procedures with long-lasting effects (allowing to adjust and improve the application), while also being able to observe real impact and improvement in the lives of this broad-based population.

DOI: 10.4018/978-1-7998-6705-0.ch005

To do so, the author will explain the need for prior work to be undertaken before the launch of the application, such as studying the results from participatory action research (PAR) and group dynamics aiming to provide safe spaces in which to create trust, a previous requirement of engagement. One of the main purposes of this new DHI is to mitigate pain. In that sense, Beristain (2007) points out:

We must avoid thinking of reparation as a fallacy. We know that nothing can replace the relatives or repair the pain of the victims, or recover lost years of life; in essence the repair speaks of a problem without solution. We are talking more about how we mitigate the damage and what commitment there is for there to be a restitution of the rights of the victims (...) and how we help victims, who have been marginalized from a history, to be reintegrated from an active position and have social recognition by the State that has violated their rights and by society. (in Escudero, 2020, pp.72-73)

Concerning the app, while it is common knowledge that mobile health apps are generally utilized to complement established treatment methods and to improve treatment accessibility (Bakker, Kazantzis, Rickwood & Rickard, 2016; Donker et al., 2013 in Sander et al., 2020) unfortunately the overall quality of such apps contrasts considerably with their quantity, demonstrating the need to create procedures for the evaluation and assessment of their usability and impact (Schellong, Lorenz & Weidner, 2019). Hence, the author believes that their previous research and work in group dynamics with the families of stolen babies could significantly contribute to the launching of a digital health intervention.

While there have been various digital interventions designed for the improvement of people's health, many of them have failed due to lack of commitment by users (Yeager & Benight, 2018) or their mistrust, as the use of apps also involves different risks and challenges where privacy and data protection are not always presented in terms of information for the client and requests for consent (van Dijck, Poell & de Waal, 2018). Also, mobile health apps have the potential to increase the quality of processes related to health remotely and on a global scale (Palos-Sánchez et al, 2019, p.125). For this reason, the author believes that for such an application to work, it is necessary to undertake a prior analysis of similar applications and also better understand the population to be given this opportunity, in order to above all create a safe space of trust, so that subsequent access to such an application be considered an agreement based on commitment from both sides.

BACKGROUND

The first cases of stolen babies in Spain occurred during Francisco Franco's regime but continued until relatively recently. The stolen babies' organizations specify that there are around 300,000 stolen babies in Spain. The theft of babies in Spain (1940-1999) was systematic (medical doctors, nurses, Catholic priests and nuns) and based on Doctor Antonio Vallejo Nágera's concept of eugenics used, during the period under Franco, as an argument for the appropriation of descent. After the dictatorship, in 1975, these practices were motivated economically (Bueno Morales & González Besteiro, 2018, in Escudero, 2020, p.71).

After the transition to democracy in 1978, the number of stolen babies decreased but the practice continued. As Aguilar specifies, the stealing of babies continued in private and public hospitals for some time after democracy and this is due to the fact that many people working for Franco's regime continued working in public offices in times of democracy (2017 in Barrenechea Lopez, 2017). But the modification of the adoption law in 1987 provided greater control and documentation of births and adoptions and therefore it was no longer possible to continue taking babies as before (Bueno Morales & González Besteiro, 2018).

There are no previous studies of this population of families of stolen babies from a social psychosocial perspective in Spain nor in how the recovery process can be implemented through a digital health intervention. As several studies point out, the psychological distress symptoms of traumatized people simultaneously call attention to the existence of an unspeakable secret and deflect attention from it. Post-traumatic Stress Disorder (PTSD) will be studied in a population that has never been studied before. Because of this, new characteristics and contributions to trauma studies may emerge.

Despite the lack of previous studies of this population, a pilot study based in Media Social Psychology was carried out in which workshops and group dynamics provided results and information on the needs of this population that may contribute to the creation of an application responding to these needs.

Issues

As for missing people in Spain, the Spanish Ministry of the Interior has registered a total of 202,529 files (1,228 prior to 2010) of missing persons since 2010, 5,529 of which were still open on 31st December 2019, in other words, 9.3% of the total, according to the 2020 report "Missing People in Spain", with figures taken from the Missing Persons and Non-Identified Mortal Remains (PDyRH) database and the Criminal Statistics System (SEC) (see inset 1). Of the 5,529 cases still open on 31st December 2019, a total of 2,556 (40% of the total) were Spanish missing persons, followed by Moroccans, counting for 1,432 files for missing persons (25% of the total) (see inset 2).

In these cases, the families who reported the disappearance of one of their relatives can access the app Alertcops, for which they are given a keyword. This app is used for cases of burglary, robbery, theft, vandalism, damages, assault, fighting, sexual assault, domestic violence, bullying, anti- extremisms, assistance to the hard of hearing and missing persons (Ministry of the Interior, 2020). Nevertheless, this app does not correspond to the idea of an app for psychological accompaniment with a view to trauma recovery; just as with families of stolen babies, this population has no psychological accompaniment or tools allowing them to get through this situation. While certain people within these groups receive or have received individual treatment, until now they have not had digital and easily accessible tools allowing them to reduce their levels of anxiety or apprehension in various moments of the day or the week. This is why the author believes it to be necessary to make the PTSD Coach app available to this population. Consequently, it should be taken into account how this app has been adapted to other countries and possible challenges in terms of cultural diversity, high expectations, amongst other issues. Although there exist antecedents in the adaptability of the app, the aforementioned challenges are not only related to questions of logistics and technological adaptability, but rather to the adaptability of a population which feels considerable contempt for institutions, having survived a coup d'état which lasted for more than four decades. For all these reasons, the author stresses the importance of gaining in-depth knowledge of the population for which the app is intended, and of choosing representatives from the organizations (of the relatives of missing persons and the relatives of stolen babies) who can provide more data to help better define this population and build trust from the very start.

The author believes that the launch of such an app will not only be the first action in the country for the reparation of this broad and diverse population, but will also be a new opportunity within society for the initiation of a process of reparation, recuperation and trust. For many reasons, a large part of this population is aware of the limitations in resources for care and accompaniment in this country, often related to logistical, financial and geographical problems. Equally, the question of age, as most are older adults, in addition to difficulties in terms of mobility, means that the launch of an app represents for them access to the possibility of improvement without having to leave their safe spaces, or without depending on someone to take them and bring them home. For this reason, it is extremely interesting to define this population in order to adapt the app in the best way possible. While it may be thought that being older adults could represent a challenge when using the app, previous studies of this population, specifically the families of stolen babies in Catalonia (SOS Bebés Robats), have shown that searching for their stolen babies motivated them to learn and transcend digital barriers, creating online profiles and uploading photos to help in their searches. This individual motivation, which also leads to a process of group support and resilience, defines them not as Digital Immigrants but rather as a group undergoing Digital Mutation (Escudero, 2020).

Population

When referring to the population of the families of stolen babies in Spain, the author is alluding to a population of older adults between the ages of 60 and 98, given that babies were stolen throughout the Spanish territory between 1940 and 1999. Most of these older adults are members of organizations/platforms that guide and support the families when beginning to search for their babies. Although over the past years the families have been gathering data and documentation proving that many of the babies did not die, as they had been told, but rather had been stolen, until now no government had recognized them as victims. Despite the numerous cases and evidence presented, the fact that these families were not recognized as victims only increased the discredit of the various democratic governments in power since the end of the dictatorship in 1974, at the death of Franco. This population, constituted by mothers, fathers, brothers, sisters, uncles, aunts and grandparents began to make their cases and their search known publicly, via campaigns such as "Te Estamos Bucando" (We are Looking for You), known as TEB (Escudero, 2017, 2018), in which the first results can be seen from group dynamics carried out as part of an introspective study of this population (Escudero, 2020). The TEB campaign gives the survivors a voice: they speak in front of the camera, tell their stories and describe who they are looking for (based on the Identified Victim theory, Ubuntu practices and resilient communication). Even if these cases were covered by the Spanish mass media from a political, historical and social perspective, through TEB a different way to treat the survivors and focus on possible actions in the present is put forward (Escudero, 2020, p.19).

Herman specifies: "Sharing the traumatic experience with others is a precondition for the restitution of a sense of a meaningful world (...) The response of the community has a powerful influence on the ultimate resolution of the trauma." (1992, p.70). For this reason, it is important to portray the traumatic event of the stolen babies and that experienced by the families of missing persons based on a combination of community and social intervention actions, and group dynamics, so as to later implement and design an application responding to the needs of this population. When accompanying and studying the traumatic event, trust and respect are very important preconditions to take into consideration. This trust and respect enable the survivors to share their perspective in depth, allowing for a better adaptation of the app and thereby a better accompaniment of both populations.

As Herman points out (1992), the story of the survivor is the crucial and active part of the investigation, because "their cases" are their lives. There are many survivors of stolen babies and families of missing people throughout the European Union, and therefore the experience of this app in Spain could be transferred to other countries and their populations with similar experiences and needs.

Regarding the organizations of the families of missing persons, the representative organizations of reference are Inter-SOS (Association of the Families of People Missing with no Apparent Cause) and QSDglobal, European Foundation for Missing Persons.

Faced with the need to create an application for a large population such as the one described, various factors emerge such as those set forth by van Dijck, Poell and de Waal:

The insights into specific health platforms help us reflect on the conflicting public values at stake in this debate: the concern for privacy versus the benefit of personalized medicine and the privatization of data by corporate owners versus the accessibility of health data and knowledge to public research. (2018, p.98)

While it is common knowledge that the purpose of most health platforms is to solicit and collect all kinds of health information from users (Adibi, 2015), the stance of the users of these applications must also be considered. More specifically, during the author's work with groups of families of stolen babies, a high level of mistrust of government institutions was observed. To be able to develop a DHI, it is essential that people feel safe when using this type of application because the greater the security when using the platform, the greater the commitment, in the prior knowledge of their rights and how the results obtained from the app will be used. In order for these people to give away their data, they need to be convinced of the personal/collective gain from the exchange and that it complies with European regulations (see consent).

Previous experiences and workshops held between 2016 and 2019 with the relatives of stolen babies in Spain were organized as part of participatory action research (PAR) based on concepts by Kurt Lewin (1946) and the experience of Fals Borda, Bonilla & Castillo.

Silenced & Traumatized Mothers

As part of the PAR research, between 2017 and 2018 twenty SOSBRC mothers from stolen babies in Catalonia were observed in terms of participation and actions, and their testimonies were analyzed separately. Different modalities were observed by which these mothers were silenced during and after labor and from which these three categories, integrated by the same component, silence, are presented:

a) Silenced through Medication (STMe): women were given medication after having had their baby - sleeping, soothing pills for a maximum duration of three months. Added to this were doctors' recommendations and instructions, such as: remove everything related to the baby from the house, take a trip as a couple, move out of the house or flat and think about the future; b) Silenced through Manipulation (faith, fear, madness) (STMa): in this case the presence of nuns was paramount with regards to faith and being a "good catholic", "devout woman" and ready to accept God's decisions; c) Silenced through medication and manipulation (STMea): in this case a combination of the categories listed above is observed. (Escudero, 2020, p.4)

From this investigation, for each of these silences induced by those who were part of what is presented as the systematic theft of babies, part of the testimonies was selected to illustrate the stories. As stated in a previous study:

The mothers belonging to the STMe group showed a constant and high degree of distrust of the institutions (political, ecclesiastical and public, private health spaces). They avoid visits to the doctor and they remember and show with their bodies the state in which their bodies were at the time of the medication, stating that "I understood everything that happened around me, I was paralyzed, immobilized, I thought the words I couldn't say because I couldn't even open my mouth." (Escudero, 2020, p.4)

Taking into consideration previous studies based on PAR, the following conclusion can be drawn:

the emergence of SOSBRC families, in particular the mothers, in the media through two campaigns (TEB 2017, 2018) is an element of breaking a paralyzing silence present during and after the dictatorship, a silence that many mothers have dared to break but that many more have not yet done so. The process of giving voice to their own cases is a way to transcend the "this is not said." (Escudero, 2020, p.6)

As part of this PAR, group dynamics make it possible to observe the various processes through which this population has progressively passed, the changes within the group due to external and internal problems, the decisions taken and the standards of operation based on shared priorities and objectives. Group dynamics and research work conducted with the group correspond to that stated by Kurt Lewin, emphasizing the power of the fluid, ever-changing forces that characterize interpersonal groups.

PAR and Group Dynamics

For various reasons, laid out below, the author believes that participatory action research (PAR) represents a coherent methodology for this population and thematic, and therefore their work will be based on specific experiences such as the TEB campaign and the establishing of typologies from the ways in which the mothers were silenced; in both cases, the author was able to get to know the population in greater depth and to organize encounters in safe spaces in which trust was primordial for them to be able to speak, remember and share experiences from the past and present.

Along the same lines, Anderson specifies that before conducting studies into novel psychosocial interventions, such as a mobile app, it is important to carry out more basic work evaluating the development and feasibility of the intervention (2008). In this case, this "basic work" as stated by Anderson, includes group dynamics that foster trust in and commitment to using the application.

The author has also resolved to use specific methods of evaluation for this population in order to improve and/or modify whatever may be necessary in the app both quickly and accurately.

For this reason, to approach this subject based on PAR has made it possible to understand the connection between the individuals within this population and their circumstances (García-Ramírez, Balcázar & Suárez-Balcázar, 2003). Hence, strategies have been developed aimed at fostering well-being and quality of life within the community, and thereby this population has been given an active role in increasing its power and its voice (Prillestensky & Nelson, 2002; Stoecker, 1999), such as through the "We are Looking for You" campaigns known as TEB ("Te Estamos Buscando") (Escudero, 2017, 2018).

One of the objectives of the PAR research was that, through these workshops and group dynamics, a safe space would be created in which people expressed themselves without the fear of reproach or judgement; "hence the group maintained the confidentiality of the workshops. This confidentiality allowed the group to see the workshops as a safe place" (Escudero, 2020, p.74).

Along these lines, the workshops carried out with this target group were divided into three phases: "the first, the process of joining the meeting, of joining the space as a safe space of trust; the second refers to the production (here the creative work); and the third refers to the reflection and sharing (here the various individual expressions and experiences arise) and corresponds to what is known in Gestalt psychology as the "realization", following which in many cases comes the question as to what can be done once we have realized" (Escudero, 2020, p.74). One example is from one of the group dynamics in which, through the questions asked, they came to the realization that the group is resilient because, despite having had a baby stolen from them (in many cases it was their first child), they tried to have other children and became mothers. To recognize this resilience makes it possible for them to become empowered, to find themselves and to see the positive side of a situation which appeared to be solely negative and to make them see themselves as victims.

More specifically, in certain group dynamics, artistic mediation was used to create a tool with which to create a space of safety and trust for the participants, called "The Bus Experience". Through this group dynamics experience, they realized that their wounds are still open and that stigmas persist in time: through the artistic mediation carried out in the workshops, the families of stolen babies taking part were able to express what was often never formally expressed. Considering that wounds need to be expressed in order to heal, when these people have expressed themselves in a literal manner they have often encountered silencing responses, such as:

'What a horrible thing to have happened to you', 'How did you manage to go through all this?', 'Please don't go on, be quiet, it makes me sick to the stomach to hear it' (...) Hence shame, stigmatization, humiliation and guilt are added to the pain of what they already went through (Escudero, 2020, p.75)

In PAR, the researcher takes the side of the social issue that represents the group (silenced, oppressed), developing research aimed at highlighting the group's strengths and minimizing its problems and limitations, generally used to identify and victimize it.

In other words, PAR is a community-centered model, in this case around the population in question, based on questions such as "what is needed, in terms of scientific information and capacity, to produce effective initiatives" (García-Ramírez, Albar-Marín, Morano Báez & Castro, 2017, p.53). For this reason, the work undertaken with this population has given rise to collaborations and has made it possible for the voice of this collective to be heard, identifying criteria and guidelines (Bradbury & Reason, 2001) and has highlighted the group's strengths (Balcázar et al., 1998).

Problems

Even if several treatments and various adaptations of the same app have been developed, many authors conclude that the impact and implications of mental healthcare for PTSD remain insufficient (Kazlauskas et al., 2016; Koenen et al., 2017; Sareen, 2014) and that even when patients are diagnosed with PTSD, there may be a shortage of available therapists specializing in psychotraumatology, leading to lengthy delays in receiving treatment (Mojtabai et al., 2011).

In the case of Spain, various challenges arise in order that a health app meet the needs of such a broad and heterogenous population, including how to reach as many people as possible respecting their needs, their identities, their knowledge of technology and the time required for them to become familiar with the app but also the factor of trust in the app and the information they put into the app, appears amongst other concerns.

The mistrust felt by this population is clear and related to their life stories and the various political contexts they have lived through. More specifically, the theft of babies in Spain began during Francoism (1940) but continued until well into the period of democracy, in other words babies where stolen for twenty years after the end of the dictatorship. This has led this population to lose faith in the country's institutions and to avoid sharing too much information as they distrust how it could be used.

Considering these aspects of mistrust, to begin adapting the app, more information about this target group is needed, and the challenge must be faced of creating new channels of contact and encounters, and new questionnaires in order to investigate in greater depth the difficulties faced by this population in terms of PTSD, anxiety and depression. To make progress in this field, it is necessary to proceed with caution, in a clear and precise manner, as this could have the opposite effect to that desired: reticence and mistrust could lead to a lack of collaboration from certain organizations.

Nevertheless, the launch of a new device answering their needs may also lead to high expectations as well as the fear of not being able to use the app due to a limited knowledge of new technologies or for fear that their answers be read by the authorities they mistrust. This is why it is necessary to present the initiative of the DHI in a specific and clear way, explaining its main objectives as well as its limitations, possible errors and needs for adjustment, trial and error that may take time and thereby not follow the schedule that was laid down.

In this sense, and taking into consideration what has been expressed within the groups of older adults, the benefits of an app for this population would be via specific tools and a connection to new means with which to improve their distress. In the same vein, Harte and colleagues state:

Older adults not familiar with technology are often afraid that they might delete something by accident or fail to save important data properly. Warning tones or symbols, such as a caution symbol, should only be used if absolutely necessary. For audio feedback, clear and low frequency tones should be used. (2017, p. 21)

From these previous experiences, it became apparent for the need to make headway in the creation of an application allowing each individual to implement new tools in order to improve their quality of life and initiate the recovery of their traumatic event in a continuous and monitored manner. Hence, the author has based this work on various studies demonstrating the significance of an app for the trauma, elaborated below.

DHI for Trauma Survivors

As in all other sectors (news, food, transportation, education), health is undergoing a process of *platformization*. Numerous platforms promoting a healthier life have emerged in which patients, doctors, IT teams and researchers interact. In this case, the aim is to develop DHI for populations with PTSD who have still not received treatment or a coordinated and linear monitoring of their cases in Spain.

Cases of post-traumatic stress disorder (PTSD) are widespread around the world (Kessler et al., 2017). PTSD is a prevalent mental health issue that commonly occurs after a person has experienced a traumatic event, which can include being threatened with death or experiencing the death of others (e.g. the death of a family member or a friend), sexual violence, or serious injury (Ptsd.va, in Rodríguez-Paras et al., 2017;

p.1). Meanwhile, effective treatment or any kind of treatment seems to be out of reach for at least half of the afflicted population and only a minority receives specialty mental health care (Kessler et al., 2017).

Evidence-based psychological treatment methods for PTSD include trauma-focused cognitive behavioral therapy, prolonged exposure therapy, cognitive processing therapy, narrative exposure therapy, and eye movement desensitization and reprocessing (EMDR) (Bisson, Roberts, Andrew, Cooper & Lewis, 2013; Charney, Hellberg, Bui & Simon, 2018; Mueser et al., 2015; National Institute for Health and Care Excellence [NICE], 2018; in Sander et al., 2020).

Also, cognitive behavioral therapy with a trauma focus typically involves psychoeducation, homework, exposure and cognitive work as well as relaxation and stress management techniques (Berliner et al., 2019). PTSD treatments are divided into two categories that are not mutually exclusive: (1) pharmacotherapy and (2) psychotherapy with two cognitive behavioral therapy (CBT) treatment methods considered the most effective: (1) prolonged exposure (PE) therapy and (2) cognitive processing therapy (CPT) (Forbes et al., 2010 in Rodriguez-Paras et al., 2017).

Digital health interventions, known as Telemedicine or e-Health, have become key for the improvement, analysis, study and development of medicine and care for patients. Electronic Health or e-Health is defined by Eysenbach as:

an emerging field in the intersection of medical informatics, public health and business, referring to health services and information delivered or enhanced through the Internet and related technologies. In a broader sense, the term characterizes not only a technical development, but also a state-of-mind, a way of thinking, an attitude, and a commitment for networked, global thinking, to improve health care locally, regionally, and worldwide by using information and communication technology. (2001, p. 1)

It is common knowledge that Mobile Health Applications / Digital Health Interventions have been the object of study in various lines of research over the last decade (Lin & Yang, 2009; Madsen, 2018), certain studies showing that health apps are a subset of those apps with a pronounced medical quality (e.g. detection, prevention, monitoring, treatment or alleviation of certain diseases or disorders) and must conform to control norms, while for medical device apps even stricter norms apply (Gregor-Haack, 2018).

While it is true that there has been a rapid proliferation of psychological interventions delivered via the Internet for a range of mental health disorders (Andersson & Titov, 2014; Arnberg, Linton, Hultcrantz, Heintz & Jonsson, 2014), most of these apps lack empirical support, i.e. their efficacy is unknown (Yeager & Benight, 2018) and the poor quality of the evidence of effectiveness as well as the absence of quality standards for the development of apps (Byambasuren, Sanders, Beller & Glasziou, 2018; Hussain et al., 2015), all show that it is therefore important to evaluate mobile apps through rigorous scientific studies (Olff, 2015).

Currently, various studies demonstrate that apps are still lagging behind in their potential benefits for people with PTSD: both people affected by PTSD and mental health providers have difficulties in identifying and recommending apps that meet the needs of their patients, "and most apps lack scientific evidence of their effectiveness" (Sander et al., 2020, p.12). On the other hand, Olff argues that the goal of PTSD apps should be to aid in the treatment and monitoring of trauma survivors with PTSD and to provide both patients and clinicians with timely remote feedback that can supplement or enhance current therapies (2015), while Hoffman and colleagues state: PTSD Coach is the most rigorously evaluated PTSD specific app (2011).

As mentioned previously, PTSD Coach is a mHealth app designed to explain PTSD concepts to patients, also providing self-management tools based on cognitive behavioral therapy (CBT). Patients can use the tools in this app to learn, perform self-assessments, manage symptoms and find support. In addition, people using the app may, through the "self-assessment" and "manage symptoms" options, track over time to see their progress and a list of mitigation techniques (Rodriguez-Paras et al, 2017). Considering this population, the mitigation techniques are a very important tool, as this will make a variety of options available to them if their distress persists or if the technique was not to their liking; more specifically: once the trauma survivors finish using each mitigation technique, they are prompted to complete the checklist again and if their level of distress is still the same or higher, then they are offered another mitigation tool to try. Tracking the users is key as, before each tool is presented, the user is asked to provide a Subjective Unit of Distress Scale (SUDS) rating their current distress from 0 to 10. The SUDS rating is presented as a thermometer (Cernvall, Sveen, Johannesson & Arnberg, 2018). Finally, another feature of this application of particular interest can be found within the section "Find support", in which users can make a previous selection and store the contact information of those they rely on during emergency or crisis situations, with direct access via the app. Considering that most of the population of the relatives of stolen babies are older adults, this option is of particular interest.

PTSD Coach Version for Spain

This study will focus on PTSD Coach, a mobile app developed by the National Center for PTSD-Dissemination and Training Division, Veterans Administration (VA), by means of participatory research, where focus groups with PTSD patients were used to generate input on features and design (Kuhn et al., 2014).

The study will be based on this app as it covers all four main goals: providing information to the patients; tracking the patients' symptoms; managing these symptoms; and making additional support available (in this field, Considering the characteristics of families looking for missing relatives, actions focusing on this type of case will be included). This app has been adapted for several different countries and, since 2011, versions of the app in English have been publicly available in Google and Apple app markets.

Experiences in Other Countries

Furthermore, each country that has used Coach as a prototype, progressively adapted it to its needs and priorities. Part of managing the symptoms is grounding and resource activation. For this purpose, mindfulness-training exercises were converted into electronic form and made available as part of the app (Kuhn et al., 2017, Miner et al., 2016).

As mentioned above, various countries have adapted the Coach app, such as the Netherlands with a version that includes a calendar function for the purpose of scheduling appointments and activities (Schellong, Lorenz & Weidner, 2019). While in Germany, the app is known as Coach PTBS and includes a mood diary and contents from CBT-I Coach for insomnia (Miner et al., 2016), as well as a tool to assess levels of anxiety and depression (Kuhn et al., 2018).

In the Swedish version, patients who used the app suggested more personalized settings such as being able to change color and voices, or create shortcuts from the home screen to specific content in the app. Participants also found specific functions of the app helpful such as relocation exercises (Cernvall, Sveen, Johannesson & Arbnerg, 2018). Meanwhile, the Canadian version also adapted the app by adding a section with information about 'operational stress injuries', related to any psychological difficulties sustained while serving in the Canadian Armed Forces or the Canadian Royal Mounted Police (Kuhn et al., 2018).

Cuijpers & Schuurmans argue that in the case of PTSD, unguided exposure without a treatment plan might increase symptom severity (2007). Furthermore, evidence suggests that mHealth apps should only be used as a supplement and not a replacement for professional therapy (Keen & Roberts, 2017). As Price and colleagues state (2014): this has to be made abundantly clear to the user, to which the author would add that this should be made known from the start, via the presentation of the app, as the high expectations of the population could lead to a negative impact in its use and objectives, which is why it is essential to know how to mediate the information and communicate its potential and limitations.

Adaptation of the Three Versions

The adaptation of the PTSD Coach app for the relatives of missing persons and the relatives of stolen babies could have a positive impact and be implemented in the short term given that there are various versions of the app which could be well adapted to the needs of this group. More specifically, initially the known versions of the app could be adapted in this way:

PTSD Coach is for part of the population currently not consulting a therapist; it is based on education about PTSD, information about professional care, self-assessment for PTSD sufferers, opportunities for finding support, and tools that can help trauma survivors with PTSD manage the stresses of daily life. Tools range from relaxation skills and positive self-talk to anger management and other common self-help strategies. Through this app it is possible to customize tools based on your preferences and you can integrate your contacts, photos and music (PubMed, table 1, 2018).

PE Coach (which deals with prolonged or repeated exposure) would be ideal for the relatives of stolen babies as many of them have spent more than 30 years coping with this situation. PE Coach is designed to be used during therapy for posttraumatic stress disorder (PTSD) with a health professional who is trained in Prolonged Exposure (PE) therapy (PubMed, table 2, 2018).

CPT Coach (related to cognitive processing therapy) would be for the families of a relative who has been missing for less than 5 years. The aim of CPT is to help patients understand and change the way they think about traumatic events, focusing on the fact that they are not at fault. CPT Coach is not intended to be used as self-help without the guidance of a professional mental healthcare provider (PubMed, table 2, 2018).

Accessibility

In order to launch the digital health intervention, practical and technological questions must be previously addressed such as the accessibility of the population to smartphones and their prior knowledge of how to use mobile apps.

The term "digital inclusion" refers not only to the skills possessed by the users, but also to their interest (Ragnedda, 2017). As stated by Bradbrook and Fisher, the "5C's" in digital inclusion are: connectivity, capability, content, confidence, and continuity. Ragnedda and Ruiu point out that: since social inclusion and digital inclusion are deeply intertwined, the emphasis has changed to a focus on digital inclusion policies and strategies and the discussion over the digital divide has thus been remapped in terms of a social inclusion agenda (2017, in Ragnedda 2018).

Digital literacy in Spain focuses its attention on the acquisition and mastery of skills for the use of information and communication, and not so much on the ability to use technology (Peñalva-Velez, Napal-Fraile & Mendioroz-Lacambra, 2018; p.3).

Firstly, centering observations on digital inclusion, it can be seen that, according to the results of the Fundación Telefónica (2019), in 2018 for 91.9% of the population in Spain their smartphone was their favorite device and various findings stand out on the use of the smartphone: for 95.1% of Spaniards, instant messaging was their preferred channel of communication, above that of face to face communication, situated at 86.60%.

Secondly, in terms of digital literacy, it can be seen that for 87.1% of older adults (who represent the large majority of this study population) over the age of 64, the use of instant messaging apps was usual (Fundación Telefónica, 2019). Hence, using an app would not be an impossible challenge for them, confirmed by the results of Statista (2018): by the year 2021, a total of 352.9 billion applications are expected to be downloaded.

Access Criteria

People with access to the application could be those registered by the police as having reported a missing relative according to the report criteria and protocols and the monitoring of the case in each autonomous region of Spain, or family members belonging to organizations/associations legally registered in Spain, searching for their stolen babies. This population must then comply with what are known as the inclusion criteria for this DHI and the use of the app: 18 years or older, Spanish speaker, access to a smartphone connected to the Google Play Store or the Apple App Store and experience of a potentially traumatic event, as defined by the DSM-5 (American Psychiatric Association, 2013), during the past five years. The exclusion criteria are as follow: people under 18, and people that present ongoing severe psychiatric problems such as psychotic symptoms, severe depression, mania, ongoing substance abuse or risk of suicide.

Consent

As specified, group dynamics are essential to obtaining greater knowledge of the population. Other practical features to be included in the app, such as questionnaires, must be re-examined for their efficacy and adapted to the population and their special needs. In this sense, various authors agree that knowledge of the needs and attributes of the app's target population is essential information that must be reflected when used digitally and then converted into a form compatible with mobile phone screens (Marcano Belisario et al., 2015; Price, Kuhn, Hoffman, Ruzek & Acierno, 2015; Van Ameringen et al., 2017; van der Meer et al., 2017 in Schellong, Lorenz & Weidner, 2019). Each intended function of the app has to be associated with one or more of the goals set at the beginning, just as every goal should be represented by the functions that are working to achieve that goal.

Moreover, before an app is released and made available to the target population, a rigorous process of evaluation and assessment has to take place. In this sense, Grundy, Wang & Bero state that: if the app is to be qualified as a medical device, it is not only important to apply all the norms concerning such devices in development, but also to establish that the app has the desired effect (2016).

In turn, it is essential to note that it is equally important for the medical team to be informed of technical limitations as it is for the technical team to have a clear understanding of what the program should be able to do, the target population's characteristics and the app's limitations.

Rights

The user rights for this app should be published from the beginning, providing those adapting the app and its users with in-depth information, and if necessary providing for a space for consultation, doubts and reservations, all of which are key to increasing the level of trust in the app. The work on an mHealth Code of Conduct started in April 2015, when a drafting team of industry members started developing the text of the code. The European Commission acted as a facilitator, provided legal and policy expertise, oversaw the development of this work and provided resources. The green book aims to be a reference framework for the exchange of information related to Mobile Health Applications, as well as to stimulate innovation among the parties and providers interested in the development of these types of applications.

The drafting team included the App Association (ACT), App Developers Alliance, Apple, COCIR, Digital Europe, ECHA, DHACA, EFPIA, Google, Intel, Microsoft, Qualcomm and Samsung. They had regular meetings, and presented their work at various events in order to obtain further feedback. The vision was that the Code should be easily understandable, also for SMEs and individual developers who may not have access to legal expertise. Based on that put forward by the EU, it is important that the app features be presented to these populations as follows:

User Consent: User consent for the processing of personal data must be free, specific and informed. Explicit consent needs to be obtained for the processing of health data. Any withdrawal of consent must result in the deletion of the user's personal data.

Purpose Limitation and Data Minimization: The data may only be processed for specific and legitimate purposes. In this specific case, for research purposes.

Privacy by Design and by Default: The privacy implications of the app must be taken into consideration at each step of development and wherever the user is given a choice.

Data Subject Rights and Information Requirements: The user has the right to access their personal data, to request corrections and to object to further processing. The app developer must provide the user with certain information on processing.

Data Retention: Personal data may not be stored longer than necessary.

Security Measures: Technical and organizational measures must be implemented to ensure the confidentiality, integrity and availability of the personal data processed and to protect against accidental or unlawful destruction, loss, alteration, disclosure, access or other unlawful forms of processing.

Advertising in mHealth Apps: There is a distinction between advertising based on the processing of personal data (requiring opt-in consent) and advertising not relying on personal data (opt-out consent). In this case, there should not be advertising.

Use of Personal Data for Secondary Purposes: Any processing for secondary purposes needs to be compatible with the original purpose. Further processing for scientific and historical research or statistical purposes is considered compatible with the original purpose. Secondary processing should be clearly informed in advance, as non-compatible purposes require new consent and their objectives and intentions must be specified.

Disclosing Data to Third Parties for Processing Operations: The user must be informed prior to disclosure and the app developer must enter into a binding legal agreement with the third party. Parties processing this information in Spain may not be linked to any political party or government, but rather to universities/research.

Data Transfers: For data transfers to a location outside the EU/EEA, there must be legal guarantees permitting such a transfer, e.g. an adequacy decision of the European Commission, European Commission Model Contracts or Binding Corporate Rules. In this case, there would initially not be data transfers outside the EU.

Personal Data Breach: The Code provides a checklist to follow in case of a personal data breach, in particular the obligation to notify a data protection authority.

Data Gathered From Children: This app will be implemented for people aged 18 or over.

Pilot Test in Spain

Although different ways of measuring the usability of any given app have been proposed (Wakefield et al., 2017), such as the eHealth literacy scale (Norman & Skinner, 2006), initially a certain agility is necessary within the development environment so that changes, made after needs become apparent through testing, can be easily implemented without creating false hopes or increasing the anxiety of the app's target population. Moreover, with regard to health research, current regulatory frameworks still lack sufficient tools with which to handle issues and situations related to the commodification of health data collections (Seife, 2013), to which may be added that put forward by Van Dijck, Poell & de Waal: "researchers from universities and hospitals have much less power than major platform owners and data firms when it comes to developing sophisticated analytic tools; and they have no leverage at all to combine health data with other personalized information" (2018, p.112). For this reason, prior work with group dynamics as well as a pilot test in which to test the app are necessary steps in order to avoid that which authors such as Giunti and colleagues (2018) described as a failure in accommodating the expectations of the targeted users that often results in the users' lack of willingness to use the app. This would be detrimental to the goal of lowering the barrier of mental health interventions for PTSD through easily accessible smartphone apps.

It is important to establish a methodology for the measurement and monitoring of the app, as suggested by Owen and colleagues (2015): Post-market communication (PMC) is an important part of the life cycle of every serious app, among other things, and involves keeping an overview of the number of down-loaded apps and their subsequent reviews and feedback by the users, as well as using this information to update the product.

For this reason, it is of interest to implement a semi-structured interview via telephone, conducted after access to the app, as part of a pilot test with ten relatives of stolen babies and ten relatives of missing persons in Spain, in order to discover and study in depth the user patterns and experiences from interacting with the app. The possible questions to be asked during this first stage should focus on the following: "Have you been able to access the application?" ; "Which parts or features of the app have been most useful to you?". Along these lines, that put forward by Kuziemsky and colleagues on how to build effective interdisciplinary communication should be taken into consideration, when "from a systems design perspective, we discussed how electronic data support, electronic process facilitation and video or web conferencing could be used to support 'e-teams'" (2009).

SOLUTIONS

Various lines of research provide growing evidence that apps are engaging, easy-to-use, and present a relative advantage over traditional care without apps (Owen et al., 2018). Furthermore, pilot studies on how end users (e.g. clinicians and patients) are interacting with the app seem promising (Cernvall, Sveen, Johannesson & Arnberg, 2018; Rodriguez-Paras et al., 2017). This is the case for the population of families of stolen babies which, at the same time, "underlines the need for supplementary low-threshold, high-coverage means of assessing and supporting the mental health of this population by either supporting professional healthcare or bridging the possibly substantial gap before therapy can begin" (Schellong, Lorenz & Weidner, 2019, p.2).

Hence, these intervention practices help them to be receptive to different perspectives in the world, encouraging their integration into and connection with the world as well as their non-traumatic introduction into diverse social areas of life.

The use of the PAR approach in the study of phenomena and problems of a psychosocial nature, such as the case of stolen babies and missing people, has two clear limitations: social conflict and the need for consensus. Consequently, as specified by Price, social consensus is essential for defining the problems, their causes and possible solutions (1989). Along the lines of that suggested by Price, part of the solution is to generate a free-flowing exchange with the organizations in which a dialogue of consensus is created and trust is strengthened, as explained below.

Proposed Solutions

The adaptation of PTSD Coach to this population would be the first health approach/solution/accompaniment for this target group, and this may lead to high expectations but also reservations. It is therefore important to communicate clearly and share with the organizations the progress made by this DHI.

Considering that the trust/mistrust factor will be latent, the fact that PTSD Coach is an app that was already developed, outside Spain, and that other countries have already adapted it to various target groups, will be a positive aspect for the families, leading to greater trust in the new challenge of adapting it to their own country. In this way, it must be explained to the organizations that European regulations on rights and consent for the use of the app will be respected, and this is another aspect which will promote calm and trust within the groups.

However, in accordance with the aforementioned, in relation to trust, as this is a population which, for various reasons, is in search of a relative, the use of terminology and words in such an app is, as has been observed in prior interventions, and will be extremely important, as part of their level of trust in the app and part of their process of recuperation in which a sense of individual work prevails within a group support system. This is why it essential to include the representatives of the families of missing persons in the adaptation process, as in this way they will take part in and bear witness to the preparation of this DHI in order to provide a space, a voice and recognition for the families' needs.

On the other hand, while an app with these characteristics will present challenges contemplated beforehand and others that may appear, and therefore the team must be prepared; nevertheless, for the adaptation process to be coherent, it must occur progressively and with direct communication with the representatives of the main organizations working in the various autonomous regions, as the aim is for this DHI to be launched at a national level and that its target population be aware of its progress in advance.

Another of the key aspects of this app is that it will be launched in Spanish to avoid confusion and/or language limitations, with the possibility that in the future it may be adapted to include the other languages of the country: Catalan, Galician, Basque, amongst others. It must not be overlooked that a high proportion of the affected families feel mistrust towards the Spanish institutions, which is why the manner in which the app is presented is key and will contribute to its progress in terms of its implementation. In this sense, regional issues and questions of identity will be assessed and taken into account, giving a voice to the representatives of the organizations making up the platforms "Te Estamos Buscando" and "X24".

Considering that these are older adults, access to this app and its launch should be coordinated by the organizations. For this reason, each representative of the associations will receive instructions and the possibility that one of the members of the team help them to know, explain and share how to use the app and the scope of its functions. An explanatory video about the app will provide support for users in which they can listen, visualize the functions and try out the app

CONCLUSIONS

The fact that, until now, no progress has been made in social policies aiming to find strategies of reparation for this population - diverse, spread throughout Spain and a majority of whom are older adults - illustrates the complexity of this issue. This group is concerned with the issue of age combined with the passing of time, and this motivates them to take action and in many cases they become avid of digital know-how, because through using and understanding social media they can share their cases, publish documents and share photos. Hence, and in accordance with other research, it is known that this group own mobile devices and use them frequently.

Nevertheless, and also based on prior research carried out with this population and the progress made in group dynamics and workshops, it may be confirmed that they recognize their needs and are open to sharing their stories, as long as this is done within spaces which are respectful and free of political or religious interests, but rather in pursuit of knowledge and the dissemination of the facts. Experiences such as the TEB campaign and the group dynamics created by this group known as "The Bus Experience", have allowed it to be observed and confirmed that the groups tend to be well-coordinated internally and show respect for their members. These various experiences of group dynamics have allowed spaces of trust to be created which could not have been created through individual interviews, which is why trust has been established with the groups but it must be maintained and adapted to new research.

Finally, after various experiences with this population, this is the moment to take a new step forward towards psychological reparation with the implementation of an app in which they feel involved, and of which they are the inspiration and the beneficiaries.

FUTURE RESEARCH DIRECTIONS

To be privy to prior information on the target population of a health app, above all when involving psychosocial interventions and group dynamics, not only makes it possible to design an adaptation responding to the needs of the population in question, but also allows for exchange and better understanding of such a broad-based and diverse population as this in Spain.

The significance of previous results, such as those from PAR, also means giving a voice and a space for reparation to a population that has not been recognized in the past and makes it possible to move forward towards the beginning of a process of reparation in which there will surely be a need to make adjustments, changes and new adaptations. Hence, future research should concentrate on the population of relatives of missing persons and on how the use of a health app can influence their health. The amplitude and diversity of this population, as previously stated, provides for a variety of interdisciplinary studies to be carried out into health, reparation and improvements for this population.

While health apps intent on gathering personal data often promise to offer personalized solutions, in this particular case, the contributions of this population will help to acquire a deeper understanding of the severity of the traumatic event undergone by this population, not solely for medical purposes but as a document of historical memory and posthumous reparation. While the information gained from the people using this type of platform tends to be seen as a contribution to the greater common good, in this particular case, this common good makes a far-reaching contribution seeing that there are no studies in which the cases of these populations are brought together in an analysis of the significance of the traumatic event, which allows the author to affirm that the study of these results will constitute an interdisciplinary contribution (in Social Psychology, Psychology, Psychiatry, Gender Studies amongst others) and possibly allow them to define new lines of research and prevention.

RECOMMENDATIONS

Nevertheless, the results of these interventions for the adaptation of a trauma application in Spain will permit a new experience in the field of digital health interventions to be established in which the monitoring and commitment of the app's users must be improved. As similar experiences to that put forward have not been found, the author believes it to be of interest to pursue this initiative creating a concrete and respectful space for exchange with the target population, with the representatives of the organizations key to the beginning of the adaptation of this app. Based on previous experience with this population, it will be essential to promote a communicational exchange on the progress of the app with the representatives of the organizations of the families of missing persons in Spain, so that they can explain them to the rest of their members via their internal platforms. In order for the exchange to work, it is recommended that an agenda of communicational exchanges be set up with these representatives, through which the progress, challenges and limitations of the app will be made public. This type of exchange should also be used to express the importance of their support, feedback and requirements to the app's adaptation process. It is equally recommended that from the start, it should be specified to the organizations, through the spokespersons, that the app does not substitute any type of medication or treatment, explaining the criteria of access to the app and that each case will be assessed individually so as to respond to the needs of each person.

The author recommends that it be clearly stated that three versions of the PTSD Coach app will be launched in order to respond to individual needs, each version passing through a pilot process which may require a certain number of adjustments, and each version should be tested, assessed and if necessary, redesigned. Publishing information in advance will make it possible from the start to count on volunteers for future tests and to confirm how important it is to have the support of the organizations throughout the whole adaptation process.

Additionally, as observed in previous studies, certain aspects may be considered of a political nature or related to matters of identity (such as flags, religious symbols or images representing a political or nationalist thinking) which may cause profound and irreversible divisions in the population; for this reason, at the launch of this app and during the entire process of communicational exchange, these aspects must be considered.

REFERENCES

American Psychiatric Association. (2013). *Diagnostic and statistical manual of mental disorders (DSM-5®)*. American Psychiatric Association.

Adibi, S. (Ed.). (2015). *Mobile Health: A Technology Road Map*. Springer. doi:10.1007/978-3-319-12817-7

Anderson, R. (2008). New MRC guidance on evaluating complex interventions. *BMJ 337*(1). *Article a, 1937–9*. Advance online publication. doi:10.1136/bmj.a1937

Andersson, G., & Titov, N. (2014). Advantages and limitations of internet-based interventions for common mental disorders. *World Psychiatry: Official Journal of the World Psychiatric Association, 13*(1), 4–11. doi:10.1002/wps.20083

Arnberg, F. K., Linton, S. J., Hultcrantz, M., Heintz, E., & Jonsson, U. (2014). Internet-delivered psychological treatments for mood and anxiety disorders: A systematic review of their efficacy, safety, and cost-effectiveness. *PLoS One*, *9*(5), e98118. Advance online publication. doi:10.1371/journal.pone.0098118

Bakker, D., Kazantzis, N., Rickwood, D., & Rickard, N. (2016). Mental health smartphone apps: Review and evidence-based recommendations for future developments. *JMIR Mental Health*, *3*(1), e7. Advance online publication. doi:10.2196/mental.4984

Balcázar, F. E., Keys, C. B., Kaplan, D. L., & Suárez-Balcázar, Y. (1998). Participatory Action Research and people with disabilities: Principles and challenges. *Canadian Journal of Rehabilitation*, *12*, 105–112. https://www.researchgate.net/profile/Christopher_Keys/publication/239921190_Participatory_Action_Research_and_People_with_Disabilities_Principles_and_Challenges/links/0deec53260cdd4e251000000.pdf

Barrenechea López, G. (2017). *Gender in Transitional Justice Contexts: Reproductive Rights in the Spanish Case of Stolen Babies* [Master Thesis]. Lund University, Lund, Scania Province, Sweden.

Beristain, C. (2007). Reparations to victims in the inter-American human rights system. In *The role of the justice system in the face of massive violations of human rights*. Abuelas de Plaza de Mayo.

Berliner, L., Bisson, J., Cloitre, M., Forbes, D., Goldbeck, L., & Jensen, T. … Shapiro, F. (2019). *ISTSS PTSD prevention and treatment guidelines methodology and recommendations*. International Society for Traumatic Stress Studies. https://www.istss.org/treating-trauma/new-istss-prevention-and-treatment-guidelines

Bradbury, H., & Reason, P. (2001). Conclusion: Broadening the band with of validity: issues and choice-points for improving the quality of action research. In P. Reason & H. Bradbury (Eds.), *The SAGE Handbook of Action Research: Participatory Inquiry and Practice*. SAGE. https://www.researchgate.net/publication/281504513_Conclusion_Broadening_the_bandwidth_of_validity_Issues_and_choice-points_for_improving_the_quality_of_action_research

Bueno Morales, M. M., & González Besteiro, J. M. (Eds.). (2018). *Bebés robados en Andalucía desde 1936. Buscando la verdad, la justicia y la reparación* [Stolen babies in Andalusia since 1936. Looking for truth, justice and reparation]. Dirección General de Memoria Democrática. Junta de Andalucía.

Byambasuren, O., Sanders, S., Beller, E., & Glasziou, P. (2018). Prescribable mHealth apps identified from an overview of systematic reviews. *NPJ Digital Medicine, 1*(1), Article 12. https://www.nature.com/articles/s41746-018-0021-9?report=reader

Cernvall, M., Sveen, J., Johannesson, K. B., & Arnberg, F. (2018). A pilot study of user satisfaction and perceived helpfulness of the Swedish version of the mobile app PSTD Coach. *European Journal of Psychotraumatology, 9*(sup1). Advance online publication. doi:10.1080/20008198.2018.1472990

Clement, J. (2020). *Annual number of global mobile app downloads 2016*–2019. Statista. Retrieved 4 April 2020, from https://www.statista.com/statistics/271644/worldwide-free-and-paid-mobileapp-store-downloads/

Cuijpers, P., & Schuurmans, J. (2007). Self-help interventions for anxiety disorders: An overview. *Current Psychiatry Reports, 9*(4), 284–290. doi:10.100711920-007-0034-6

Dalvit, L. (2018). From Access to Proficiency: Reconceptualizing Digital Inclusion in a Rural Area in South Africa. In M. Ragnedda & B. Mutsvairo (Eds.), *Digital Inclusion: An International Comparative Analysis*. Lexington Books.

Donker, T., Petrie, K., Proudfoot, J., Clarke, J., Birch, M. R., & Christensen, H. (2013). Smartphones for smarter delivery of mental health programs: A systematic review. *Journal of Medical Internet Research, 15*(11), e247. Advance online publication. doi:10.2196/jmir.2791

Escudero, C. (2020). Stolen Babies in Spain: Mediated Stories for Recovery. Mothers' activism through online campaigns. *International Journal of Business and Social Science, 11*(3). Advance online publication. doi:10.30845/ijbss.v11n3a3

Escudero, C. (2020). Giving voice to the traumatic event, Spanish mothers of stolen babies. Three strategies to silence mothers during and after the dictatorship. *International Journal of Humanities and Social Science, 3*(10). doi:10.30845/ijhss.v10n3p1

Escudero, C. (2020). Digital Mutation, a Result of Motivation and Resilience. Stolen Babies in Catalunya. *Review of Journalism and Mass Communication, 1*(8), 1–8. doi:10.15640/rjmc.v8n1a1

Eysenbach, G. (2001). What is e-health? *Journal of Medical Internet Research, 3*(2), e20. Advance online publication. doi:10.2196/jmir.3.2.e20

European Commission. (2018) Privacy Code of Conduct on mobile health (mHealth) apps. In *Shaping Europe's digital future.* https://ec.europa.eu/digital-single-market/en/privacy-code-conduct-mobile-health-apps

Fals Borda, O., Bonilla, V., & Castillo, G. (1972). *Causa popular, ciencia popular: una metodología del conocimiento científico a través de la acción* [Popular cause, popular science: a methodology of scientific knowledge through action]. Publicaciones de La Rosca de Investigación y Acción Social.

Fundación Telefónica. (2019). *Sociedad Digital en España 2018.* https://www.fundaciontelefonica.com/cultura-digital/publicaciones/sociedad-digital-en-espana-2018/655/

García Ramírez, M., Albar-Marín, M. J., Morano Báez, R., & Castro, V. P. (2017). Metodología de la intervención social: implicaciones para la investigación y la práctica de la psicología comunitaria [Methodology of social intervention: implications for research and practice in community psychology]. In I. Maya-Jariego, M. García Ramírez, F. J. Santolaya -Soriano (Eds.). Estrategias de Intervención Psicosocial [Strategies for Psychosocial Intervention]. Psicología Pirámide.

García-Ramírez, M., Balcázar, F., & Suárez-Balcázar, Y. (2003). Investigación acción participativa en la psicología comunitaria: Una herramienta para entender la diversidad humana [Participatory Action Research in community psychology: a tool for understanding human diversity]. *Apuntes de Psicología, 21*(3), 409–417.

Giunti, G., Mylonopoulou, V., & Romero, O. R. (2018). More stamina, a gamified mhealth solution for persons with multiple sclerosis: Research through design. *JMIR mHealth and uHealth, 6*(3), e51. doi:10.2196/mhealth.9437

Gregor-Haack, J. (2018). Erstattung von Health-Apps durch die gesetzliche Krankenversicherung [Reimbursement of health apps by the German statutory health insurance]. *Bundesgesundheitsblatt, Gesundheitsforschung, Gesundheitsschutz, 61*(3), 328–333. doi:10.100700103-018-2689-z

Grundy, Q. H., Wang, Z., & Bero, L. A. (2016). Challenges in assessing mobile health app quality: A systematic review of prevalent and innovative methods. *American Journal of Preventive Medicine, 51*(6), 1051–1059. doi:10.1016/j.amepre.2016.07.009

Harte, R., Quinlan, L. R., Glynn, L., Rodríguez-Molinero, A., Baker, P. M., Scharf, T., & ÓLaighin, G. (2017). Human-Centered Design Study: Enhancing the usability of a mobile phone app in an integrated falls risk detection system for use by older adult users. *JMIR mHealth and uHealth, 5*(5), e71. Advance online publication. doi:10.2196/mhealth.7046

Herman, J. (1992). *Trauma and Recovery. The Aftermath of Violence - From domestic abuse to political terror*. Pandora Edition.

Hoffman, J. E., Wald, L., Kuhn, E., Greene, C., Ruzek, J., & Weingardt, K. (2011). *PTSD Coach* [mobile application]. US Department of Veterans Affairs. https://itunes.apple.com/de/app/ptsd-coach/id430646302?mt=8

Hussain, M., Al-Haiqi, A., Zaidan, A. A., Zaidan, B. B., Kiah, M. L. M., Anuar, N. B., & Abdulnabi, M. (2015). The landscape of research on smartphone medical apps: Coherent taxonomy, motivations, open challenges and recommendations. *Computer Methods and Programs in Biomedicine, 122*(3), 393–408. doi:10.1016/j.cmpb.2015.08.015

Kazlauskas, E., Javakhishvilli, J., Meewisse, M., Merecz-Kot, D., Şar, V., Schäfer, I., Schnyder, U., & Gersons, B. P. R. (2016). Trauma treatment across Europe: Where do we stand now from a perspective of seven countries. *European Journal of Psychotraumatology, 7*(1), 29450. Advance online publication. doi:10.3402/ejpt.v7.29450

Keen, S. M., & Roberts, N. (2017). Preliminary evidence for the use and efficacy of mobile health applications in managing posttraumatic stress disorder symptoms. *Health Systems (Basingstoke, England), 6*(2), 122–129. doi:10.1057/hs.2016.2

Kessler, R. C., Aguilar-Gaxiola, S., Alonso, J., Benjet, C., Bromet, E. J., Cardoso, G., ... Ferry, F. (2017). Trauma and PTSD in the WHO world mental health surveys. *European Journal of Psychotraumatology, 8*(sup5). *Article, 1353383.* Advance online publication. doi:10.1080/20008198.2017.1353383

Koenen, K. C., Ratanatharathorn, A., Ng, L., McLaughlin, K. A., Bromet, E. J., Stein, D. J., Karam, E. G., Meron Ruscio, A., Benjet, C., Scott, K., Atwoli, L., Petukhova, M., Lim, C. C. W., Aguilar-Gaxiola, S., Al-Hamzawi, A., Alonso, J., Bunting, B., Ciutan, M., de Girolamo, G., ... Kessler, R. C. (2017). Posttraumatic stress disorder in the world mental health surveys. *Psychological Medicine, 47*(13), 2260–2274. doi:10.1017/S0033291717000708

Kuhn, E., Greene, C., Hoffman, J., Nguyen, T., Wald, L., Schmidt, J., Ramsey, K. M., & Ruzek, J. (2014). Preliminary evaluation of PTSD Coach, a smartphone app for post-traumatic stress symptoms. *Military Medicine, 179*(1), 12–18. doi:10.7205/MILMED-D-13-00271

Kuhn, E., Kanuri, N., Hoffman, J. E., Garvert, D. W., Ruzek, J. I., & Taylor, C. B. (2017). A randomized controlled trial of a smartphone app for posttraumatic stress disorder symptoms. *Journal of Consulting and Clinical Psychology, 85*(3), 267–273. doi:10.1037/ccp0000163

Kuhn, E., van der Meer, C., Owen, J. E., Hoffman, J. E., Cash, R., Carrese, P., Olff, M., Bakker, A., Schellong, J., Lorenz, P., Schopp, M., Rau, H., Weidner, K., Arnberg, F. K., Cernvall, M., & Iversen, T. (2018). PTSD Coach around the world. *mHealth, 4*(5). Advance online publication. doi:10.21037/mhealth.2018.05.01

Kuziemsky, C. E., Borycki, E. M., Purkis, M. E., Black, F., Boyle, M., Cloutier-Fisher, D., ... Tschanz, C. (2009). An interdisciplinary team communication framework and its application to healthcare 'e-teams' systems design. *BMC Medical Informatics and Decision Making, 9*(1), 43. Advance online publication. doi:10.1186/1472-6947-9-43

Lewin, K. (1946). Action research and minority problems. *Human Relations, 1*(2), 34–46. doi:10.1111/j.1540-4560.1946.tb02295.x

Lin, S. P., & Yang, H. Y. (2009). Exploring key factors in the choice of e-health using an asthma care mobile service model. *Telemedicine Journal and e-Health, 15*(9), 884–890. doi:10.1089/tmj.2009.0047

Madsen, W. (2018). History in health: Health promotion's underexplored tool for change. *Public Health*, *154*, 118–122. doi:10.1016/j.puhe.2017.10.028

Miner, A., Kuhn, E., Hoffman, J. E., Owen, J. E., Ruzek, J. I., & Taylor, C. B. (2016). Feasibility, acceptability, and potential efficacy of the PTSD coach app: A pilot randomized controlled trial with community trauma survivors. *Psychological Trauma: Theory, Research, Practice, and Policy*, *8*(3), 384–392. doi:10.1037/tra0000092

Ministerio del Interior. (2020). *AlertCops* [mobile application]. https://alertcops.ses.mir.es/mialertcops/

Mojtabai, R., Olfson, M., Sampson, N. A., Jin, R., Druss, B., Wang, P. S., Wells, K. B., Pincus, H. A., & Kessler, R. C. (2011). Barriers to mental health treatment: Results from the National Comorbidity Survey Replication. *Psychological Medicine*, *41*(8), 1751–1761. doi:10.1017/S0033291710002291

Norman, C. D., & Skinner, H. A. (2006). eHEALS: The eHealth Literacy Scale. *Journal of Medical Internet Research*, *8*(4), e27. Advance online publication. doi:10.2196/jmir.8.4.e27

Olff, M. (2015). Mobile mental health: A challenging research agenda. *European Journal of Psychotraumatology*, *6*(1), 27882. Advance online publication. doi:10.3402/ejpt.v6.27882

Owen, J. E., Kuhn, E., Jaworski, B. K., McGee-Vincent, P., Juhasz, K., Hoffman, J. E., & Rosen, C. (2018). VA mobile apps for PTSD and related problems: Public health resources for veterans and those who care for them. *mHealth*, *4*(7), 28. doi:10.21037/mhealth.2018.05.07

Owen, J. E., Jaworski, B. K., Kuhn, E., Makin-Byrd, K. N., Ramsey, K. M., & Hoffman, J. E. (2015). mHealth in the wild: Using novel data to examine the reach, use, and impact of PTSD coach. *JMIR Mental Health*, *2*(1), e7. Advance online publication. doi:10.2196/mental.3935

Palos-Sánchez, P., Saura, J. R., & Álvarez-García, J. (2019). Innovation and Creativity in the Mobile Applications Industry: A Case Study of Mobile Health Applications (e-Health Apps). In M. Peris-Ortiz, M. R. Cabrera-Flores, & A. Serrano-Santoyo (Eds.), *Cultural and Creative Industries. A path to Entrepreneurship and Innovation*. Springer. doi:10.1007/978-3-319-99590-8_7

Peñalva-Velez, A., Napal-Fraile, M., & Mendioroz-Lacambra, A. M. (2018). Competencia digital y alfabetización digital de los adultos (profesorado y familia) [Digital skills and adult literacy (faculty and family)]. *International Journal of Nursing Education*, *1*(1). Advance online publication. doi:10.24310/IJNE1.1.2018.4892

Price, R. H. (1989). Bearing witness. *American Journal of Community Psychology*, *17*(2), 151–167. doi:10.1007/BF00931004

Price, M., Yuen, E. K., Goetter, E. M., Herbert, J. D., Forman, E. M., Acierno, R., & Ruggiero, K. J. (2014). mHealth: A mechanism to deliver more accessible, more effective mental health care. *Clinical Psychology & Psychotherapy*, *21*(5), 427–436. doi:10.1002/cpp.1855

Prilleltensky, I., & Nelson, G. (2002). Doing Psychology Critically. Making a Difference in Diverse Settings. *Journal of Community & Applied Social Psychology*, *13*(4). Advance online publication. doi:10.1002/casp.711

PubMed Central. Table 1. (2018). *Mhealth, 4*(28). http://mhealth.amegroups.com/article/viewFile/20524/20219/123518

PubMed Central. Table 2. (2018). *Mhealth, 4*(28). http://mhealth.amegroups.com/article/viewFile/20524/20219/123519

Ragnedda, M. (2017). *The Third Digital Divide: A Weberian Approach to Digital Inequalities*. Routledge. doi:10.4324/9781315606002

Ragnedda, M., & Mutsvairo, B. (2018). Digital Inclusion: Empowering People through Information and Communication Technologies. In M. Ragnedda & B. Mutsvairo (Eds.), *Digital Inclusion: An International Comparative Analysis*. Lexington Books.

Rodriguez-Paras, C., Tippey, K., Brown, E., Sasangohar, F., Creech, S., Kum, H. C., Lawley, M., & Benzer, J. K. (2017). Posttraumatic Stress Disorder and Mobile Health: App Investigation and Scoping Literature Review. *JMIR mHealth and uHealth, 5*(10), e156. Advance online publication. doi:10.2196/mhealth.7318

Sander, L. B., Schorndanner, J., Terhorst, Y., Spanel, K., Pryss, R., Baumeister, H., & Messner, E. M. (2020). 'Help for trauma from the app stores?' A systematic review and standardized rating of apps for Post-Traumatic Stress Disorder (PTSD). *European Journal of Psychotraumatology, 11*(1), 1701788. Advance online publication. doi:10.1080/20008198.2019.1701788

Sareen, J. (2014). Posttraumatic stress disorder in adults: Impact, comorbidity, risk factors, and treatment. *Canadian Journal of Psychiatry, 59*(9), 460–467. doi:10.1177/070674371405900902

Seife, C. (2013, November 27). 23andMe Is Terrifying, but not for the Reasons the FDA Thinks. *Scientific American*. https://www.scientificamerican.com/article/23andme-is-terrifying-but-not-for-the-reasons-the-fda-thinks/

Schellong, J., Lorenz, P., & Weidner, K. (2019). Proposing a standardized, step-by-step model for creating post-traumatic stress disorder (PTSD) related mobile mental health apps in a framework based on technical and medical norms. *European Journal of Psychotraumatology, 10*(1), 1611090. doi:10.1080/20008198.2019.1611090

Bebes Robados Catalunya, S. O. S. (2017, June 16). *Te Estamos Buscando* [We are looking for you]. Created by Escudero, C. [Video]. YouTube. https://www.youtube.com/watch?v=MfOMKmqKgsM

Bebes Robados Catalunya, S. O. S. (2018, November 29). *Te Estamos Buscando* [We are looking for you]. Created by Escudero, C. [Video]. YouTube. https://www.youtube.com/watch?v=G0D6Txc08hU

Stoecker, R. (1999). Are Academics Irrelevant?: Roles for Scholars in Participatory Research. *The American Behavioral Scientist, 42*(4), 840–854. doi:10.1177/00027649921954561

Van Dijck, J., Poell, T., & de Waal, M. (2018). Healthcare and health research. In *The platform society. Public values in a connective world*. Oxford University Press.

Wakefield, B. J., Turvey, C. L., Nazi, K. M., Holman, J. E., Hogan, T. P., Shimada, S. L., & Kennedy, D. R. (2017). Psychometric Properties of Patient-Facing eHealth Evaluation Measures: Systematic Review and Analysis. *Journal of Medical Internet Research*, *19*(10), e346. Advance online publication. doi:10.2196/jmir.7638

Yeager, C. M., & Benight, C. (2018). If we build it, will they come? Issues of engagement with digital health interventions for trauma recovery. *mHealth*, *4*, 37. doi:10.21037/mhealth.2018.08.04

Chapter 6
The Immune Space:
Monitoring Narratives in Contemporary Pandemic

Chiara Davino
Architecture University of Venice (IUAV), Italy

Lorenza Villani
Architecture University of Venice (IUAV), Italy

ABSTRACT

In the context of SARS-CoV-2 health emergency, strongly framed in the normalization of logic of risk, the authors analyze three digital platforms of contagion containment and population tracking in order to investigate, through a comparative-descriptive analysis, the relationship between different socio-political-cultural contexts and the respective responses adopted—the Chinese government tool Health Code, the South Korean app Corona100, and the Italian app, Immuni—to counter a single global emergency. The objective is to investigate the framing operations that introduced on a global scale the use of apps in bio-security and immunity regime for which individual privacy increasingly collides with collective security. The authors consider central the opening of a debate on how the logic of risk and worst case scenario are paradigmatic nowadays in the development of increasingly sophisticated systems, potentially invasive of privacy, even in function of complex threats interconnected on a global scale.

INTRODUCTION

In the contemporary social space, emergency, exception, control, security and risk structure practices and policies that, in function of increasingly frequent states of crisis, introduce technological systems and digital platforms aimed at containing the emergency, defining security spaces and monitoring flows of goods and people. Through these systems there is a functional confluence between physical and digital dimensions, in which the latter is used to regulate the former.

DOI: 10.4018/978-1-7998-6705-0.ch006

Post September 11, more and more states of crisis, complex threats extended to the global scale and widespread individual insecurity have led to the consolidation of the *normalization of the state of emergency* (Agamben, 2003, p. 11). This condition is identified, at the legal level, with the permanent *state of exception*: a governmental and legislative strategy of *extrema ratio* increasingly adopted today, which normalizes the exception, to the detriment of ordinary law, converting the "contingent fact" into "law" (Agamben, 2003, p. 40). At a spatial level, the permanent emergency and exception are translated both in the frequent militarization of public space and heterogeneous territories considered "at risk", and in the diffusion and strengthening of technological systems capable of collecting data and monitoring areas extended to national and international scale, borders, people. Thanks to these systems, there is the above mentioned functional confluence between the physical and digital dimensions. In this order, which has been consolidated in recent decades, especially in the more economically developed countries, there is the idea that society is constantly going through a condition of crisis and catastrophe, within a temporal dimension without a precise moment when the *disaster occurs* (Fisher, 2009, p. 27). Thus, *risk factor* becomes dominant device capable of ensuring behaviour, conduct and opinions as a function of an action that institutionalizes forecasting and outlines, from time to time, *worse and worse scenarios* that consolidate the coincidence between the concept of *freedom* and *security* (Comitato Invisibile, 2019, p. 63).

All these facts are defined by Didier Bigo as "politics of discomfort": a condition that affects and regulates all levels of society (2008, p. 8) and that places on the same level extremely heterogeneous phenomena (migration, terrorism, epidemic, climate or economic crisis), against which equally heterogeneous mechanisms of monitoring and socio-spatial organization are equally developed (2008, p. 19).

In the last instance it is outlined state of security, within which every level of security corresponds to a projected threat and in which every political subject to defend is opposed to a subject (individual, collective, biological) to fight, dangerous and a-political, understood as a reproduction, in the opposite sense, of the former (Cavalletti, 2005, pp. 17-18, p. 59, pp. 222-223). In this sense, the concept of security is extremely *mobile*, and, similarly to this, so is the concept of risk; in function of this *mobility* is defined, from time to time, the degree of power of the *regulatory authority* and the devices promoted by it (Lam, 2012, p. 167).

The relationship between the concepts of emergency, exception, control, security and risk is now evident in the orders of bio-economy, bio-politics and bio-security – the latter clearly dominating today. To articulate these orders, Donna Haraway's concept is adopted: "human beings, like any other component or subsystem, must be localized in a system architecture whose basic modes of operation are probabilistic, statistical. [...] organisms have ceased to exist as objects of knowledge, giving way to biotic components" (1991, pp. 163-164).

The condition of bio-politics, today extremely evident, for example, along border and frontier areas, makes the latter real interactive architectures that change according to the citizenship of those who cross them. Borders, understood as pure prototypes of more or less porous bio-political spaces according to citizenship, are built and deconstructed as devices of regulation between birth and nation (Petti, 2007, p. 6). *Necropolitic* spaces, taking up the concept coined by Achille Mbembe (2016), in which the violent security policy of border control takes shape by deciding directly on who has the right to live and who to die, within a state of exception that unequivocally questions inalienable human rights.

To this order is added today's neo-liberal regime, in which the individual collaborates in the construction, on a global scale, of *surveillance capitalism*. This economic system is a bio-economy that exploits human experience as raw material, useful to optimize and standardize market logics, and that imposes a new collective global order based on *absolute security* (Zuboff, 2019, p. 13).

To these two orders, diffused on a global scale, today, in 2020, bio-security one, sophisticated evolution of the bio-political regime, returns to be added. In this condition of biological security policy, the *right to health* is transformed into an *obligation to health* (Zylberman, 2013; Agamben, 2020, par. 1) since each individual, inserted within a system which imposes directives and regulates spaces, is legally obliged to adopt certain behaviours and digital devices in function of collective health security.

Against these considerations, today's SARS-CoV-2 health emergency is investigated by the authors not as an *event* but rather as a *phase of a longer-lasting social process*, in which the concepts listed above, closely related for some time, now emerge within a single global crisis. In this particular juncture, emergency, exception, control, security and risk are simultaneously widespread and their effect is the establishment of a global bio-security regime aimed at containing a pandemic that affects indiscriminately the different countries of the globe.

To cope with the pandemic crisis, sophisticated technological systems of monitoring, tracking and socio-spatial reorganization have been employed at the same time. These systems record each individual and collective activity, storing information to allow authorities to intervene in risk contexts or to produce statistical models. Although monitoring modalities were already possible through extremely widespread digital platforms such as, for example, social networks, it is only thanks to the pandemic that these modalities were institutionalized, with the specific and declared aim of tracking and monitoring the population, through government platforms. While in the social platforms it is the individual himself who autonomously supplies data and information tied to his own person, to the circle of contacts and also to the states of mind and health – in a context which is not "declared" aimed at social monitoring – with the governmental platforms of tracking, the citizen is asked, through precise narrative formulas, to communicate information to collaborate actively in the containment of the contagions. Such collaboration is, however, extremely diversified from country to country, with different degrees of "invasiveness" being voluntary or obligatory depending on the cultural and political context within which each system is inserted.

In order to make the diffusion of such digital systems possible in crisis contexts – both in the specific case of health emergencies and in different emergency states – each governmental institution adopts specific languages and modes of communication capable of speaking simultaneously to several segments of the population or to individual groups. In this way there is a multiplicity of speeches, conveyed in heterogeneous digital platforms and media channels, capable of building the *logic of risk* as a whole. This logic contributes to the consolidation of certain attitudes, behaviours and opinions necessary to re-organise the population socially according to a danger to be eradicated.

The authors' objective is to investigate the processes through which certain technological systems, framed by legal measures and accompanied by heterogeneous narratives, are introduced to counteract states of crisis and emergency, specifically the SARS-CoV-2 emergency, and how, consequently, tracking and monitoring systems are institutionalized more easily in emergency conditions than in *normal times*.

In the following section, with the aim of reconstructing today's emergency context from a socio-spatial point of view, and also using a historical perspective, the concepts of emergency and normalization of the latter, risk and worst case scenario, bio-security and finally digital surveillance are investigated. In the section *Security narratives at the base of digital systems in time of SARS-CoV-2* it is argued that, in the current pandemic context, the biosecurity regime has easily allowed the introduction of digital systems both for virus monitoring and socio-spatial organization, whose implications, in both cases, are closely related to the possible invasion of users' privacy. On the relationship between crisis states, digital platforms adopted to counteract the latter and consequent privacy implications, particular emphasis is placed on a different digital platform, iBorderCtrl, an experimental European border control system. Within the same section is given an overview of the different digital platforms introduced in recent months by government institutions and private agencies to counter the development of the pandemic, paying particular attention to the implications of these apps with respect to the protection of user privacy. As discussed in the introduction to the case studies, the survey takes into account the cultural differences that are reflected in the systems analyzed from time to time, highlighting how there is, to some extent, a direct proportionality between the emergency response and the governmental system. The comparative survey conducted in the following sections – referring to the Chinese, South Korean and Italian case studies – links some digital systems, adopted at institutional level to combat the SARS-CoV-2 pandemic, with pre-existing governmental systems of socio-spatial organization or with the more general degree of digitization of the investigated socio-political contexts. The final aim is to show how, in today's bio-security regime, tracking and monitoring have become constituent practices of contemporary times and how these can consequently influence, in the near future, the introduction of further technological systems and digital platforms that until now have only been evaluated on an experimental basis.

BACKGROUND

By it very nature, emergency is, as the Israeli historian Harari argues, an accelerator of social processes (2020, par. 2). Furthermore, the measures and rules introduced temporarily in emergency conditions tend to become permanent (Bigo, 2008, p. 31). An emblematic example are the measures issued by the State of Israel in 1948, the year in which the state of emergency was declared due to Arab-Israeli conflict. Despite the end of the conflict, the state of emergency continues to be permanent, as do some of the temporary measures introduced in 1948 (Harari, 2020, par. 6). A further and more recent example of a *tendency to permanence* is the Italian operation *Strade Sicure* introduced by Law No. 125 of 24th July, 2008, which provides, for specific and exceptional needs of crime prevention, the employment of military personnel in areas of national territory where an increase in control is opportune (Italian Army, 2020, par. 1, 2, 3). Following this provision, renewed year after year, stations, urban centres, squares and main streets are still, in 2020, subject to capillary controls by the Italian armed forces.

The normalization of the emergency and its provisions is closely linked to the *narrative policy* adopted to consolidate it in the population, here considered not only in its biological characteristics but also in those of the grip surface it offers as *audience*, therefore in its behaviour, prejudices and fears that can be oriented in a precise direction (Foucault, 2005, p. 60). The inclusion of the emergency in precise narrative frameworks has the power to consolidate specific provisions in the conscience of the population that accepts and justifies them in favour of security, protection and wellness (Tsoukala, 2008, p. 78). This framing operation consists in contextualizing specific fragments of narrative within a context that can isolate aspects and exalt others, attribute qualities or radically change their meaning so as to present a specific event, person, subject through precise categories. This narrative process has been adopted, together with the metaphor – which condenses a story in very few words – to frame the SARS-CoV-2 emergency in warlike terms. The metaphor summarizes complex concepts that, synthesized in it, determine how to react to a situation. In this sense, the war metaphor, adopted for the contextualization of the biological agent, has the capacity to activate directional social processes, based on the strategy of fear and threat. If on the one hand the formula of war generates union, mobilization and solidarity of communities and resources for the collective good against a common enemy, obscuring the emergency as a product of complex social and cultural problems (Sontag, 1979), on the other hand it activates processes of acceptance of the rules introduced as necessary for the protection of the entire community (Tsoukala, 2008, p. 52). The metaphor of the war threat, characterized by its unpredictability and its potential range of action, which make the enemy as difficult to *identify* as to *quantify* (Tsoukala, 2008, p. 55), has meant that the SARS-CoV-2 emergency has been managed through narrative strategies of fear that have consolidated new models of life for entire communities (Clément, 2020, par. 3).

In war there are no realistic evaluations, the resources are used in their totality because one is in pure emergency and, therefore, every sacrifice is considered necessary (Cassandro, 2020, par. 5). A similar process occurred after the attacks of September 11, when George W. Bush announced the "War on terror" declaring the state of war and emergency based on the safeguarding of the security, rights, freedom and lifestyle of entire communities. In this context, as in the SARS-CoV-2 emergency, although of a different nature, the preliminary communication played a central role in consolidating and propagating the state of emergency and the measures adopted for the management of the *war* (Godard, 2018, p. 124).

Within this emergency order, many social groups are in the phase of dislocation, i.e. due to a break they move from a state of harmony to a dislocated state compared to the previous one, characterized by a precarious social, economic and working condition in which *simple thinking* can prevail. The latter does not make it possible to grasp the complexity of the problems as much for lack of cultural tools as for the psychological refusal to confront the complexity, and leads entire social groups to embrace specific ideals and opinions or to adopt behaviors considered immediate for the resolution of the problem (Orsini, 2019, p. 49).

Within this social order the *logic of risk* is mostly adopted as a governmental tool of emergency of the so-called risk society by which we mean, taking Ulrich Beck, a social-spatial structure of a reflexive type – since it is society itself that produces its own risks – and catastrophic since the state of emergency is the norm and the unknown and unintended consequences rise to the role of dominant force in history and society (2000). This socio-political organization, based on the logic of risk, is aimed at preventing the *worst-case scenario*, according to Zylberman, divided into three points (2013). The first consists in the construction, on the basis of a possible risk, of a fictitious scenario; in this phase data are presented in order to favour specific behaviours that allow to govern in emergency condition. The second point provides for the adoption of the worst-case scenario logic as a regime of *political rationality*. In the third, action is taken according to the organization of the body of citizens to strengthen as much as possible the adherence to government institutions (Zylberman, 2013). Risk is therefore conceived as an all-encompassing term through which the State acts in anticipation of the worst-case scenario. In this sense, the authority that holds the power and control over a particular risk varies according to the manifestation of the risk itself. Therefore, the authority-risk system is mobile: as the level of risk increases, the level of authority with power and control over it increases (Lam, 2012, pp. 166-167).

Risk materializes in particular forms of *mediation* – scientific, political, economic, popular – essential to its own understanding: risk is "produced" only in terms of "knowledge", the latter functional to the growth of awareness of what is unknown and unpredictable – the risk itself (Luhmann, 1993). The social and political meaning of "knowledge" is therefore central to the conceptualization of "risk": those who have the power to establish, through legal and intellectual resources, what is or is not "risk" and to have the media both to structure a certain type of "risk knowledge" and to spread risk itself, have real power over the social production of risk. In this sense, the authority that holds the power and control over a certain risk varies according to the manifestation of the risk itself. Therefore, the authority-risk system is mobile: as the level of risk increases, the level of authority that holds the power and control over it increases.

From a perceptual point of view, according to Adams, risk can occur through three manifestations. In the first instance, the risk can be directly perceptible - for example, of the continuous projection of the videos of the attack on Twin Towers which made terrorist threat *tangible* for the majority of global population. In the second instance, the risk can be perceived through science: the worst-case scenario perspective is managed by technicians, experts and professionals; with reference to the post-9/11 period, for example, TIA (Total Information Awareness) program to monitor and prevent terrorist action, managed by the US Army and experts (Lam, 2012, p. 175). Ultimately, the risk can occur virtually by anticipating a collective threat that urges the population to involve the catastrophic imagination as a means of action. The paradigm of risk, understood in this way, is the product of a cultural process both "citizen" and "political"; in fact, the threat can be considered real only if a sufficiently large group of people recognizes it as such, therefore only if the logic of risk is internalized by the population itself (Adams, 2003, pp. 87-103).

In contemporary societies, security measures are not adapted to the level of risk; on the – turning protection into one of the greatest risks (Esposito, 2002, p. 19).

The government's risk-based strategy facilitates the consolidation of measures introduced in response to the state of emergency. These provisions are designed and presented as evidence of altruism and civic participation, based on a precise narrative framework and the internalisation of the threat itself by the population. In this sense, the obligations imposed on the global population during the SARS-CoV-2 emergency have translated the *right to health*, health safety, into a legal *obligation to health*, bio-security (Zylberman, 2013; Agamben, 2020, par. 2). In this sense, the concept of immunity becomes the protective response to a risk – specifically, in this case, a biological agent – which establishes an individual, collective or systemic body, activating a boundary between *internal* and *external*, the *individual* and the *common*. Today's interconnection and uninterrupted socio-political-economic chain that binds together the global order makes the contagion – the alteration, transformation and corruption of a body by means of a risk – uncontrolled (Esposito, 2002).

In this order the concept of "security" acts as a narrative machine capable of shaping individual and collective spaces and behaviors that, by influencing each other, produce contemporary social space, seen today as a result of the mutual confluence of biopolitics and bio-security (Zylberman, 2013).

In this scenario, health institution takes on a real political power:

Under the stress of crisis, the professional who is believed to be in command can easily presume immunity from the ordinary rules of justice and decency. He who is assigned control over death ceases to be an ordinary human. [...] Because they form a charmed borderland not quite of this world, the time-span and the community space claimed by the medical enterprise are as sacred as their religious and military counterparts (Illich,1975, p. 34).

As Ivan Illich argues, the ability of medical science to "decide on the exception", thus acquiring decision-making power, undoubtedly establishes the logic of risk in favour of that of danger. While the second one is an empirical judgement, based on experience, the first one is a statistical construct detached from an experienced context. The claim of power in emergency suspends hourly evaluations (1975).

Furthermore, the sovereignty capable of creating law regardless of existence of law itself, as emerged in Schmitt's thought (1984), is replaced, in bio-security, by professional category.

Each street is placed under the authority of a syndic, who keeps it under surveillance. [...] On the appointed day, everyone is ordered to stay indoors: it is forbidden to leave on pain of death. [...] Each family will have made its own provisions. [...] If it is absolutely necessary to leave the house, it will be done in turn, avoiding any meeting. Each individual is fixed in his place. And, if he moves, he does so at the risk of his life, contagion or punishment (Foucault, 1995, p. 195).

The surveillance regime described by Michel Foucault – the context is that of a 17th century public order adopted against the plague – is based on the permanent registration of activities and social conditions. The registration of disease is centralized and, as Foucault argues, the relationship of each individual with death or disease "passes through the representatives of power, the registration they make of it, the decisions they take on it" (1995, p. 197). The plague is interpreted by the philosopher as a "political dream" because it is a device capable of implementing clear social divisions and entering with capillary forms of regulation even in the most hidden aspects of everyday life; a perfect disciplinary condition.

The centralization of power and gaze finds its spatial counterpart in the architecture of the Benthamite Panoptic based on visibility and transparency adopted as correction devices. According to Bentham, in fact, the moral improvement of society is linked to the constant surveillance of each citizen. In this sense, the architecture of Panoptic is understood both as a concrete disciplinary institution, "the old simple schema of confinement and enclosure – thick walls, a heavy gate that prevents entering or leaving – began to be replaced by the calculation of openings, of filled and empty spaces, passages and transparencies" (Foucault, 1995, p. 172), as well as ideological, "the theme of the Panoptic – at once surveillance and observation, security and knowledge, individualization and totalization, isolation and transparency – found in the prison its privileged locus of realization [...] It was the most direct way of expressing the intelligence of discipline in stone" (Foucault, 1995, p. 249).

The Panoptic constitutes a principle of centralized political order, organized in function of a one-way view that dominates and watches from above aimed at guiding human behavior; this logic is applicable to all public buildings, and, as demonstrated previously, also to the entire city (Foucault, 1995, p. 206).

The control paradigm described by Foucault, although dated and culturally distant from contemporary times, describes a spatial and *permanent recording* regime very similar to the system used in the management of SARS-CoV-2 health emergency. However, dynamics linked to the internalization of the emergency, the reaction of the population to specific narrative frameworks and the introduction of measures that, in different forms, limit the freedom or privacy of individuals, are also linked to a new socio-political order.

To introduce the contemporary social paradigm the authors adopt the Orwellian spot *1984* for the launch of the first Macintosh personal computer. In the sequences of the commercial, shown in Figure 1, there is a transition from a situation of alienation to a turning point, announced by the advent of the new digital media that sanctions the passage of power from *few* to *many*: each individual becomes the master of his own actions, his own time and his own gain.

The process is further consolidated with the advent of Internet, which erodes power, moving it from the centre to the periphery, from institutions to ordinary people, giving them back the possibility of managing their own lives (Dyson as cited in Pariser, 2012, p. 52). The new *system of individualization* (Zuboff, 2019, p. 43) promotes direct experience thanks to network, apps and social networks, devices through which everyone can act and access multiple services according to his will and desires (Pariser, 2012, p. 53).

Figure 1. Collage made by the authors with the video sequences of the Macintosh commercial "1984". Alienated workers march controlled by cameras while a young woman in sportswear, chased by guards, runs towards a big screen on which a man - explicit reference to Orwellian Big Brother – says: "Today we celebrate the first glorious anniversary of the Directives on the Purification of Information. We have created, for the first time in all of history, a paradise of pure ideology where each worker can realize himself safe from destabilizing invasions of contradictory truths and causing confusion. Our Unification of Thoughts is a more powerful weapon than any fleet or army on earth [...]". The heroine destroys the screen by throwing a hammer at it. The commercial ends with the ruling "On January 24, Apple will introduce Macintosh. And you'll understand why 1984 won't be like 1984"

Within this order, the prospective Panoptic becomes the a-prospective Panoptic (Han, 2017, p. 93), within which groups of technologized individuals manage their public and private time supported by digital devices. These, on the one hand, facilitate the multiple daily actions, on the other, make the different profiles protocollable and monitorable both by media owners and by other profiles (Pariser, 2012, p. 53), storing information within *big data* systems:

highly efficient psychopolitical instrument that makes it possible to achieve comprehensive knowledge of the dynamics of social communication. This knowledge is knowledge for the sake of domination and control (Herrschaftswissen): it facilitates intervention in the psyche and enables influence to take place on a pre-reflexive level. [...] Big Data is making it possible to predict human behaviour (Han, 2017, p. 30).

It is the same inhabitants of the a-prospective digital Panoptic who structure it and keep it voluntarily active according to the comfort, the economic and personal profit and the security it ensures.

Unlike the Benthamite Panoptic, control is therefore achieved not by external imposition but rather by the self-produced need of each individual who finds in the digital dimension a space through which to express and represent himself (Han, 2017, pp. 101-102) .

The digitalization of society also emerges at a participatory and representative political level with *digital citizenship*, a new form of social participation that allows citizens, thanks to a series of tools and services, to exercise their rights institutionally (Italian Republic, 2017).

Digitization is therefore today a typical condition of *states of normality* and of *emergency*; in both, the State acts to optimize the management of the social space in order to make both ordinary and exceptional responses more effective.

SECURITY NARRATIVES AT THE BASE OF DIGITAL SYSTEMS IN TIME OF SARS-CoV-2

The authors' aim is to investigate, in the context of SARS-Cov-2 emergency, the process through which specific digital systems have been introduced in the common use of heterogeneous populations. Specifically, the authors argue that a precise narrative framework, related to the state of crisis, facilitates the adoption of devices and measures that, while limiting freedom and privacy of individuals, become indispensable tools for the population itself to counter the threat.

In order to manage SARS-Cov-2 emergency, digital technologies for monitoring of the biological agent and, consequently, of the population have been employed all over the world and in different forms – as will be analyzed below referring to the extremely heterogeneous panorama of apps currently in use globally. Today's bio-security regime has resulted in digital systems being introduced as devices for the socio-spatial organization of a new normality – opening, in some cases, more possibilities to monitoring devices that are still experimental. The use of digital technologies is not limited to virus monitoring, but plays a central role in the management of multiple spheres of life, both public and private; it is emblematic how digital platforms soon became essential devices in response to social distancing, work and distance education, as well as telemedicine.

In the regime of bio-security, *health* and *privacy* can potentially enter into conflict because the State adopts nationwide monitoring systems that, through a targeted narrative policy, are assumed by the population in its daily use as they are considered indispensable. Although several tracking apps do not act in a more invasive way than other platforms daily used by the majority of population (such as social networks, Google or Amazon), they introduce into people's consciousness the paradigm that "healthcare" cannot fully coincide with "privacy".

The digital technologies used to combat states of crisis and the consequent implications for the privacy of those who use them voluntarily or compulsorily, are conditions that emerge frequently, however, beyond the specific context of the health emergency in question.

In *prolonged emergency regimes*, such as the migration emergency in Europe, several digital platforms have been developed to counteract illegal arrivals of third parties, some of which are potentially detrimental to the privacy of users. Although this is a different field of investigation than the health sector investigated in the chapter, the authors believe it is appropriate to focus on it for several reasons, ranging from the considerable number of heterogeneous digital systems used to combat states of crisis to the policies adopted along the borders – an issue that returns with emphasis in today's national policies in an attempt to contain the number of infections – to the ethical implications of systems that are still experimental today.

Some of the platforms used in border control management, such as the *ETIAS* system, have officially entered into operation; others, as in the case of *iBorderCtrl* project, have instead remained experimental because they are invasive of privacy and data protection. *iBorderCtrl* is an experimental system, funded from 2016 to 2019 by European Union's *Horizon 2020* research and innovation programme and active for about nine months along parts of the Schengen external borders of Hungary, Serbia, Latvia (European Union [EU] – iBorderCtrl, par. 1). The main objectives of the system, in order to deal effectively with the migration emergency, are to improve the security and reliability of controls and speed up the crossing of the external borders of the Schengen area (EU – iBorderCtrl, Objectives par. 3). Within this narrative framework, focused in particular on the latter aspect, *iBorderCtrl* introduces a pre-registration phase for third country nationals who intend to cross the Schengen border, during which, in one of the various steps, the user is asked to undergo an interview from home. Filmed from the cameras of his webcam, the user must answer questions asked from the *avatar* of a border agent. As shown in Figure 2, during the video interview, biometric systems are used in order to determine whether the user is lying or telling the truth. This is possible by detecting facial microexpressions, i.e. non-verbal behaviours.

Figure 2. Collage made by the authors with the video sequences of "La macchina della verità" (Ludovica Jona, 2019). The images taken during the first interview with the iBorderCtrl avatar are used to create the user's biometric model and detect, through color codes, whether the user is telling the true or false

Through biometric technology, integrated with the cross-checking of multiple and heterogeneous information contained in digital servers, a *risk factor* is attributed to the user on the basis of which it is established whether the third country citizen should be subjected, once at the border, to further controls by the "real" border police (EU – iBorderCtrl, The two phase procedure). However, the introduction of the project as a stable control system conflicts with fundamental rights on the processing of personal data and privacy (EU – iBorderCtrl, Scenarios Results).

The studies on privacy and protection of fundamental rights on the processing of personal data, in the case of iBorderCtrl carried out by the Institute of Legal Informatics (IRI) of the University of Hanover in 2017, highlight the need for reflection on the application of some systems, personal data collectors, considering in the evaluation the tendency to use data already present for new purposes. In fact, once established, a database can be consulted, for example, for security or health reasons, regardless of the reason for which the data were initially collected (Stoklas, 2017).

If, therefore, in sudden and global states of emergency, such as the pandemic, "immediate" population monitoring platforms are adopted on a national scale, the authors argue that digital securitarian experimental systems, currently considered invasive, can be more easily introduced in the near future depending on greater openness on the part of governments and the population for increasingly connected, complex and global securitarian reasons.

The topicality of the SARS-CoV-2 emergency makes it difficult to define the precise long-term consequences that the introduction and consolidation of digital platforms and tracking systems bring with them. However, the authors consider the issue to be central because the global scale of pandemic has made a biological agent the common denominator against which to mobilise globally. This condition, albeit in different forms and degrees and in heterogeneous socio-political contexts, has accelerated the introduction and consolidation of technological devices. In this sense, the technological and digital dimension has assumed a scientific value, making digital platforms real "medical" forms of virus prevention.

Dozens of tracking, monitoring and telemedicine apps have been promoted and introduced since the beginning of the pandemic by government institutions or private companies to counter the spread of contagion and ensure a new socio-spatial organization. From the survey conducted by Sharma and Bashir, published in March 2020, it emerges that, on a sample of fifty platforms analyzed, the main features concern the formulation of real-time maps of confirmed cases and the sending of location alerts to users, the monitoring, by the competent authorities, of those who are in domestic isolation or quarantine, the possibility to send reports to health authorities even self-disclosed symptoms; more generally, these apps spread news about the pandemic and education to it (2020, par. 3). To these features are added both the self-assessment of the daily physiological status of the *user-patient* and the monitoring of his vital parameters and the possibility of virtual medical consultations – as in the case of the Italian platform of *connected health* or *digital health* ADiLife Covid-19 that connects patients under active surveillance or fiduciary isolation with medical staff to manage home care according to institutional directives (ADiLife – Covid-19). On the sample of fifty apps, thirty require access to contacts, photos, location, camera, microphone, wi-fi and only sixteen explicitly refer to data anonymity (Sharma & Bashir, 2020, par. 4, 5). With specific reference to the theme of privacy, the current tracking apps therefore define an extremely varied landscape due to cultural, technological, political instances, making it clear that certain platforms are not applicable in different countries (Cho, Ippolito, Yu, 2020, par. 3); some of these apps work with GPS, others with Bluetooth and through different degrees of centralization of the system, thus defining extremely heterogeneous possibilities of protection or invasion of privacy (Hart et al, 2020); finally, the same institutional authorities have adopted different pandemic management and containment policies, more or less damaging to the privacy of citizens and users. In the case of Israel, for example, the digital epidemiological investigation has been entrusted to the Internal Security Agency ISA, normally employed in terrorist espionage (Amit et al, 2020, par. 2), approving a law to trace the cell phone data of persons with suspected infection; in South Korea, the government has set up a public database of known patients in which data on age, gender, occupation and travel itineraries are provided for each of them; in Taiwan, medical institutions have access to the data of those who travel or are in quarantine; In Singapore, government institutions have introduced the *TraceTogether* app that, through Bluetooth, associates each user with a token that changes every thirty minutes or so and is anonymously recorded both by other mobile devices with which the citizen comes into close contact and by the government itself - only in the event of an established infection or potential infection does the government access all the user's data (Cho, Ippolito, Yu, 2020, par. 2).

METHODOLOGY AND INTRODUCTION OF CASE STUDIES

In order to read, in the context of SARS-CoV-2 pandemic, the current orientation of digital tracking and monitoring systems, three case studies, belonging to different socio-political contexts, are analyzed: the governmental tool *Health Code* created specifically by the Chinese government to be included in existing applications for mobile devices, *Alipay* and *WeChat*; the South Korean governmental app *Corona100*, active since February; the Italian governmental app *Immuni*, currently active in experimental phase only in four regions. With specific reference to the latter case study, further institutional arrangements, launched on different digital platforms in order to monitor the national health and social situation, will also be analyzed.

Since, as mentioned above, this is a health emergency still in progress, the analysis of the case studies will proceed through a comparison between the current provisions on contagion containment (government apps) and the more general situation of digitization, also in relation to specific systems, of each of the socio-political contexts analyzed. The comparison between the previous digital systems, already adopted or experimental, with those recently introduced to combat the health emergency allows to identify the degree of proportionality between the responses introduced in an emergency regime and those adopted under normal conditions. For the Chinese context, a comparison will be structured between the governmental tool *Health Code* and the experimental project *Social Credit System* (SCS) – designed in 2007 and which should officially enter into force throughout Chinese mainland in 2020. For the South Korean case-study, reference will be made to the highly technologized urban and social context that has largely favoured the diffusion and use not only of the government app but also of other private apps also aimed at tracking and monitoring contagions. Finally, for the Italian case-study, some public urban monitoring systems already activated in recent years on the territory and their implementation will be considered in order to better respond to the needs related to the health emergency. The unprecedented introduction of the contact tracing system will also be related to the development of more precise apps for potential future epidemics.

The narrative policies, adopted by the institutions in order to promote the use of various apps by population, will also be investigated.

The analysis is conducted from a diverse array of sources including government protocols and regulations, official documentation, speculative investigations, recent articles and surveys and social communication. The heterogeneity of the documents taken into consideration is, according to the authors, functional to return the different languages used by institutions in their official communication.

The analysis shows that there is a certain degree of proportionality between the tracking and monitoring systems introduced during the SARS-CoV-2 emergency and those "prior" to the outbreak of the pandemic, although the sudden introduction of these systems has been the subject of much criticism in the socio-political democratic realities of Europe.

The analysis aims to highlight how different government institutions, in the emergency context, adopt digital security systems that, while not drastically changing directions already taken, represent the acceleration of decision-making processes, the introduction and activation of specific new technological models. This aspect highlights the fact that the global scale of current emergency and consequent diffusion of monitoring systems at opens up, at the same time, many further questions linked to the possible subsequent developments of a *global security process* that has been going on for a long time.

The Chinese Case

As previously mentioned, the government tool *Health Code*, designed by the Chinese government to counter the development of possible outbreaks of SARS-CoV-2 and to monitor infected, potentially infected and healthy people, is an extension of the main Chinese mobile apps, *WeChat* and *Alipay* (Zunino, 2020, par. 10). The *Health Code* tool attributes a daily electronic health code to each citizen, depending on his travel history, the time spent on public transportations and places frequented and the more or less prolonged contacts individual has with other people. This, in three colours (green, yellow, red) and in the form of a QR-code, must be shown to the authorities from time to time (Ranucci, 2020, par. 35). As shown in Figure 3, every time one goes up and down a public vehicle or crosses or enters a space, the electronic code must be scanned so as to inform the authorities of every movement. The green code allows movements; the yellow code and the red code impose quarantine: the first one according to the prolonged exposure to potential virus carriers, the second one if the person is infected (Ranucci, 2020, par. 10).

Table 1. Categories of analysis of case studies investigated

Privacy	Sources	Digitalization	Socio-political context
• Impact of SARS-CoV-2 digital pandemic control systems on the collection of personal data by government and private institutions.	• Availability of documents and data at the governmental and institutional level.*	• Level of digitization of the population	• Modes of introduction of digital systems in contrast to the SARS-CoV-2 pandemic compared to the socio-political context.
• Analogies between pre-existing governmental digital systems and new platforms introduced in response to the SARS-CoV-2 emergency.	• Availability of documents and data from agencies and private companies.	• Pre-existing digital systems projects at institutional level	• Formulas and narrative policies used to introduce digital systems to contrast the SARS-CoV-2 pandemic.
		• Impact of tracking and pandemic control apps compared to already developed digital systems.	• Impact that, in each socio-political contexts, the narrative policies adopted to introduce digital systems aimed at counteracting the SARS-CoV-2 emergency have on the population and the use of the apps.

Notes
* In case governmental sources are not available, institutional sources referring to similar already active digital systems will be examined.

Figure 3. Collage made by the authors with the video sequences of "Wuhan open city?" (Sigfrido Ranucci, 2020). By scanning the QR-code the user makes all movements traceable and, consequently, the authorities are informed of user's movements

The use of the tool is mandatory, therefore the decision to implement, depending on the emergency, *WeChat* and *Alipay* applications has made the introduction of the contrast system immediate – since these are extremely widespread among the population and essential to carry out transitions and daily actions, digital and not. *WeChat*, together with the integrated payment system *WeChat Pay*, makes it possible to communicate, publish content, make purchases, rentals and bookings; *Alipay* is a payment system of Alibaba group (Zunino, 2020, par. 10). The operating system of two apps is also integrated with the management systems of the main Chinese smart cities and with that of *Gaode Maps*, the Chinese equivalent of *Google Maps* (Zunino, 2020, par. 10). The monitoring of the population was therefore capillary from the start.

Integrated systems of digital platforms and apps for mobile devices are at the basis of the Chinese socio-spatial organization both in states of normality, the case of smart cities, and in states of emergency, such as today's SARS-CoV-2. The *interoperability* between different platforms is, more generally, particularly evident in the governmental plans promoted by Chinese institutions to optimize the social organization and economic competitiveness of the country, as shown, for example, in the experimental *Social Credit System* (SCS) project launched in 2007. Since then, it is constantly improved by the authorities, in collaboration with private companies, and it will officially enter into force in 2020. Analyzing the SCS is functional to the understanding of interoperability between digital platforms mentioned above and to show the similarities between the mechanisms of monitoring systems both in crisis and normality scenarios.

The SCS is a tool for sharing existing public and private big data, collected for other purposes, and reused by the Chinese government to assess the reliability of citizens and companies. With the aim of standardizing individual and collective behaviors and optimizing economic, time and resources, the system promotes precise behavioral patterns and gives social credit scores, which can be viewed online in *blacklist* or *red list*, depending on on commercial, judicial, social reputation (State Council of the People's Republic of China, 2014). Figure 4 shows how, in the context of a train journey, the information message refers to the individual social credit score to deter possible offenders from committing offences on board the wagons.

Figure 4. Collage made by the authors with video sequences published by James O'Malley on his Twitter profile on October 29th 2018 during a train trip. "Dear Passengers / People who travel without a ticket / behave disorderly, or smoke in public areas / will be punished according to regulations / and the behaviour will be recorded in individual credit information system / To avoid a negative record of personal credit / please follow the relevant regulations / and help with the orders on the train and at the station."

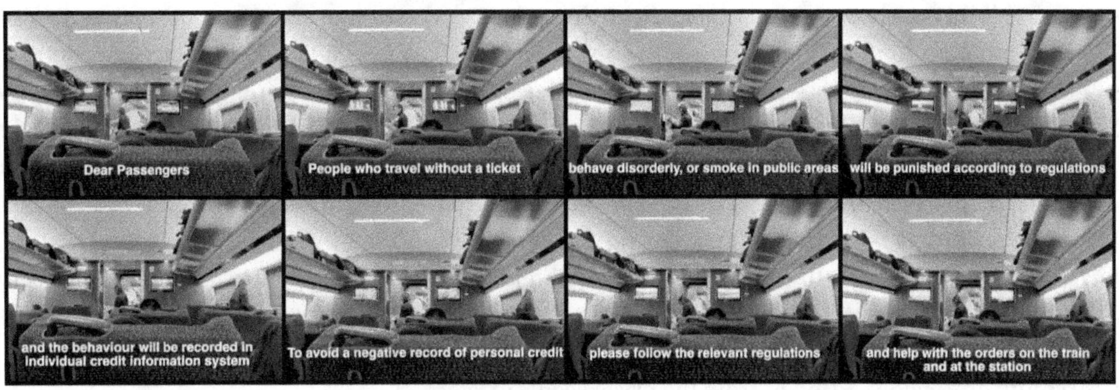

Although the government protocols of 2007 and 2014, containing the project guidelines, are sometimes vague and not functional to an exact operational understanding of the system – highlighting an *uncertainty* and *linguistic ambiguity* attributable to a greater extent to contemporary Chinese law itself (Cao, 2012, p. 15) – the project is currently operational in the credit pilot cities Suzhou and Fuzhou. According to the definition expressed during the "Credit Cities Construction Summit", organized in 2017 by the National Development and Reform Commission (NDRC), in *credit cities* the local government and several technology companies share data with each other to determine the level of reliability of companies and individuals (Ahmed, 2019, p. 56). These cities are structured through two different levels of collaboration: that between the national government and technology companies, for the construction of online platforms to monitor credit data, and that between municipalities and technology companies, for the creation of local rating systems (Ahmed, 2019, p. 57). In the cities of Suzhou and Fuzhou these collaborations take shape in the *Xinyi+* project in which the NDRC collaborates to share data with big Chinese companies, each of which is dominant in its respective sector: Ant Financial (of the Alibaba group), Didi Chuxing and Ctrip.

The collaboration, in the mutual sharing of data, between the Chinese government and the Alibaba financial group appears to be solid both at local level, in the context of credit cities, and at national level, in the context of the digital *Health Code* platform to combat the health emergency. The collection and use of big data, functional to the social organization, tracking and monitoring of the population, appear paradigmatic both in crisis and normality scenarios. In addition, as is the case in credit cities, the contagion containment system also includes rewards. In the pilot cities, in fact, the level of credit, in addition to influencing each other between companies and between citizens, may rise as a function by reporting system transgressors to the authorities (Ahmed, 2019, p. 58). In the same way, in the context of SARS-CoV-2 emergency, in order to encourage both general compliance with the rules and reports of possible infected persons, transgressors and violations of directives, the government has provided for rewards; *WeChat* already "normally" has a reporting section (Zunino, 2020, par. 14). Again, a relationship of analogy emerges between different systems, prepared or in conditions of normality or emergency, which concerns, in both contexts, the consolidation of the systems themselves in individual and collective habits.

The introduction of governmental tool *Health Code* is part of a socio-political and cultural context already strongly digitalized and centralized, at state level, in function of a widespread and capillary monitoring of the population operated not only by the institutions but also, reciprocally, between citizens. Total accessibility to data, information and people emerges as a structuring aspect of SCS and *Health Code*, contributing to define Chinese society as a *society of transparency*. This society consolidates itself according to states of crisis, risk and threat and uses the digital space to regulate the physical one, determining the accessibility to specific spaces in contexts of emergency and normality. This society acts on the population and the latter contributes to consolidate it according to the narrative policy that institutions themselves disseminate. In fact, in society of transparency decision-making processes are speeded up according to the *logic of increasing risk* (Davino & Villani, 2020).

Health Code, coming into force, has formalized and therefore anticipated many of the provisions of the *Social Credit System*. This aspect helps to argue how serious states of crisis can facilitate the introduction of digital systems that were initially only examined on an experimental basis in *situations of normality* profoundly different from the crisis scenario.

The South Korean Case

As in China, the South Korean government has developed its own tracking system for SARS-CoV-2 contagion containment, *Corona100* app, officially launched on February 11th 2020. However, unlike the Chinese app, the download and use of the South Korean app is on a voluntary basis. By cross-referencing geolocation data of users and video-surveillance cameras with government data on the national health situation, the system communicates the physical distance of users from verified Covid-19 patients and sends a notification each time the former enter within 100 meters of a potentially infected or sick person (Zunino, 2020, par. 6). The app is the tool through which communications with health authorities take place and through which both infected and potentially infected are traced. The system sends additional notification messages, also intended as part of health service, to describe in detail movements and behaviour of those who are infected. In this way, forms of social stigma and uneasiness are potentially induced in infected people and in those who could be recognized in the described behaviours (Zunino, 2020, par. 7). In this way, the system acts, to counteract the emergency, both spatially, by geolocalizing the users, and socially, by leveraging the moral and social reputation.

Thanks to highly developed technological infrastructures, Internet has been widely used in South Korea for civic actions and political campaigns since the 1990s, becoming a consolidated tool of action for institutions, traditionally understood, and for social movements and activism – the case, for example, of *net-izen movement* spread in the early 2000s (Lee, 2013, pp. 346-348). In the context of a highly technologized society, such as South Korea, several digital tracking systems for SARS-CoV-2 contagion containment were designed in the first weeks of February by private app developers – the best known are *CoronaMap* and *Corona Doctor*, as well as *Corona100*. The latter has been spreading rapidly among the population – 1 million installations in less than three weeks since launch with "20,000 installations per hour" (Bae as cited in Watson & Jeong, 2020, par. 3, 4) – and optimal functioning, also thanks to the integration with the *physical* health monitoring system based on swabs.

The South Korean system has become a worldwide reference model, in the containment of the epidemic, also for different socio-political-cultural contexts – recurring the formula "South Korea model" in the Italian narrative policy, for example. This assumption was also reiterated in the speech of the President of the Republic of Korea, Jae-in Moon, on May 10th 2020, for the three years of his mandate: "we have already become a leading country in the world in the prevention and control of epidemics [...], Korea has become a global standard. The status of the Republic of Korea and the pride of the people are higher than ever" (Moon, 2020). In addition to leveraging national pride, the narrative policy of the Republic of Korea, adopted to frame the pandemic situation, largely relates the virus to the war context: "we won a war against the virus", "the government will meticulously manage the Covid-19 situation, bearing in mind that we are in a prolonged state of war", "this is undeniably an economic situation in wartime" (Moon, 2020). This interpretative framework, as mentioned in the previous paragraphs, consolidates in the collective consciousness of the *audience-population* the use of digital tracking systems that reorganize the physical social space.

The Italian Case

Digital platforms, as previously argued by the authors, have assumed a central role on a global scale in response to SARS-CoV-2 emergency. The following will examine the case of Italy as a democratic State strongly affected by the pandemic, and, for the authors, representative of how digital technologies have been used, in democratic contexts, in the management of emergency state. In this sense, in addition to the introduction of new tracking and monitoring devices, platforms already active in the territory have been implemented to respond to health needs, as in the case of SUS (Unique Reporting System) platform of the Municipality of Rome, activated in 2005 in order to collect suggestions, complaints and reports from citizens (Municipality of Rome, 2020, par.1). With the emergency and the ministerial provision relating to social distancing, the platform of the Italian capital was implemented to allow citizens to report any groups. The initiative, which was followed by numerous protests by citizens for "incitement to hatred", was announced to the population through a post on Facebook profile of "Roma Capitale" through a post, displayed in Figure 5:

Are there groups of people that you believe to be in conflict with the rules of the health emergency? You can report them directly to the competent Authority with SUS (Unique Reporting System) active on the institutional portal of Roma Capitale. It's simple, just follow the instructions (Roma Capitale, 2020, March 26).

Figure 5. Post from Facebook profile "Roma Capitale" on the bulletin board. In the post are indicated all the steps to follow to make a report on SUS platform

Other Italian municipalities, even minor ones, have made the same request on their social profiles, inviting citizens to take charge of public order through the virtual channels available to them. Although these public invitations have also been followed by numerous protests, there have been a considerable number, throughout Italy, of reports of groups photographed and sent to institutions through the social channels of the municipalities themselves – despite the fact that this violates the privacy rights of the individuals filmed.

These requests are part of a narrative framework that sees each citizen involved, as the authors previously articulated, in the management of public order and emergency. It is within this narrative order that the *Immuni* tracking app is included; the app is a unique national platform for the management of the alert system of those who install, on a voluntary basis, the application on mobile devices (Italian Republic, 2020, Art.6). The app, established by the Italian Government in collaboration with the Ministry of Health and the Ministry for Technological Innovation and Digitization, uses only public infrastructures located within national borders (Presidency of the Council of Ministers [PCM], 2020, par. 1); moreover, the operating methods of the system are complementary, as stated in the decree, to those provided by the National Health Service to deal with the emergency (Italian Republic, 2020, Art.6).

The app uses the Bluetooth Low Energy system that allows to detect contacts within a few meters without tracking the position of the user.

The IDs of each user, which vary daily, register each other when they make contact locally. Subsequently, if one of the two users turns out to be positive for the virus, based on a risk assessment, which varies depending on the duration of exposure and the distance between the two devices, the other is notified of the possibility of infection. In addition, to ensure that only users who tested positive for SARS-CoV-2 upload their keys to the server, the upload procedure can only be finalized with the collaboration of an authenticated health care professional. However, it should be noted that distance estimation is subject to errors. In fact, attenuation of a low energy Bluetooth signal depends on factors such as the orientation of the two devices – one relative to the other – and the obstacles (including human bodies) in between (PCM, 2020, How it works). For the development and design of the national platform, institutions collaborated with Bending Spoons, the Milanese company in charge of its implementation, pursuing the objective of maximum accessibility of the app to the largest number of users – as a public platform (PCM, 2020, Principles par. 2). However, one of the criticalities expressed by the creators concerns the fact that mobile device must have specific system requirements: the app uses technology for exposure notifications provided by Google and Apple and is not compatible with some previous versions of iOS, Android and Google Play Services (Presidency of the Council of Ministers [PCM] – Immuni, 2020, Questions).

Moreover, one of the fundamental principles on which *Immuni* app was built is privacy in accordance with Articles 13 and 14 of Regulation (EU) 2016/679, on the free movement of personal data and the processing of such data, in order to protect those involved. Therefore, before the activation of the application "users must receive clear and transparent informations in order to achieve full awareness, in particular on the purposes and processing operations, pseudonymisation techniques used and data retention times". In addition, pursuant to art. 25 of Regulation (EU) 2016/679 "the personal data collected by the application are only those necessary to alert users of the application to be among the close contacts of other users found positive to Covid-19" (Italian Republic, 2020, Art.6). Therefore, the application does not collect any personal data that could detect the user's identity such as name, address, e-mail address, telephone number or age, nor does it use any geolocation data – the app is not able to say where contact with a potentially infected user took place (PCM, Executive summary par. 4.5). However, the geolocation service must be enabled on Android smartphones that incorporate GPS and Bluetooth to track Bluetooth signals and save the ID codes detected nearby (PCM – Immuni, 2020, Questions). Although the app is not authorized to access the geolocation services, its use involves the constant activation of GPS with consequent data collection, by Google, on the user's position.

In order to protect privacy, moreover, the duration of the exposure is measured in five-minute increments and limited to thirty minutes so that even in function of always different identification code, *Immuni*, while establishing the risk of contagion, is not able to determine whether exposures occurring on different days may have involved the same user and for how long (PCM, Epidemiological information par. 3). All data, for which the Ministry of Health will be responsible only for the containment of epidemic, must be eliminated, by ministerial order, when they are no longer necessary and in any case no later than December 31th 2020 (PCM, 2020, Privacy, par. 8). The design of the app is therefore based on the confidentiality and accuracy of the information collected on individual users.

To the substantial technological limitations, as stated by the institution itself, are added the already existing gaps in the *physical* health system – lack of tests and swabs. Within this context it is evident, and it is stated several times in the official documentation, that the app is a technological system to be implemented considerably and that to do so it is necessary to collect more information (PCM, 2020, How it works par. 5). In this regard, the development of tracking system, as seen above, largely examines the protection of user data in order to encourage greater public acceptance and, consequently, greater adoption of the app – an indispensable condition for its improvement (PCM, 2020, Privacy par. 1). The various national institutions promote a narrative that leverages the civic conscience of each citizen who, using the app, supports the National Health Service in the collection of data. The latter, in fact, thanks to the collection of data offered by citizens, can "make better decisions" to maximize the effectiveness of the app and, consequently, patient care (PCM, 2020, Operational Informations par. 4). In this sense, on Immune app site, there are formulas such as "help yourself, your family and your country" or "let's protect ourselves and our loved ones. [...] In this way, they help to contain the epidemic and to promote a rapid return to normality" – as shown by screenshots of the app in Figure 6. Through these narrative pictures, *Immuni* app becomes a real device through which each citizen can demonstrate civic awareness and a strong bond with relatives.

Figure 6. Immuni app infographic. The third screen reads "Let's slow down the epidemic together"; the fourth is dedicated to data protection, "Your privacy is protected"

In parallel, the war framing adopted makes the app indispensable in the fight against epidemics (PCM, 2020, Vision and goals par. 2) such as Covid-19, a threat to health and economies on a global scale (PCM, 2020, Executive Summary par. 1). The use of the word "epidemic" in the plural, adopted in the official documentation, highlights how the development of the technological system is based on logic of risk and worst-case scenario previously articulated by the authors. In the same documentation, in fact, it is made explicit that many experts agree that, in the future, there will be the concrete possibility of new epidemics potentially more dangerous than the one "we are "fighting" – *we are* as a collective formula, *fighting*, as a war formula. In this context it is stated that the contribution of technological innovation can be decisive (PCM, 2020, Introduction par. 2, 3).

Immuni is designed to address the current crisis, but the vision behind it is for the tools that are being developed to make us all better prepared in addressing similar threats that may arise in the future (PCM, 2020, Vision and goals par. 4).

Table 2. Comparison of case studies investigated. With "privacy guarantee" we refer to the one concerning the degree of accessibility, through the platforms, to the personal information of the users.

	Government regime	**Launch data**	**Use**	**Operating system**	**Guarantee of privacy from contacts**		**Guarantee of privacy government from authorities***		**Feedback**
					Exposed users	Diagnosed users	Exposed users	Diagnosed users	
Health Code	One-party socialist republic	?.02.2020	Mandatory	GPS. Centralized operating system	Yes	Yes	No	No. Personal information revealed.	Download is not necessary (the app is an extention of the most used app).
Corona100	Unitary presidential constitutional republic	11.02.2020	Voluntary	GPS. Centralized operating system	Yes	Partial. Open government database with some personal information on infected persons.‡	No	No. Personal information revealed.	1 mln downloads in 2 weeks.
Immuni	Unitary parliamentary constitutional republic	15.06.2020	Voluntary	Bluethooth Low Energy. Decentralised privacy-preserving proximity tracing.	Yes	Yes	Yes	No. Personal information, all tokens and all contact tokens revealed.	2,2 mln downloads in 10 days.†

Notes

* However, the companies that create the apps collect anonymous information and only the health autorities collect and store personal information.

‡ Identification number with date of positive report, age, gender, citizenship, travel made, places visited, persons with whom you have come into contact, profession.

† Downloads stopped at 4,6 mln (05.08.2020) – requires at least 36 mln for optimal operation.

CONCLUSION

The three countries to which the case studies refer are characterized by a high number of infections, especially in the first phase of epidemic: China is the first country that has seen the virus spread in its territory; South Korea is the Asian country, after China, with the highest number of infections; Italy, finally, is one of the most affected nations in the world.

Although the three contexts are different from a socio-political-cultural point of view, the authors have decided to investigate them together as representative of the different responses adopted to fight SARS-CoV-2 epidemic. In fact, the three different digital systems adopted show three different forms of tracking.

In the People's Republic of China there is such a sense of collectivity that the concept of privacy cannot be contemplated in Chinese thought, as it is understood in Western thought (Han, 2020, par. 6, 7). This condition is, moreover, contextualized and generated by a totalitarian governmental regime, of which the modes of use of apps and digital technologies described are representative. Through the analysis of Chinese case, in the context of SARS-CoV-2 emergency, it has been demonstrated how the pandemic has institutionalized and activated, in a very short time, total and mandatory surveillance devices similar to some not yet active ones such as *Social Credit System*. In this sense, unlike other socio-political contexts, no narrative policies have been adopted to consolidate the choices, provisions and devices introduced by the government in the population. The authors, in fact, starting from testimonies of people who have lived in China, have noted the absence, for cultural reasons, of a critical and oppositional political discourse to the government and its dispositions. In the Chinese socio-political and cultural context, the governmental digital technologies act as devices of moral correction which see employed forms of surveillance ascribable to the Benthamite perspective Panoptic.

South Korea promotes, as seen before, the use of digital technologies different from those of the Chinese regime. The Korean Republic, in addition to the war metaphor, introduces the new technological devices as platforms which are born with the emergency but which are part of an already highly digitalized system; according to this order, South Korea assumes the role of guide and model of world reference in the fight against Covid-19. It is emblematic that in the highly digitized social context, to cope with the SARS-CoV-2 emergency, private digital technology agencies have been employed to develop tracking systems, using data provided by the government and self-financing or being financed by the users themselves – this is the case of *Corona100* app which was only later adopted by the government keeping its use voluntary.

Ultimately, the app adopted by the Italian government was considered emblematic by the authors to highlight the more general policies on privacy, personal data processing and transparency that characterize the democratic countries of the European Union. In a scarcely digitized country like Italy, *Immuni* app, as mentioned above, is active in an experimental phase, from 1st to 15th June, in only four Italian regions. Moreover, in order to introduce the future contact tracking system on a national scale, specific narrative formulas have been used which, although declaring the need to collect information and data, frame the adoption of the app in a community process of solidarity and unity in the fight against the enemy, the virus.

Finally, the three apps are characterized by different degrees of guarantee of privacy – by contacts and government authorities – regarding the personal information of users exposed to the virus and those infected. Summarizing what is reported in Table 2, it emerges that privacy is more guaranteed by contacts and almost completely denied by government authorities when you are infected or exposed to the virus; moreover, it emerges that the Italian app is, among the three, the one that guarantees the highest levels of privacy.

The development of increasingly complex digital systems, carried out and funded by States around the world, is related to the complexity of contemporary threats that involve, on a global scale, multiple contexts – in addition to health. In this sense, the documentation relating to tracking apps states that these devices, if adopted by a significant number of people, are indispensable tools to deal with the economic and social crisis, initiated or exacerbated by the health crisis.

Increasingly complex emergencies have made the crisis management mechanism *transparent*. As seen through the analysis of the case studies, transparency can be used in different forms and degrees: total and compulsory, as in the Chinese system; on a voluntary basis in a context already based on digital comfort, such as South Korea; and finally as pseudonymized transparency, which nevertheless introduces a new normality within a poorly digitalized system such as the Italian one.

In this sense, the health emergency has led to a reorganization on a global scale of daily life on the basis of a potential state of crisis extremely prolonged in time that, introducing itself as a new category of normality based on bio-security, has made technological devices, more or less invasive of privacy, new institutional forms.

FUTURE RESEARCH DIRECTIONS

This extremely topical research context has concerned technological systems used to deal with global emergencies that make society increasingly transparent – as articulated previously. The authors believe, however, that a scientific literature that brings together heterogeneous disciplines should be developed. In the analysis of the case studies, in fact, sources dealing with specific aspects of technological systems have been examined, but not considering more general repercussions on society and space.

In investigating these issues it is central, according to the authors, to highlight different values of transparency itself. In fact, the latter, used as a capillary monitoring and surveillance device, consolidates the creation of big data, even anonymous ones, which, on the one hand, make the action of the population protocollable and predictable, and on the other, consolidate speeches and thoughts based on logic of risk. At the same time, however, the concept of transparency can be the basis of a *social pact*, as understood by Rousseau, in which each person and all his powers are united, under the supreme direction of the general will, in a form of association that protects and defends, with all the common force, the person, goods and rights of each member (Rousseau, 1997, p. 24).

In a highly evolving context such as today's pandemic, the concepts of risk, uncertainty and precaution – respectively understood as the possibility, linked to almost always undesired consequences, that such an event may occur, the indeterminateness as an imperfect quality of our knowledge of the world and the set of ethical implications linked to scientific choices – have more or less greater impacts on the social space according to the evaluation processes that concern them and those inherent in their own communication and return to society by the scientific committees involved in their management. The revival of the need for greater transparency also in risk management processes, and therefore the inclusion of scientific uncertainty within the political and public debate (Strand, Oughton, 2009, p. 16), emerges even more clearly with the health emergency in progress. The solutions, both physical and digital, proposed by researchers and professionals involved in the management of the crisis are in fact, at least in Europe, within a public debate extremely conflictual, *a priori*, with them because the degree of scientific uncertainty in which the solutions are framed is always interpreted differently depending on the context of who evaluates – industrial sector, government, health system, cultural institutions. Greater cohesion and cooperation by the different social sectors would make it possible to address the issue of scientific uncertainty in a systemic way, thus facilitating coordinated and as unambiguous as possible assessments on ethical, social and economic issues that are essential in facing an emergency crisis – assessments, for example, regarding the threshold of tolerance of certain risks or the proportionality of damages and benefits generated, as in this specific case, by the adoption of digital systems and platforms of contrast. In this way, the technological solutions adopted, in response to increasingly systemic crises in their manifestations, would be more integrated within the different social sectors, since the latter are collectively assessed against a real inclusion of scientific uncertainty in the social debate. The space, understood by the authors as a product of the actions and thoughts of society, in this scenario would be really shared and common, since, based on the conscious individual conscience, it would open a transparent collective debate.

REFERENCES

Adams, J. (2003). Risk and morality: three framing devices. In R. V. Ericson & A. Doyle (Eds.), *Risk and morality* (pp. 87–103). University of Toronto Press. doi:10.3138/9781442679382-006

ADiLife – Covid 19. (2020). *Covid-19*. https://www.adilife.net/covid-19/

Agamben, G. (2003). *Stato di eccezione*. Bollati Boringhieri.

Agamben, G. (2020). *Biosicurezza e politica*. https://www.quodlibet.it/giorgio-agamben-biosicurezza

Ahmed, S. (2019). Credit Cities and limits of the Social Credit System. In N. D. Wright (Ed.), Artificial intelligence, China, Russia and the Global Order (pp. 55-61). Air University Press.

Amit, M., Kimhi, H., Bader, T., Chen, J., Glassberg, E., & Benov, A. (2020). *Mass-surveillance technologies to fight coronavirus spread: the case of Israel.* doi: 10.1038/s41591-020-0927-z

Beck, U. (2000). *La società del rischio. Verso una seconda modernità*. Carocci. (Original work published 1986).

Bigo, D., & Tsoukala, A. (Eds.). (2008). *Terror, insecurity and liberty. Illiberal practices of liberal regimes after 9/11*. Routledge. doi:10.4324/9780203926765

Cao, D. (2012). Linguistic uncertainty and legal transparency: statutory interpretation in China and Australia. In V. K. Bhatia, C. A. Hafner, L. Miller, & A. Wagner (Eds.), *Transparency, power and control. Perspective on legal communication* (pp. 13–30). Ashgate.

Cassandro, D. (2020). *Siamo in guerra! Il coronavirus e le sue metafore*. https://www.internazionale.it/opinione/daniele-cassandro/2020/03/22/coronavirus-metafore-guerra

Cavalletti, A. (2005). *La città biopolitica. Mitologie della sicurezza*. Mondadori.

Cho, H., Ippolito, D., & Yu, Y. W. (2020). *Contact Tracing Mobile Apps for COVID-19: Privacy Considerations and Related Trade-offs*. https://arxiv.org/pdf/2003.11511.pdf

Clément, G. (2020). *Strategia della paura*. Academic Press.

Comitato Invisibile. (2019). *L'insurrezione che viene – Ai nostri amici – Adesso* (M. Tarì, Trans.). Nero. (Original work published 2007, 2014, 2017).

Davino, C., & Villani, L. (2020). La società del controllo. Lettura di spazi e fatti urbani attraverso il paradigma della trasparenza. In S. Casini, F. Di Blasio, & G. Perletti (Eds.), Trasparenze – Elephant&Castle, 22. https://cav.unibg.it/elephant_castle

Esposito, R. (2002). *Immunitas. Protezione e negazione della vita*. Einaudi.

European Union – iBorderCtrl. (2019). *iBorderCtrl: Intelligence Portable Control System*. https://www.iborderctrl.eu/

Fisher, M. (2018). *Realismo capitalista* (V. Mattioli, Trans.). Nero. (Original work published 2009).

Foucault, M. (1995). *Discipline and punish. The birth of the prison* (A. Sheridan, Trans.). Vintage Books. (Original work published 1975).

Foucault, M. (2005). *Sicurezza, territorio, popolazione. Corso al Collège de France (1977-1978)* (P. Napoli, Trans.). Feltrinelli. (Original work published 2004).

Godard, F. (2018). *Il consenso nell'epoca del terrorismo* (A. L. Carbone, Trans.). Eleuthera. (Original work published 2016).

Han, B. C. (2017). *Psychopolitics. Neoliberalism and new technologies of power* (E. Butler, Trans.). Verso. (Original work published 2014).

Han, B. C. (2020, April 7). La cura al virus è lo Stato di polizia? *Avvenire*.

Harari, Y. N. (2020). *The world after coronavirus*. https://www.ft.com/content/19d90308-6858-11ea-a3c9-1fe6fedcca75

Haraway, D. J. (1991). *Simians, cyborgs and women: the reinvention of nature*. Routledge.

Hart, V., Siddarth, D., Cantrell, B., Tretikov, L., Eckersley, P., Langford, J., Leibrand, S., Kakade, S., Latta, S., Lewis, D., Tessaro, S., & Weyl, G. (2020). *Outpacing the Virus: Digital Response to Containing the Spread of COVID-19 while Mitigating Privacy Risks*. Academic Press.

Illich, I. (1975). *Medical nemesis. The exploration of health*. Pantheon Books.

Italian Army. (2020). *Operazione "Strade Sicure"*. http://www.esercito.difesa.it/operazioni/operazioni_nazionali/Pagine/Operazione-Strade-Sicure.aspx

Jona, L. (2019). *La macchina della verità*. https://www.rai.it/programmi/report/inchieste/La-macchina-della-verita-27b730ca-6997-419d-a74c-deea8237e4e3.html

Lam, A. (2012). Visualizing the terrorist risk in president Bush's War on Terror and Peter Jackson's "The two towers". In V. K. Bathia, C. A. Hafner, L. Miller, & A. Wagner (Eds.), *Transparency, power and control. Perspective on legal communication* (pp. 165–181). Ashgate.

Lee, J. (2013). The netizen movement: a new wave in the social movements in Korea. In H. Y. Cho, L. Surendra, & H. J. Cho (Eds.), *Contemporary South Korean Society. A critical perspective* (pp. 346–397). Routledge.

Luhmann, N. (1993). *Risk: a sociological theory*. Walter de Gruyter. (Original work published 1991).

Mbembe, A. (2016). *Necropolitica* (R. Beneduce, Trans.). Ombre Corte. (Original work published 2003).

Moon, J. (2020). *Discorso "Tre anni di Mandato" Presidente della Repubblica di Corea Moon Jae-in.* http://overseas.mofa.go.kr/it-it/brd/m_8792/view.do?seq=760669&srchFr=&srchTo=&srchWord=&srchTp=&multi_itm_seq=0&itm_seq_1=0&itm_seq_2=0&company_cd=&company_nm=&page=1

Municipality of Rome. (2020). *Sistema Unico di Segnalazione.* https://www.comune.roma.it/web/it/di-la-tua-segnala.page

Orsini, A. (2019). *Viva gli immigrati! Gestire la politica migratoria per tornare protagonisti in Europa.* Rizzoli.

Pariser, E. (2012). *Il filtro. Quello che Internet ci nasconde* (B. Tortorella, Trans.). Il Saggiatore. (Original work published 2003).

Petti, A. (2007). *Arcipelaghi e enclave.* Mondadori.

Presidency of the Council of Ministers. (2020). *Immuni's High-Level Description.* https://github.com/immuni-app/immuni-documentation

Presidency of the Council of Ministers – Immuni Italia. (2020). *Aiuta te stesso, la tua famiglia, il tuo paese.* https://www.immuni.italia.it/? gclid=Cj0KCQjwoPL2 BRDxARIs AEMm9y9 YLjUzg32 CT-v5rH7GX usHnHmrS4O5 uAEvh-KI2FBTt EIzKVA_1vv8aAs5JEALw_wcB

Ranucci, S. (2020). *Wuhan città aperta?* https://www.rai.it/dl/doc/1586270571878_wuhan_citta_aperta_report.pdf

Repubblica Italiana. (2017). *Disposizioni integrative e correttive al decreto legislativo 26 agosto 2016, n. 179, concernente modifiche ed integrazioni al Codice dell'amministrazione digitale, di cui al decreto legislativo 7 marzo 2005, n. 82, ai sensi dell'articolo 1 della legge 7 agosto 2015, n. 124, in materia di riorganizzazione delle amministrazioni pubbliche* (GU n.9 del 12-1-2018). Gazzetta Ufficiale.

Repubblica Italiana. (2020). *Misure urgenti per la funzionalità dei sistemi di intercettazioni di conversazioni e comunicazioni, ulteriori misure urgenti in materia di ordinamento penitenziario, nonche' disposizioni integrative e di coordinamento in materia di giustizia civile, amministrativa e contabile e misure urgenti per l'introduzione del sistema di allerta Covid-19.* Gazzetta Ufficiale.

Roma Capitale [@RomaCapitaleOfficialPage]. (2020, March 26). *Hai notato un assembramento nella tua zona?* [Status update]. Facebook.

Rousseau, J. J. (1997). *Il contratto sociale* (M. Garin, Trans.). Laterza. (Original work published 1762).

Schmitt, C. (1984). *Le categorie del "politico": saggi di teoria politica* (G. Miglio & P. Schiera, Eds., SchieraP., Trans.). Il Mulino. (Original work published 1934).

Sharma, T., & Basir, M. (2020). *Use of apps in the COVID-19 response and the loss of privacy protection.* https://www.nature.com/articles/s41591-020-0928-y.pdf

Sontag, S. (1979). *Malattia come metafora* (E. Capriolo, Trans.). Einaudi. (Original work published 1978).

State Council of the People's Republic of China. (2014). *Planning Outline for the Construction of a Social Credit System (2014-2020)*. https://chinacopyrightandmedia.wordpress.com/2014/06/14/planning-outline-for-the-construction-of-a-social-credit-system-2014-2020/

Stoklas, J. (2017). *Sicherheit im Schengen-Raum: Eine unendliche Datensammelei?* https://www.repo.uni-hannover.de/bitstream/handle/123456789/4587/Datensammelei.pdf?sequence=1&isAllowed=y

Strand, R., & Oughton, D. (2009). *Risk and Uncertainty as a Research Ethics Challenge*. National Committees for Research Ethics in Norway.

Watson, I., & Jeong, S. (2020). *Coronavirus mobile apps are surging in popularity in South Korea*. https://edition.cnn.com/2020/02/28/tech/korea-coronavirus-tracking-apps/index.html

Zuboff, S. (2019). *Il capitalismo della sorveglianza. Il futuro dell'umanità nell'era dei nuovi poteri* (P. Bassotti, Trans.). Luiss. (Original work published 2018).

Zunino, G. (2020). *Coronavirus, app e sistemi per tracciare i positivi: come funzionano (nel mondo, in Italia)*. Academic Press.

Zylberman, P. (2013). *Tempêtes microbiennes. Essai sur la politique de sécurité sanitaire dans le monde transatlantique*. Gallimard.

ADDITIONAL READING

Agamben, G. (2006). *Che cos'è un dispositivo?* Nottetempo.

Baudrillard, J. (2002). *L'altro visto da sé* (M. T. Carbone, Trans.). Costa & Nolan. (Original work published 1987).

De Leonardis, O. (2001). *Le istituzioni*. Carocci.

Foucault, M. (1969). *La nascita della clinica. Una archeologia dello sguardo medico* (A. Fontana, Trans.). Einaudi. (Original work published 1963).

Han, B. C. (2014). *La società della trasparenza* (F. Buongiorno, Trans.). Nottetempo. (Original work published 2012)

Latour, B. (2020). What protective measures can you think of so we don't go back to the pre-crisis production model? (S. Muecke, Trans.).

Le Bon, G. (1996). *Psicologia delle folle* (G. Villa, Trans.). Longanesi. (Original work published 1895).

Lefebvre, H. (1976). *La produzione dello spazio* (M. Galletti, Trans.). Moizzi. (Original work published 1974).

Zafra, R. (2012). *Sempre connessi: spazi virtuali e costruzione dell'io* (E. C. Vian, Trans.). Giunti. (Original work published 2010).

KEY TERMS AND DEFINITIONS

Bio-Security: A regime in which the right to health becomes, through governmental provisions, an obligation to health and in which the State of security – spatial institutionalization of the logic of security – legislates on life itself.

Corona100: Tracking and reporting system, on a voluntary basis, developed in the context of the SARS-CoV-2 outbreak by a private company using geolocation data from users and government health data and then officially taken over by the Republic of Korea.

Health Code: Chinese governmental mandatory tracking tool that expands existing mobile apps, WeChat and Alipay. It assigns a daily electronic health code that allows travel or imposes quarantine in the context of the SARS-CoV-2 outbreak.

Immuni: Digital contact tracing system, on a voluntary basis and currently in the testing phase, introduced in Italy by the Presidency of the Council of Ministers in the context of the SARS-CoV-2 epidemic. The system assigns a daily ID that ensures compliance with privacy.

Narrative Policy: An operation of framing, used on a discursive or visual level in institutional communications, which attributes specific meanings to certain facts also using metaphor as a conceptual device.

Risk: A conceptual device, produced by political vision and consolidated by city's one, through which authority uses power in function of an increasingly worse future scenario.

SARS-CoV-2: An epidemic occurred between 2019 and 2020, understood by the authors as a prototype of potential other global crises that, while starting from a health context, affect heterogeneous sectors, requiring the definition of new regimes of normality on a global scale.

State of Exception: A condition through which ordinary law is suspended in function of exceptional directives indispensable in the resolution of an emergency. It is spatialized in securitized areas where the power of the authority varies according to the risk.

Tracking System: A heterogeneous systems of data collection of mobile device developed in multiple ways according to the socio-political-cultural context in which they are introduced. In the context of the SARS-CoV-2 epidemic they allow the monitoring of the population and possible risk spaces.

Chapter 7
Forest Fire Scenarios in Digital Platforms:
The Case of Portugal

Liliana Gonçalves
Digital Media and Interaction Research Centre, University of Aveiro, Portugal

Lídia Oliveira
https://orcid.org/0000-0002-3278-0326
Digital Media and Interaction Research Centre, University of Aveiro, Portugal

ABSTRACT

Forest fires are widespread in Portugal, particularly in the summer. Recently, in 2017, Portugal had two great fires. As a result, more than 120 people died, hundreds suffered injuries and registered significant economic and environmental losses. Since then, and due to the evolution and democratization of the internet and technology devices, forest fire content is much more common in cyberspace. Thus, to understand this issue, the authors propose to outline a profile of the digital platforms used in forest fire situations. The goal is to understand the uses and commitment arising from forest fires' issues in digital platforms by presenting a conceptual framework in Portugal's specific case. The authors analyzed webpages, mobile apps, Facebook pages and groups, and YouTube channels, focusing on forest fires contents. By understanding the kind of digital platform, its contents, uses, and interaction, this chapter contributes to understanding digital platforms' role in crisis and disaster scenarios such as wildfires.

INTRODUCTION

A multiplicity of social networks and the increase in the use of new technologies, such as smartphones, have transformed digital platforms in privileged communication stages in situations of natural disasters. This increased the trend of connection, communication, and mobilization through the virtual world.

DOI: 10.4018/978-1-7998-6705-0.ch007

In the context of crisis communication, digital platforms – particularly social media – present some advantages: the speed in dissemination and access to information; to bring together institutions and citizens; with essential features to communicate in emergency situations (Eriksson & Olsson, 2016; Reuter, Stieglitz, & Imran, 2020). The potential of digital platforms to crisis communication is vast. However, there are also some challenges, like potential communication constraints – organizational, cultural, stigma, denial, hostility, lack of trust in institutions and/or science, learning difficulties, Etc. – hat should be recognized and adequately address. This implies an in-depth knowledge of the audience and public engagement in crisis communication. Here the preparation phase, as well as the use of social media, are essential (Lundgren & McMakin, 2018). In the specific context of social media, the challenges to crisis communication arise at several levels: to understand the use and utility sense that the different users make of these platforms; to continually update the information between institutions and with the public, to develop new features to improve communication flow; or even methodological issues of real-time data analysis (Eriksson & Olsson, 2016; Reuter et al., 2020).

Keeping in mind these key issues in crisis communication, in this chapter, the authors intend to understand the uses and the commitment arising from the issue of forest fires in digital platforms by presenting a conceptual framework in the specific case of Portugal. In the Portuguese scenario, forest fires are one of the major natural catastrophes that occur every year. As expected, the presence of this topic on digital platforms has been gaining preponderance. Thus, this work aims to outline a profile of the digital platforms used in forest fire situations. The chapter presents a quantitative and content analysis of web pages, mobile apps, and social media (Facebook and YouTube) related to Portugal's forest fires theme. To understand the presence of the forest fires theme' on digital platforms in the Portuguese context and to create a conceptual framework, the following categories were analyzed: type of digital platform, type of content presented, number of users, and territorial scope of influence.

Academic and technical literature refers that prevention should be the focal point to minimize forest fires situations. However, this research's main conclusions suggest that the forest fires theme in Portuguese cyberspace focuses mainly on real-time events alert based on geographic information and maps visualization. The echo of forest fires on digital platforms in the Portuguese context is high, mobilizing stakeholders and citizens around it. Thus, despite the potential of digital platforms to engage users to prevent forest fires situations, the information about forest fires in digital platforms for the Portuguese contexts remains dispersed, unstructured, and underuses prime platforms such YouTube, the most used social media in Portugal. These are some of the main findings of this research. To understand them, it is essential to enlighten the context of forest fires in Portugal, the use of digital platforms in this European country, and the data gathered to perform this research.

Forest Fires: a Reality Powered by Climate Change

Forest fires are a natural catastrophe that affects several areas in the world. Recently, the world witnessed violent fires in Australia and the Amazon. Thousands of forest hectares were destroyed, ecosystems, fauna, and flora were compromised. Those forest fires generated significant economic losses and put human lives at risk. The scenario happens all over the world, and the European continent is no exception. In recent years, Europe has suffered from forest fires every summer, particularly in the Mediterranean regions (de Rigo, Libertà, Houston Durrant, Artés Vivancos, & San-Miguel-Ayanz, 2017, p. 8; Turco et al., 2016, p. 2).

Climate change, combined with other environmental and socio-economic factors, seem to be at the root of this natural catastrophe (Turco et al., 2016, p. 2). The local environment, topography, vegetation, socio-economic patterns in the region, as well as climate and human behavior play a fundamental role in the propensity to forest fires (Costa, de Rigo, Libertà, Houston Durrant, & San-Miguel-Ayanz, 2020, p. 5; de Rigo et al., 2017, p. 29; Turco et al., 2016, p. 15). On the other hand, in a long-term perspective, the literature points out climate change as one of the main risk and vulnerability factors to forest fires, decreasing the ecological resilience of regions (Costa et al., 2020, p. 8; de Rigo et al., 2017, p. 29). In this context, climate change enhances forest fires in two different ways: due to weather conditions (dry weather, low precipitation, winds, Etc.) that increase the ignition and spread of fires; and indirectly, by the effect they produce in vegetation and natural fuel (Schmuck et al., 2018, p. 9). This is particularly significant in rural areas. On the one hand, because several forest fires have human causes. On the other hand, due to the threat that fire poses to rural populations, on economic and human levels (Costa et al., 2020, pp. 8–9, 32). Thus, forest fires are a complex natural disaster "caused by a combination of territorial management, human activity, cultural traditions, and climatic conditions" (Schmuck et al., 2018, p. 9).

The solutions to decrease natural disasters caused by forest fires include awareness campaigns and prevention programs to reduce fires and health implications. Those should be combined with forest management practices, such as fuel management, forest reorganization, and building ecosystems with less flammable species (Feyen L., Ciscar J.C., Gosling S., Ibarreta D. & Soria A., 2020, p. 43).

FOREST FIRES: THE PORTUGUESE SCENARIO

Forecasts for Europe are pessimistic: climate change will severely intensify the risk of forest fires, will increase the number of days with high and extreme fire risk, heat waves, and periods of drought ("Forest fires", 2020). This means an increased impact for people and ecosystems of the most vulnerable regions, a growth of the fire season, areas at risk and their progressive desertification, and also the occurrence of large forest fires, particularly in southern Europe (Iberian Peninsula, Turkey, part of Greece and Balkans, and areas of central and southern Italy and France) (Costa et al., 2020, pp. 5–6; de Rigo et al., 2017, p. 40; Schmuck et al., 2018, p. 9). In this context, Portugal, Spain, and Turkey present the highest risk levels of forest fires ("Forest fires", 2020).

The forecasts presented for this group of countries are alarming: in the worst climate change scenarios, the risk of fire can increase between 40% and 100%, leading to more frequent forest fires, changes in ecosystems, biomass and biodiversity reduction, increased drought, and impacts on citizens' lives (de Rigo et al., 2017, p. 40; "Forest fires" 2020). Solutions indicated include adaptive strategies such as controlled burning, fire management of breaking zones, or prevention activities (*Climate change impacts and adaptation in Europe - JRC Science for Policy Report*, 2020, p. 43; "Forest fires", 2020)

Portugal is one of the European Mediterranean countries that presents the highest risk of forest fires. It is also the only country in this group that does not follow the trend of decreasing the burnt area in the last decades ("Forest fires", 2020). In 2018, the burnt area reached about 45000 hectares, mainly due to Monchique's fire in the south. This big fire contributed with almost 28000 hectares of burnt area, and it was the most significant fire registered in Europe, Middle East, and North Africa. The 2018[1] numbers represented a decrease of about 48% compared to the last decade average and were the lowest recorded since 2014 (San-Miguel-Ayanz et al., 2019, pp. 63, 141). Even so, Portugal was the European Union (EU) country with the highest number of fires, with the most burnt area in 2018 among the 22 EU countries, only Turkey surpassed Portugal in 2018 (San-Miguel-Ayanz et al., 2019, pp. 134, 141).

Two thousand eighteen numbers correspond only to 7% of those recorded in 2017 (San-Miguel-Ayanz et al., 2019, p. 141), which is already considered a national catastrophe. In 2017, Portugal started the fire season in June, with a fire known as the Great Fire of Pedrogão. It was a single fire, started in Pedrogão Grande (a small town in the Center Region), quickly spreading to neighboring locations. In the end, this fire had more than 200 hectares of burnt area (Schmuck et al., 2018, p. 59). In addition to the environmental loss, the fire caused high material losses, as it passed through several villages, but it was especially tragic due to the 66 deaths. It was the deadliest fire in the history of Portuguese fires (Schmuck et al., 2018, p. 118). In the same year, the catastrophe repeated in mid-October, after the typical fire season. In this great fire, more than 500 ignitions devastated several places in the Center Region. Again, this fire had more than 200 hectares of burnt area, significant environmental and economic losses, and another 48 deaths (Schmuck et al., 2018, p. 59; Viegas et al., 2019). In this particularly tragic year, the forest fires resulted from severe weather conditions, with hot and dry weather and strong winds. Two thousand seventeen totaled 560000 hectares burnt across the country, more than in the previous six years together and representing 498% of the burnt area compared to the previous decade average (INE - Instituto Nacional de Estatística, 2019; Schmuck et al., 2018, p. 56). These numbers naturally placed Portugal at the top of European countries with the largest burnt area in 2017 (47%). In terms of forest fires, it was the third most critical year for Europe since 1980 (INE - Instituto Nacional de Estatística, 2019; Schmuck et al., 2018, pp. 56, 82, 118).

Although climate change effects are still uncertain, weather conditions are one of the main factors to happen to forest fires. Soil moisture and, as in Portugal, high fuel risk forest areas are also determinants to spread fire (de Rigo et al., 2017, p. 4,8).

In Portugal, the firefighting phase is the one that mobilizes more resources, involving several entities: National Authority for Civil Protection (Prociv), Special Fire Fighting Unit, Special Firefighters Force, National Guard Protection Group (GIPS), Institute for the Conservation of Nature and Forests (ICNF), Nature Protection Service and Republican National Guard (GNR) (Schmuck et al., 2018, p. 59). The operations still focus on training for firefighting. In 2017, around 13500 operatives were engaged in firefighting, supported by more than 2200 terrestrial and aerial means (INE - Instituto Nacional de Estatística, 2019; Schmuck et al., 2018, p. 59).

COMMUNICATION IN EMERGENCY SITUATIONS

In the context of forest fires, particularly those that affect areas with human communities, the challenges arise in many ways: fighting fires, the heterogeneity of fire factors, human behaviors and values, and external factors, such as climate changes (Spies et al., 2014, p. 4). Thus, the forest fire scenario is characterized by the complexity of interrelationships between different ecosystems that hinder communities' resilience from learning and adapting to the external environment (Spies et al., 2014, p. 10). In this framework, social networks and institutions have a crucial role in structuring this resilience, which integrates social concepts, knowledge, values, interpersonal and local relationships (Spies et al., 2014, p. 4). Also, stakeholder networks can collaborate with local communities searching for solutions, defense mechanisms, and social and political strategies to deal with the ecological processes that trigger forest fires. Working together as a network and focusing on communication and knowledge and resource sharing allows communities to improve resilience to deal with forest fire (Spies et al., 2014, p. 4).

Communication is essential to prepare communities to deal with disaster situations (Brengarth & Mujkic, 2016, p. 589; Haworth, Whittaker, & Bruce, 2016, p. 115). The literature demonstrates that the focus has been the response to emergencies, neglecting the preparedness, prevention, and mitigation phases (Haworth, 2018; Haworth et al., 2016, p. 117). However, this trend is changing progressively. It has been recognized that a simple spread of information is not enough to improve communities' preparedness to cope with disasters. Also, the need for participatory approaches involving the community has been identified as a principle for disaster risk reduction and building resilient communities (Haworth et al., 2016, p. 115).

Usually, communication in emergency situations assumed the vertical model, flowing from the authorities to the citizens, through the traditional media. Until just over a decade ago, horizontal communication (citizens-citizens and citizens-authorities) was almost non-existent (Spinsanti & Ostermann, 2011, p. 2, 2013, pp. 36–37). However, widespread access to mobile devices, the democratization of web 2.0, and particularly the advent of social media have created new opportunities for citizens' collaboration and participation. Consequently, it changes communication in emergency situations such as forest fires. In this context, recognizing information is useful. Local citizens' knowledge allows a greater decentralization of risk and disaster management processes and creates a closer relationship between authorities, specialists, professionals, and citizens. This approach allows members' cohesion and inclusion, leading to greater community empowerment when dealing with forest fire situations (Haworth et al., 2016, pp. 118; 125).

By being a low-cost communication channel, the new web 2.0 media, particularly social media, and mobile apps, play a crucial role in supporting and responding to emergencies. This new media arena allows users to collect and control information about events and people. It allows citizens' engagement to map the emergency, to give emotional support, and express solidarity. Also, digital media can enhance relations between citizens and authorities (Appleby-Arnold, Brockdorff, Fallou, & Bossu, 2019; Brengarth & Mujkic, 2016, pp. 590; 595; Reuter et al., 2020, p. 242). According to Brengarth and Mujkic (2016), in situations of forest fires, the content generated by users shared on digital platforms allows the public to stay informed in real-time, stimulates the collaboration with the rescue authorities, and mutual assistance among the community. Nevertheless, authorities must be able to communicate with citizens through digital platforms. Here, the literature notices a lack of preparedness of authorities to understand the uses, perceptions, and features of digital media, particularly social media, and also lack of coordination between different official services to communicate in digital arenas, which may arise some confusion and uncertainty among users (Eriksson & Olsson, 2016; Reuter et al., 2020). On the other hand, authorities should not entirely replace traditional media for digital media in forest fires; otherwise, they could exclude those disconnected from the web. 2.0 (Brengarth & Mujkic, 2016, p. 595; Haworth, 2018, p. 234).

THE POTENTIAL OF DIGITAL PLATFORMS FOR COMMUNICATION IN EMERGENCY SITUATIONS

In recent decades, citizens' participation has increased to collect and share volunteered geographic information (VGI) and local knowledge through user content generation (UCG), mainly through social media such as Facebook, Twitter, Weibo, or mapping platforms (open street maps - OSM). This allows a real-time update of information (Ahmouda, Hochmair, & Cvetojevic, 2018, p. 209; Bhavaraju, Beyney, & Nicholson, 2019, p. 1; Sachdeva, McCaffrey, & Locke, 2017, p. 1147). This occurs mainly due to the ubiquity of mobile devices that allow users to share, in real-time, the emergency situations they are witnessing or engaged in (Athanasis et al., 2018, p. 75; Liu, Zhu, Shao, Adusumilli, & Wang, 2020, p. 2; Pekar, Binner, Najafi, Hale, & Schmidt, 2020, p. 43). Thus, when they are in places experiencing first-hand emergency situations, users collect and share information, even before the reaction of emergency services or traditional media (Boulianne, Minaker, & Haney, 2018, p. 699; Tavra, Racetin, & Peroš, 2019, p. 415). In the form of images (photographs), texts, or emergency mapping data, this information can be scrutinized by emergency response organizations (Riccardi, 2016, p. 127). Thus, this crowdsourcing trend allows information sharing on multiple platforms by organizations and citizens, becoming a powerful tool for information dissemination in emergency contexts and disaster management (Riccardi, 2016, pp. 123; 127; Sachdeva et al., 2017, p. 1148).

In this sense, VGI increases the communication between affected communities and authorities, enables the public to collect and share a large volume of information, and allows citizens to participate actively in disaster management (Haworth et al., 2016, p. 116). The use of digital platforms to share VGI is seen by communities as a form of active participation in emergency situations, and by the authorities as an opportunity to improve emergency management, with benefits in terms of preparing and increasing the resilience of communities when participation is encouraged through sharing (Haworth, 2018, p. 234; Liu et al., 2020, p. 18).

Regarding social media (such as Facebook, Twitter, or YouTube) in the context of emergencies, the literature points out several uses and roles: in the pre-emergency phase, they can be used for sharing protective information, alert messages, and disasters detection. In this phase, social media can enhance the users' self-confidence, motivation, and hability to learn and memorize practices, behaviors, and preventive measures. Social media are used in the emergency phase to exchange real-time emergency information (location, conditions, assistance, Etc.), to warn users and improve response by the emergency management authorities. These digital platforms are also useful to users to seek information, reduce uncertainty and isolation, increase self resilience and collaboration, and make sense of the emergency. In the mitigation phase, social media plays a key role to debate, connect communities, increase emotional and material support, and generate solidarity among users, which are rarely achieved by traditional media (Ahmouda et al., 2018, p. 196; Appleby-Arnold et al., 2019; Athanasis et al., 2018, p. 75; Boulianne et al., 2018, pp. 697–708; Çolak & Sunar, 2020, p. 10; Jurgens & Helsloot, 2018, p. 86; Sachdeva et al., 2017, p. 1157).

Thus, when it comes to strategies supported by VGI or by crowdsourcing, it is essential to guarantee the training and motivation of users to collect relevant data (Riccardi, 2016, p. 127; Tavra et al., 2019, p. 418). In the same sense, authorities must have the ability to adapt and prepare for this new communicational reality provided by the new digital media. They need to understand users, to listen to citizens, and even to privilege the web influencers in the dissemination of reliable information (Eriksson & Olsson, 2016, p. 10; Liu et al., 2020, p. 19; Lundgren & McMakin, 2018, p. 45).

In this context, it is also important to consider that VGI or other types of data in collaborative platforms and social media are only valid as "sensors" of emergency situations if this information can be extracted and analyzed in real-time (Bhavaraju et al., 2019, pp. 1; 9). Here, technology presents a set of resources such as Big Data, social media analytics, text mining, or machine learning (Athanasis et al., 2018, p. 79; Pekar et al., 2020, p. 44; Reuter et al., 2020, pp. 243–244). These kinds of technology resources can process a large volume of information at great speed. They allow automated analysis; incorporate different information sources and data types - satellite data, digital data, crowdsourcing data, real-time weather data, Etc. - bringing possibilities such as visualizing and forecasting natural disasters more efficiently and, consequently, enhancing of decision processes (Athanasis et al., 2018, p. 79; Çolak & Sunar, 2020, p. 11; Pekar et al., 2020, p. 44).

THE CHALLENGES OF DIGITAL PLATFORMS TO IMPROVE EMERGENCY COMMUNICATION

Citizens' participation in emergency contexts and valuing the local communities' knowledge is central to narrowing the temporal information gap between authorities and citizens. Also, participatory approaches in digital contexts reveal quality results (Craglia, Ostermann, & Spinsanti, 2012, pp. 401, 405). However, there are obstacles. Considering that digital platforms focus on information sharing, disinformation is a danger, particularly in emergency contexts, where it can mean significant losses (Brengarth & Mujkic, 2016, p. 590; Haworth et al., 2016, p. 117; Liu et al., 2020, p. 19). Here, the key to dealing with this challenge may be to raise the awareness that each user is responsible for the information posted online (Brengarth & Mujkic, 2016, p. 590).

Associated with this obstacle, and particularly regarding VGI, other issues arise: data quality has direct consequences on its usability in emergency scenarios, lack of credibility and sources reputation leading to questions of trust in the information made available in digital contexts (Brengarth & Mujkic, 2016, p. 591; Haworth, 2018, p. 232; Haworth et al., 2016, p. 117; Yang & Tian, 2017, p. 5). Regarding the use of crowdsourcing in emergency contexts, it is necessary to filter and control the information quality, particularly those collected in social media, where data can present much noise, overlaps, and unrelated information (Zhong, Duckham, Chong, & Tolhurst, 2016, p. 2). These challenges do not mean that VGI or other information made available on digital platforms is not reliable. However, the solution to coping with these issues may involve treating it as a complementary data source, to be integrated with other official sources, in emergency and disaster contexts (Haworth et al., 2016, p. 117). According to Zhong et al. (2016), in the context of forest fires, those sources may include official data about risk factors of the regions, data from sensors in the territory or satellites such as combustion conditions, topography, temperature, and humidity in time -real, monitoring population movements in real-time and mobile network coverage, among others. These, plus VGI and crowdsourcing data, based on the community contribution on digital platforms, are essential to assess emergency situations. Although, none of them, individually or jointly, are useful to determine the perimeter of fire in extreme forest fires (Zhong et al., 2016, pp. 1; 6).

Other challenges arise from data's unstructured nature, which hinders its storage, management, and processing. The heterogeneity of data and information sharing formats, the noise around digital platforms, or the massive volume of data in emergency situations are also issued (Athanasis et al., 2018, pp. 76–78; Haworth, 2018, p. 232; Haworth et al., 2016, p. 117). These pose challenges to processing and analyzing Big Data, particularly from VGI and social media, since authorities seem to lack resources to manage data and, consequently, limiting acceptance and innovation (Haworth et al., 2016, p. 117; Reuter et al., 2020, p. 247). One of the solutions proposed in the literature may seem contradictory at first glance. It refers to the massive engagement of people in VGI processes, which allows reducing errors, misinformation, and spread of rumors associated with the credibility of data shared by non-expert users (Haworth, 2018, p. 232).

The literature also points out data privacy threats, personal information security, or malicious use of data (Haworth et al., 2016, p. 117). Likewise, data protection guaranteed by recent legal regulations (for example, the GRDP in European Union) or changes to access data from digital platforms can bring barriers to retrieve data available for the management of emergency situations (Ahmouda et al., 2018, p. 209).

Finally, demography is also an obstacle to the use of shared digital data because not all people use digital platforms similarly, nor does everyone have the same access conditions or usage abilities (knowledge, time, or financial and equipment conditions). This can contribute to the digital divide, marginalizing those excluded from the empowering technologies to the community (Bhavaraju et al., 2019, p. 9; Brengarth & Mujkic, 2016, p. 590; Eriksson & Olsson, 2016, p. 4). This can also mean territorial inconsistencies in data collection, i.e., usually, the most densely populated areas are those where technologies' use and information sharing occurs most, which influences data usefulness, particularly when emergency situations are located in less densely populated areas (Yang & Tian, 2017, p. 5; Yue, Dong, Zhao, & Ye, 2019, p. 2). The solution to these issues is in the hands of government institutions. They must, simultaneously, adapt to the new web 2.0 formats and keep using traditional channels and media, ensuring communication reaches all groups (Brengarth & Mujkic, 2016, p. 590; Eriksson & Olsson, 2016, p. 8; Lundgren & McMakin, 2018, p. 45).

In the specific context of forest fires, the literature shows that this thematic echo on digital platforms is high and must be contemplated. However, it should consider the different challenges about social digital data and use the relevant information to the forest fire context (Craglia et al., 2012, p. 414).

Forest fires in Portugal have also been taking up digital platforms. Thus, it is important to understand how and what kind of subjects this theme has been disseminated, whether by authorities, formal and non-formal institutions, or citizen networks. Thus, the present research aims to map Portuguese forest fires on digital platforms - web, apps, and social media - to profile the pages and networks found regarding objectives, interaction, and use. This work aims to understand the kind of uses and engagement that arise from the forest fires thematic and the digital platforms used for that to present a conceptual framework about the digital platforms used in situations of forest fires in Portugal.

THE DIGITAL PROFILE OF PORTUGAL

With just over 10 million inhabitants and an urbanization rate of 65% (higher than the world average of 55%), Portugal follows the growth trends of the digital footprint, based on data provided by datareportal.com, which collects the primary official sources of statistical data (Hootsuite & We Are Social, 2020).

Currently[2], Portugal has 8.52 million internet users, corresponding to an internet penetration rate of 83%, above the world average of 59%, and only 1% below the European average (84%). These numbers have been growing progressively around the world. Portugal follows this trend, as well. Portuguese spend 6h38m surfing the internet, a value very close to the world average of 6h43. Regarding the internet speed available in the national territory, the average points out to 101.58 Mbps, above the world average of 73.58 Mbps (Hootsuite & We Are Social, 2020; INE - Instituto Nacional de Estatística, 2020). The websites most visited by the Portuguese are Google, Youtube, and Facebook, and, according to Google Trends (Google, n.d.), in 2017, the terms "fogos", "Pedrogão Grande"[3] and in 2018 the term "Fogo de Monchique"[4] joined the top 10 most searched expressions on Google search engine. Regarding online activities, the most consumed content by the Portuguese is videos (96%), streaming music (51%), and online radio (43%). Concerning electronic devices, 87% of Portuguese have a computer, 93% have a smartphone, and 52% have a tablet. (Hootsuite & We Are Social, 2020).

About social media, Portugal has 7 million active users, which corresponds to 69% of the population. This is an increasing number that beats the world (49%) and European (55%) averages. Portuguese users spend about 2 hours on social networks, slightly less than the world average of 2 hours and 24 minutes. It should also be noted that 94% of social network users also use mobile devices (smartphones) to access it through an app. In Portugal, the top 5 most used social media are YouTube, Facebook, Facebook Messenger, WhatsApp, and Instagram (Hootsuite & We Are Social, 2020).

Regarding mobile connection, Portugal occupies the 6th place in the world ranking, with 155% of mobile connections (number of registered users related to the total population), compared to the 103% world average and the 128% European average. However, Portuguese values show a downward trend. The internet speed on mobile devices reaches 37.94 Mbps, slightly above the world average of 32.01 Mbps, which shows a global growth trend. About the use of apps on mobile devices, the Portuguese top 5 include social media apps (94%), messaging/chat apps (93%), video and entertainment apps (82%), geolocation apps/maps (80%), and shopping apps (60%) (Hootsuite & We Are Social, 2020).

Thus, considering the digital use profile in Portugal, this study focuses on understanding the presence of forest fires theme in three major types of digital platforms: web, mobile apps, and social media, applied to the Portuguese context.

METHODOLOGY

Different types of digital platforms were analyzed to perform the present study: web pages, mobile apps, and social media (Facebook and YouTube). The goals were to understand what kind of digital platforms are being used in forest fire situations in the portuguese scenario and what type of content they focus on. To accomplish those goals and create a conceptual framework, the following dimensions were analysed: typology, use, and interaction. These dimensions integrate categories of analysis, such as:

- Type of digital platform: web pages, blogs, mobile apps, Facebook pages, Facebook groups, YouTube channels were the types of digital platforms found in our research.
- Type of content presented: the content of each validated result was categorized in terms of the page's objectives. Five main results were:
  - Education and Prevention: includes all the results focused on education and forest fire prevention
  - Specialized Information: assembles the results with specialized content, namely technical, legal and legislative, scientific, Etc.
  - General Information: aggregates the results that focus on fires from a general perspective
  - Alert and Information – Maps: includes the results that present information and fire events alert through maps navigation
  - Causes: gathers the results related to support to victims, firefighters, animals, Etc., focusing essentially on the help content.
- Territorial scope of influence: this category seemed relevant due to the content and the platforms' target. International, national, regional, and local were the main categorizations. Here, the international scope refers to platforms from international institutions that directly relate to forest fire

Portuguese context. National pages are those from national institutions and/or who cover all the Portuguese territory. Regional platforms regard to a specific region of the country. Local platforms refer to a specific municipality.

- Number of users, posts, comments, reactions: those categories were considered only in quantitative terms to those platforms which present this data publicly.

The research and data collection for the present study took place in May 2020. Since the goal was to map the forest fires theme on digital platforms, the following words/expressions were searched: "fires" and "fire emergencies"[5]. In all scenarios - web, apps, and social media - the platform's sensitivity to the accentuation was considered. Terms/expressions were searched in its plural version to integrate the singular form as well, extending the possibility of results to the unspoken words "forest/forestry" associated with the term "fires". All scenarios were searched manually, except YouTube, where automatic digital data extraction tools were used, since, for the other digital platforms, no automatic search tool to cope with data collection criteria was found.

The web search was performed using the encrypted search engine DuckDuckGo, to limit the results bias due to the personal settings of the most common search engines such as Google Chrome, Bing, or Mozilla Firefox. In terms of settings in the search engine, the Portuguese language and Portugal's location were defined for the search criteria. For each of the search terms, DuckDuckGo returned 500 results, making a total of 1000. After collecting all the results, they were manually selected to identify the digital platforms explicitly focused on forest fires in Portugal. This selection process excluded media web pages with published news about forest fires. The authors considered that media sites do not exclusively focus on forest fires. Also, the digital news about forest fires will be the subject for future research on this topic. Results from social media pages (Facebook) were also disregarded, as these will be analyzed in the Social Media topic. Also, results related to Brazil and those not related to forest fires (for example, results related to urban fires, insurance companies, fire management equipment companies, fireworks, games, or music) were dismissed.

Thus, for the search term "fires", 58 results were selected, and for the expression "fire emergencies" 36, resulting in a total of 94 webpages. Of these, there were also 17 overlaps, i.e., results that appeared simultaneously in two or more searches. In the end, data analysis is focused on 77 web pages.

Regarding mobile apps, considering that the Portuguese population mainly uses the Android operating system (77.8%) (Hootsuite & We Are Social, 2020), the authors decided to limit the search to Google Play Store for Androids. The search was performed in the apps tab. Five hundred four results were found for the searched term "fires" and 374 for the searched expression "fire emergencies". The selection process was identical to the previous one: manually, apps directly related to forest fires were identified, particularly for the Portuguese digital universe. All the results related to simulation/role-playing games, fire detection software, sounds/alarms, wallpapers, and fireworks were excluded. Twenty-one results for the term "fires" and 15 for "fire emergencies" were found. Of the 36 selected results, there were also 14 overlaps. Finally, we found 22 mobile apps for analysis.

Regarding social media, we integrated Facebook and YouTube into this research, as they are the two most used social media in Portugal (Hootsuite & We Are Social, 2020). WhatsApp and Facebook Messenger were dismissed in this research due to the impossibility of collecting personal message data. About Instagram, which was in the top 5 social media most used by the Portuguese until January 2020 (Hootsuite & We Are Social, 2020), we concluded that it is unfeasible to integrate this social media after intense searching in this research. In 2019, Facebook changed the Application Programming Interface

(API) to limit data collection on its different platforms, including Instagram. Since then, the existing data extraction tools have ceased to function. On the other hand, searching for data on Instagram in a web environment does not allow the standardization of search criteria such as location or search type (accounts, publications, Etc.). Thus, facing the impossibility of collecting systematized data on this platform, and considering that research with these conditions would bring more noise than it would add to this work, Instagram was left out, limiting social media to Facebook and YouTube.

Despite the restriction of data collection with automatic digital tools, provided by the platform's APIs alteration, systematic research is still possible for Facebook. However, the search was done through a personal user account. This represents a bias confirmed by the platform's terms and conditions: "Facebook's search results are based on your activity from Facebook. (...) Your Facebook search results are also based on the activity of the Facebook community". Thus, the Facebook platform presents a data bias impossible to avoid. Likewise, data collected through the Facebook search engine is limited to publicly available data since it is impossible to access Facebook statistics, which are only available to pages administrators. Despite all these factors conditioning data collection, Facebook is the second most used social media by Portuguese social media users (Hootsuite & We Are Social, 2020). Because of that, it was considered essential to this research.

Thus, Facebook pages and groups were searched on the Pages tab and the Groups tab, with the Any Category setting. Geographic limiters were not used because they can only be applied to cities and not to countries on Facebook. The total results (pages and groups) for each term were 404 for "fires" and 379 for "fire emergencies". After a manual selection, 48 results were selected to the term "fires" and 20 to "fire emergencies", excluding the results for Brazil, Spain, and Latin America, the services of companies and fire detection equipment in buildings, the results related to fireworks and music, converging in a total of 68 results. After this manual selection, we also find ten overlaps. So, we kept 58 pages/groups for the final analysis.

Finally, for YouTube data, a search was conducted through YouTube Data Tools. We made four searches with the terms "fires Portugal", and "fire emergencies Portugal"[6]. On YouTube, it was necessary to specify the country in the search terms since, in the first attempts, the results obtained focused entirely on Brazil and Latin America. After this adjustment in the search terms, the search was done in the Channel Network module with parameters defined for two interactions (the first 100 results for each search terms), with no seeds indication and 0 defined in the parameter depth. The results were presented by relevance and extracted using a CSV file. Two hundred results were identified to the terms "fires Portugal" and 44 for "fire emergencies Portugal". The result selection was made manually. All results with metadata for other countries than Portugal (PT) were eliminated, and personal channels not centered or related to the theme of fires. We reached 19 results for "fires Portugal" and 5 for "fire emergencies Portugal". After disregard the four overlaps, 19 YouTube channels were analyzed.

After collect data and select results specifically focused on forest fires, the authors performed a content analysis of the results according to type, use, and interaction. We found similar categories of analysis for these three major analysis dimensions to work for all digital platforms. However, the heterogeneity of the available data, the missing data in some cases, and the nature of the different digital platforms - web, mobile app, and social media - did not allow us to do a systematic analysis. Thus, the study uses only the available data to analyze each of the dimensions mentioned.

ANALYSIS AND DISCUSSION OF RESULTS

After checking each digital platform's validated results, they were analyzed according to the categories such as typology, use, and interaction. The results were categorized according to the type of page, its geographical scope, and the content presented for typology. About the use, we found indicators of users/followers and navigation. In the interaction category, we found publications, comments, and reactions as indicators. It should be noted that the digital platforms analyzed have different characteristics and features. Therefore, not all platforms present data for each of the categories mentioned.

1. Web

Seventy-seven validated results were found on web pages; 56 of these are websites, 20 are blogs, and one forum page. Most of the pages found fit in the Portuguese scope (68), i.e., they presented general content, written in Portuguese from Portugal. This is mainly due to the specific search criteria for the Portuguese/Portugal digital universe. Only 1 page of international scope was found (European Commission - fires) with content written in Portuguese and related to the European context. In addition to these, we classified five results as regional and three as local. Here, the contents referred directly to regional or local contexts.

Regarding the use, most of the results presented links, i.e., the forest fires' theme did not fill the web pages' home page (58). Ten results returned PDF documents, and finally, only nine results presented the theme of forest fires on the home page. This allows us to understand that only a minority of the webpages have forest fires as their central theme.

In terms of content, they were categorized into five major types of contents, as shown in Graph 1. The majority of results fits in the "specialized information" category (37). This included webpages regarding technical, legal, and governmental information (pages with legislation about fires and forests, technical reports, or political decisions on the subject), statistical and academic information, official funding programs regarding forest fires, and firefighting training. The second category was "general information" (25). These are pages of generic information about forest fires, such as wikis and blogs. "Alert and information – Maps" (7) reached webpages where the public can find out the active fires in any region of the country through a map navigation tool. With the same expression, the "Education and Prevention" category (7) presented webpages with content about alert information and prevention measures. Finally, there was also specific content about support for fires victims, which fit the "causes" category.

Due to their nature, it was not possible to find indicators of engagement on the webpages. The next table presents the categories analyzed to webpages.

Regarding the most frequent results, the National Authority for Emergency and Civil Protection (Prociv) website was the only one that appeared in all searches[7]. This is the webpage of the official government source of information on forest fires, which presents mostly real-time forest fires maps information. The Wikipedia page on Forest Fires as well as the Forest Fires (Fogos Florestais) and Deco Protest page were frequent results in most of the searches. The first two webpages present general information about forest fires, and Fogos Florestais is an informational blog. The Deco Proteste page is not dedicated to the forest fires' theme but presents a set of contents about fires, forests, properties, and life prevention in forest fires. In the following graph (Figure 2), it is possible to identify the pages found most frequently in web searches.

Figure 1. Webpages contents

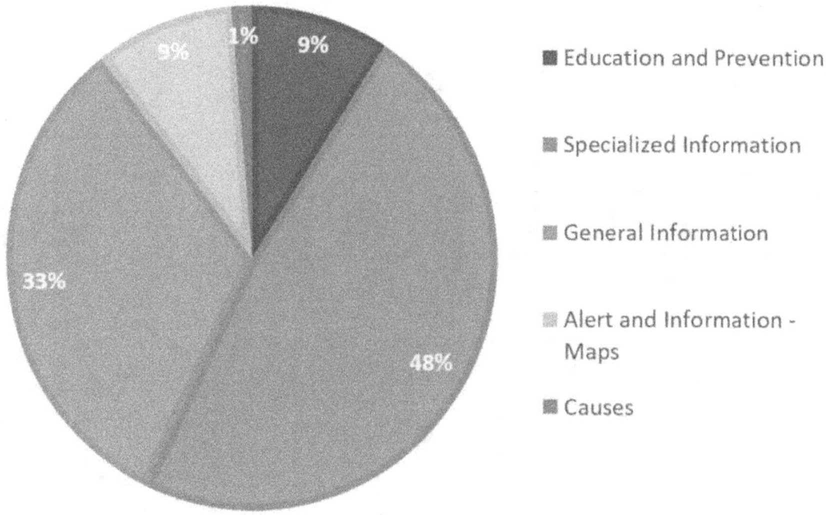

Table 1. Web analysis

Web analysis								
Tipology						**Use**		**Interaction**
Type of page		Geographical scope		Content		Navigation		
Site	56	Internacional	1	Education and Prevention	7	Home Page	9	No categories found to this dimension of analysis
Blog	20	Nacional	68	Specialized Information	37	Link	58	
Forum	1	Regional	5	General Information	25	PDF	10	
		Local	3	Alert and Information - Maps	7			
				Causes	1			
Total webpages analyzed: 77								

2. Mobile Apps

The mobile apps' geographic scope is mostly directed to the national context (7). Also, international apps with application to the Portuguese context (7) were identified. International apps with content in English (4), regional (2), and local scope apps (2) were also listed.

Concerning the use, the category found for this analysis dimension was the number of downloads of the apps, considering that they may correspond to the apps' users. Thus, only one app was identified with more than 100,000 downloads (fogos.pt). It is an app for real-time fire visualization through map navigation. The vast majority is at the level of 10,000 downloads, and there is still an application that did not present these numbers because it is in the pre-release phase.

Figure 2. Webpages most frequently found in searches

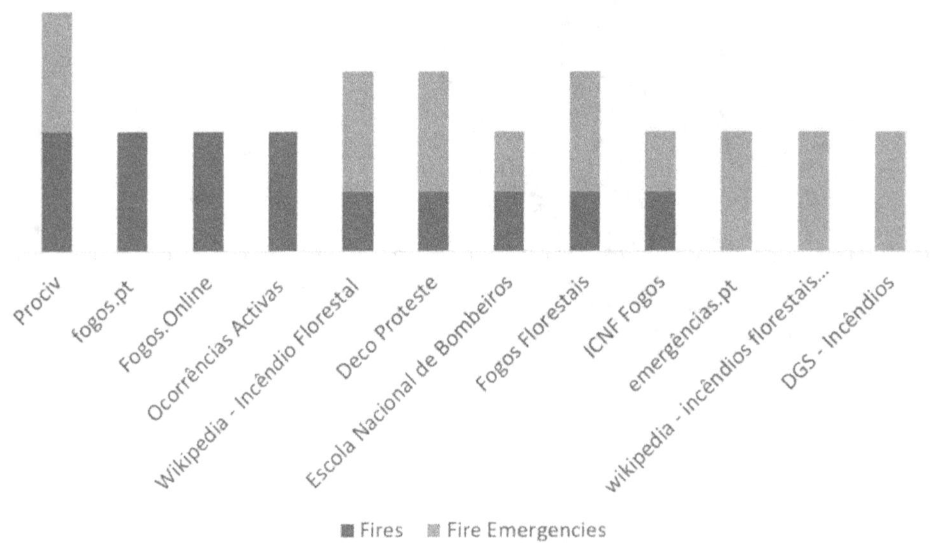

Regarding the interaction, categories such as comments and reactions were analyzed quantitatively with the Google Play Store portal's information. Most apps had up to 100 comments, and only one exceeded 1000 user comments, the same that had more than 100,000 downloads (fogos.pt). The classification of mobile apps (1 to 5 stars. 1 = minimum and 5=maximum) was collected to analyze the reactions category. Users scored most apps with 4 to 4.9 stars. Only one app had 5 stars. This app – Loulé Civil Protection – was the one with the fewest downloads (50) and was considered a local app. Here, the proximity factor can justify both data: it has a lower use, but it also has a proximity use, so it is possible that users related more to this app because it serves their needs and interests directly. In the next table, it is possible to understand the quantitative analysis of mobile apps.

About the content of the apps, the majority of them present content about "education and prevention" (12). Here are included mobile apps with warning services and protection measures, as well as disaster survival guides. These last ones are international apps, with contents in English. There are also interactive apps with general information about fires and forest fires events in this main category. The next category with more apps is "Alert and Information – Maps" (9), which includes apps with real-time forest fires maps to visualize, and they present contents very similar to one other. Here are also included one app with a map showing weather information. Finally, it was also found one app included in the "specialized information" category, that targets specifically firefighters with analysis and real-time information about the fire behavior

Regarding the mobile apps that most appeared in the searches[7] (graph 4), Fogos.pt, Incêndios Rurais - Portugal and Ponto de Situação stood out. All of them appeared in all search terms, and all of them are national apps presenting real-time forest fires maps for visualization. The one with the highest score (4.4 stars), the largest number of comments (1378), and the largest number of downloads (+100,000) is Fogos.pt.

Table 2. Mobile app analysis

Mobile apps analysis										
Tipology					Use		Interaction			
Type of digital platform	Geographical scope		Content		Users		Comments		Reactions	
mobile apps	Internacional applied to Portugal	7	Education and Prevention	12	100 000 +	1	> 100	14	2* to 2,9*	2
	Internacional in english	4	Specialized Information	1	10 000 +	11	100 +	3	3* to 3,9*	7
	Nacional	7	General Information	0	5 000 +	5	200+	2	4* to 4,9*	10
	Regional	2	Alert and Information - Maps	9	1 000 +	2	1000 +	1	5*	1
	Local	2	Causes	0	500 +	1	no comments	2	no reaction	2
					50 +	1				
					pre-release	1				
Total mobile apps analyzed: 22										

Figure 3. Apps types of contents

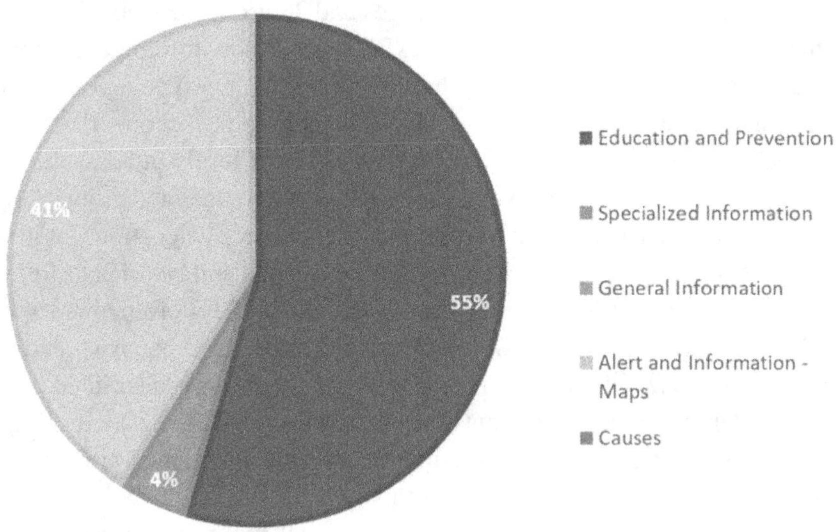

Figure 4. Apps most frequently found in searches

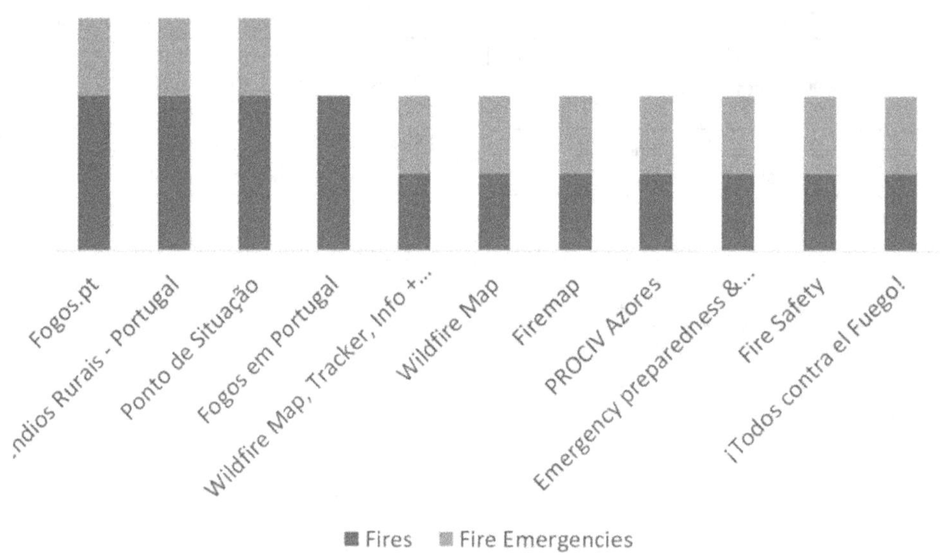

3. Facebook

On the social network site Facebook, we found 30 pages and 28 groups. Facebook's categorization was used regarding the pages' typologies, which identifies 22 pages, 6 nonprofit organizations, 1 community service, and 1 multimedia content. Concerning Facebook groups, the typologies are divided between public groups (20) and private groups (8). Regarding the geographical scope, the majority are pages/groups with national scope (43), 12 were recognized at the local level, and 3 at the regional level. It should be noticed that results at the local level are mainly based on pages and groups devoted to support victims and share information and alerts on forest fire events in specific locations, or in specific regions in groups/pages classified with a regional scope. In terms of use, the number of page followers and group members was used as a category. Most have between 1000 and 5000 followers/members, with only one page exceeding 100,000 members. This is the National Authority for Emergency and Civil Protection (Prociv) page, which has 182,776 followers. This page is not exclusively focused on the theme of forest fires. It presents general information about events and emergencies of different kinds. Regarding the posts of the pages and groups, a category of interaction dimension, as mentioned previously, the Facebook platform settings did not allow for the collection of this type of data, so it was only possible to identify the periodic average of posts in some groups. Although not all groups presented this information, we found that Ocorrências Activas Nacionais e Internacionais (Active National and International Fire Events) group presents the highest activity, with an average of 240 daily posts, compared to other groups.

About the contents (graph 5), the highlight is information and alert, with more than half of the results fitting into this category (32). Here it was found mostly pages and groups with specific information and warnings about forest fires events. There is also a trend for content to support victims (13), with pages and groups dedicated, for example, to collect goods to donate to the victims who lost their houses in forest fires, or to support firefighters organizations. The third type of content most found was information

Table 3. Facebook analysis

Facebook analysis											
Tipology								Use		Interaction	
Type of digital platform		Type of page		Geographical scope		Content		Followers/ Members		Frequency of posts	
pages	30	non-profit organization	6	Nacional	43	Education and Prevention	0	> 100	10	Daily	9
groups	28	page	22	Regional	3	Specialized Information	1	101-1000	14	Weekly	4
		comunity service	1	Local	12	General Information	10	1001-5000	22	Monthly	1
		multimedia content	1			Alert and Information - Maps	32	5001-10000	4	Anually	6
		private group	8			Causes	15	10001-15000	5		
		public group	20					15000-50000	2		
								+100000	1		
Total Facebook pages and groups analyzed: 58											

and discussion. These are mostly public and private groups to spread general information about forest fires and promote discussion. All results that fit into this category have a national scope.

The most frequent results (see graph 6) in the various searches[7] were the pages Fogos.pt and GNR UEPS - Emergency Protection and Relief Unit of National Republican Guard. Both of them with national scope focused on sharing alerts about forest fire events and, in the case of the GNR UEPS page, for other emergencies. The GNR UEPS page is also a page from an official source of information on forest fire contexts since they are the police authority. These are also the pages with the highest number of followers – Fogos.pt has 83,742, and GNR UEPS has 42,278. Most of the other pages/groups did not exceed 5,000 followers/members.

4. YouTube

For YouTube analysis, 19 channels were selected. In this analysis, it was not possible to distinguish any territorial scope for the channels. All the validated channels were from Portugal, and the contents were in Portuguese. Regarding channel contents (graph 7), we found mostly channels related to alert and information. Here are included channels from corporations and groups of firefighters, with videos focused mainly on forest fires scenarios and training techniques, channels from the army's special forces with videos about forest fire operations firefighting. Another theme highly observed was prevention, with videos related to fire prevention training, fire ignition alerts, and public campaigns to prevent forest fires. Also, there were channels related to protection materials/goods, particularly emergency and professional equipment used in forest fires. Some channels, included in the general information category, only share news about fires, with videos from the traditional media. Finally, there was also a channel focus on animal protection, with some content related to animals' rescue in forest fire situations, categorized as "cause".

Figure 5. Facebook types of contents

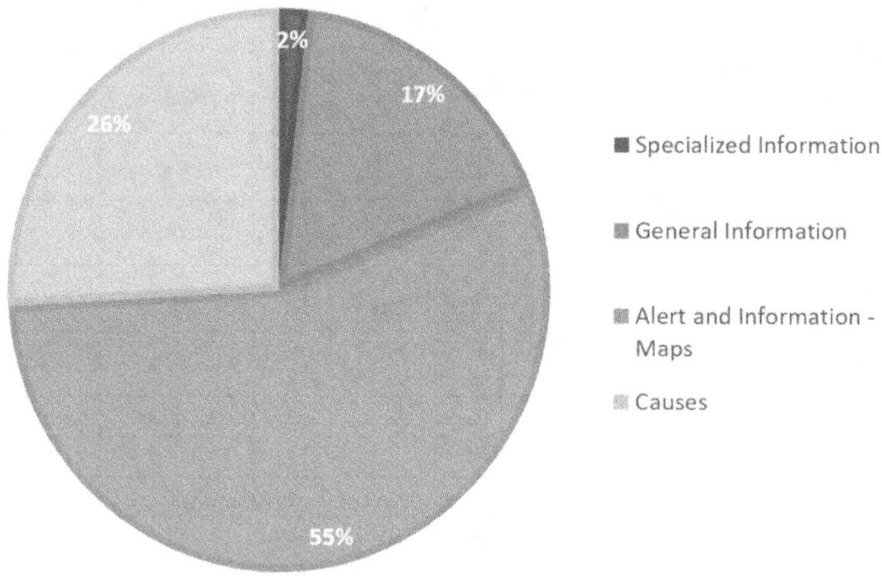

Figure 6. Facebook pages and groups most frequently found in searches

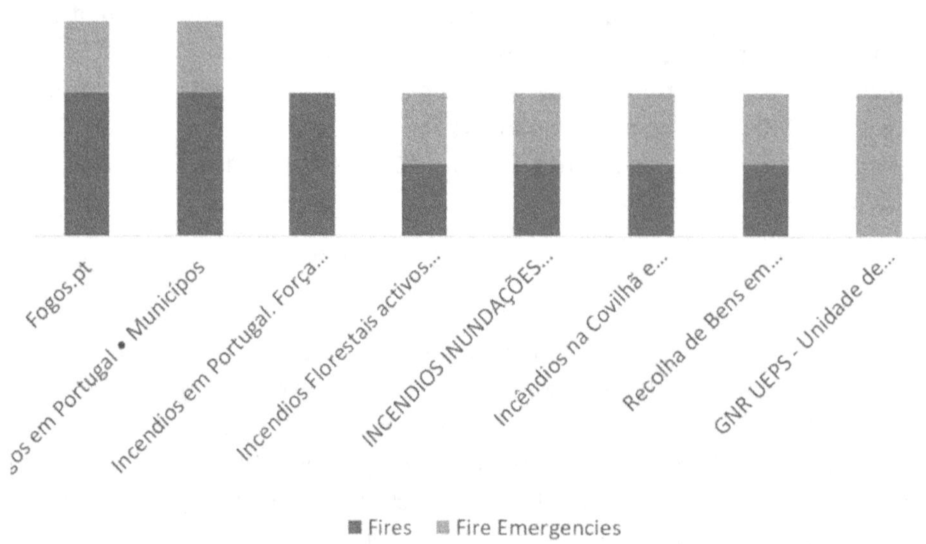

Figure 7. YouTube channels contents

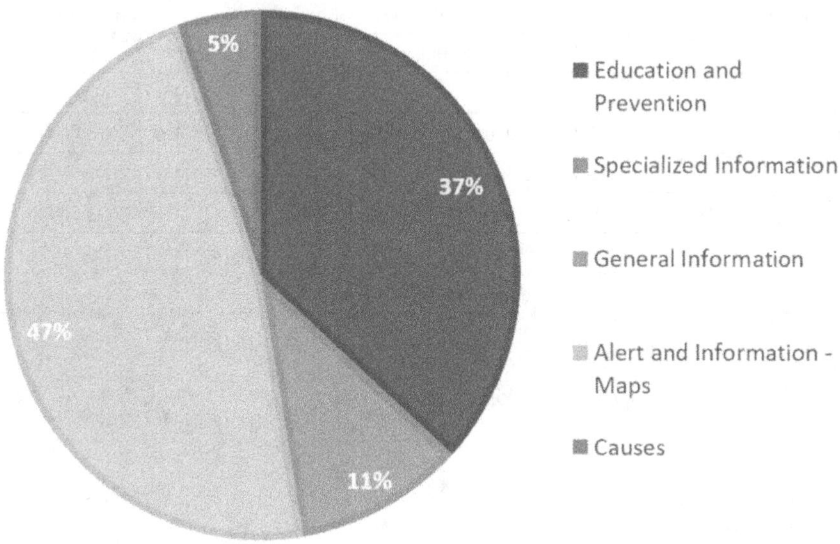

Regarding the use of the YouTube platform, the number of subscribers was used as an indicator (see table 4). Here the weak capacity to attract audiences for channels with forest fire themes is highlighted since the majority gathers only up to 100 subscribers, and only 1 channel has more than 10,000 subscribers. This one is the Forças Armadas Portuguesas (Portuguese Armed Forces) channel, one of the official government sources to operate in forest fires contexts. Despite not focusing specifically on the theme of forest fires, it includes playlists and lots of content related to the GNR Emergency Protection and Relief Unit, managing to gather a total of 18,800 subscribers. It should be noted that this unit (GNR UEPS) also has its own YouTube channel, with only 408 subscribers.

The number of videos and views for each channel was used as indicators for the interaction category. Here, it is noteworthy that most channels have only a maximum of 10 videos, followed by channels that count between 10 to 30 videos and channels with a maximum of 100 videos. Only 1 channel counts more than 100 videos. This is the Portuguese Armed Forces channel, with 175 videos. The same channel also gathers the largest number of views (33,906), while most channels gather 100 views at the most. The data can be seen in the following table.

Regarding the channels that most appeared in the searches (graph 8), there are only 3. Alto Minho FireCamp channel stands out in this category since it is the only one who appears in all searches. It is a channel dedicated to forest management specialists and firefighting professionals with contents about forest fire prevention. It is a small channel, with only 70 subscribers, 26 videos, and 23 views. Also, on the most frequent channels is the National Emergency and Civil Protection Authority (Prociv). Although it is not a channel with a significant expression on YouTube – it has only 876 subscribers, 99 videos, and 1549 views – this profile stands out in this analysis, as we will see below.

Table 4. Youtube analysis

YouTube analysis							
Tipology		**Use**		**Interaction**			
Content		**Subscribers**		**Videos posted**		**Visualizations**	
Education and Prevention	7	>100	12	> 10	7	> 100	12
Specialized Information	0	101-1000	4	10-30	6	101-1000	1
General Information	2	1001-5000	1	31-100	5	1001-5000	4
Alert and Information - Maps	9	5001-10000	1	+100	1	5001-10000	1
Causes	1	+ 10000	1			10000 +	1
Total Youtube channels analyzed: 19							

Figure 8. YouTube channels most frequently found in searches

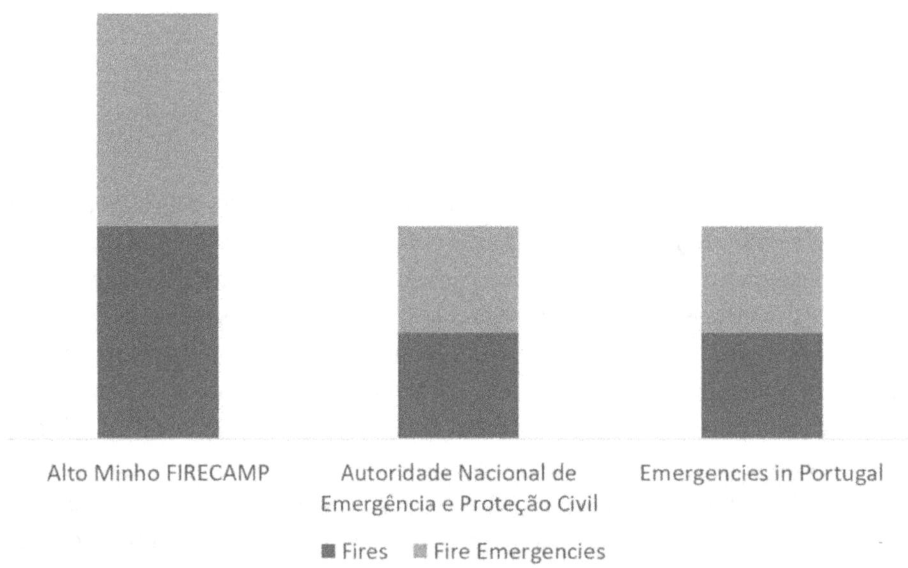

After analyzing data and outlining a profile of the forest fires theme on digital platforms, the most outstanding results on all platforms were iditified and related to each other. In this context, the graph below (9) presents the results that stand out most in each analyzed category. This graphic allows us to understand the diversity of the data collected, its heterogeneity, and the limited possibility of systematization. It was worth to notice that 3 of the 7 results that stand out are official digital platforms from governmental sources (Portuguese Armed Forces, Prociv and Proteção Civil Loulé – Prociv Loulé). The others emerged from the community (Facebook groups) and private initiative (Fogos.pt).

In terms of contents, by analyzing all the digital platforms, the majority of results were integrated the category alert and information - maps (57), followed by the categories of specialized information (39), general information (37), education and prevention (26) and causes (17) (see graph 10).

Figure 9. Outstanding results in each digital platform

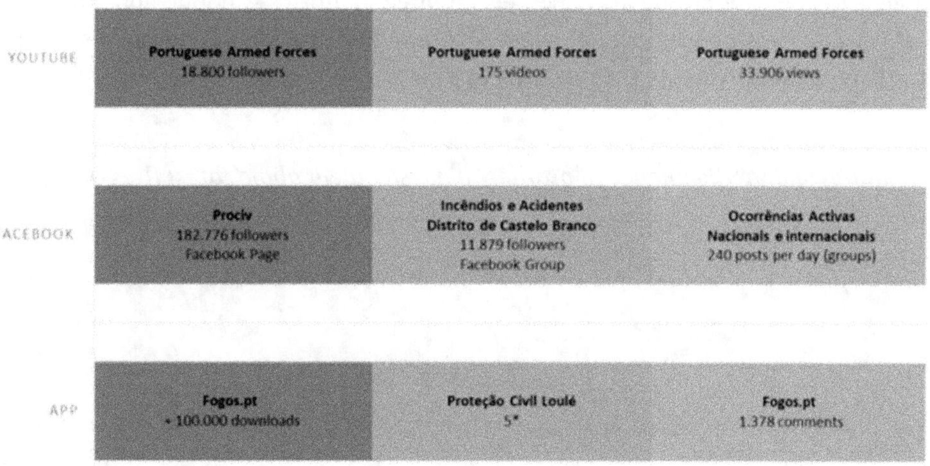

Figure 10. results distributed by major content themes

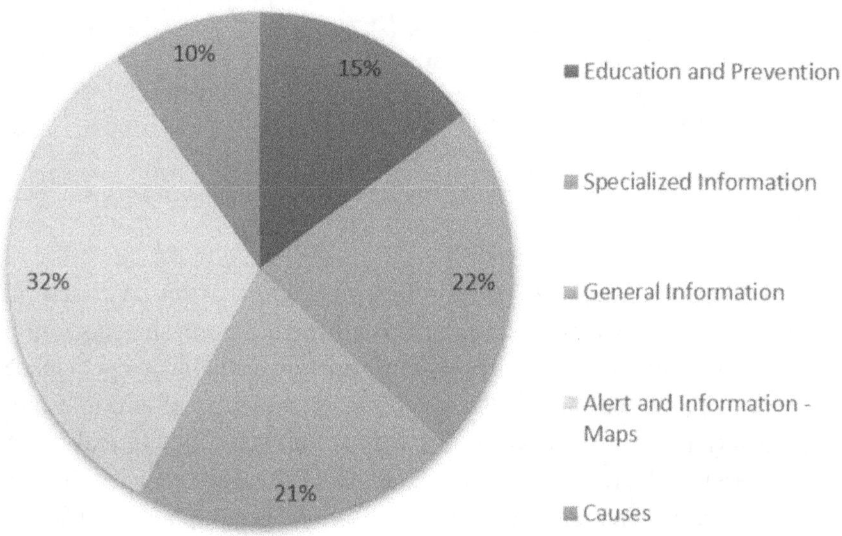

The authors also identified the profiles that appeared the most on all digital platforms. Here, the National Emergency and Civil Protection Authority (Prociv) is the one that stands out the most, being present on the web, on Facebook, and YouTube. Although, on each platform, it presents different content. On the web, the Prociv page fits into the category of alert and information through maps; on Facebook, it discloses general information about emergencies; on YouTube, it presents prevention content. Here it is also noted that, despite not having its app, there are four apps of this institution with regional and local profiles (Prociv Azores, Prociv Madeira, Prociv Albufeira, and Prociv Loulé), all of them focused on prevention. The Fogos.pt profile also operates on three platforms - mobile app, Facebook, and web - acting in all of them with content about event alert and information through maps. In the following graph (11), it is possible to identify the relationships between the different platforms.

Figure 11. Relation and predominance of profiles/themes/content about forest fires on digital platforms

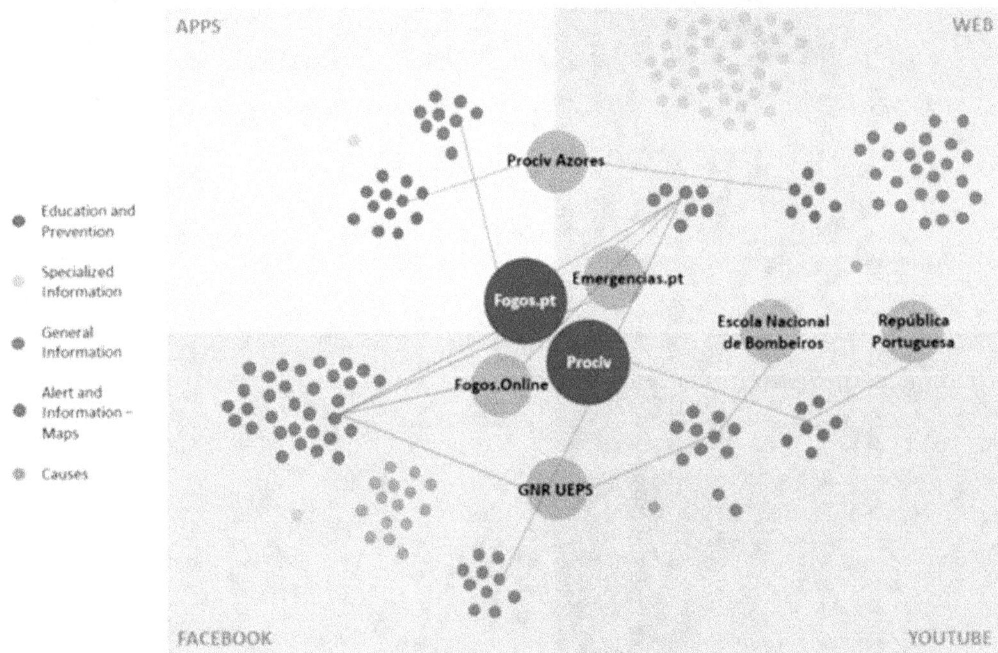

By reading this graph, we can identify patterns that profile forest fires on digital platforms. First, it appears that events alert and information through maps are the most prevalent content in different types of digital platforms, which allows us to assume that information available in cyberspace focuses mainly on the event/firefighting phase. This trend also consolidates an overlap of multiple apps and web pages with the same features (maps where forest fire occurrences are identified in real-time) or with several Facebook pages centered on the fire events' alert. The main difference between the platforms is the information origin. In the case of apps and web pages, forest fires' geographic identification comes mainly from institutions and does not rely on public collaboration (they are unidirectional information vehicles). In the case of Facebook pages/groups, the public has an active role in sharing information about alerts and fire events. This is partly in line with the literature trend, where communities are mobilized to share information about emergency situations on social media. Nevertheless, any institutional page/profile who encourage citizen VGI sharing on OSM platforms was found, nor any of them mentioned information from crowdsourcing, as it was found as a trend in the literature review.

If we relate pages with a higher presence online and the topics addressed, the focus is on forest fires alerts. This also fits in with the literature trends: the use of digital platforms to share information in emergency situations has been growing, particularly with geographic information through maps. It was also noted a high presence of official emergency management institutions online, which means Portugal organizations are following the trends and adapting to the new digital communication formats. However, it seems they are struggling with the horizontal communication format since the citizens did not contribute to the information in their different digital platforms and can only comment on Facebook posts.

Also, the literature shows that prevention is one of the leading solutions to mitigate forest fires. However, this is an area to be developed in Portuguese cyberspace. Even in mobile apps, where there are more profiles dedicated to the prevention, the truth is that many of these are international apps, which may have barriers to penetrate in the Portuguese digital environment. In the same sense, prevention does not focus only on the training of professionals and specialists in the field but also on communities' mobilization. Here, social media play a crucial role in fostering a more participatory approach. However, graph 11 shows a little presence of prevention and education contents in social media, particularly on Facebook, where this content is not found.

It is also on Facebook that support profiles for causes were found. Despite this being the category with the least presence in Portuguese cyberspace, 15 Facebook profiles focused on this theme were found, which reveals forest fires raise solidarity between community, as shown in the literature as a trend.

CONCLUSION

This research intended to outline the forest fires theme on digital platforms in the Portuguese context. It was found that forest fires are an increasing natural catastrophe, mainly due to climate change impacts. Countries like Portugal, located in the Mediterranean region, are most affected by this trend, where forecasts point out an increase in forest fires scenarios. It was also noticed that Portugal leads the European rankings in the number of fires and burnt areas, having been significantly affected by massive fires in 2017 and 2018.

Throughout the work, the authors analyzed the forest fires theme in the Portuguese digital universe. The data support the trends verified in the literature, with a high echo of this theme on digital platforms. However, as mentioned earlier, the digital platforms' heterogeneous nature and the constraints to collect data did not allow a systematic analysis. Even so, after analyzing the contents of mobile apps, web pages, Facebook pages and groups, and YouTube channels, it is possible to conclude that the forest fires theme in Portuguese cyberspace focuses mainly on real-time events alert based on maps visualization, as shown in graph 10. A high number of results on different platforms with similar features focus on sharing information about occurrences with geographic information was found, which, once again, meets the trends pointed out by the literature. However, in terms of features and contents, most digital platforms focused on fire events alert do not integrate a participatory approach. These are particularly notorious on mobile apps and web pages, since they do not allow the public to contribute to communicate fire events and present data in a unidirectional way. This is particularly relevant in official governmental sources since none of its digital platforms perform a participatory approach. In this case of forest fire alert with geographical information, the exception was on Facebook (graph 5). Here, several pages and groups rely on public information about forest fire events, mostly with regional or local scopes.

Literature also suggests that in the context of forest fires, the mitigation of these emergency situations must focus on prevention, either through public awareness, forest and fire management, or the community mobilization towards an active attitude about this issue. In this context, digital platforms play a key role, allowing new forms of communication and participatory approaches, where citizens and stakeholders can collaborate in forest fires situations. Nevertheless, in the Portuguese digital scenario, as graph 10 and 11 shows, there is an underuse of digital platforms regarding prevention. This is highlighted particularly on Facebook. This social media did not offer any page or group focused on forest fire prevention. Since Facebook is one of the most used social media in Portugal, this social network site should be used to mobilize communities for preventive practices in forest fires.

The information spread based on knowledge co-creation, and community mobilization takes place mainly on Facebook. It is mostly found during the emergency phase, with several Facebook pages and groups that allow citizens to communicate fire events. It also appears in the post-fire phase by supporting victims and professionals. Here, profiles devoted to "causes" appear mainly on Facebook (graph 5), showing that forest fires arise solidarity between community, as shown in the literature as a trend. This indicator may serve for a future further reflection about the social mobilization generated by the forest fires theme in social media, particularly through an in-depth analysis of the contents of these profiles.

In general terms, graph 11 allows concluding that information about forest fires in digital platforms is dispersed and unstructured in the Portuguese cyberspace. General and specialized information can be found mainly on web pages, although also in a scattered form. Also, the data analyzed shows the underuse of prime platforms, such as YouTube. This is the most used social media in Portugal. However, the use of this digital platform for sharing content about forest fires is still at an early stage. Here, information is dispersed, poorly systematized, with a few institutional or personal channels specifically focused on forest fires. This means that YouTube is being neglected by the public and by organizations. In the forest fires theme, YouTube could serve as a collective mobilization platform for citizens, as well as a vehicle to reach a wider audience for institutions that could even use it to foster this public mobilization.

This chapter integrates extensive research, focused on knowledge sharing and increasing literacy on forest fires. The main goal of this comprehensive research is to create a conceptual model of a digital platform specifically focused on the issue of forest fires. Characterizing the presence of the forest fires theme in Portuguese cyberspace is essential to understand the digital scenario, the limitations, and assets that must be integrated into the conceptual model under development. Nevertheless, this work itself contributes to understanding digital platforms' role in emergency scenarios such as forest fires. There is a sizeable communicational potential in cyberspace. There is a high echo of forest fires in the Portuguese context, and there are interest and mobilization through communities and stakeholders to alert for forest fires events and support victims of forest fires. However, participatory approaches are underused, particularly by the official government digital profiles, where it was noticed a unidirectional way of information spread. Also, prevention is essential to minimize forest fires events and mitigate its consequences, and regarding this content, there is still a lot to explore on Portuguese digital platforms. This content is poorly approached in every digital platform, particularly on social media (Facebook and YouTube), where it has a great potential to increase community engagement to prevent forest fires. These thematics should be developed in future studies.

ACKNOWLEDGMENT

This research was supported by the Portuguese Foundation for Science and Technology (FCT) through the FCT fellowship [SFRH/BD/140247/2018].

REFERENCES

Ahmouda, A., Hochmair, H. H., & Cvetojevic, S. (2018). Analyzing the effect of earthquakes on OpenStreetMap contribution patterns and tweeting activities. *Geo-Spatial Information Science*, *21*(3), 195–212. doi:10.1080/10095020.2018.1498666

Appleby-Arnold, S., Brockdorff, N., Fallou, L., & Bossu, R. (2019). Truth, trust, and civic duty: Cultural factors in citizens' perceptions of mobile phone apps and social media in disasters. *Journal of Contingencies and Crisis Management*, *27*(4), 293–305. doi:10.1111/1468-5973.12282

Athanasis, N., Themistocleous, M., Kalabokidis, K., Papakonstantinou, A., Soulakellis, N., & Palaiologou, P. (2018). The emergence of social media for natural disasters management: A big data perspective. *International Archives of the Photogrammetry, Remote Sensing and Spatial Information Sciences - ISPRS Archives*, *42*(3W4), 75–82. doi:10.5194/isprs-archives-XLII-3-W4-75-2018

Bhavaraju, S. K. T., Beyney, C., & Nicholson, C. (2019). Quantitative analysis of social media sensitivity to natural disasters. *International Journal of Disaster Risk Reduction*, *39*(July), 101251. doi:10.1016/j.ijdrr.2019.101251

Boulianne, S., Minaker, J., & Haney, T. J. (2018). Does compassion go viral? Social media, caring, and the Fort McMurray wildfire. *Information Communication and Society*, *21*(5), 697–711. doi:10.1080/1369118X.2018.1428651

Brengarth, L. B., & Mujkic, E. (2016). WEB 2.0: How social media applications leverage nonprofit responses during a wildfire crisis. *Computers in Human Behavior*, *54*, 589–596. doi:10.1016/j.chb.2015.07.010

Climate change impacts and adaptation in Europe - JRC Science for Policy Report. (2020). doi:10.2760/171121

Çolak, E., & Sunar, F. (2020). The importance of ground-truth and crowdsourcing data for the statistical and spatial analyses of the NASA FIRMS active fires in the Mediterranean Turkish forests. *Remote Sensing Applications: Society and Environment*, *19*(March), 100327. doi:10.1016/j.rsase.2020.100327

Costa, H., de Rigo, D., Libertà, G., Houston Durrant, T., & San-Miguel-Ayanz, J. (2020). *European wildfire danger and vulnerability in a changing climate : towards integrating risk dimensions.*, doi:10.2760/46951

Craglia, M., Ostermann, F., & Spinsanti, L. (2012). Digital Earth from vision to practice: Making sense of citizen-generated content. *International Journal of Digital Earth*, *5*(5), 398–416. doi:10.1080/17538947.2012.712273

de Rigo, D., Libertà, G., Houston Durrant, T., Artés Vivancos, T., & San-Miguel-Ayanz, J. (2017). *Forest fire danger extremes in Europe under climate change: variability and uncertainty*. JRC Science Hub., doi:10.2760/13180

Eriksson, M., & Olsson, E. K. (2016). Facebook and Twitter in Crisis Communication: A Comparative Study of Crisis Communication Professionals and Citizens. *Journal of Contingencies and Crisis Management*, *24*(4), 198–208. doi:10.1111/1468-5973.12116

Forest fires. (2020). Retrieved from https://www.eea.europa.eu/data-and-maps/indicators/forest-fire-danger-3/assessment

Google. (n.d.). *Google Trends*. Retrieved May 25, 2020, from https://trends.google.pt/trends/?geo=PT

Haworth, B., Whittaker, J., & Bruce, E. (2016). Assessing the application and value of participatory mapping for community bushfire preparation. *Applied Geography (Sevenoaks, England)*, *76*, 115–127. doi:10.1016/j.apgeog.2016.09.019

Haworth, B. T. (2018). Implications of Volunteered Geographic Information for Disaster Management and GIScience: A More Complex World of Volunteered Geography. *Annals of the Association of American Geographers*, *108*(1), 226–240. doi:10.1080/24694452.2017.1321979

Hootsuite & We Are Social. (2020). *Digital 2020 Portugal. Global Digital Insights*. Author.

INE - Instituto Nacional de Estatística (Ed.). (2019). *Estatísticas do Ambiente 2018. Instituto Nacional de Estatística*. INE - Instituto Nacional de Estatística.

Jurgens, M., & Helsloot, I. (2018). The effect of social media on the dynamics of (self) resilience during disasters: A literature review. *Journal of Contingencies and Crisis Management*, *26*(1), 79–88. doi:10.1111/1468-5973.12212

Liu, Y., Zhu, J., Shao, X., Adusumilli, N. C., & Wang, F. (2020). Diffusion patterns in disaster-induced internet public opinion: Based on a Sina Weibo online discussion about the 'Liangshan fire' in China. *Environmental Hazards*, *0*(0), 1–25. doi:10.1080/17477891.2020.1758608

Lundgren, R., & McMakin, A. (2018). *Risk Communication A Handbook for Communicating Environmental, Safety, and Health Risks* (6th ed.). IEEE Press Wiley.

NE - Instituto Nacional de Estatística (Ed.). (2020). *Anuário Estatístico de Portugal 2019 / Statistical Yearbook of Portugal 2019*. INE - Instituto Nacional de Estatística.

Pekar, V., Binner, J., Najafi, H., Hale, C., & Schmidt, V. (2020). Early detection of heterogeneous disaster events using social media. *Journal of the Association for Information Science and Technology*, *71*(1), 43–54. doi:10.1002/asi.24208

Reuter, C., Stieglitz, S., & Imran, M. (2020). Social media in conflicts and crises. *Behaviour & Information Technology*, *39*(3), 241–251. doi:10.1080/0144929X.2019.1629025

Riccardi, M. T. (2016). The power of crowdsourcing in disaster response operations. *International Journal of Disaster Risk Reduction*, *20*(November), 123–128. doi:10.1016/j.ijdrr.2016.11.001

Sachdeva, S., McCaffrey, S., & Locke, D. (2017). Social media approaches to modeling wildfire smoke dispersion: Spatiotemporal and social scientific investigations. *Information Communication and Society*, *20*(8), 1146–1161. doi:10.1080/1369118X.2016.1218528

San-Miguel-Ayanz, J., Durrant, T., Boca, R., Libertà, G., Branco, A., Rigo, D., & de, … Leray, T. (2019). Forest Fires in Europe. *Middle East and North Africa, 2018*. Advance online publication. doi:10.2760/1128

Schmuck, G., San-Miguel-Ayanz, J., Durrant, T., Boca, R., Libertà, G., Petroliagkis, T., … Schulte, E. (2018). *Forest Fires in Europe, Middle East and North Africa 2017. Scientific and Technical Research series*. doi:10.2788/1082

Spies, T. A., White, E. M., Kline, J. D., Paige Fischer, A., Ager, A., Bailey, J., ... Hammer, R. (2014). Examining fire-prone forest landscapes as coupled human and natural systems. *Ecology and Society, 19*(3), art9. Advance online publication. doi:10.5751/ES-06584-190309

Spinsanti, L., & Ostermann, F. (2011). Retrieve volunteered geographic information for forest fire. *CEUR Workshop Proceedings*, 704.

Spinsanti, L., & Ostermann, F. (2013). Automated geographic context analysis for volunteered information. *Applied Geography (Sevenoaks, England)*, *43*, 36–44. doi:10.1016/j.apgeog.2013.05.005

Tavra, M., Racetin, I., & Peroš, J. (2019). Combining social media and authoritative data for crisis mapping: A case study of a wildfire reaching Croatian city of split. *International Archives of the Photogrammetry, Remote Sensing and Spatial Information Sciences - ISPRS Archives, 42*(3/W8), 415–420. doi:10.5194/isprs-archives-XLII-3-W8-415-2019

Turco, M., Bedia, J., Di Liberto, F., Fiorucci, P., Von Hardenberg, J., Koutsias, N., Llasat, M.-C., Xystrakis, F., & Provenzale, A. (2016). Decreasing fires in mediterranean Europe. *PLoS One*, *11*(3), e0150663. Advance online publication. doi:10.1371/journal.pone.0150663 PMID:26982584

Viegas, X., Almeida, M. F., Ribeiro, M., Almeida, M. A., Raposo, J., Viegas, M. T., … Viegas, C. X. (2019). *Analysis of the Forest Fires Ocurred in October 15, 260*. Retrieved from https://www.portugal.gov.pt/download-ficheiros/ficheiro.aspx?v=c2da3d7e-dcdb-41cb-b6ae-f72123a1c47d

Yang, C., & Tian, W. (2017). Social media geo-sensing services for EO missions under sensor web environment: Users sensing information about the Ya'an earthquake from Sina Weibo. In *2017 6th International Conference on Agro-Geoinformatics* (pp. 1–6). IEEE. 10.1109/Agro-Geoinformatics.2017.8047032

Yue, Y., Dong, K., Zhao, X., & Ye, X. (2019). Assessing wild fire risk in the United States using social media data. *Journal of Risk Research*, *0*(0), 1–15. doi:10.1080/13669877.2019.1569098

Zhong, X., Duckham, M., Chong, D., & Tolhurst, K. (2016). Real-time estimation of wildfire perimeters from curated crowdsourcing. *Scientific Reports*, *6*(1), 1–10. doi:10.1038rep24206 PMID:27063569

KEY TERMS AND DEFINITIONS

Crowdsourcing Data: Digital data gathered from different digital sources with various formats or types.

Digital Platforms: Any kind of platform based on an online environment. It could be a website, a social media page, a mobile app, a digital wearable, a digital game, an artificial intelligent object, etc.

Forest Fires: A natural or human-made disaster that burns significant areas of forestry. It usually happens in the summer season and can also affect rural populations, causing environmental, economic, and social losses. Climate changes increase forest fires events.

Local Knowledge: Informal knowledge that people have about their places, territory, local uses, and local culture. It is passed from generation to generation in an informal way.

Mobile Apps: A type of digital platform specifically developed to mobile environments and used in smartphones, tablets, or other digital mobile devices.

Social Media: Digital platforms used to share digital contents and connect people through a network. Usually, social media support different multimedia content and have various features that increase connection and communication.

Volunteered Geographic Information (VGI): Information shared voluntarily by digital users on digital platforms with metadata regarding geo-localization. Could be as images, text, or another type of data. Usually, it is shared in disasters, catastrophes, or emergency situations to help the authorities and spread information among users.

ENDNOTES

1. 2018 is the most recent data to forest fires available by European Union.
2. Data for January 2020.
3. "fogos" is the Portuguese word for "fires" and "Pedrogão Grande" is the name of the municipality where the big forest fire has occurred in 2017.
4. "Fogo de Monchique" means "Monchique fire". Monchique is the name of the region affected by a big forest fire in 2018.
5. The search was made in Portuguese and four expressions were used: "incêndios", "fogos", "emergências incêndios" and "emergências fogos". Both first words means "fire" and both last expressions means "fire emergencies" The results are present in English, so the English expressions congregate the searches in their two terms correspondents in Portuguese.
6. In the Youtube search the search expressions/terms were also the Portuguese expressions used in the other digital platforms. The only difference was the reference to the country. So the four searched expressions were "incêndios Portugal", "fogos Portugal", "emergências incêndios Portugal" and "emergências fogos Portugal".
7. The diferences visible in the graphic bars are due to the number of searches. In the original research, searches were made in Portuguese and it was necessary to make four searches with the Portuguese terms mentioned in endnote 5.

Chapter 8
Evaluation of Research Trends in Social Media Crisis Communication

Daniel Ikesinachi Nwogwugwu
https://orcid.org/0000-0001-8873-4014
Bowen University, Iwo, Nigeria

ABSTRACT

An organization's survival during a crisis often depends on its speed of response. The introduction of social media into crisis communication discourse has meant that organizations must revisit their crisis communication strategies. This chapter explores a content analysis of the integration of social media platforms into crisis communication based on a comprehensive review of eight purposively selected crisis studies conducted globally. Findings revealed that Facebook and Twitter are increasingly employed as platforms for crisis communication. It was also discovered that responding to crises promptly, and engaging with the publics before, during, and after crises are crucial to managing organizational reputation. Social media platforms are also capable of spreading mis(information) about crises. Thus, organizations are advised to fully integrate and adopt social media into their crisis communication plans. This chapter extends our understanding of how social media platforms contribute to crisis communication discourse.

INTRODUCTION

The introduction of social media into crisis communication discourse calls for concerted organisational efforts in militating against its effects. This call becomes necessary as social media has the potential to spread (mis)information about crises within a short time and placing organisations at higher risks of exposure to crises. Also, the organisational reputation – described as intangible assets - are at stake and could be damaged during crises. As Veil, Sellnow & Petrun (2012) posit, social media are channels for stories to go viral within a short period. With the *share* option on Facebook and *retweet* option on Twitter, citizens can disseminate (mis)information, while organisational activities could be closely monitored

DOI: 10.4018/978-1-7998-6705-0.ch008

online. Thus, to avoid misinformation during crises, organisations need to fully optimise their abilities to handle crises in the social media era.

Social media crises are a common organisational phenomenon that keeps evolving. As Coombs, Claeys & Holladay (2017) observed, the application of social media to crisis communication is the most dramatic evolution of the practice in need of greater understanding. Hence, researchers have sought to understand the use of social media during airline crises (e.g., Adebayo, 2017; Benoit, 2018; Brataas, 2018), social media's use during natural disasters (Cho, Jung & Park, 2013), university crisis on social media (Snoeijers, Poels and Nicolay, 2014), food crisis discussions on social media (Shan, Regan, Brun, Barnett, van der Sanden, Wall & McConnon, 2013), health crises on social media (Lawrence, 2016), brand crisis communication through social media (Wang, 2016), among other studies. Findings from these studies revealed several dynamics in social media crises. These studies also revealed the growing interests and involvement of the publics in social media crisis discourse. Considering the increasing amount of crisis research, it becomes necessary to comprehensively present an analysis of the incorporation of social media by crisis managers to manage social media crises.

Therefore, this chapter explores a content analysis of the integration of social media platforms into crisis communication. The objective of this study is to understand how crisis managers employed social media platforms in responding to crises through an analysis of eight (8) purposively selected social media crises conducted globally and published from 2006 to 2019. Thus, this chapter aims to understand the strategies, trends, and paradigms that have characterised social media crisis studies. For instance, the chapter will analyse social media crisis strategies, theoretical explanations, and an investigation (if any) of new approaches that have emerged from the discourse. Findings from this study will highlight the relevance of digitalisation and its significance on different crisis types. Furthermore, it will also be useful to crisis communication and social media students as it would be a reference point for crisis communication studies.

BACKGROUND

What is Crisis?

Several definitions have been given to the word 'crisis' by various scholars and practitioners. These diverse definitions arose since there has not been any universally accepted definition of crisis. More so, these scholars and practitioners come from various disciplines, research backgrounds, with diverse experiences. While these definitions are diverse, they are yet similar. An attempt is made to provide some definitions of crisis, deduce similar crisis characteristics, and provide a working description for this study.

The word crisis originated from 'krisis,' a Greek word which suggests a moment of decision. These decisions could represent turning points where the choices made could have fundamental impacts in the future (Sellnow and Seeger, 2013). Crises present moments where critical decisions that would affect an organisation or country (positively and/or negatively) are made, which in return, could make or mar their future. It is quite true as, during crises, public relations practitioners and crisis managers seek the best possible alternatives in dealing with crises such that there is minimal or reduced reputational damage. Also, crises could be turning points or opportunities for organisations. On the one hand, when organisations poorly manage crises, it could signal the end of such organisations while on the other hand, when crises are well managed, it could present positive opportunities for growth and development. Either way, they are turning points in the history and life of such organisations or countries, as the case may be.

In one of the earliest definitions of crisis by Herman (1963), crises are surprising situations that threaten the high-priority goals of organisations and come with time-constraint in the decision-making process. According to Herman, crises are capable of threatening critical organisational goals and usually have a short response time. Herman's definition laid the foundation for other crisis ideas as other scholars built their ideas around Hermans' idea of crisis as threats to high-priority organisational goals and having short response time. For instance, Fearn-Banks (1996) defined crises as major occurrences with potentially adverse outcomes that could affect organisations or industries, publics, products, services, or good names. It interrupts routine business transactions and can sometimes threaten the existence of organisations. This definition further describes crises as threatening the existence of organisations and their goals, such as the publics, products, reputation, among others.

In another definition, Seeger, Sellnow & Ulmer (1998) describe an organisational crisis as "a specific, unexpected and non-routine event or series of events that create high levels of uncertainty and threaten or are perceived to threaten an organisation's high-priority goals" (p. 233). This definition aptly describes and combines the earlier identified characteristics of a crisis. Crises are specific. They are events that are identifiable and easy to distinguish. Similarly, Coombs (1999) describes crises as events that are unpredictable, capable of causing significant threats, and can have negative effects on organisations, industries, or stakeholders, when they are not adequately handled. Thus, Coombs (1999) makes a case for the proper management of crisis as a panacea to handling negative crisis effects.

These definitions of crises reveal certain characteristics. Some of these characteristics include incidents that are surprising, unpredictable, with short response time, threats to high-priority goals, disruption of routine activities, usually followed by media involvement as well as negative/positive publicity. Thus, for this chapter, a crisis is defined as an unexpected (and sometimes expected) occurrence that disrupts high priority goals with short response time and attracts high media attention. Thus, to mitigate the effects of crises, organisations are expected to respond by communicating effectively with all stakeholders.

Crisis Communication

Crises are unexpected incidents with potentially negative outcomes. It is assumed that a day hardly goes without the news of a new crisis or the escalation of a previous one. Since crises are surprising, unpredictable, and sometimes predictable events with the potentials of negatively affecting organisational goals, concerted efforts should be channelled towards managing them. As Fearn-Banks (2007) observed, a critical feature of managing crises is communication, which is the exchange of meaningful information that organisations have with stakeholders before, during, and after crises. Organisations plan to mitigate the (adverse) effects of crises by having constant interactions with its stakeholders before, during, and after crises. Thus, communication is a continuous and all-encompassing activity for organisations in dealing with crises. Ray (1999), cited in Stephens, Malone & Bailey (2005), believes that an essential objective of communicating during a crisis is to influence the perception of the stakeholders towards the organisation and to maintain a positive or restore damaged image among stakeholders. Communication is necessary during crises because stakeholders/publics have formed opinions about the crises and the organisation. Thus, organisations can communicate effectively and change these (negative) perceptions.

As enunciated by Coombs (2009), crisis communication focuses on how organisations use communication to manage information and meaning during crises. It provides information that helps stakeholders to make meaning out of the crisis. In essence, communication during a crisis is necessary as it will assist stakeholders with the necessary information which they need to make sense of the crisis. Thus, Coombs (2010) defines crisis communication as "the collection, processing and dissemination of information required to address a crisis situation" (p. 20). Communication involves a two-way interactive process that organisations can employ in promoting change and influencing the perceptions that stakeholders will have about the organisation. To this end, crisis communication with stakeholders, including victims of the crisis, potential victims, and family and friends of these victims, could assist organisations to mitigate or contain the adverse effects associated with crises.

Crisis Communication (Response) Strategies

Crises, as earlier discussed, are capable of threatening important organisational goals. It becomes more worrisome as no organisation is said to be immune to crises. This is because when organisational crises occur, their effects could be devastating, including having negative impacts on the financial performance, return on investment, competitive advantage, and could eventually lead to bankruptcy (Coombs & Holladay, 2008 as cited in Gerken, Van der Land & van der Meer, 2016). Thus, during crises, organisations make efforts to ensure that they minimise these effects. One significant organisational effort during crises is to respond by employing strategies aimed at protecting their image and reputation. Thus, crisis communication (response) strategies are organisational approaches in response to crises.

A crisis response strategy is a form of crisis communication that focuses on how organisational responses are presented to the public and how messages are narrated to repair their reputation (Coombs, 2007). Crisis response strategies refer to the specific approaches taken by organisations to respond to crises by presenting their narration of the crises. Crises narrations are necessary because, during a crisis, several versions of the cause(s) of such crises would be available (especially with the advent of social media). Thus, organisations respond by providing their own stories in a timely and accurate manner to influence the perceptions of stakeholders. As Coombs, Frandsen, Holladay & Johansen (2010) simply stated, crisis response can "either improve or make the crisis situation worse for a corporation and its various stakeholders" (p. 337), as the response strategy can be the difference between success and failure. Hence, appropriate crisis response strategies can assist organisations to shape attributions of a crisis, influence stakeholder perceptions about the crisis and organisation, and also reduce adverse crisis effects.

Several strategies, in the form of frameworks or theories, have been developed by scholars that are helpful to crisis managers when responding to crises. An early crisis scholar, Sturges (1994) explained a model that could assist organisations in responding to crises. He proposed that organisations should provide instructing and adjusting crisis information in mitigating the effects of crises. Instructing information involves crisis communication content from organisations in crises to update the public about what to prepare for, how to react, and the further actions needed to be taken. Adjusting information, on the other hand, involves addressing stakeholders with messages of sympathy and reassurance and updates on measures taken to avert the reoccurrence of crisis.

Similarly, Benoit (1995) developed the *image restoration/repair theory,* which presents message strategies available to organisations to protect their reputation. Also, the *situational crisis communication theory*, one of the most central and widely tested theories in crisis communication discourse, examines the reputational threats accompanied by various crises, and provides strategies in dealing with these threats. The introduction of social media in crisis communication discourse also enabled the development of the Social Mediated Crisis Communication (SMCC) Model. These theories will be explained further in the theoretical framework section.

Social Media

Social media represent an entirely new platform for crisis communication discourse. The advent of social media has meant that crisis and crisis information can spread faster and reach a wider audience within a short period. Also, with social media, citizens act as watchdogs and can engage in "citizen journalism" with the ability to share (real and fake) news around the world. Social media, also known as social networking sites (SNS), are computer-mediated tools such as Facebook, LinkedIn, Twitter, YouTube, WhatsApp, among others. SNS tools enable users to interact and communicate with one another. Kaplan & Haenlein (2010) opined that social media are groups of Internet-based applications that build on the ideological and technological foundations of Web 2.0. Social media allows for interaction among people irrespective of distance and time.

The introduction of social media has changed the way journalists, public relations practitioners, and the publics communicate. As earlier observed, it has facilitated citizen journalism where many citizens are self-imposed journalists who can share insights, opinions, experiences, and personal perspectives using pictures and videos of events with others. Coombs (2010) observed that one crucial feature of social media networks is its capacity to facilitate bi-directional communication or dialogue. In other words, social media have enabled a two-way flow of communication between users. Hence, organisations can communicate directly with stakeholders, and producers can communicate directly with consumers using social media platforms.

In the next section, an attempt is made to examine the importance of social media in crisis communication discourse.

Social Media and Crisis Communication

Organisations have employed social media platforms as tools for marketing and public relations. However, in recent years, social media have been employed as platforms for the management of crises. Crisis communication discourse reveals that the focus of crisis communication shifted from examining crisis strategies as reported by the traditional media to examining crisis communication on social media. As Marcias, Hilyard & Freimuth (2009) posit, social media may be better matched to crises than traditional media because technologies allow for the quick flow of information with the capabilities of uploading and downloading of content. Thus, social media has completely transformed how crisis information is presented and disseminated as information can spread faster and reach broader audiences. To this end, Wigley and Zhang (2011), as cited by Hilding-Hamann (2012:12), assert that "a key to successfully handling crises is controlling the flow of information to the media and one's publics." Handling information flow from social media with characteristics such as immediacy, viral news tendencies as well as the ability to also pass across (mis)information to the publics is necessary for public relations practitioners. This chapter responds to the call for research on the impact of social media on crisis management.

As Coombs and Holladay (2012), cited in Coombs and Holladay (2014) observed, "social media represent a cluster of media that facilitate the growth of crises as well as provide channels that can be used to communicate during a crisis" (p. 43). Through this, public relations practitioners and crisis managers have an array of platforms to respond to crises. Social media thus becomes a powerful crisis channel for the dissemination of crises news and information, for facilitating the chances of crises and also a necessary media in managing crises. As crisis channels, social media messages can likely trigger different questions and reactions that require timely responses. The introduction of social media also enables the active participation of the publics and thus, exposing organisations to the vulnerability of crises as citizens around the world can watch as crises and disasters unfold. Furthermore, Veil et al. (2012) opine that social media are channels for a story to go viral in a short period. Thus, within a split second, social media users can post happenings about a crisis for the consumption of other internet users. Studies have revealed that people rely more on the information made available online as the primary means of receiving crisis information. The implication is that even when the information is negative, it is still received by the publics. Therefore, the essence of organisations communicating during a crisis is to disseminate their crisis narrative to the publics to avoid rumours.

As a crisis communication platform, social media has been described as a double-edged sword that could be employed to create and escalate crises, and could also be employed in managing crises. That is, social media presents both advantages and disadvantages for organisations during crises. No doubt, social media could be beneficial as a crisis management tool when used correctly by organisations. Perry, Taylor & Doerfel (2003), in their discussions on the advantages of social media to crisis management, argued that social media could assist organisations in their effective management of crises as strategies could be searched for and are readily available online. Second, social media also assists organisations to communicate decisions quickly through various social media platforms during a crisis and also get feedback from the stakeholders. Similarly, Auer (2011) posited that another advantage of social media is in its ability to eliminate middle-men as crises managers and organisations can reach and convince the publics directly and vice versa. Indeed, in recent times, social media platforms have assisted both organisations and stakeholders to make sense of crises. It has assisted organisations to interact with stakeholders and to respond quickly and accurately by giving out crisis information. Also, the publics' reactions to crises can be gotten online. Thus, organisations and crises managers are under pressure to respond to crises by providing information.

However, a disadvantage of social media (platforms) in crisis communication discourse may be that the era of information control by organisations during crises could be limited. Since social media enable the immediate transfer of information, organisations who do not respond swiftly will find it difficult to control misinformation. Thus, crises managers must always be alert and proactive during events that could lead to crises. It is necessary to be alert and proactive as studies have revealed the role of social media in fostering the emergence of fake news as many stakeholders sometimes find it difficult to initially distinguish fake news from accurate ones (Shu, Mahudeswaran, Wang & Liu, 2019; Ogundoyin, Olagunju & Nwogwugwu, 2020).

METHODOLOGY

The exploratory qualitative research design was employed for this study. This design is employed when researchers seek to have an in-depth understanding of an existing issue. Hence, the design will serve as a guide to deeply understand trends in social-mediated crisis communication. Also, the exploratory design is suitable for studies that seek to answer "what" and "why" research questions (Yin, 2009). The findings from exploratory research designs could reveal some patterns that could assist in further research. In gaining insights into the objectives of this study, this chapter adopted a content analysis of eight purposively selected social media crisis studies. These crises cases were purposively selected because they possessed specific characteristics. These characteristics include wide reportage via social media platforms; started via social media; focused on various crises types; the cases must have occurred from 2005 when the social media platforms were launched and became prominent till 2019, and the organisations/government involved employed social media in responding to and managing the crises.

A literature key search was first conducted to select online journal articles on social media and crisis. Articles were then selected based on the characteristics mentioned above with keywords such as "digital media," "digital public relations," "social media crises," and "social media crisis communication." A total of 147 articles were identified, while only eight (8) were purposively selected for analysis. The selected articles are empirical analysis of social media crises undertaken by academic researchers and crisis professionals. Thus, secondary data was collected from only published journal articles. These articles comprised crises from government and non-profit organisations. The unit of analysis was one article. The instrument was coded for three categories. First was the social media employed in responding to the crisis, crisis strategies, theoretical explanations, and an investigation into new findings. Table 1 presents a summary of cases:

Table 1. Summary of Social Media Crisis Cases

S/N	Name of Organization	Crisis Year	Crisis Type	Social Media Involved	Crisis Response Strategies	Theories
1.	Domino's Pizza, US	2009	Human error	YouTube, Twitter	-Mortification (delayed Apology) -Telling the truth -Accepting responsibility -Corrective measures	IRT SCCT
2.	Research in Motion (RIM)	2011	Technical error	Twitter; Facebook; YouTube	-Mortification (Apology)	- IRT - SCCT - Social Media Theory
3.	Japanese Government	2011	Natural disaster	Twitter	-Instructing and adjusting information	
4.	US Centre for Disease Control and Prevention (CDC); World Health Organization (WHO)	2015	Preventable	Twitter; Facebook	Instructing and adjusting information	SCCT
5.	Malaysian Airline MH17	2014	Accident	Facebook	Stealing thunder Instructing and adjusting information Diminish Deny Excuse Victimage	SCCT
6.	University Crisis	2017	Transgression	Facebook; Twitter	Instructing and adjusting information	SCCT
7.	Nigerian government	2012	Wrongdoing	Facebook; Twitter	Instructing and adjusting information	Framing
8.	School shooting crisis	2017	Preventable	Facebook; Twitter	Instructing and adjusting information	SCCT

SOCIAL MEDIA CASE STUDIES

This section presents an analysis of each crisis. A background into each crisis is presented before an in-depth analysis of the relevant trends.

a. Crisis Case One: Social Media and Food Crisis

In April 2009, Domino's Pizza was involved in a crisis that was started and escalated through YouTube. This study conducted by Young & Flowers (2012) examined social media's use in managing the crisis. Domino's Pizza, regarded as a world-leading food delivery company, is present in more than 90 countries worldwide (Domino's, 2020). The crisis started when two employees of Domino's in the US who were bored, made and posted a video of themselves engaging in disgusting activities while preparing sandwich ingredients to be delivered to a customer. Five videos were created in total, with one of the videos showing "an individual sticking mozzarella cheese up his nose and then blowing the cheese on a sandwich, among other unsanitary and stomach-turning activities" (Young & Flowers, 2012 p. 24). These videos were posted online and were viewed by an estimated one million viewers before it was removed two days later. Both employees were relieved of their jobs at Domino's while "the franchise owner discarded all open containers of food and sanitised the location" (Hosseinali-Mirza, Marcellis-Warin & Warin, 2015 p. 11).

Meanwhile, the video had already been downloaded and reposted before YouTube removed it on the request of Domino's (Hosseinali-Mirza et al., 2015). This act led to a severe reputational crisis for Domino's Pizza on social media (YouTube, Twitter, and Facebook) as they received several queries from stakeholders, including consumers and regulatory agencies, who were seeking to authenticate the videos. Specifically, customers took to Twitter to vent their anger for the delayed response from the management of Domino's in responding to the viral video. Questions were raised on the continuous silence of the organisation in responding to the act.

In response to the crisis, however, a statement was first made by the Vice President of Corporate Communications, Tim McIntyre, within the first 24 hours after the video was posted. McIntyre confirmed that, indeed, the incidents took place. Also, in an attempt to further salvage the already damaged reputation, the President of Domino's Pizza, Patrick Doyle, recorded an apology which was uploaded on the same YouTube platform, three days after the incident (Young and Flowers, 2012). In his apology, Doyle highlighted steps taken to correct the mistake, including sanitising the entire organisation and reviewing hiring processes and also promised that such incidents would never happen again.

b. Case Analysis Two: Social Media Use in Technical-Error Crisis

This study, conducted by Hilding-Hamann (2012), examined the social media crisis response strategies of the mobile company, Research in Motion (RIM), during a four-day network outage. RIM, a global leader within wireless technology, was involved in the export of smartphones, tablets, and software and was also the manufacturer of the popular Blackberry smartphones. The crisis occurred in October 2011 when users of Blackberry smartphones suddenly experienced both browsing and messaging delays leading to further network collapse caused by a core switch failure from RIM's infrastructure. The incident first occurred around 10 am on 10th October 2011 and lasted till the evening on 13th October 2011, as the outage was felt among users in Europe, Africa, the Middle East, India, and in many parts of South

America. This network collapse generated negative publicity for RIM as users moved to Facebook to express their frustrations.

Analysis of RIM's response strategies revealed that its first message was posted on its Twitter platform on the evening of 10th October, followed by another notification on its Facebook page almost ten hours later. Also, the CEO of RIM, Mike Lazaridis, was featured in a YouTube message posted on 13th October, where he apologised to customers. Furthermore, there was a Blackberry Service Update through which updates from the organisation were uploaded. In trying to salvage the situation, RIM responded promptly to the crisis with its first Facebook post seeking to reassure the publics that the Blackberry service had been restored. However, the post turned out to be the beginning of more reputational damage for RIM at that time as the crisis had just begun. RIM employed Facebook daily in responding to the concerns of the publics. Facebook posts were employed to continually apologise, explain the progress of the crisis, and update the publics on steps taken by in restoring the network.

The YouTube apology video gathered an estimated 400,000 views and shared on all social media platforms of RIM. In the video, Lazaridis addressed stakeholders about the crisis and sought to redeem the image and reputation of the organisation. Conversely, the stakeholders voiced their concerns on Facebook. Stakeholders took to RIM's Facebook page to express their frustrations at RIM's inability to fix the issues after three days. Full service to the network was finally resolved on the fourth day (13th October) after the crisis began. RIM also employed its Facebook platform to announce the full-service restoration. However, although RIM regularly provided updates on its activities in restoring the network, the organisation chose not to answer any of the threads and comments made by stakeholders, not until 17th October, when the network was finally restored. Their refusal not to respond to stakeholder comments online was in itself, a cynical crisis strategy on social media as best practices reveal that social media platforms in crises should be employed as two-way communication tools in responding to stakeholder concerns.

c. Crisis Case Three: Social Media Use During Natural Disasters

Natural disasters are types of crises that have also attracted scholarly attention. Natural disasters could happen at any time in any country – developed, developing, or underdeveloped. For crises resulting from natural disasters, there are usually fewer attributions of crisis responsibility and control. Cho, Jung & Park (2013) examined how Twitter was used as an interactive and information media by the Japanese government and citizens during Japan's 2011 earthquake. This study was instructive as it provided insights into the social media crisis management strategies by the Japanese government.

When the study was conducted, Japan had an estimated 78.9 percent internet access rate with a 95 percent mobile phone usage rate and approximately 15 million daily visitors on Facebook and Twitter. Thus, it was expected that the number of social media engagement would be high. As Cho et al. (2013) noted, the earthquake started at 2.46 pm on 11th March 2011, in the South Kanto area. During this period, fixed and mobile telephone connections were interrupted, thereby leading to the loss of communication among the citizens. However, citizens broke the news via Twitter, approximately 20 minutes before the mainstream media. Thus, before citizens in other parts of Japan recognised the damage caused by the earthquake, Twitter users had already posted pictures and shared their "earthquake experiences." Through Twitter, the earthquake news spread fast, such that some citizens of Korea, a neighbouring country, had also begun retweeting (sharing) the news while some other launched search teams for their loved ones.

Findings from this study revealed a high use of Twitter in "initiating search and rescue operations, fundraising, providing emotional support, and creating, delivering and sharing information during Japan's 2011 earthquake" (Cho et al. p. 37). However, despite the use of Twitter by citizens and organisations, findings revealed that before the earthquake, the Japanese government made little use of Twitter in interacting with citizens. However, the incident awakened the Japanese governments' interest in using social media to interact with citizens. This interest was a result of the interactions of tweets from the government to the citizens. Some governmental agencies provided useful information on special measures for controlling gas damage, aftershocks, and nuclear power plants rather than the regular economic-related topics. Thus, Twitter's use by the Japanese government after the earthquake increased as it enabled local governments to create Twitter accounts to interact with citizens.

d. Crisis Case Four: Social Media's Role in a Health Crisis

Crises can also result from health pandemics. For instance, studies have examined crises that resulted from the deadly SARS outbreak, the Ebola outbreak, and, more recently, new studies are examining the Coronavirus (COVID-19) pandemic. Health crises present challenges for citizens as crisis information is needed to protect them from harm. Lawrence (2016) examined social media's use (Twitter and Facebook) in reporting measles outbreak in the US. This study, unlike previous studies conducted in the US, was significant as it examined the use of social media by health organisations as well as reactions from the public towards crisis information posted online.

In 2015, there was an outbreak of measles across several states in the United States from February till April 2015. The outbreak was reported to have likely started from an infected traveller who visited the amusement park. Thus, this crisis was preventable only if necessary measures were taken. News and information on the measles outbreak spread quickly on social media with pictures and videos posted online in real-time. Thus, Lawrence (2016) gathered data from 1st February 2015 till 1st April 2015, from the official Twitter and Facebook pages of the relevant government health agencies - Centre for Disease Control and Prevention (CDC); World Health Organization (WHO), and relevant hashtags (#measles #outbreak #vaccine) from Twitter and Facebook to examine public responses to the crisis.

Findings revealed that the Centre for Disease Control and Prevention (CDC) and the World Health Organization (WHO) actively employed social media (specifically Facebook and Twitter) in disseminating health information during the crisis. However, both organisations employed Twitter more than Facebook. Their messages helped citizens to protect themselves and their children from getting infected with the measles. They also kept citizens updated with growing measles cases.

Also, citizens, to a great extent, employed social media in responding to health information from the health agencies. They employed the use of relevant health hashtags in making health information to trend. Also, citizens mentioned the social media accounts of the CDC and WHO as sources for their health information and retweeted (shared) information from both organisations. Also, citizens revealed that they closely adhered to the advice from the health organisations in protecting themselves and families from the measles outbreak.

e. Crisis Case Five: Social Media Use in an Airline Crisis

Social media has also been employed in examining crises within the airline industry. Adebayo (2017) examined Facebook's use by Malaysian Airlines (MAS) after the MH17 Boeing 777-200 crash, which departed from Amsterdam to Kuala Lumpur. The aircraft crashed after a missile was reportedly fired from the Ukrainian airspace on 17th July 2014. This incident came just four months after another missing aircraft, owned by Malaysian Airlines MH370 from Kuala Lumpur, Malaysia to Beijing, China with 12 crew members and 227 passengers were declared missing. In response to this crisis, MAS swiftly reported the crisis news to the public through its Facebook page with a promise to update the public on its Twitter handle. MAS employed the *stealing thunder* approach, where an organisation self-discloses crisis information before the public gets wind of such information from other sources. Thus, within the first 24 hours of the occurrence, MAS had updated its social media pages (Facebook and Twitter) and website with the news of the crash. MAS mostly employed Facebook in communicating with the publics. For instance, between 17 – 20th July 2014, the first four (4) days after the incident, MAS posted messages that provided instructing information and responded to public enquiries about the crisis.

Also, stakeholders employed Facebook to communicate with and react to the crisis. Some of the reactions include expressing grievances, showing empathy with the organisation, supporting the organisation, and blaming the organisation for the crisis. However, MAS also employed the traditional media in responding to the crisis as this was to cater to stakeholders with no social media presence or accounts. This study also revealed that MAS had a robust online relationship and engagement with stakeholders before and during the crisis. Thus, it was easy to garner support for the organisation due to its previous online relationship with the public.

f. Crisis Case Six: University Social Media Crisis

Social media crises have also received scholarly attention from colleges and universities. As earlier discussed, crises can happen to any organisation irrespective of status. Colleges and Universities in the US have been reported to be vulnerable to students' protests, racial discrimination, and violence, potentially impacting their reputation. Thelen and Robinson (2019) examined the social media crisis communication strategies employed by the President of the University of Florida (UF), Kent Fuchs, when a controversial speaker, Richard Spenser, planned to speak at UF. The study aimed at examining how crisis leaders interacted with publics during a crisis and how publics also responded to these crisis communication messages.

Richard Spenser, head of the National Policy Institute, was described as a controversial nationalist speaker due to his white-supremacist propaganda and was disallowed to speak at several universities in the US due to his supposed agenda. Thus, his refusal to be granted permission to speak created severe crises for those institutions. His supporters had engaged or threatened to engage in lawsuits to enable his "controversial messages" to be delivered to them on campus. Therefore, the threats of lawsuits enabled many universities to take steps in addressing the impending crises. It was in the midst of all these issues that Fuchs had invited Spenser to speak and share his ideas.

On 12th August 2017, Fuchs informed the University community of a request granted to reserve space at UF in Gainesville for a speaking event featuring Spencer. Four days later, UF denied Richard Spenser's request to speak, citing potential risks and violent conduct that occurred where Spenser had last spoken. However, findings from the study revealed that Spenser was later cleared to speak. On the day of Spenser's speech, two arrests were reportedly made for minor incidents while three men were arrested and charged with attempted murder. Thelen and Robinson (2019), therefore, analysed statements and information provided by the University's leader on his official Facebook and Twitter pages between 12th August (when the first public statement was made) and 19th November (a month after the event).

Findings from the study revealed that President Fuchs utilised information-giving strategies on social media. He always provided both instructing and adjusting information on Facebook and Twitter to assist students and other stakeholders in making sense of the situation and to protect themselves from harm. Findings also revealed that students responded and reacted to Presidents Fuchs' statements through Facebook and Twitter. Students who clicked the like/love options on Facebook were more than those who clicked on the angry option, implying that more students supported the President's responses. Also, there were more responses from the public when the President gave out factual and concrete information on social media.

Also, Facebook and Twitter platforms were employed by President Fuchs, the management of UF, as well as students and other stakeholders in responding to crisis messages. The crisis was reported to have also escalated via social media as immediate happenings were posted online. Also, similar crises involving Spenser in the past were posted online, which further escalated the situation. Thus, social media contributed to the escalation of the crisis.

g. Crisis Case Seven: Social Media and Protest Against Government Policies

Social media's use by organisations, governments, political actors, employees, and citizens has also attracted scholarly attention. Within governments, citizens have employed social media platforms to show support for favourable policies and also express grievances against unfavourable government policies. An instance of citizens' show of disapproval of a government policy that escalated through social media was the #OccupyNigeria protests against fuel subsidy removal in Nigeria in January 2012. In a study by Egbunike (2015), the #OccupyNigeria was a hashtag by Nigerian online citizens (netizens) who revolted against the sudden increase of Premium Motor Spirits (PMS) from N65 to N140. This sudden PMS increase was announced on 1st January 2012, causing transportation fares across the country to rise to as high as 120% and an astronomical rise in the cost of living.

The resultant effect of this increment was nationwide protests on 2nd January, which first started online through social media platforms (precisely, Twitter and Facebook), before moving to physical (offline) protests. The physical protests were orchestrated via social media as netizens formed groups from different locations, planned and executed on protest strategies. Also, celebrities, religious leaders, civil society groups, labour unions all lent their voices against the increment via social media and were all in support of the protest.

Although the physical protests led to the death of some protesters, the Nigerian government was, however, forced to rescind on its earlier decision by withdrawing the subsidy it had earlier put on the price of PMS. This act, which was made possible by social media, showed the powers of collective actions on social media.

h. Crisis Case Eight: Social Media Use During School Shooting Crisis

This last case focuses on the use of social media during an active school shooting incident in the US. This study, conducted by Thompson, Mazer, Payne, Jerome, Kirby & Pfohl (2016) examined, among other things, the social media challenges faced by school crisis teams during active shooting incidents. This study was noteworthy since it was reported that school shootings in the US increased from 6.4 to 16.4 per year within twelve years, from 2000 till 2012 (Blair & Schweit, 2014 cited in Thompson et al. 2016). Thus, these incidents opened up challenges faced by schools, especially on social media, as it revealed their level of crisis preparedness. Thus, this study examined the challenges associated with information emanating from social media during school shooting incidents.

Findings revealed the emergence of three salient social media challenges faced during active school shooting crises. The first challenge was a lack of social media control, followed by message content issues and social media strategies. As a crisis incident, information about school shootings are difficult to control social media while the struggle to get ahead of social media posts were significant challenges faced by district schools. Besides, the independent nature of social media made it difficult to control online information as there was no basic control or censorship over what the public could say and do, thereby leading to misinformation. While it was difficult to prevent the spread of rumours and misinformation on social media, it was also difficult to stop the online criticisms that arose during active shooting incidents.

Furthermore, responding through message strategies on social media was difficult as students spread misinformation about the shooting incidents. This difficulty arose from the unpreparedness and aftermath of shooting incidents as schools had fewer plans to respond to the crisis on social media. Therefore, respondents identified the lack of social media strategies as areas of need to enhance the management of social media crises. Issues around misinformation and fake news have been fundamental issues arising from social media studies. For instance, Ogundoyin et al. (2020) revealed that social media's influence had increased in Nigeria, such that they were significant carriers of fake news. Hence, findings from this study revealed that students, parents, and teachers found social media use as a difficult challenge during active school shootings.

Social media wields enormous powers in escalating and reducing crises. People take to social media platforms to disseminate, receive, and discuss crisis information. For school districts in the US, it was difficult to counter crisis narratives online because they were not adequately prepared for the communication challenges associated with social media crises. Hence, controlling social media information through adequate social media crisis strategies is an important requirement for organisations.

ANALYSIS AND DISCUSSION

Social Media Platforms in Crisis Cases

One objective of the study was to examine the social media platforms that characterised the crisis studies. Findings revealed that while Facebook, Twitter, and YouTube were the social media platforms employed in managing crises, Facebook and Twitter were dominant platforms. Since the introduction of social media platforms, crisis communication studies have continued to examine the continuous importance of social media to crises. With the knowledge of the global rate of social media acceptance and use, public relations practitioners and crisis managers have incorporated social media platforms as communication channels with the publics.

Statistics as of August 2020 from an online business intelligence portal statistic, *Statista*, currently ranks Facebook as the platform with the highest number of global users while Twitter and YouTube, on the other hand, are among the top ten platforms (Clement, 2020). For crisis discourse, however, Facebook, Twitter, and YouTube are among the top five (5) social media platforms used as channels of communication between organisations and the public. Facebook and Twitter offer easy to navigate processes, the ability to upload videos, and user interactivity in real-time, while YouTube makes use of videos in disseminating messages to the publics. Findings from this study are similar to those of Cheng

(2016) and Huang, Wu & Huang (2016), who also found Twitter, Facebook, YouTube, and blogs as the most frequently employed social media platforms in digital public relations research. Also, Apuke and Tunca (2018) reported similar findings in their study.

Theoretical Explanations

Findings revealed that the Situational Crisis Communication Theory (SCCT), Image Restoration Theory (IRT), social media theory, social-mediated crisis communication (SMCC) theory, and framing theory were employed in the studies. However, findings revealed that the SCCT was the dominant theory, followed by the IRT. As earlier discussed, crises are capable of damaging the reputation of organisations. Hence, these theories present strategies available to public relations practitioners and crisis managers when responding to crises.

The SCCT, introduced by Coombs (1995), focuses on protecting organisational reputation following crises. Social media crises are threats to organisational reputation. Hence, the SCCT presents strategies that crisis managers can employ in managing crises and also stakeholders. The SCCT strategies include denying the existence of the crisis and attacking/blaming the accusers - persons or groups outside of the organisation - as the cause of the crisis. Crisis managers can also seek to reduce (diminish) the organisation's connection to the crisis by providing excuses or justifications for the crisis. Furthermore, crisis managers can seek to rebuild reputation by paying compensation and/or apologising for acts that led to the crisis. Finally, crisis managers, seeking to bolster their reputation, can remind stakeholders of favourite aspects of their past before the crisis or present themselves as victims of the crisis.

The IRT, on the other hand, also provides strategies that help to shape the public's immediate perceptions about crises. The IRT, developed by Benoit (1997), focuses on communication message strategies – what crisis managers can do or say – when faced with crises. The IRT, as opposed to the SCCT, presents communication actions also aimed at maintaining favourable organisational reputation. Some strategies include denial of the crisis incident; evasion of crisis responsibility, but subtle admittance to the crisis incident, reduction of perceived offensiveness of the crisis, corrective actions by promising ways to solve the crisis; and mortification, which focuses on confessing, admitting, and apologising for acts that led to the crisis. Findings from this study on the dominant use of IRT in crisis communication discourse are similar to findings from Avery, Lariscy, Kim & Hocke (2010) as cited by Cheng (2016), and Nwogwugwu (2018).

Social Media Crisis Communication Strategies

An array of crisis response strategies were found in the cases discussed. Findings revealed that instructing and adjusting information are necessary strategies in handling social media crisis. Sturges (1994) posited that instructing information includes content that emanates from organisations in crises to update the public about what to prepare for, how to react, and what further action to be taken. Coombs (2007) also adds that instructive information involves informing stakeholders about what to do to protect themselves physically during crises. Instructing information is necessary as stakeholders need information that could reduce tension and rumours that could lead to reputational damage. Adjusting information, on the other hand, involves addressing stakeholders with messages of sympathy and reassurance and updates on measures taken to avert the reoccurrence of crisis (Sturges, 1994; Coombs, 2007).

The Japanese government employed instructing and adjusting information during the 2011 earthquake crisis. There was necessary instructing information as well as information on measures through which citizens could protect themselves. Similarly, findings from MAS revealed the use of instructing and adjusting information. MAS provided information about the crash and how the stakeholders could cope psychologically with the crisis. Similarly, the health agencies in the US employed both instructing and adjusting information in handling the measles pandemic. Necessary information on the pandemic, the number of casualties, and the need for children to be vaccinated were disseminated. Adjusting information on protective measures against measles was also provided. Also, the University of Florida employed instructing and adjusting information to students on how to protect themselves from harm during the protests. However, the crisis at RIM only reported the use of instructing information in handling its crisis

Another crisis response strategy observed was apology – of the SCCT (known as mortification in the IRT) by some crisis managers in responding to crises. For instance, Domino's crisis reported the use of the apology strategy in managing the crisis. The President of Domino's, Patrick Doyle, had used YouTube and Facebook platforms to apologise to customers for the wrongdoings of two of its employees. Similarly, RIM's response strategy reported the use of apology. However, it was found that the choice of words contained in RIM's apology on Facebook worsened the crisis. Thus, apologising for crisis acts may necessarily not reduce the anger from stakeholders. Rather, apologising and using the right selection of words could be a better strategy.

When organisations in crises genuinely apologise for crisis acts, matched with strategies that reveal sympathy for stakeholder concerns, they could be absolved from the crises. On the one hand, apologising for crisis acts could mean admitting and accepting of crisis responsibility. On the other hand, however, Pace, Fediuk & Botero (2010) believe that sometimes organisations apologise as acts to show stakeholders sympathy but not to take the blame. Nonetheless, in severe situations, apologising for crisis acts could lead to litigation.

Furthermore, MAS employed the deny, diminish, and victimage strategies in its response to the MH17 plane crash. The deny strategy attempts to remove any connection between the organisation and the crisis because, if the organisation is not directly involved in the crisis, it will not suffer any damage from the crisis (Coombs, 2007). MAS employed the deny strategy when they announced that the plane was fired by a missile, thereby shifting the blame (IRT) away from them to the missile that brought down the plane. Also, organisations use the diminish strategy to make the public believe that crises are not as bad as they think or that the organisation lacked control over the crisis. The diminish strategy was employed by MAS when they framed the loss of contact with the plane as something that was beyond their control. The victimage strategy is used when crisis managers remind stakeholders that the organisation was a victim of the crisis (Coombs 2007). MAS played the victim by portraying itself as a victim of the plane crash since they also lost crew members.

Furthermore, another strategy found in this study and worthy of mention is the stealing thunder approach. Stealing thunder is a concept in crisis communication that explains how organisations in crisis self-disclose crisis information to the publics before it becomes known from other sources, especially the (social) media. By employing the stealing thunder strategy, organisations are in control of crisis information. Also, available research has revealed that organisations that employ the stealing thunder approach suffer less reputational damage than organisations where the media first reports crisis information (Coombs et al. 2017). MAS employed the stealing thunder strategy by self-disclosing the crisis before stakeholders got wind of such information. By doing this, they took control of the information that was spread about the crisis. Through this, they received support from stakeholders.

Finally, the corrective action strategy of the IRT was also recorded in the reviewed studies. Benoit (1997) argued that organisations correct their actions by promising and planning to solve or correct it. Domino's employed this strategy when its CEO promised that such an incident would not repeat itself, including the sack of the employees involved in the act, discarding all open containers of food, sanitising the entire organisation, and a review of its hiring process.

New Findings

Crisis response timing and its consequences were observed in this study. While some organisations responded within the first six hours after crises broke out, others responded after 24 hours. Domino's initial response to the viral video incident came after 24 hours. Within the 24 hours before Domino's response, customers had already taken to Twitter to post negative news, which damaged Domino's reputation. However, Hosseinali-Mirza et al. (2015) argued that the reason for Domino's delayed response was because they did not want to attract more attention to the case. Nonetheless, for a crisis of that magnitude, there ought to have been an immediate response within the first few hours, especially since it involved food.

Also, RIM first responded to online customer complaints within the first six hours after the incident. While the response timing was not delayed, RIM's choice of words in its apology further aggravated the crisis. Their response only acknowledged the loss of network and did not provide information on corrective measures to calm the stakeholders. The lack of adjusting information ensured that some customers abandoned the services of RIM. Thus, in crisis communication discourse, while responding to crises early enough is necessary, it is essential to pay attention to the content of messages.

MAS was also found to have responded to the crisis within the first 24 hours of the incident by updating its social media (Facebook and YouTube) pages. An early response, instructing and adjusting information, and prior stakeholder relationship, enabled stakeholders to be supportive, empathetic, and grieved with MAS after the MH17 plane crash. This example shows how a delayed response with the right selection of strategies could yield positive results. Also, early responses were observed in the violent protests in Nigeria and the school shooting incident in the US Citizens responded quickly after the increase in PMS in Nigeria by calling for protests while prompt responses were recorded within minutes of school shootings in the US.

Another finding worthy of mention is the introduction of the social-mediated crisis communication model into social media crisis studies. The SMCC model is the first theoretical model developed to test social media crisis communication (Austin, Liu & Jin, 2012). Although still being widely tested, the SMCC model, developed from the blog-mediated crisis communication, highlights the importance of integrating social media, including blogs and traditional media, into the media mix for crisis communication (Liu, Jin, Austin & Janoske, 2012). The SMCC model addresses how crises spread across social media and identify critical elements that interact to share and produce information before, during, and after crises. These elements include influential social media creators, social media followers and social media inactives (Liu, Jin, Austin & Janoske, 2012).

Social media creators are those who develop and post-crisis information online. They recognise the gravity of crises and thus, discuss and disseminate information online. These could include crisis managers and public relations practitioners. Social media followers comprise people who consume crisis information from the creators and share information online and offline. Offline media sharing includes traditional media, while social media inactives receive crisis information through other channels – traditional media and word-of-mouth communication. Thus, active players in the SMCC model are organisations, the publics' social media, traditional media, and other media that could be employed before, during, and after crises. The SMCC model thus discusses the interaction of these elements to produce the desired results.

SOLUTIONS AND RECOMMENDATIONS

Crisis communication studies have revealed that the publics prefer to utilise social media platforms during crises to not only disseminate crisis information but to discuss, debate, and to seek support from one another. Thus, Facebook, Twitter, and in some cases, YouTube were mostly employed by organisations, governments, and the public when responding to social media crises. Facebook and Twitter provide opportunities for organisations and the publics to express themselves through the use of texts, pictures, and videos, while YouTube provides visuals. While these platforms were popular, crisis managers must ensure that they own and control the information posted on the organisation's social media platforms.

This chapter presents the following solutions and recommendations:

1. Organisations and governments of nations are advised to be prepared for crises by integrating and adopting social media into their crisis communication plans and monitor all online activities. Monitoring online activities are necessary as times had changed from the times when traditional media influenced the public. Thus, organisational reputation could be easily damaged in an online crisis if social media is not fully integrated into crisis communications plans.
2. Since social media platforms have altered the way organisations communicate with stakeholders and vice versa, public relations practitioners and crisis managers are expected to engage the publics before, during, and after crises actively. The pre-crisis online dialogue is essential as it could come in handy during crises. Research has shown that a strong pre-crisis relationship with the publics can assist in reducing perceived reputational threats or negative perceptions during a crisis.
3. It is not enough to create social media accounts. It is expedient to be aware and monitor comments from the publics about an organisation. Thus, it is crucial to be fully aware of comments made by the publics online as these comments are capable of shaping opinions, attitudes, and behaviours of other publics. Not responding or late response to crises could have devastating consequences.
4. One of the best practices in crisis communication is to respond in a timely, accurate, and consistent manner. Thus, in a social media crisis, crisis managers must ensure that they respond, usually within the first six (6) hours after a crisis has been made public. When responding, crisis managers must also ensure that they are honest and open.
5. Communication with stakeholders during a crisis can prove to be an effective strategy. Providing informative and adjusting information during crises could help stakeholders to stay informed and protected with coping mechanisms against the crisis.
6. Crisis managers are expected to be mindful of and respect the cultural, religious, and racial backgrounds of the publics because they come from diverse ethnic, religious, and racial backgrounds. Thus, responding to one race or religion and ignoring others could lead to another crisis.
7. Finally, while social media platforms are increasingly becoming poplar crisis response tools, traditional media should not be jettisoned. Instead, unique blends of both traditional and social media during crisis communication are recommended. Traditional media could be employed to complement efforts from social media platforms

CONCLUSION

The introduction of social media into crisis communication discourse significantly changed the communication dynamics between organisations and the public. From being a useful tool in public relations and marketing, social media has grown to become crisis management tools. Social media has also proven to be useful tools in escalating and reducing crises. Findings from this study have assisted in extending the current knowledge of social media crisis communication. The use of social media during crisis provides a platform for stakeholders to seek and share crisis information. Findings revealed that Facebook and Twitter are increasingly employed as platforms for crisis communication discourse, with YouTube receiving only a few mentions in studies. It was also discovered that responding to crises promptly and engaging with the publics before, during, and after crises are crucial to managing organisational reputation. Issues around crisis timing response are particularly important when responding to a crisis. Given the independent nature of social media, it is possible for fake news on crises to escalate. It is no longer sufficient for a company to follow up on a crisis several hours after the outbreak. Hence, organisations and crisis managers are advised to fully integrate and adopt social media into their crisis communication plans. Also, crisis managers must respond to crises by providing information promptly and with strategies that will help stakeholders to make sense of the crisis.

REFERENCES

Adebayo, O. (2017). *The application of Facebook to crisis communication management: a case study of Malaysia airlines* [Unpublished doctoral dissertation]. Salford Business School, University of Salford, UK.

Alexander, G. (2010). *Defining crisis communication.* www.sagepub.com/upm.37705_1.pdf

Anthonissen, P. F. (2008). No thrillers, but hard reality. Crisis communication, practical PR strategies for reputation management and company survival, 7 – 23.

Auer, M. R. (2011). The policy sciences of social media. *Policy Studies Journal: the Journal of the Policy Studies Organization, 39*(4), 707–736. doi:10.1111/j.1541-0072.2011.00428.x

Baron, G. (2010). *Response suggestions and public participation: the new norm in response management.* http://www.emergencymgmt.com/emergency-blogs/crisis-comm/Response-suggestions-and-Public-111510.html

Benoit, W. L. (1997). Image Repair Discourse and Crisis Communication. *Public Relations Review, 23*(2), 177–186. doi:10.1016/S0363-8111(97)90023-0

Cheng, Y. (2016). How social media is changing crisis communication strategies: Evidence from the updated literature. *Journal of Contingencies and Crisis Management*, 1 – 11. DOI: doi:10.1111/1468-5973.12130

Cho, S. E., Jung, K., & Park, H. W. (2013). Social Media use during Japan's 2011 Earthquake: How Twitter transforms the locus of crisis communication. *Media International Australia, 149*(1), 28–40. doi:10.1177/1329878X1314900105

Clement, J. (2020). *Global social networks ranked by number of users.* Retrieved from https://www.statista.com/statistics/272014/global-social-networks-ranked-by-number-of-users/

Coombs, W. T. (1999). *Ongoing crisis communication: Planning, managing, and responding*. Sage.

Coombs, W. T. (2002). Deep and surface threats: Conceptual and practical implications for "crisis" vs "problem.". *Public Relations Review, 28*(4), 339–345. doi:10.1016/S0363-8111(02)00167-4

Coombs, W. T. (2007). Protecting organisation reputations during a crisis: The development and application of situational crisis communication theory. *Corporate Reputation Review, 10*(3), 163–176. doi:10.1057/palgrave.crr.1550049

Coombs, W. T. (2009). Conceptualising crisis communication. In R. L. Heath & H. D. O'Hair (Eds.), *Handbook of crisis and risk communication*. Routledge.

Coombs, W. T., Claeys, A., & Holladay, S. (2017). Social Media's Value in a Crisis. Channel Effect or Stealing Thunder? In *Social Media and Crisis Communication* (pp. 159–167). Routledge. doi:10.4324/9781315749068-12

Coombs, W. T., Frandsen, F., Holladay, S. J., & Johansen, W. (2010). Why a concern for apologia and crisis communication? *Corporate Communications, 15*(4), 337–349. doi:10.1108/13563281011085466

Coombs, W. T., & Holladay, S. J. (2014). How publics react to crisis communication efforts. *Journal of Communication Management (London), 18*(1), 40–57. doi:10.1108/JCOM-03-2013-0015

Coombs, W. T. (2010*)*. Parameters for crisis communication. In W. T. Coombs & S. J. Holladay (Eds.), *The handbook of crisis communication* (pp. 17 – 53). Wiley Blackwell.

Domino's. (2020). *Domino's 101: Fun Facts*. https://biz.dominos.com/web/public/about-dominos/fun-facts

Egbunike, N. (2015). Framing the #Occupy Nigeria Protests in Newspapers and Social Media. *Open Access Library Journal, 2*(05), e1486. doi:10.4236/oalib.1101486

Fearn-Banks, K. (1996). *Crisis communication: A casebook approach.* Lawrence Erlbaum Associates.

Gerken, F., Van der Land, S. & van der Meer, T. (2016). Crisis in the air: an investigation of AirAsia's crisis-response effectiveness based on frame alignment. *Public Relations Review,* 1 – 14. doi:10.1016/j.pubrev.2016.09.002

Hermann, C. F. (1963). Some consequences of crisis which limit the viability of organisations. *Administrative Science Quarterly, 8*(1), 61–82. doi:10.2307/2390887

Hilding-Hamann, S. (2012). *The effects of social media on crisis communication: A case study of the response strategy performed by Research in Motion* [Unpublished B. A Thesis]. Aarhus University, Denmark.

Hosseinali-Mirza, V., Marcellis-Warin, N., & Warin, T. (2015). Crisis communication strategies and reputation risk in the online social media environment. *International Journal of Business and Social Science, 6*(5), 7–21.

Huang, Y. C., Wu, F., & Huang, Q. (2016). Does research on digital public relations indicate a paradigm shift? An analysis and critique of recent trends. *Telematics and Informatics*. Advance online publication. doi:10.1016/j.tele.2016.08.012

Jin, Y., & Liu, B. F. (2010). The blog-mediated crisis communication model: Recommendations for responding to influential external blogs. *Journal of Public Relations Research*, *22*(4), 429–455. doi:10.1080/10627261003801420

Kaplan, A. M., & Haenlein, M. (2010). Users of the world, unite! The challenges and opportunities of social media. *Business Horizons*, *53*(1), 59–68. doi:10.1016/j.bushor.2009.09.003

Lawrence, A. K. (2016). A study of the media's portrayal of the measles outbreak. University Honors Program Theses. *Paper*, *175*, 1–38.

Lerbinger, O. (2012). *The crisis manager. Facing disasters, conflicts and failures* (2nd ed.). Routledge. doi:10.4324/9780203222133

Liu, B. F., Austin, L. L., & Jin, Y. (2011). How publics respond to crisis communication strategies: The interplay of information form and source. *Public Relations Review*, *37*(4), 345–353. doi:10.1016/j.pubrev.2011.08.004

Macias, W., Hilyard, K., & Freimuth, V. (2009). Blog functions as risk and crisis communication during Hurricane Katrina. *Journal of Computer-Mediated Communication*, *15*(1), 1–31. doi:10.1111/j.1083-6101.2009.01490.x

Nwogwugwu, D. I. (2018). Influence of crisis communication strategies on stakeholders' perception of organisational reputation: A review of research trends. *Journal of Communication and Media Research*, *10*(2), 125–138.

Ogundoyin, O., Olagunju, L., & Nwogwugwu, D. (2020, February). *Influence of Digital Technologies on Journalistic Practices and Quality of News Reporting in Nigeria*. Paper presented at the @frica Digital Media Conference, J. Valenti School of Communication, University of Houston.

Perry, D. C., Taylor, M., & Doerfel, M. L. (2003). Internet-based communication in crisis management. *Management Communication Quarterly*, *17*(2), 206–232. doi:10.1177/0893318903256227

Pratt, C. B. (2012). Theoretical approaches to and sociocultural perspectives in crisis communication. In A. George & C. Pratt (Eds.), *Case studies in crisis communication. International perspectives on hits and misses* (pp. 3–27). Routledge.

Seeger, M. W., Sellnow, T. L., & Ulmer, R. R. (1998). Communication, organisation, and crisis. *Annals of the International Communication Association*, *21*(1), 231–276. doi:10.1080/23808985.1998.11678952

Sellnow, T. L., & Seeger, M. W. (2013). *Theorising crisis communication*. Wiley.

Shan, L., Regan, A., Brun, A., Barnett, J., van der Sanden, M., Wall, P., & McConnon, A. (2013). Food crisis coverage by social and traditional media: A case study of the 2008 Irish dioxin crisis. *Public Understanding of Science (Bristol, England)*, *23*(8), 911–928. doi:10.1177/0963662512472315 PMID:23825281

Shu, K., Mahudeswaran, D., Wang, S., & Liu, H. (2019). *Hierarchical propagation networks for fake news detection: Investigation and exploitation*. Retrieved from https://www.semanticscholar.org

Snoijers, E. M., Poels, K., & Nicolay, C. (2014). #universitycrisis: The impact of social media type, source, and information on student responses towards a University Crisis. *Social Science Computer Review*, *32*(5), 647–661. doi:10.1177/0894439314525025

Stephens, K. K., Malone, P. C., & Bailey, C. M. (2005). Communicating with stakeholders during a Crisis. Evaluating message strategies. *Journal of Business Communication*, *42*(4), 390–419. doi:10.1177/0021943605279057

Sturges, D. L. (1994). Communicating through crisis: A strategy for organisational survival. *Management Communication Quarterly*, *7*(3), 297–316. doi:10.1177/0893318994007003004

Thelen, P. D., & Robinson, K. L. (2019). Crisis Communication in Institutions of Higher Education: Richard Spencer at the University of Florida. *Communication Quarterly*, *67*(4), 444–476. doi:10.1080/01463373.2019.1616586

Thompson, B., Mazer, J. P., Payne, H. J., Jerome, A. M., Kirby, E. G., & Pfohl, W. (2016). Social Media and Active Shooter Events: A School Crisis Communication Challenge. *Qualitative Research Reports in Communication*. Advance online publication. doi:10.1080/17459435.2016.1247111

Utz, S., Schultz, F., & Glocka, S. (2013). Crisis communication online: How medium, crisis type and emotions affected public reactions in the Fukushima Daiichi nuclear disaster. *Public Relations Review*, *39*(1), 40–46. doi:10.1016/j.pubrev.2012.09.010

Veil, S. R., Sellnow, T. L., & Petrun, E. L. (2012). Hoaxes and the Paradoxical Challenges of Restoring Legitimacy Dominos' Response to Its YouTube Crisis. *Management Communication Quarterly*, *26*(2), 322–345. doi:10.1177/0893318911426685

Wang, Y. (2016). Brand crisis communication through social media: A dialogue between brand competitors on Sina Weibo. *Corporate Communications*, *21*(1), 56–72. doi:10.1108/CCIJ-10-2014-0065

Yin, R. K. (2009). *Case study research: Design and methods*. Sage Publications.

Young, C. L., & Flowers, A. (2012). Fight viral with viral: A case study of Domino's Pizza's crisis communication strategies. *Case Studies in Strategic Communication*, *1*, 93–106.

KEY TERMS AND DEFINITIONS

Crisis: Crises are unexpected (and sometimes expected) occurrence that disrupt high priority goals with short response time and attracts high media attention.

Crisis Communication: Information given by organisations/nations in crisis to help stakeholders make sense of the crisis.

Crisis Response Strategies: Organisational tactics in response to crisis aimed at repairing and/or rebuilding reputation

Organisation: Organised entities comprising groups of people who are poised to achieve specifically defined objectives.

Organisational Publics: These are people that can affect or can be affected by activities of organisations

Reputation: Valuable and intangible organisational assets which are formed based on how well stakeholder expectations are met and could be damaged during crises.

Social Media: These are groups of internet-based applications that build on the ideological and technological foundations of Web 2.0 and allow for interactivity among users.

Chapter 9
Communication Crisis Management of the Public Security Policy:
The Social Media Landscape of the Police in Portugal

José Gabriel Andrade
https://orcid.org/0000-0002-9778-5865
University of Minho, Portugal

Nuno Jorge de Lima Ferreira
Higher Institute of Police Sciences and Internal Security, Portugal

ABSTRACT

This chapter discusses crisis communication management in the Portuguese public security police digital landscape, mainly focused on social media. This is an exploratory investigation it is intended to understand how crisis communication management methodologies in the digital environment can be applied in police intervention, so its legitimacy is reinforced. The study is divided into two parts, the literature review, and the empirical research. The data was gathered together through an interview survey, with the participation of four public security police officers and four investigators. The interviews were transcribed, and their content was analyzed. In an emergency situation, most of the actions are communication activities, so its management in the digital landscape—above all social media—is critical for the police success, as an institution under strong scrutiny and whose actions depend on the citizen cooperation. Communication is essential for building reputation and should be used in threats to legitimacy, and as a tool for reaching out to the community.

DOI: 10.4018/978-1-7998-6705-0.ch009

INTRODUCTION

Communication and new technologies are essential cornerstones of the actual society we live in. Based on networks with a dynamic and flexible social structure, today's society is characterized by the extinction of certain ideals, as well as relationship artificiality and individualism. Currently, we live in a context, where a break with tradition is perceived (Bauman, 2001). Technology has become part of the present and will enhance time and the compression of space, which leads to a replacement of context, of worlds, where the conventional is to be unconventional. With the title of Information, this Society is dependent on the new information and communication technology (Ilharco, 2003), largely motivated by the emergence of social networks and the strong implementation of the internet, "the great transformer of the global world" (Andrade, 2009, p.3952) who, together, implemented major changes in the way Men behave and interacts in society. Due to their size, public exposure, and predictability of disorder, these events can affect the Police reputation and credibility. The institutional image is, in today's society, one of the main assets of organizations, being influenced by the context and the society in which it operates. Like other institutions, the Police must adapt to this new reality and must be attentive to factors that may put their character and values at stake. Otherwise, incidents may arise, if the Institutional response is not adequate and within a limited period of time for accurate decision making, will be perceived as crises threatening its fundamental values, its reputation, and will assemble a negative public opinion (Claeys & Coombs, 2020; Coombs, 2007; Coombs, 2010; Mendes & Pereira, 2006). The causes of a crisis can be diverse (Gama 2000; Orduñã 2004) and, therefore, its classifications or typologies of crisis can also vary. All types can be converted into optics (or reputational) crises, with the respective consequences on the credibility and trust of the organization's stakeholders (Andrade, 2008; Andrade, 2015). However, since most crises follow a certain development cycle, institutions can prepare themselves to deal with and manage them, implementing crisis management processes.

Thus, our study is focused on the Police Sciences and connected with a new area within the Communication Sciences - Crisis Communication Management - in the new technological environment in which we live in. The literature on crisis communication has increased in recent years with the construction of its own doctrine and methodology, which shows a growing interest in Crisis Communication Management. However, we found that it is still an undiscovered field of study in the Police Sciences universe, highlighting the studies already done in the area of Communication and Public Relations in a Police context. We, therefore, consider that the theme of communication as prevention and response to crises in the digital landscape is of the utmost relevance and needs further investigation studies.

CRISIS MANAGEMENT

Growing public activism, technological advances, and neglect of organizational planning contribute to the increased value and need for crisis management, forcing organizations to be prepared to deal with and to manage them effectively (Coombs, 2007; Coombs, 2010; Claeys & Coombs, 2020). It is essential that crises are understood naturally and that institutions create mechanisms to deal with them, "either in reducing the likelihood of them happening or in the need to deal with an occurrence" (Jorge, 2010, p.47). It is in this context that Crisis Management begins, as a structured procedure, configuring itself as a recent and constantly changing science, in line with the development of communication technologies (Andrade, 2008).

As a process, crisis management must begin before the crisis occurs. Prevention and monitoring are the best ways to manage crises (Caetano, *et. al*, 2006; Coombs, 2010), it is necessary "to take special care with monitoring situations and information about critical or potentially critical events, with training and construction of a strong and healthy corporate and institutional culture" (Caetano, *et. al*, 2006, p.19).

In this liquid modernity, we have seen a break with tradition and the valorization of the new replacing the old (Bauman, 2001). In the same sense, Ilharco (2005) considers that we are in the presence of a renewal of context, of worlds, stating that:

"Today the conventional is to be anti-conventional. We all have something to say against tradition, customs, and the heavyweights of history. We got used to a new world where the history of the recent past, namely with regard to two of its most relevant events, the cold war, and left-right polarization, appears outdated and strangely distant" (Ilharco, 2005, p.68).

This "another world that now appears to us along the paths of new technology" is something that opens up new possibilities and questions and changes the way we act on a daily basis and "forgets tradition, routines, and customs, creating new involvement and new projections of ourselves towards ourselves" (Ilharco, 2005, p.62).

With the Information and Communication Technologies and the almost instantaneity of time, the devaluation of space is announced, since all parts can be reached at any time (Bauman, 2001). As a result, "localities are stripped of their cultural, historical, and geographical sense and are reintegrated into functional networks (….) providing a space for flows that replaces the space for places" (Castells, 2011, p.492). This was not true in heavy modernity (that of hardware) as there was an obsession with conquering space and its delimitation (Bauman, 2001). The conquest of spaces (territorial conquest) was the motto and size was the power. Time and space, fundamental dimensions of human life, have been radically transformed with the current communication system (Castells, 2011), in which we see the "transition from traditional mass media to a system of horizontal communication networks, organized around the Internet and wireless communication" (Castells, 2011, p. XXXVIII). This factor leads to "a fundamental cultural transformation as virtuality becomes an essential dimension of our reality (Castells, 2011, p. XXXVIII).

In addition "modern societies rely on complex technological systems that are deeply intertwined with other complex systems that stretch across geographical, judicial and administrative borders" (Boin, 2019, p. 94). This is a time of volatility, with constant changes and movements, in which Man is more and more individual and less citizen (Bauman, 2001). All of those aspects create uncertainties and insecurity, with the resulting consequences - "If the individual is the worst enemy of the citizen, and if individualization announces problems for citizenship and for politics based on citizenship, it is because the care and concerns of citizens individuals as individuals fill the public space to the top, asserting themselves as their only legitimate occupants and expelling everything else from public discourse" (Bauman, 2001, p.46).

This communication system and its changes, motivated by ICT, "aggravate the feeling of disorientation" (Castells, 2011, p. XXXVIII), and this networked social structure creates "new forms and channels of communication, shaping life and, at the same time, being shaped by it" (Castells, 2011, p.3).

The "Transboundary Crisis effortlessly exceeds geographical, pol-icy, cultural, public–private and legal boundaries that normally enablepublic managers to classify, contain and manage a crisis" (Boin, 2019, p. 94).

THE NEW CRISIS MANAGEMENT TOOLS

The society in which we live is constantly changing, largely due to the development of new technologies (with a strong contribution from the massive implementation of Web 2.0) and, on the other, due to the emergence of social networks. These platforms have implemented major changes in the way we act and how we interact in society and have helped to create a world that connects in a large network that encompasses individuals and all entities in our society (Castells, 2011; Pinho, 2015).

With the technological revolution, and in the current context of computer-mediated communication, Public Relations and image crises have also undergone major changes. The problems that were previously limited to the geographic area of the organization today are followed up all over the world and in real-time (Andrade, 2009; Fearn-Banks, 2011).

These platforms (social media and social networks) give users the possibility to produce and manipulate content quickly and to take control of what interests them, at the same time they become a kind of opinion leaders and creators of a viral word-of-mouth marketing (buzz and word of mouth), which may or may not be harmful to an organization (Pinho 2015; Sebastião, 2009).

In this context, institutions also take advantage of social networks to target and connect with their audiences. These networks are an important communication tool between institutions and their audiences, during all phases of their work. These platforms can also play a very important role when organizations go through the darkest phases of their existence (Pinho, 2015). This issue also applies to the use of mass alert and notification tools.

Derived from the growing use of social networks and the evolution of digital devices, there is a trend for organizations to adopt social networks and start to integrate them into their corporate and communication strategies. The question that arises when thinking about strategic communication and new media is to understand how organizations are going to integrate the internet and social networks in their own strategic communication plans (Pinho, 2015), because, as Alves defends (2007), crises are usually caused by failures, simultaneously, with interactions between technology, people and organizations.

By allowing massive access to information from all over the world, new technologies bring advantages and setbacks when it comes to managing a crisis. Although the use of social networks is important during the day-to-day life of an organization, it is during the period of crisis that they can become a great ally, since they allow a quick and almost instant response to problems that may arise (Fearn-Banks, 2011). As stated by Sónia Sebastião:

"Public relations find on the internet another vehicle that allows them to easily, quickly and effectively: communicate and get to know the opinion of targeted audiences; publicizing events (...) you can find the information you need to know the characteristics and trends of the public online and, thus, position yourself strategically" (Sebastião, 2009, p.147).

With regard to the drawbacks, we can immediately list two. One side a simple fact can become an event of unusual notoriety, and on the other, many issues are created based on rumors conceived in forums (Ordunã, 2004). To minimize these situations, social networks can function as a channel of monitoring the environment that surrounds the organization and detecting problems or failures. Monitoring the environment and what is happening on the internet should be one of the responsibilities of Public Relations and crisis managers, in order to control the mentions that are made to the organization, as well as to avoid the creation and spread of rumors (Pinho, 2015). This idea is corroborated by Santos (2006, p. 193) when he states that "there is a need for constant identification, monitoring, and management of issues". It is also important to note that, despite the growing and large implementation of computer-mediated communication and the communication tools made available over the internet, the PR professional and organizations should not replace or underestimate face-to-face communication, since not all individuals access the Internet. Therefore, the information does not reach the entire population (Sebastião, 2009). The author also states that "the Internet does not replace other means of communication, but complements them" (Sebastião, 2009, p. 148).

Digital technologies provide easy and constant access to an enormous amount of information, instantly, anywhere in the world. "This context has advantages and disadvantages that relate to crisis management. The main disadvantages are the possibility of viral spread of the contents, and the ease of handling them, and the unprecedented speed and reach, which are reflected in the notoriety (negative in the event of a crisis)" (Dias & Andrade, 2015, p. 94). In addition, there are numerous problems that result from content without rigor or even rumors that are produced and published by users of forums, virtual communities and other social platforms.

CRISES AS OPPORTUNITIES

Some crises can result from a fulminating event (unexpected and unpredictable), however, organizational crises tend to result from the interaction between a large number of causes. Organizations must learn from these situations, even though they appear to be harmless since attention to small causes can avoid the big consequences since most crises result from the incremental accumulation of (micro) causes (Cunha, 2006).

Most of the time, crises are associated with difficult situations and negative consequences. Nevertheless, and depending on how they are approached and managed, crises can translate into opportunities for organizations. Opportunities arise in the sense of learning from the crisis and improving so that, if repeated, the answer is more efficient (Pinho, 2015). Thus, organizations have the opportunity to analyze incidents and accidents, not so much for the purpose of identifying the culprits, but with the aim of preventing future occurrences (Cunha, 2006).

In Chinese culture, a crisis is always seen as a "dangerous opportunity", since if it is anticipated and worked on it can bring positive results for the organization (Andrade, 2008; Pinho, 2015; Sellnow & Seeger, 2013), when a company suffers a crisis, is subjected to a free public exhibition that is unrepeatable. If the crisis is well managed, it is possible to have the opportunity to mitigate the negative effect of media coverage and it is feasible to send positive messages about the company, its products, and services (Andrade, 2008; Orduña, 2004). Within each crisis, there is always an opportunity in a latent state where before the crises it is essential to cultivate the right values, those that are perceived as positive by society. This aspect is fundamental during the crisis and can cause the crisis to "cool down".

In managing a crisis, it is not just about what you do, but above all how you do it, as perception often counts more than the actions themselves. To reach the public, especially in crisis situations, a symbolic demonstration of the values expressed by the leader or by the company must be given (Andrade, 2015). Regester and Larkin also agree that a crisis can be an opportunity:

"Surviving a crisis guarantees a great opportunity for the organization to reexamine and reorganize itself, to ensure that it is not in a similar position again. The crisis can represent a turning point in organizational life, presenting opportunities to establish a reputation, attacked, but reinforced and more competent" (Regester & Larkin, 2008, p.231).

It is through Public Relations and communication that a crisis can become an opportunity, "the opportunity for the company to be seen and known" (Sebastião, 2009, p.175). The author argues that the management of communication with the media is preponderant, as they are usually the ones who give media visibility to information about the crisis. Miguel Pina and Cunha (2006, p.146) considers that an organization that critically reflects on its own activity and rethinks its responses and behaviors, although previously effective, is an organization that learns and tends to avoid and minimize its potential problems. For the author "these organizations will tend to see crises as opportunities for learning". On the opposite side, crises are seen as threats by organizations with learning difficulties, which repeat the same behavior even when it becomes manifestly dysfunctional (Cunha, 2006). The same author presents the following factors as reasons for the difficulty in organizational learning: organizational complexity, excessive information, a highly controlling hierarchical culture, and bureaucracy. This preparation is inherent in the way organizations face change and face crises - as threats to their own activity or as opportunities for learning before, during, and after the crisis (Mendes & Pereira, 2006). The first is inherent to more bureaucratic organizations, which do not see change naturally and who have more difficulty in learning from mistakes, eventually repeating their behavior even when it becomes dysfunctional, capable of generating crises and making it difficult to resolve. In the second, the organization learns when it critically reflects and changes its behavior in order to avoid or minimize problems (Mendes & Pereira, 2006). "As tragic as a disaster is when nothing can be learned from its occurrence" (Cunha, 2006, p.159).

COMMUNICATION AS A RESPONSE TO CRISES

In addition to the most current definition a crisis, "serious, complicated, difficult phase, a moment of tension or stalemate in the life of a person, a social group, or the evolution of certain situations" (Mendes & Pereira, 2006, p.107), is also an evaluation of options because when you start your stakeholders are forced to give an answer. As we have already mentioned, a crisis is a situational reality with different interpretations and repercussions. The community and the whole environment of the (crisis) phenomenon have their own perception and interpretation of the degree of guilt and the involvement of each actor, and therefore, demand different responses and different reactions from the different parties involved (Mendes & Pereira, 2006, p.107). As Sónia Sebastião (2009, p.175) teaches us, "crises can be the result of different factors, among them public perception". For this author, audiences consist of:

"(...) groups of individuals with certain common characteristics who directly or indirectly contact the organization influencing its activity. They are internal when they contribute to the company's existence and organizational functioning. They are external when they influence the existence of the organization in terms of meeting in the market of its sector of activity" (Sebastião, 2009, p.91).

For this reason, "the relationships established between the organization and its different audiences require differentiation of messages (...), each audience is treated as a specific target, with distinctive characteristics" (Sebastião, 2009, p.91). In conclusion, "a crisis is mainly a perceptive phenomenon developed by third parties about an agent who sees himself or is involved in an unpredictable event or action" ((Mendes & Pereira, 2006, p. 108). Given this framework, it is now important to understand how an organization should act in the face of a crisis, that is, what is its structure for responding to the crisis. The existence of a crisis implies the activation of the reactive mode of Crisis Communication Management, which is translated into the operational and symbolic response. Furthermore "organizations in crisis often fail to select the optimal crisis response strategy, preferring strategies that avoid short-term losses over the ones that offer long-term gains" (Claeys & Coombs, 2020, p. 290). Its critical tasks are the identification of the crisis and stakeholders, the selection of the response strategy, and the activation of the response system. As for the identification of crises and their typologies, we will separately address the remaining critical tasks, namely, the crisis factors, the identification of stakeholders and the audiences involved, and the selection and preparation of a crisis response strategy. Consequently, "crisis management organizations should ultimately learn to independently evaluate and reorganise their own crisis management performance on-the-fly." (van Laere & Lindblom, 2019, p. 48) In addiction, social media represented a paradigmatic shift for Strategic Communication. On these collaborative platforms, based on the creation of horizontal links and content created by users, the reputation of companies and brands is much more exposed to scrutiny, comment, and word of mouth, all with a potentially viral scope. "In order to adapt to this new context, Strategic Communication went from a dissemination paradigm to a dialogue paradigm, and social media gained an increased importance for building reputation" (Dias & Andrade, 2017:55).

THE DIMENSION OF COMMUNICATION IN THE POLICE

To be successful in their duties and to fulfil their purpose, to ensure public order and security, the Police need the support and voluntary cooperation of citizens (Denef *et al.*, 2013; Lourenço, 2012). Such support is only possible if citizens recognize the Police interventions as legit. Police legitimacy is sustained by the citizens based on "their perception on how the Police exercise their authority and the trust they place in them" (Lourenço, 2012, p.182). However, the environment in which the Police activity takes place is complex, diversified and limited, since, in addition to normative legitimacy, Police still lacks social legitimacy, that comes from citizens who voluntarily accept their authority, realizing the "Police activity as necessary and useful" (Valente, 2014, p.169). In this sense, the Police's command cannot be exhausted in its activity in combating crime. It must relate, simultaneously, to a social commandment, that must perform functions of reassurance and strengthening the community (Brown, 2014). The Police also have a symbolic function, in the sense that they must represent what is right and symbolize the nation's efforts to guarantee order and preserve peace (Brown, 2014).

From the study of the aforementioned authors, we realized the Police's relationship with the community is truly important for its mission. The way in which citizens perceive and evaluate the Police's action can influence (directly or indirectly) the way and how they will react, as well as their attitudes towards the Police's requests for support and collaboration, thus adopting behaviors accordingly (or not) with the law (Tyler & Fagan, 2008). This idea is corroborated by authors like Denef when they conclude "the way how the public reacts to the actions of the Police depends on the relationship between the Police and the public, and more specifically on the image of the Police in (within) society" (Denef *et al.*, 2013, p.9). This notion of legitimacy depends on the citizens' perception of the Police procedures (whether they consider them fair or not), as well as on the knowledge they have about the Institution and its objectives and strategies. Lourenço (2012, p.184) considers that:

"Obedience is thus not rendered to the person who has the authority, but to the social and impersonal order that is the guarantor of the position. The legitimate authority of a Police officer lies in the institution to which he belongs and it is this that is perceived as deserving of legitimacy and not the individual.

On the other hand, when there is a sense of rejection towards the Institution, an "anti-police" feeling is created, verifiable in neighborhoods in many cities" (Lourenço, 2012, p.185).

That said, it is essential for the Police to ensure that citizens recognize the legitimacy of their actions and, at the same time, achieve a high level of trustworthy relationships from individuals and communities, which will increase the efficiency of the Police actions (Lourenço, 2012).

This leads us to question how crisis situations regarding Police actions should be considered. Whether they should be considered as a crisis for the Police involved or as a crisis for the whole community? Should a crisis be treated as an organizational event or as a community event? (Heath, 2010).

Crises occur in organizations within the community and, for this reason, Heath (2010) considers that communication is more than sharing and transmitting information and that new crisis management paradigms must be based on causal attribution to organizations. From the above, the importance of communication for the involvement of organizations with the community in which they operate is fundamental, making the Police no exception.

The Police must always maintain a relationship with the community, keeping Robert Peel's[1] "maximum" authenticity, "the Police is the community and the community is the Police" (Her Majesty's Chief Inspector of Constabulary [HMIC], 2008; Kingshott, 2011; Williams, 2003; Stott & Gorringe, 2014). This is one of the "Peelian" principles that help us understand that the Police's ability to perform its duties depends on its public approval. The Police must ensure voluntary cooperation and obedience on the part of the community, and the degree of cooperation obtained proportionally decreases the need to use physical force (Kingshott, 2011; Williams, 2003; Stott & Gorringe, 2014).

According to Robert Peel's principles, dated 1829, the Police's power to fulfill its duties do not lie in its capacity to generate fear and deterrence, but in the public's approval and its collaboration. The Police must be impartial and use force only when persuasion fails. For the same author, the role of the Police is not to replace the community, but to promote the ability to self-regulate (Stott & Gorringe, 2014).

DIALOGUE AS A GUARANTEE OF POLICE LEGITIMACY

Violence causes are not always struck by crowds. Sometimes, violence originates from less tough Police actions (Santos, 2015). As already mentioned, in order to legitimize Police action in the citizen scrutiny, it is essential that the Police establish a positive relationship with all the members of the different groups. To do this, you must know the values and objectives of the stakeholders, aiming to differentiate between the different groups (Gorringe *et al.*, 2012; Santos, 2015). Only then, with an informed and effective performance, will it create a positive impact on the dynamics of the crowd (Gorringe *et al.*, 2012; Santos, 2015). This implies a change effort, regarding crowd policing, being necessary to abandon the eminently reactive and repressive posture, the escalation model of coercive means, and adopt a sustained methodology approach to the management of levels of violence, which it implies communication, dialogue, and facilitation by the Police (Felgueiras, 2015).

According to Robert Reiner, policing should be understood as a process and not as a result and, as such, policing strategies should extend the period of the Police operation, transforming it into an (almost) continuous operation, in order to be able to develop functions intelligence, dialogue, and communication, in order to build lasting relationships with participants in collective action and community events (Reiner, 1998 *apud* Lourenço, 2012; Felgueiras, 2015). In these models, "Police intervention is intended to be transparent and based on conflict-reducing principles, allowing policing to be simultaneously facilitating, differentiating and informed" (Felgueiras, 2015, p.29). The main objective should be to bring (through techniques of dialogue, communication, and negotiation) the frameworks of collective action to the behaviors considered legitimate from the Police perspective, and, above all, of community standards (Felgueiras, 2015). Thus, tactical action (intervention and the use of physical force) must be targeted and segmented, that is, surgically directed at individuals who exhibit behavior considered illegitimate, increasing the perception of the legitimacy of Police intervention (Felgueiras, 2015).

In Police activity, especially in interaction with the citizen, communication has the function of clarifying the information available, in order to avoid any rumors or misinterpretations and to reduce the surprise factor (no surprises) for people (Felgueiras, 2015).

"In order to achieve this intention: It is essential to have a communication plan that is implemented before, during and after the event, to inform participants, Police officers and the general public about the event's characteristics, expected behaviors, circulation constraints, advice for the promoting security, general lines of Police action, avoiding rumors in a crowd" (Felgueiras, 2015, p.23)

Due to the complexity of the issues of its activity and the diversity of its audiences, it is very important that the Police are able to effectively communicate. In policing collective action events, communication gains a greater dimension, being vital to contain and control crisis situations (Denef *et al.*, 2013).

However, in the events studied by Donatella della Porta and Herbert Reiter, one of the main causes of misunderstanding between the Police and protesters, and the consequent escalation of levels of violence, was the lack of cross-cutting communication in all phases of the events (Porta & Reiter, 2006 *apud* Santos, 2015).

DIALOGUE POLICING - A MODEL OF DIALOGUE POLICE

Sweden was the first country to take Police dialogue as a tool for policing collective events. This measure came after the European Summit in 2001 in Gothenburg, as a result of the high number of injured officers and protestors (HMIC, 2008; Santos, 2015; Stott, 2009; Stott *et al.*, 2016). These events were a "trauma" for the country and made the event a turning point that led the Swedish government to recognize the need to reformulate the policing tactics that were until then used (Santos, 2015; Stott *et al.*, 2016). They decided to abandon the policing style based on a law and order model (not very flexible and based on the escalation of force levels) and adopt dialog method, facilitation, and the principle of non-confrontation (Holgersson, 2010; Santos, 2015). As a result, as part of the Special Police Tactics idea, dialogue team officers or dialogue polis were created. These are special management crowd units that encourage dialogue with extremist movements in order to promote restrained and non-confrontational protests (Institute for Strategic Dialogue, 2014).

In a similar context, in 2008, the Danish Police, after realizing the confrontational stance (which was previously adopted in situations of risk) led to delegitimized and negative views of their activity, introduced Police dialogue in policing major events (Havelund *et al.*, 2013 *apud* Santos, 2015). The new form of policing leads to a posture of greater openness and differentiation, always based on dialogue (Havelund *et al.*, 2011). These values, inherent to the ESIM model, are applied in sports policing and protest by dialogue teams, called Event Police (Santos, 2015).

In 2009, like its Nordic counterparts, the UK Police also decided to change their model of policing major events. The decision comes after four nights of turmoil, during the G20 summit in London, in which the Police used more harmful and unusual means of coercion in their performance (Gorringe *et al.*, 2012; HMIC, 2008; Santos, 2015; Stott, 2009). The measure adopted was translated, in practice, by the creation of teams - liaison officers, Police Liaison Officers (Gorringe *et al.*, 2012; Santos, 2015) in order to be more easily identifiable and visible, uniformly different, that gives the Police a less ostentatious approach. Usually, these elements are used to mediate situations of emerging tension and establish a dialogue with the different protesters (Santos, 2015).

Despite the organization and structure of the teams they vary depending on the context of the country and the Police itself, "the principles that guide them and the actions they develop are similar" (Osterling, 2013; Santos 2015, p.55) in general their activities "are, regardless of the nature of the event they take into account the information collected about it, seek to establish relationships of trust with the different organizers, group leaders and/or spokesperson present" (Santos, 2015, p. 55).

HOW DO THE POLICE COMMUNICATE?

Police activities come to life in a "24x7" world, where events happen at a high pace where everyone wants to be the first one to know about it. In this reality, it becomes even more complicated for the Police to have effective communication due to changes in the "how, when and where" people access information because with social media evolution and the exponential power of the internet it is very difficult to follow what is being said or shown (Stephens *et al.*, 2011).

The Police communication policies should aim to make information available to the public (Felgueiras, 2015) and to the policing community, in order to reduce the fear of crime and enhance a community policing philosophy (Kingshott, 2011).

In an incident situation, the Police must transmit facts to the public, appeal to witnesses and reassure community security and crime prevention advice, as well as give alerts on traffic, thus reducing the fear associated with this incident (Braziel *et. al.,* 2016; Kingshott, 2011; Velde *et. al.,* 2015).

Police communication, in addition to this informative function and the purposes already mentioned, also seeks "guarantee the transparency of police action"(Godiac, 2013a, p.9). A philosophy of communication and policing should be adopted, no surprises, in order to avoid surprises for the Police, protesters and protest targets (Felgueiras, 2015; HMIC, 2008; Joint Committee on Human Rights, 2009; Santos, 2015). The dialogue between the Police and its stakeholders must be continuous to ensure that the protesters and the rest of the public are aware of the probable police actions so that their decision making is informed (HMIC, 2008)

a. The New Media

In order to be able to communicate effectively with the public, the Police must manage not only their relationship with traditional media but also know and embrace new communication technologies. They must understand how the citizen obtains information and use the methods most likely to reach his target audiences (Police Executive Research Forum, 2014; Stephens *et al.*, 2011; Velde *et al.,* 2015).

Authors like Reicher suggest the use of various means that facilitate communication between the Police and participants in the protests (Reicher *et al.,* 2007). Before the events, the authors give us an example of the use of social networks and the use of flyers, in addition to the traditional media. During the event, loudspeakers, LED screens and other technological means can be used to transmit and receive messages easily and instantly.

During the events, there has been a recent trend by the Police to use social media (Police Executive Research Forum, 2014, 2016; Reicher *et al.*, 2007; Stephens *et al.*, 2011; Velde *et al.,* 2015). These platforms are used in the events by the participants to plan and discuss the possibility of generating riots and are used by the spectators to record the events. By monitoring these platforms, the Police can obtain information about possible threats (among other information), which can help to understand the number of groups that may be involved; which of these groups are already known; if they are from the city itself or if they come from outside; and what are their intentions (Police Executive Research Forum, 2014, 2016).

During the events, the Police can use social media to disseminate information to the public, to refute false information, and to correct misleading information. It is important that the monitoring of social media is maintained throughout the event, since, events take place, plans and conditions may change (Braziel *et al.*, 2016).

At the end of the event, the Police can use social media for criminal investigation purposes (Office of Community Oriented Policing Services [COPS], 2013; Police Executive Research Forum, 2016; Procter *et al.*, 2013).

In short, social networks are presented as a powerful and valuable information collection tool to keep the public informed and to transmit warnings and advice (Police Executive Research Forum, 2014, 2016; Procter *et al.*, 2013).

b. A Three-stage Communication Model

Like a crisis, policing a crowded event can also be divided into three phases, each of which has particularities with regard to communication (Holgersson, 2010; Santos, 2015).

Before the events, taking into account that this is a policing philosophy based on a no surprise approach, it is essential that the Police establish a dialogue with the event organizers (when they are known) and with the possible participants, as soon as they receive information from the holding an event (Holgersson, 2010). In the preparation phase of policing, police dialogue is essential, since the information collected allows them to evaluate and plan each policing as a unique event, taking into account the particular contexts of the same (Osterling, 2013; Santos, 2015). The aim is to establish agreements with the different parties involved in it and, from the outset, to combine the places of concentration, the routes, the schedules, and the destination locations. It also seeks to demonstrate which facilitation strategies will be put into practice taking into account the (legitimate) objectives of the protest (Holgersson, 2010; Santos, 2015). It should be noted that, in order to legitimize and accept police action, it is crucial to clarify what will be the actions to be taken, if the protesters want to assert unlawful intentions (Holgersson, 2010; Santos, 2015).

During the events, contact (as a rule, at the place of concentration) with the different entities must be made by the people who were involved in the preparation phase (where the initial agreements were established), in order to facilitate the construction of a relationship of trust (Holgersson, 2010; Santos, 2015). The agents responsible for the dialogue must be easily recognized and identifiable by those present (Holgersson, 2010; Santos, 2015), since it is paramount that the Police are able to ensure that there is someone identified and ready to negotiate.

After the events, it is important to arrange meetings with the main parties involved in the demonstrations (Office of Community Oriented Policing Services [COPS], 2015). These debriefings allow the Police to have a better knowledge of the groups' identity, thus reducing the risk of creating stereotypes, namely with regard to the reasons for the adoption of certain behaviors. This feedback can contribute to a better interpretation of group phenomena in future events (Granström 2002 *apud* Holgersson, 2010).

As a crisis, these events can also be seen as an opportunity and not "just be seen as a problem or threat" (Godiac, 2013b; Reicher *et al.,* 2007, p. 10). Kingshott (2011) also considers that Police should view and use an incident as an opportunity to build bridges with the community, to do so, they must harness the power of the media and establish a cooperative relationship with them so that both OCS and the Police can achieve their goals (Kingshott, 2011) To reach this level, the police must assume themselves as the main actor in collective events (Godiac, 2013a; Santos, 2015).

METHODOLOGICAL OPTION AND RESEARCH DESIGN

For interpretivists, there are no "special" research methods that automatically and inevitably lead to the truth. Their research techniques and procedures are the same as those of other researchers. The big difference is that the majority of researchers who opt for qualitative methods (and all those who resort to quantitative methods) consider that there are certain methods that have to be used to obtain a valid study, while interpretivists do not accept this premise. For these, studies are situational (Smith, 2008) since people's social activities are contingent and not universal (Vannini, 2009). For this methodology, qualitative research is the most appropriate when it is intended to understand the meaning of the phenomenon under study and not the phenomenon itself as an external and objective reality. This implies understanding the context in which the phenomenon is experienced and interpreted, and understanding the process through which it occurs (Maxwell, 2005).

Regarding the interviewees' choice process, considering that it must be adequate to the research objectives, we opted for a personalized adaptation, translating into an "intentional sample, in which a set of qualified informants is sought" (Carmo & Ferreira, 2008, p.150). According to Carmo and Ferreira (2008, p.209) "qualitative research typically focuses on relatively small samples, or even single cases selected intentionally". The idea is the purposeful choice of the participants and, in addition to the small number that characterizes these types of studies, there is no ideal number defined for them (Creswell, 2014).

Bearing in mind the objectives of our research work, we opted for a non-probabilistic sampling technique (Carmo & Ferreira, 2008) and included in our study eight participants - four PSP officers and four researchers. As for the PSP elements of Portugal, the following were interviewed: Pedro Clemente (Chief Superintendent; Doctor of Political Science and Director of ISCPSI); Hugo Palma (Director and Director of the GIRP of the DN); Paulo Flôr (Subintendent; Commander of the 1st Police Division of COMETLIS; and former Director of GIRP); and Sérgio Soares (Commissioner; Chief of NIP; former Chief of NIRP of COMETLIS). As for the researchers, the following were interviewed: António Marques

Mendes (Professor at the London College of Communication - University of the Arts); Clifford Stott (Professor at the School of Psychology at Keele University; he is co-director of the Keele University Police Academic Collaboration (K-PAC) and was a consultant to the Portuguese Police at Euro 2004); Mariana Victorino (PhD in Communication Sciences at the Catholic University of Lisbon, and Director of the Porter Novelli Communication Agency); and Timothy Coombs (PhD in Public Relations and Affairs Management, and Professor at College Station, Texas, and Crisis Management consultant).

a. Corpus

The corpus concerns the set of materials collected during scientific research. For Bardin (1977) the corpus comprises the set of documents subject to analytical procedures. In our study, the corpus consists of eight interviews, applied using two different scripts. The average recording duration was approximately 39 minutes, for a total of 29987 words.

b. Data Collection Instruments

Interviews are fundamental instruments when an investigator needs to represent the practices, beliefs, and values of social universes that are not, properly, clarified (Duarte, 2004). An interview makes it possible to collect evidence of the way in which each of the subjects (in this group) perceives and attributes meaning to their reality, as well as to collect the explanatory data present in the subject's own language (Duarte, 2004). The interview is configured as an adequate instrument for the analysis of the interpretation that the actors give to the practices and events with which they are confronted, and allows to analyze the reading they make of their personal experiences (Quivy & Campenhoudt, 1998).

c. Data Analysis Tools

In the treatment of the data collected in the interviews, we resorted to content analysis with the purpose of "making inferences, based on an explicit logic, about the messages whose characteristics were inventoried and systematized" (Vala, 1986, p.104).

The analysis of the interviews is a complicated task and requires a lot of attention, essentially, in the interpretation and construction of the categories, in order to avoid the natural tendency of the researchers to extract the elements that confirm their research hypotheses. A category is a representation of the raw data, that is, it gathers, after the detection of indicators, a certain group of elements (under a generic title), the category (Bardin, 1977; Vala, 1986).

In the analysis of the interviews, it is necessary to take into account the interference of our subjectivity, and therefore we must be aware of it and assume it as part of the investigation process (Duarte, 2004, p.216). One of the methods suggested by Alberti to avoid partiality concerns the transcription of the interviews (Alberti, 1990 *apud* Duarte, 2004). These were transcribed and went through the "trustworthiness conference" process, that is, the recording was heard (with the text transcribed in hand) accompanying and checking each sentence. According to Alberti, transcribing and reading each interview helps to correct errors, to avoid induced responses, and to reevaluate the direction of the investigation (Alberti, 1990 *apud* Duarte, 2004). The interviews were edited in order to remove excessively colloquial phrases, repetitions, incomplete sentences, slang, and grammatical errors. However, according to Duarte, we maintained an original version and an edited version of all transcripts (Duarte, 2004).

THE ROLE OF POLICE COMMUNICATION IN THE INFORMATION SOCIETY

Communication is "absolutely vital for the Police" (Inv3) and "essential in the management process of any public entity" (Pol4), having, nowadays, "a preponderant role in all the activities that need to be performed" (Inv1).

Society is now marked by the digital context (which has evolved much more rapidly than organizations) and by the "rapid dissemination of information through 24-hour news and also through social media" (Inv3). These factors lead to "an acceleration, greater speed and the need for a quick response or to react" (Inv1), however, the institutions "are not prepared, due to their hierarchical structure and all the obstacles that are encountered along the way, to have a quick answer" (Inv1).

Communication must be seen as "a critical success factor" (Pol4), especially "when we talk about crises, this proves to be the most essential" (Pol4), since "decision-making and the response process is extremely fast" (Inv1). "At certain times, [the institution] survival, is at stake but the main problem is related to the institutional competence and its legitimacy of action" (Inv2).

According to the information collected, we found that nowadays the social context "is much more fluid, much more liquid. The fluidity with which things evolve makes it difficult for institutions, at times, to keep up with this pace" (Pol4), "we live in a running world" (Pol2). However, we also realize that "more important than technology, it was what technology did to people" (Inv4), is, "more important than the information society in a purely technological perspective, are the sociological consequences of these changes technological" (Inv4).

Currently, "people have a much higher degree of demand and influence, but a lower degree of tolerance, especially new generations have very high levels of anxiety, have much higher levels of expectations regarding the interactions they enter" (Inv4). "This causes social relations to change and, when they change, all social processes also change, namely the relationship of more formal institutions, with these new audiences and new realities" (Inv4). As a result, "there is much greater superficiality in terms of knowledge of public things because people have stopped internalizing and reflecting on things since information is always available" (Inv4). In relation to the Information Society, the question remains: "Is our audience more and more informed?" (Pol2).

a. Police and Presence on Social Media and Social Networks

The presence of the Police on social media can be seen as an opportunity and as a challenge. The interviewees refer that "all the amplifications of our image have positive and negative questions" (Pol1) and that "adherence to these new communication tools, like social media and social networks, is a double-edged sword" (Inv4) because "they are an opportunity but they have to be well managed" (Pol3).

From the interviews conducted, we realized that "the Police cannot live closed in their cocoon, they have to modernize and adapt to these new methods of communication" (Pol3). Thus, "the Police use social networks to reach their public directly" (Inv1), which is one of the characteristics of social networks and the "new digital context in which we move" (Inv1), that is, "the possibility of there are no intermediaries, the possibility of us speaking directly and, of the institution, being the producer of the information, without this information being mediated by anyone" (Inv1). Therefore, social media allows the police to "communicate very quickly and better control what is put out of the institution" (Pol4).

We also understand that "mass communication and notification tools help the Police to distribute information accurately and quickly. [What] helps the Police to establish itself as an information authority, which quickly distributes information to people who need it" (Inv2). Also at the operational level, it is a field with advantages, namely, "the ability to collect intelligence" (Inv3) and, for example, in the demonstrations the Police have the possibility of "being where the communications are, which helps to predict what will happen happen" (Inv3).

However, there is a common idea, "The Police must be present on social networks, but they cannot be exclusively present only on social networks. Social networks are a good tool, but they are not the only ones (…) especially in a country that is aging" (Pol2). The police "cannot think only of the unconscious young people who spend their lives clinging to the computer. You have to think about the other types of citizens who still only know the newspapers, the church leaves (....)" (Inv4). "The Police must use digital channels as an integral part of the "mix" of the communication process" (Inv2).

As for the challenges arising from joining this new world, the fact that "membership (…) may be opening the door to there is a space for free, anonymous, and incendiary criticism" can be highlighted there. (Inv4). This idea is reinforced by the question of rumors "In social media, everything is propagated. The effect of the rumor and the way in which the rumor (...), which is not true, manages to assume an absolutely extraordinary dimension, poses an extra challenge" (Pol4). This aspect gains greater relevance because nowadays people "spend a large part of their day on social networks" (Pol1). Another aspect to highlight is the constant control of police activity by citizens, "Nowadays, the Police and any other institution is exposed, because there can always be a cell phone recording a situation and, after a second, this video is on a journalist's cell phone, or on "youtube" and after a few minutes it goes viral" (Inv1). This aspect is directly linked to police activity because "it was clearly realized that every time there was a demonstration with solid Police interventions, this had repercussions on social networks. People used the PSP chronology to criticize and dishonor the PSP´s reputation" (Pol1).

b. The use of Facebook

In general, the interviewees consider Facebook as a valuable asset because it enhances and promotes Police's image and allows "catapulting all our operational results" (Pol3). It is a useful tool that "can help to manage people's understanding and understanding of police action, and create a connection to the community, and thus demonstrate the nature of police action as well as what challenges police officers face" (Inv3).

Adhering to Facebook "the PSP shows some transparency, some openness and shows that it has already embraced new ways of communicating, in a more open way" (Inv1) because the discourse used in these platforms allows "a little more human vision in the day-to-day life of the Police, a different view from the traditional bodies" (Inv1). It is intended, therefore, to "carry the logic of policing proximity to the virtual community, the community of social networks" (Pol4).

Facebook also has advantages for the police in terms of crime prevention, "because we are able to pass on safety, proactive and self-protection advice" (Pol3). However, "capital gains have to be perceived not in the short term, but in the medium / long period. The way we communicate today on social networks has a medium and long-term logic that is institutionally legitimized" (Pol4). In this sense, Facebook is configured as "a form of two-way communication. What is particularly important with regard to crisis situations" (Inv3).

Regarding the fact that it is the government page with the most followers, the interviewees consider that "the fact that there are a lot of people there, allows these people to clarify the situation, and give the Police's vision" (Inv1). Thus, the PSP has a means, direct and without intermediaries, that allows communicating to society about a certain incident and "helping people to understand what is happening and to create appropriate responses and give feedback" (Inv3). Facebook, if it is "strategically thought through, can be a force to reinforce the image and reputation of the Police" (Inv4).

Thus, "the negative side is the possibility of misuse of the official page. To avoid this situation, it would be ideal to provide an email so that these people can discuss these issues and concerns outside the public sphere" (Inv2). Researcher 1 considers that "having many followers and many people interacting with the page, also makes it more exposed, which I think is inevitable nowadays. If you do not stand out from the crowd (social media) you are missing out.

c. The Potential of Twitter

Twitter is configured as "a way of bringing the Police closer to the population, but it is also a way of managing the conflict, since we are giving people timely information, even to know which are the transit routes that they are cut off, they can avoid it, they know what is expected to happen and what is important" (Pol3). The benefits of using Twitter were evident during the attack on the Boston Marathon, in which the Police used Twitter "as an official channel to keep people informed" (Inv2).

In Portugal, Twitter, although without a wider widespread "is a tool that, due to its characteristics, can be useful for the Police, in the sense of quickly alerting to certain situations, drawing attention to security procedures that people must have, even to position police officers as opinion leaders in certain security matters" (Inv1). When there is a big event or an incident, "the first task is to understand what's happening" (Inv3), so the Police must monitor Twitter, "to quickly and accurately find out what's going on. happening, [since] there are people who are actually in the place and who publish video content showing what is happening" (Inv3).

We found that one of the advantages of sharing information, in real-time on Twitter, is the ability to verify the population's acceptance of the PSP's intention in a given occurrence. From the experience of the PSP in this context, it was possible to "understand what society, in general, was waiting for and what they were thinking about police intervention (….) and this also helped to shape the decision of who was in charge (Pol3), even "because any public order intervention always has unpopularity with the police" (Pol4). In this sense, the use of this tool allows better management of staff, as well as the level of force to be used to legitimize the intervention of the Police.

CONCLUSION

The analysis made it possible to ascertain that a crisis in terms of the image is a public reality, interpreted by the citizen and exposed and mediated, not only by traditional media but especially in social media. Its classification as a crisis depends on the damage caused and the fault (the causal attribution) attributed to the Institution. The crisis represents a controversial phase in which the Institution is under high media attention and, therefore, disturbs its daily life. For the Police, crises can result from an unpredictable event or phenomenon, but also from a predicted, poorly resolved event that calls into question its legitimacy and leads to the intervention of the Institution's stakeholders.

Crises require quick decision-making and urgent response, and in policing, if possible, this should be taken live and live. As a consequence, a crisis affects the structure of the organization, with repercussions for its members (at least psychologically, in terms of absenteeism, demoralization, and disunity amongst them).

This work showed the most used tools for communication with different audiences, in addition to traditional media, are social networks and social media. In this field, we conclude that the Police essentially uses Facebook for social interaction and approximation. On the other hand, Twitter is used in emergency situations. These two tools should be used as different channels, for different situations, with a distinction in terms of the form and content of the communication.

With regard to the relationship between the Dialogue Policing and Crisis Communication, we can say that these complement each other. On the one hand, the Police, by using dialogue at events, reduces conflict, prevent the occurrence of potential crisis phenomena. On the other hand, the better the image and legitimacy attributed to the Police, the more open people are in participating in these events, to hear what officers have to say. This aspect facilitates the mission of the dialogue teams, as advocated by Dialogue Policing. We believe, therefore, that the two concepts, when interconnected, contribute to the Police improving its capacity for persuasion and cooperation with the community, in order to achieve acceptance of its mission and objectives.

The relationship that is established between an organization and its audience cannot be seen as a simple exchange. It should be seen as social interaction, marked by the history of previous impressions. In the case of the Police, in addition to the daily interactions between its members and the citizen, it is the organizational discourse (strategic and institutional communication) that helps to shape the way the citizen views the police mission. The Police, as a member of the community, must ensure voluntary cooperation and obedience on the part of society, and the degree of cooperation obtained proportionally decreases the need to use physical force. Thus, communication has the function of informing and clarifying procedures, in order to avoid any rumors or misinterpretations, since society has the right to know organizations and to comment on them. Following this reading, the Police must see crises and policing major events not only as potential problems, but also as an opportunity to build bridges with the community, and, to that end, should harness the power of the media (the public attention that is dedicated to a phenomenon and its characteristics influences the framework of the phenomenon and leads to the formation of values and judgments) and establish a cooperative relationship with them so that both can achieve their goals.

On the other hand, and because no country is immune, the police must reflect on their ability to manage, in terms of crisis communication, the phenomenon of terrorism. It is important that institutions with responsibility at that level establish mechanisms that can inform people about what is happening so that they can proceed with their own security measures. In the Portuguese case, this could be challenging due to the high number of entities involved.

Taking into account all the analysis options, we believe this work may be the starting point for new research both in the area of crisis management and in the area of crowd policing, based on a dialogue philosophy, where the Police can help society to face its problems.

REFERENCES

Alves, C. (2007). *Comportamento Organizacional: a Gestão de Crise nas Organizações*. Escolar Editora.

Andrade, J. (2008). *A sociedade da informação e as organizações em momentos de crise: uma investigação interpretativista do fluxo de comunicação* (Dissertação do Mestrado em Ciências da Comunicação). Lisboa: Universidade Católica Portuguesa.

Andrade, J. (2009). Gestão de crises organizacionais e a Web 2.0. In *6º Congresso SOPCOM - Sociedade dos Media: Comunicação, Política e Tecnologi*, (pp. 3948 - 3953). Associação Portuguesa de Ciências da Comunicação.

Andrade, J. (2015). Portugal no "mensalão" - um estudo de caso sobre o envolvimento da Portugal Telecom na crise brasileira de compra de votos parlamentares. *Comunicação & Marketing*, *6*(4), 54–71.

Bardin, L. (1977). *Análise de Conteúdo*. Lisboa: Edições 70.

Bauman, Z. (2001). *Modernidade Líquida*. Rio de Janeiro: Jorge Zahar Editor.

Boin, A. (2019). The Transboundary Crisis: Why we are unprepared and the road ahead. *Contingenciesand Crisis Management*, 94–99. doi:10.1111/1468-5973.12241BOIN|99

Braziel, R., Straub, F., Watson, G., & Hoops, R. (2016). Bringing calm to chaos: A critical incident review of the San Bernardino public safety response to the December 2, 2015, terrorist shooting incident at the Inland Regional Center. U.S. Community Oriented Policing Services, Department of Justice.

Brown, J. (Ed.). (2014). *The Future of Policing*. Routledge.

Caetano, J., Vasconcelos, M., & Vasconcelos, P. (2006). *Gestão de crise*. Editorial Presença.

Carmo, H., & Ferreira, M. (2008). *Metodologia da Investigação: guia para a auto-aprendizagem* (2nd ed.). Universidade Aberta.

Castells, M. (2011). A era da informação: Economia, Sociedade e Cultura. Volume I: A sociedade em rede (4th ed.). Lisboa: Calouste Gulbenkian.

Claeys, A., & Coombs, T. (2020). Organizational Crisis Communication: Suboptimal Crisis Response Selection Decisions and Behavioral Economics. *Communication Theory*, *30*(3), 290–309. doi:10.1093/ct/qtz002

Coombs, T. (2007). *Crisis Management and Communications*. Institute for Public Relations. http://www.instituteforpr.org/crisis-management-and-communications/

Coombs, T. (2010). Parameters for Crisis Communication. In T. Coombs & S. Holladay (Eds.), *The Handbook of Crisis Communication, (pp. 17 - 53)*. Wiley-Blackwell. doi:10.1002/9781444314885.ch1

Creswell, J. (2014). *Research design: qualitative, quantitative, and mixed methods approaches* (4th ed.). SAGE.

Cunha, M. (2006). Crises Organizacionais: Ameaças ou oportunidades? In A. Mendes & F. Pereira (Eds.), *Crises: de ameaças a oportunidades - gestão estratégica de comunicação de crises* (pp. 143–159). Edições Sílabo.

Denef, S., Bayerl, P., & Kaptein, N. (2013). *Social media and the Police - Tweeting Pratices of British Police Forces during the August 2011*. Fraunhofer Institute for Applied Information Technology. doi:10.1145/2470654.2466477

Dias, P., & Andrade, J. (2015) Desafios das RP na Era dos Media Sociais. Novos media e novos públicos.

Dias, P., & Andrade, J. (2017). O papel dos social media nas crises reputacionais. In *A Comunicação Organizacional e os desafios tecnológicos: estudos sobre a influência tecnológica nos processos de comunicação nas organizações* (pp. 55–69). CECS - Centro de Estudos de Comunicação e Sociedade Universidade do Minho.

Duarte, R. (2004). Entrevistas em pesquisas qualitativas. *Review of Education*, *24*(24), 213–225. doi:10.1590/0104-4060.357

Fearn-Banks, K. (2011). *Crisis Communication: a casebook approach* (Vol. 4). Routledge.

Felgueiras, S. (2015). Ação Policial face à ação coletiva: teoria para uma estratégia de policiamento de multidões. Lição Inaugural da Abertura Solene do ano letivo 2015/2016. Lisboa: ISCPSI.

Gama, M. G. (2000). Quando o inferno desce à terra: A gestão de crises e a sua problemática. *Comunicação e Sociedade*, *14*, 535–542. doi:10.17231/comsoc.2(2000).1419

GODIAC. (2013a). *Field Study Handbook: GODIAC - Good practice for dialogue and communication as strategic principles for policing political manifestations in Europe.* http://www.polisen.se

GODIAC. (2013b). *Recommendations for policing political manifestations in Europe. Good practice for dialogue and communication as strategic principles for policing political manifestations in Europe.* http://www.polisen.se

Gorringe, H., Stott, C., & Rosie, M. (2012). Dialogue Police, Decision Making, and the Management of Public Order During Protest Crowd Events. *Journal of Investigative Psychology and Offender Profiling*, *9*(2), 111–125. doi:10.1002/jip.1359

Havelund, J., Ilum, J., Jensen, M., Nielsen, B., Rasmussen, K., & Stott, C. (2011). *Event Policing - Dialogue in the policing of mass events in Denmark*. European Police Science and research. *Bulletin*, *4*, 3–7.

Heath, R. (2010). Crisis Communication: defining the Beast and De-marginalizing Key Publics. In T. Coombs & S. Holladay (Eds.), *The Handbook of Crisis Communication, 1 - 14*. Wiley-Blackwell. doi:10.4324/9780203891629

Her Majesty's Chief Inspector of Constabulary (HMIC). (2008). *Adapting to protest– nurturing the British model of policing*. Author.

Holgersson, S. (2010). *Dialogue Police: experiences, observations and opportunities*. Swedish National Police Board.

Ilharco, F. (2003). *Filosofia da Informação: uma introdução à informação como fundação da acção, da comunicação e da decisão*. Universidade Católica Editora.

Ilharco, F. (2005). A Tecnologia como Contexto ou a Ordenação Informacional e Comunicacional do Mundo (Prisma.Com, Ed.). Porto: Faculdade de Letras da Universidade do Porto.

Institute for Strategic Dialogue. (2014). *Case Study: Dialogue Police.* Author.

Joint Committee on Human Rights. (2009). *Demonstrating respect for rights? A human rights approach to policing protest.* House of Commons & House of Lords.

Jorge, N. (2010). *Reputação: um elemento diferenciador e protector face a crises organizacionais.* Escola Superior de Comunicação Social.

Kingshott, B. (2011). Effective police management of the media. Criminal Justice Studies: A Critical Journal of Crime. *Law and Society*, *24*(3), 241–253. doi:10.1080/1478601X.2011.592728

Lourenço, N. (2012). Legitimidade e confiança nas polícias. *Revista do Ministério Público*, 181-198.

Maxwell, J. (2005). *Qualitative research design: an interactive approach.* Sage.

Mendes, A. M., & Pereira, F. C. (2006). *Crises: de ameaças a oportunidades - gestão estratégica de comunicação de crises.* Edições Sílabo.

Office of Community Oriented Policing Services (COPS). (2015). After-Action Assessment of the Police Response to the August 2014 Demonstrations In Ferguson, Missouri. Washington, DC: Institute for Intergovernmental Research.

Orduña, O. (2004). *A comunicação em momentos de crise.* de Biblioteca on-line de Ciências da Comunicação (BOCC): http://www.bocc.ubi.pt/pag/orduna-octavio-comunicacao-em-momentos-de-crise.pdf

Osterling, O. (2013). Dialogue police work: a balancing act on a slack. In The anthology: GODIAC -, Good practice for dialogue and communication as strategic principles for policing political manifestations in Europe, (pp. 54-60). Swedish National Police Board and GODIAC.

Pinho, C. (2015). *Repensar a Gestão de Crises em Redes Sociais: o caso do voo MH370.* Universidade do Minho.

Police Executive Research Forum. (2016). *Recommendations on advancing Community Policing in the Pasco Police Department: Critical Response Initiative.* Office of Community Oriented Policing Services.

Procter, R., Crump, J., Karstedt, S., Voss, A., & Cantijoch, M. (2013). Reading the riots: What were the police doing on Twitter? *Policing and Society*, *23*(4), 413–436. doi:10.1080/10439463.2013.780223

Quivy, R., & Campenhoudt, L. (1998). *Manual de Investigação em Ciências Sociais* (2nd ed.). Gradiva.

Regester, M., & Larkin, J. (2008). *Risk issues and crisis management in public relations: a casebook of best practice* (4th ed.). Kogan Page.

Reicher, S., Stott, C., Drury, J., Adang, O., Cronin, P., & Livingstone, A. (2007). Knowledge-Based Public Order Policing: Principles and Practice. *Policing*, *1*(4), 403–415. doi:10.1093/police/pam067

Santos, R. (2006). Crises no e-world. In A. Mendes & F. Pereira (Eds.), *Crises: de ameaças a oportunidades - gestão estratégica de comunicação de crises, (pp. 191 - 244).* Edições Sílabo.

Santos, R. (2015). *Dialogue Policing: uma nova abordagem à gestão de multidões* (Dissertação do Mestrado integrado em Ciências Policiais). Lisboa: ISCPSI.

Sebastião, S. (2009). *Comunicação Estratégica: as Relações Públicas*. Instituto Superior de Ciências Sociais e Políticas.

Sellnow, T., & Seeger, M. (2013). *Theorizing Crisis Communication*. Wiley-Blackwell.

Smith, J. (2008). Interpretive Inquiry. In L. Given (Ed.), *The SAGE Encyclopedia of qualitative research methods* (Vol. I & II, pp. 459–461). SAGE.

Stephens, D., Hill, J., & Greenberg, S. (2011). *Strategic Communication Practices: A Toolkit for Police Executives*. The Office of Community Oriented Policing Services (COPS).

Stott, C. (2009). *Crowd Psychology & Public Order Policing: An Overview of Scientific Theory and Evidence*. School of Psychology - University of Liverpool.

Stott, C., Havelund, J., Lundberg, F., Khan, S., Joern, L., Hoggett, J., & Vestergren, S. (2016). *Policing football in Sweden: enabling an evidence-based approach*. ENABLE.

Tyler, T., & Fagan, J. (2008). Legitimacy and Cooperation: Why Do People Help the Police Fight Crime in Their Communities? *Journal of Criminal Law*, *6*, 231–275.

Vala, J. (1986). A análise de conteúdo. In *Metodologia das Ciências Sociais, 101-128*. Edições Afrontamento.

Valente, M. M. (2014). *Teoria Geral do Direito Policial* (4th ed.). Almedina.

van Laere, J., & Lindblom, J. (2019). Cultivating a longitudinal learning process through recurring crisis management training exercises in twelve Swedish municipalities. *Contingencies and Crisis Management*, 38–49, doi:10.1111/1468-5973.12230

Vannini, A. (2009). Interpretive Theory. In Encyclopedia of Communication Theory, (pp. 557-562). SAGE. doi:10.4135/9781412959384.n209

Velde, B., Meijer, A., & Homburg, V. (2015). Police message diffusion on Twitter: Analysing the reach of *social media* communications. *Behaviour & Information Technology*, *34*(1), 4–16. doi:10.1080/0144929X.2014.942754

Williams, K. (2003). Peel's Principles and their acceptance by American Police: Ending 175 years of reinvention. *The Police Journal*, *76*(2), 97–120. doi:10.1177/0032258X0307600202

KEY TERMS AND DEFINITIONS

Dialogue Policing: Proposal for emphasis on oral and informal communication with the aim of preventing confrontations and violence connected to events like demonstrations or football matches.

Facebook: Social media and virtual social network launched on February 4, 2004, operated, and privately owned by Facebook Inc. In 2012 it reached the mark of 1 billion active users, making it the largest virtual social network in the world.

New Media: Broad term referring to the sum of new technologies and communication methods to differentiate from traditional communication channels such as TV, broadcasting, press, etc.

Public Relations: The set of information activities, systematically coordinated, related to the exchange of information between an individual, organization (public or private), or a non-governmental organization and its public.

Social Media: The collective of online communication channels dedicated to inputs, interactions, content sharing and community-based collaboration.

Twitter: Social media and a microblogging server, which allows users to send and receive personal updates from other contacts (in texts of up to 280 characters, known as "tweets"), through the service's website, by SMS and by specific software management.

Viral Marketing: Marketing strategy that aims to explore the connections between people to spread and go viral. It is considered a technique with lower costs than traditional actions, since the media used is the target audience itself.

ENDNOTE

1 Sir Robert Peel is quoted as one of the founding fathers of the modern Police and his nine principles ("Law Enforcement Principles" of 1829) are the foundations for today's Police. His teachings were also at the basis of what today´s community policing (Kingshott, 2011; Williams, 2003, Stott & Gorringe, 2014).

Chapter 10
Setting the Aware Agenda of the COVID-19 Health Emergency:
The Italian PAs Social Media Coverage

Marica Spalletta
Link Campus University, Rome, Italy

Dario Fanara
Independent Researcher, Italy

Paola De Rosa
Link Campus University, Rome, Italy

ABSTRACT

Among its main goals, crisis management aims at promoting people awareness in respect of the crisis which they are going to face or in which they are already involved. In order to do that, it uses a wide range of communication tools, among which, over the last decade, social media have proved to be of paramount importance. Based on these premises, the chapter analyses a very meaningful case of crisis communication, which consists of the social media coverage of the early stages of the COVID-19 emergency coming from Italian national and local institutions. The media content analysis carried out on Facebook and Twitter confirms a communication strategy aimed at creating people awareness in respect of the health emergency, suggesting citizens which conducts they need to stop or adopt. However, the analysis also shows that the goal of crisis awareness represents the first step of a wider agenda coming from the Institutions' social posting, which aims at transferring their awareness to citizenship and, as a consequence, inspiring citizens' own responsibility.

DOI: 10.4018/978-1-7998-6705-0.ch010

INTRODUCTION[1]

In the final season of *Scandal* (episode 7x12), the popular US political thriller television series, the main character Olivia Pope, a well-known crisis manager, after her dismissal from the White House starts to serve as a guest lecturer at a local university. During her first class, she asks which is "*the biggest mistake people make in a crisis*" and students answer quoting "*a failure to plan*" or "*an ineffective communication strategy*", or – provoking the lecturer – "*telling the truth*"; however, all these statements don't seem able to satisfy Olivia. After a short silence, a voice from the latest desks suggests that the biggest mistake could consist in *"not knowing people are in a crisis"*. The voice belongs to Annalise Keating, the main character of another popular TV series (*How to Get Away with Murder*), and this meeting kicks off the crossover between the two series created by Shonda Rhymes.

Even though very short, the dialogue between Olivia and Annalise points out a feature of paramount importance aiming to introduce crisis communication, that is the central role played by *crisis awareness*.

We are living through a major period of historical transition and constant transformation defined by authoritative scholars as a "crisis society" (Fraser, 1981; Heide & Simonsson, 2019), marked by a persistent exposure to multiple and different risks (Beck, 1992; Giddens, 1990; Adam *et al.*, 2000) that lead both to endemic uncertainty (Bauman, 1999) and the need to handle, also from a communication point of view, the disrupting events breaking normal routines (just like crisis events) (Perry, 2007). Nevertheless, as Annalise Keating suggests, and Olivia Pope confirms, managing a crisis is the second step of an upstream process based on the primary need to personally gain (and even more let people gain) awareness of being in a crisis situation, and this happens regardless of the nature of the crisis (natural disasters, terrorist attacks, health emergencies, etc.) (Grant & Mack, 2004; Quarantelli, 2005).

Indeed, the very first step in managing a crisis is clearly understanding what is going on and then predicting how the crisis may evolve (Endsley & Garland, 2000) which is essentially about the so called "situation awareness" relying on three main components: «the perception of the elements in an environment within a volume of time and space, the comprehension of their meaning, and the projection of their status in the near future» (Endsley, 1995, p. 36).

What is more, situation awareness is considered as a key precondition to develop an efficient and accurate crisis communication and turns to be necessity both for citizens (and their pressing need for information during a crisis event) and for Public Authorities (PAs), who seek to get control over high-risk environments – from public safety to unpredictable events, such as natural disasters or health emergencies – and to monitor how they develop, carrying out at the same time the indispensable tasks to spread (and get) information, to interact with the public and to perform an effective and real-time decision-making process (Coombs, 2004; 2007a; 2007b).

Over the last years, the process of achieving and maintaining situation awareness has been making evident that social media can play a strategic role on the occasion of the main crisis affecting modern societies (Bruns & Burgess, 2014; Kaufmann, 2015), because of their attitude to act as "awareness system" (Maireder & Ausserhofer, 2014), responding to emotional needs and making people feel as they are contributing and coping with the crisis events (Liu *et al.*, 2013). Of course, the social media strategic role emerges also referring to PAs, which bear a significant responsibility for managing crisis events (Coombs, 2014; Austin & Jin, 2017).

In fact, in an increasingly connected world where citizens are often victims of information overload, PAs have the duty and the need to attract the attention of individuals. Modern decision-making processes do not end in the chambers of power but go on with communicating the implementation of certain public policies or initiatives correctly and in the most widespread way. The spread of social media has radically transformed the ways in which users access the news they are interested in and this is a mechanism that PAs have had to take into account over time, promoting a new communication pact in which the citizen/user is central (Comunello, 2014; Comunello & Mulargia, 2017).

Shifting from "everyday communication" to the crisis one, Comunello and Mulargia (2017) explain how increasing number of citizens turn to social media to seek or provide information during emergencies and natural disasters: on one hand, official accounts and institutional pages on social networks can be an efficient way to disseminate emergency alerts or information to reach a large number of people; on the other hand, social networks are the first media to be used by the population to find news concerning an emergency or to produce/share new information.

Therefore, even in emergency communication, the use of social networks represents a challenge for the PAs to overcome, in order to organize the communication flow and act as a filter to provide citizens verified information with timely updates (Coombs, 2002; van der Meer & Jin, 2020). A quick advice dissemination is one of the most important things for reducing risks in disaster scenarios. Social media has become an efficient way of sharing information between government organizations and citizens during times of emergencies (Coombs & Holladay, 2014; Hornmoen & Backholm, 2018).

AIMS & METHODOLOGY

Based on these premises, the chapter focuses on a very emblematic case of crisis communication, which consists of the social media management of the early steps of the Covid-19 health emergency coming from some Italian national and local PAs. The research aims at understanding how the analysed institutional actors have managed the communication concerning the so-called "coronavirus" promoting people awareness (Coombs & Holladay, 2010; Svedin, 2012) and even setting the agenda upon which people are called to build their own responsibility in everyday life (McCombs & Shaw, 1972; McCombs, 2004), also highlighting the gradual transition from an initial *risk communication* to a more properly *crisis communication*, (i.e. from a mainly informative communication to a more persuasive one, aimed at transforming the "declaration" into "action") (Morgan *et al.*, 2002; Reynolds & Seeger 2005; Littlefield & Sellnow, 2015).

From a methodological point of view, the research carried out a qualitative media content analysis (Neuendorf, 2002; Krippendorff, 2004; Altheide & Schneider, 2013) on a database consisting of the Facebook posts and tweets shared from 12 social accounts of Italian national and local PAs, taking into account both the institutional social profiles and the ones related to their political representatives: on one hand (national PAs), Presidency of the Council of Ministers (*Palazzo Chigi*) and Prime Minister (*Giuseppe Conte*); Ministry of Health and Health Minister (*Roberto Speranza*); on the other hand (local PAs, that is the Regions), *Lombardy*, *Veneto*, *Campania* and *Sicily* and their related Presidents (*Attilio Fontana*, *Luca Zaia*, *Vincenzo De Luca*, *Nello Musumeci*). The choice to focus the media content analysis on these actors, within the wider Italian institutional scenario, is closely related to either the role they played during the health emergency (by virtue of their leading function and technical expertise) or their degree of involvement in the coronavirus pandemic (both in terms of incurred infections and containment measures adopted), with also a look toward a geographical distribution, reflecting the unusual polarization (North *vs.* South/Islands) that marked the pandemic's course in Italy[2].

The choice to focus both on national and regional PAs relies on the specific nature of the Italian system, which provides for a Prime Minister – who is the head of government – and is characterized by a territorial division in 20 Regions, each of which is ruled by its own President or Governor. The relations between national and local Institutions is set on the basis of the division of legislative competences between the State and the Regions established by the Italian Constitution (Art. 117), according to which the Regions hold their own competence about health issues.

As to the choice to take into account both institutional and political actors, since Law 150/2000, all Italian PAs have been equipped with institutional communication tools with the aim of managing, developing and improving the relations of Institutions with the citizens. This is a typical feature of the Italian scenario, which needs to distinguish *institutional communication*, considered as the activity aimed at favoring and promoting PAs, from *political communication*, that is an activity carried out by individual political subjects through their own communication channels; both of them are part of a wider *public communication* (Faccioli, 2006; Grandi, 2013). Starting from the 2010s, this communication activity has also been carried out through social institutional channels, which joined those already used by political actors who *pro-tempore* hold the related Institutions, without overlapping them (Lovari, 2013; Ducci 2017).

Referring to the evolution of the Covid-19 health emergency, the research focuses on five topical moments, selected in respect both of the "crisis timeline" suggested by the Italian Department of Civil Protection (that is the institutional authority which, due to its role, "set the crisis", standing out as the main primary source of information both for citizens and PAs) and the most important containment measures issued by the involved PAs. The five moments considered in this analysis are the following:

1) January 30: the so called "time zero" of the pandemic when, after the first two cases of coronavirus involving a Chinese couple of tourists in Rome, the Department declared the health emergency;
2) February 22: the set-up of the first two "red-zones", one centred on 10 towns in Lombardy and another one in Veneto;
3) March 4: the decree about the closure of all schools and universities in the Country, coincident with the beginning of the Department's daily appointment at 6 p.m., which points out the crisis evolution;
4) March 8: the decree expanding the coronavirus quarantine red-zone across the whole Lombardy and parts of other Northern Regions;
5) March 9: the decree *#iorestoacasa* placing the entire Country under quarantine and declaring the whole Italy as a "protected zone".

The analysed corpus consists of 600 original social posts, coming from Twitter (304) and Facebook (296), to which are added the most relevant retweets and re-shared Facebook posts. The choice to focus on both social networks derives from their different distinguishing features, as Twitter is the most suitable channel for providing frequent, immediate and quickly accessible information while Facebook allows to build a more in-depth relationship based on dialogue, discussion and audience participation.

The overall contents (posts/tweets) shared by political and institutional actors have been reported in a suitable matrix and codified according to: the main topics on which they focused (contagions, health emergency, guidelines for dealing with the emergency, school/university, invitation to civic sense and change of habits, containment measures/ministerial decrees, announcement of press conferences or other institutional appointments); the main goals of institutional communication (informing, empowering, blaming, alarming, reassuring); the different "tones" used for communication purposes (institutional, aseptic, ironic, participatory, peremptory).

In order to answer the research question, the chapter is structured in two sections: the first one focuses on the social posting coming from the analysed PAs and more specifically on the use they make of their own Facebook and Twitter accounts, in terms of social media coverage as a whole as well as of distinguishing features contributing to define their communication strategy; the second one focuses mainly on the role played by the above mentioned actors, underlining the degree of *situation* (and consequently *crisis*) *awareness* displayed during the emergency and the way in which it affected their role of "crisis agenda builder" aimed at promoting responsibility among their publics.

THE SOCIAL POSTING CONCERNING THE COVID-19 EMERGENCY

The first phase of the analysis focuses on the social posting coming from the examined accounts, aiming at understanding the relationship between the institutional actor and the political one (contrast/integration) as well as analogies or differences between the Facebook posting and the Twitter one, the newsworthiness of the Covid-19 crisis in respect of the other covered topics (meaningful or not, and in respect of which other issues), the main features characterising posts and tweets (goals, tones, recurrent tools, as well as photos, videos or hashtags).

Palazzo Chigi/Giuseppe Conte

The joint analysis of the "self-produced" communication about Covid-19 performed by Palazzo Chigi and the Italian Prime Minister Giuseppe Conte allows us to identify a clear communication strategy: on one hand, it shows a "content overlapping", aimed at drawing people's attention to the same main issues; on the other hand, it serves the targeted purpose of differentiating the methods through which the actual contents have been created and shared with the public.

This ideal "division of tasks" is detectable starting from the different use they make of the examined social media: Palazzo Chigi's social posting develops mainly on Twitter while Conte's social communication is more prominent on Facebook.

In the timeframe analysed, it is from February 22 that both accounts start sharing only information about Covid-19, avoiding reference to any other matter unrelated to the health emergency. Those "*no-Covid*" issues are indeed circumscribed at the "time zero" of the pandemic, coinciding with institutional and representative events, already scheduled by Italian Government, that only by chance go hand-in-hand with the first news about a critical event, whose nature is by definition unexpected.

Both tweets and Facebook's posts published by Palazzo Chigi serve mainly an informative function, providing constant updates about Government decrees and the related containment measures introduced from time to time, with a major use of links redirecting to Italian Government website, as well as a strong preference for the use of videos rather than photos. In the few cases where the aim is to reassure citizens or to empower them, the overall communication preserves its institutional tone, limiting itself to share the claim of a specific institutional campaign (*"If you love Italy, keep your distance"*), or to replicate the main statements spoken by the Italian Prime Minister.

Conte's social communication, for its part, serves several functions combining the more basic informative aim with the trickiest functions to reassure Italian people and call them for responsibility through the use of different tones that gradually turn from an institutional and detached level (January 30) to an increasingly involved one (starting from February 22 and even more after March 4). On one hand, his social communication aims at reinforcing the dissemination activity about decrees and containment measures already carried out by Palazzo Chigi; on the other hand, its main interest is to empower Italian citizens making large use of patriotic appeals and calling for civic sense and collective responsibility, in its literal meaning of "sum of individual responsibilities". The large use of videos, especially on Facebook, may be considered as the distinctive feature of his communication strategy, in its dual form of recorded video message (as it happens on March 4, with his heartfelt message starting with the statement "*The whole of Italy is called to do its part*") and live broadcasts, which allow the public to benefit from both a real-time view and a delayed view in the following hours (which comes in handy since many of his press conferences take place late at night).

Both actors do not make use of particular hashtags, other than the more generic "#coronavirus" which may be found in almost all tweets. Starting from March 9, Premier Conte starts using the hashtag #iorestoacasa (#stayathome), which also gives the name to the decree that will be issued at the end of the same day.

Ministry of Health/Roberto Speranza

The communication flow that characterizes the platforms of Italian Ministry of Health and its political leader, Minister Roberto Speranza, clearly shows the role of informant par excellence played by the Ministry. Considering the two platforms examined, the most used social channel is Twitter.

On a communicative level, Speranza's role seems to be that of "amplifier" of the messages of the Ministry: apart from a few posts and some tweets in which he reaffirms personal statements or produces the report of previous interventions, the Minister is used to share the contents both of the Ministry of Health, the Premier and, of course, the Italian Department of Civil Protection. Attention is also paid to other types of actors, such as the mainstream media. One of the most significant examples is the relaunch of a video initially presented on Twitter by Raitre in which the intervention of the Minister within the program *Mezz'ora in più* is proposed (March 8).

Therefore, the Twitter account of the Ministry of Health dominates the communication arena in this phase of the emergency. The most retweeted actors are Palazzo Chigi, Premier Conte and the Department of Civil Protection. The sharing of the messages of these three accounts seems to follow a precise strategy: the first and the second ones are useful for informing the population about the new rules to be followed, through a narration marked by the sharing of the videos of the press conferences of the Prime Minister or by links aimed at redirecting the user to "other pages" and allow to deepen the contents of the various decrees. The third one, on the other hand, is useful for updating the user about the progress of the infections. The attention towards these three actors also goes along with the gaze turned to Lombardy and Lazio, through the sharing of tweets launched by the two Regions containing updates on the evolution of the epidemic in the respective Regions. It almost seems that the gaze towards these two "physical spaces" is a mirror that outlines the extent of the emergency at national level. Obviously, the relaunch of a greater number of tweets of Lombardy Region, compared to the Lazio one, reflects the worse position of North than Central Italy during the emergency.

The Ministry also pays attention to the media as its "political leader": Raitre's videos showing the television interviews of Minister Speranza are re-broadcasted and, at the same time, it contributes to the greater dissemination of the video promoted by the Ministry itself with the participation of Michele

Mirabella, the well-known anchorman of scientific information programs on public service television, with the aim of preventing inappropriate behaviours and promoting its own website (www.salute.gov.it) for correct information. In this sense, the retweeting amplifies the information, by empowering the mission embodied by the Ministry of Health. A mission that becomes clear through a careful analysis of the tweets of this institutional account. These mainly consist of recommendations that indicate how to behave and the behaviours to avoid in order to slow the spread of the epidemic. Among the recommendations, the most widespread is "staying at home" promoted through the #iorestoacasa campaign, aimed at raising public awareness about the need to avoid gatherings as much as possible and implement social distancing by remaining in one's own home. This campaign was supported by infographics and a hashtag that allowed the contents to go viral. It is no coincidence, in fact, that even the big names in music, cinema and entertainment have re-launched the campaign with an appeal, addressed especially to young people. The commitment of the showbiz characters was highlighted through a tweet on March 9, once again demonstrating the awareness about the importance of networking in order to make contents more viral through the Net.

Lombardy Region/Attilio Fontana

The Lombardy Region (*Lombardy*) develops its social communication mainly on Twitter; by contrast, President Attilio Fontana is practically absent on Twitter on these specific dates, relegating to his Facebook page the main task of "social intermediary" between his concerned public and the institutional role he embodies.

At the "time zero" of the pandemic, the social posting of both actors is mainly focused on *"no-Covid"* issues, dealing with institutional events or various updates about different topics of public interest, and just a few mentions about the coronavirus emergency with the clear aim to deny any alleged case of infection within Lombardy. A growing interest starts developing from February 22, when the social posting begins to focus exclusively on the health emergency, leaving no room for any other issue.

From a content point of view, the information disseminated about Covid-19 is almost the same on both social media. The main focus is about practical tips for preventing coronavirus contagion and diffusion shared through explanatory *vademecum* divided into different episodes (one for each tweet/post) and accompanied by several pictures referring to national or regional awareness campaigns. The same "episodic" structure is actually replicated in the same way on Facebook, together with the same short and incisive tweet texts, even where there would be more room for a wider communication. The overall aim to provide citizens with clear examples of good practices also relies on the common use of videos involving several testimonials, from virologists (called to provide short and clear explanations on some scientific issues), to Italian show business personalities and professional athletes who display themselves in the guise of ordinary citizens, asking people to follow their own example in facing the health emergency. Alongside the practical "to do", the social posting of *Lombardy* also focuses on the containment measures adopted at both national and regional level. While the reference to Government decrees is marked by an informative tone, the communication about regional provisions takes on the character of pressing "need", and hence is marked by the use of peremptory tones that call for strict compliance with the rules, pointing out the risk of penalties.

The hashtag #LNews, already used by *Lombardy* for everyday communication on Twitter, remains unchanged also during the emergency, opening almost all posts in combination with the more generic hashtags #coronavirusitalia #covid19. On the other hand, the hashtag #coronavirus opens all Facebook posts focused on that subject. From March 4, the hashtag #fermiamoloinsieme (#LetStopItTogether) makes its entrance both on Twitter and Facebook posts, together with the launch of a suitable communication campaign carried out by *Lombardy* aimed at strengthening the prevention measures.

As far as President Fontana, through his Facebook page he partially retraces the communication performed by *Lombardy* making use of short, concise texts and institutional tones (consider, for example, the posts through which he merely announces further press conferences and upcoming statements). On the other hand, he also makes use of longer and detailed posts in order to address heartfelt and participatory messages to his people, as well as frequent praise to medical and health personnel engaged in the front line against Covid-19 emergency, together with his personal and institutional support to private fundraising activities. The various posts, as well as his many television interventions, are equally distributed among the various topics, with a greater focus on containment measures adopted at a regional level (first of all the ones about the schools closure, marked as a priority in Lombardy even before the national decree), as well as on the empowerment of citizens, which he seeks to achieve through recurring appeals to civic responsibility, frequently enhancing the feeling of "regional pride" of Lombard citizens.

Both social accounts bring out the presence of different multimedia contents, which are translated into the use of videos and photos (more or less in equal measure) bringing out a professionally managed and multi-channelled communication.

Veneto Region/Luca Zaia

The analysis of the communication flows of Veneto Region (*Veneto*) shows several differences in respect of the other examined actors, which concern the different use of social media, and – more specifically – the covered issues and the related goals, tones and strategies of crisis communication related to the health emergency. If the Twitter coverage of *Veneto* is equally distributed between the two actors, on Facebook it seems more difficult to identify which actor plays the leading role of the communication process, since both of them tend to take turns on the social arena.

At the "time zero" of the pandemic (January 30) the only reference to "coronavirus" can be found in a single Facebook post shared by Zaia announcing an upcoming press conference focused on that issue. With exception of this minimal reference, there is no other hint to the Covid-19 emergency, which is indeed largely overtaken by institutional communications on priority themes related to ongoing events and services of public interest (issued by the Region), rather than political actions in support of the autonomy of the Region (promoted by Zaia). The social media campaign promoting the autonomy of Veneto is indeed one of the most widespread *"no-Covid"* issues reported by Zaia in the very first topical moments analysed, together with the wish of a re-election for the leadership of his Region, and it develops mainly on his Facebook page, where he is used to share television or press interviews focused on these topics, even calling for an active interaction of his followers.

It is from February 22 that a crisis communication about the emergency starts to take shape, showing a clear "division of tasks" between the two actors and their related social media profiles. At that time, the social posting of Veneto on Twitter consists of just "service tweets" announcing the cancellation or postponement of scheduled events, aimed at informing citizens about the toll-free numbers to call in case of need, or just sharing the links to Facebook posts without any additional detail, while the substantial information about the health emergency is relegating mainly to retweets (posted by Zaia or social accounts belonging to Provinces and Municipalities of Veneto). On the contrary, in the same dates, *Veneto* gains a much greater visibility than the President on its Facebook page, providing constant updates about the course of the pandemic as well as practical instructions about how to behave in the event of suspected infection, making large use of infographics, and even more of videos focused on Zaia's interventions and press declarations.

As of March 4, and to a greater extent of March 8/9, Zaia starts dominating the social arena, focusing his social posting on timely updates about the spread of contagions, reassuring information about the economic aid in favour of workers and companies and severe criticism towards the Government decrees. It is exactly this last issue to become a "trend topic" of his crisis communication, with the aim to underline that the containment measures adopted at national level are *"scientifically disproportionate if compared to the epidemiological trend"*, as well as harmful to the economy of Veneto. The criticism towards the Government decrees goes hand-in-hand with the proposal of autonomous regional intervention plans, aimed at emphasizing the skills and preparation of the technical staff of his Region as well as underlining the self-recognition of a very strong leadership. Another key feature of Zaia's social posting relies on the use of emotional communication aimed at promoting regional pride and cohesion, that can be summarized with the claim *"I ♡ VENETO"*, frequently mentioned in his posts, with the main goal to rehabilitate the tourist vocation of the Region.

As a whole, the media coverage analysis that comes from Zaia's social accounts points out a crisis communication which makes large use of videos and is especially marked by an overproduction of contents, spread through different channels, showing a deep use of crossmediality which turns to be its very strong point, if compared to other institutional actors analysed so far.

One last issue to point out is the one related to the overall social posting "tones". Although the main aim of both actors is often to reassure people, but also to inform and empower them, the large use of "worrying symbols" they make in their tweets/posts (red crosses and alert icons, multiple exclamation marks) draw recipients' attention but also tend to alarm them (regardless of the more neutral content they intend to convey).

Campania Region/Vincenzo De Luca

Contrary to what happens with other institutional actors, the social posting about Covid-19 emergency carried out by Campania Region (*Campania*) and its President Vincenzo De Luca develops mainly on Facebook. The Twitter account of *Campania*, indeed, does not return any result over the analysed days (the latest tweet is dated back to February 2, 2020) while De Luca performs a very limited Twitter coverage, marked by frequent references to the Facebook page, which is thus appointed as "social arena" par excellence. The social posting about the health crisis goes hand-in-hand with the *"no-Covid"* issues, not only at the first stage of the emergency but also in the following days, although the communication flow gradually becomes more and more focused on coronavirus updates highlighting some important differences between the two actors.

While *Campania* carries out a communication mainly focused on informative issues, and only at the moment the contagion spread is well-known and perceived as a potential danger at a national level, De Luca's social posting starts as early as the news about the first confirmed cases of coronavirus in Italy hits the headline (on January 30), clearly announcing his decision to *"centralize information at regional and ministerial level"* with the main goal of preventing the spread of fake news and psychosis. On one hand, *Campania* provides updated daily reports about the number of infections, at a regional level, as well as the results of the overall swabs carried out (almost always combined with the launch of press releases and links redirecting to the institutional website). On the other hand, its main aim is to provide citizens with news of public interest and practical instructions (toll-free numbers, regional services), marked by institutional tones aiming at informing people as well as empowering them about how to deal with the emergency.

For his part, De Luca manages to provide his communication with a strong "personal imprint" which seeks to differ from both the institutional-aseptic tones of regional information, and the expressive modes (in terms of *themes, goals* and *tones*) featuring all other actors examined. In doing so, even the same report of contagions (elected as "trend topic" of the Region) plays a different role in De Luca's social media coverage, where it stands for the most visible result and most obvious outcome of the successful containment measures adopted at regional level, becoming the heart of his policies and dissemination activities. Taken as a whole, his Facebook posts, as well as his live broadcasts and radio-TV interviews, are primarily focused on regional decrees (as opposed to Government ones) and especially on those measures dealing with the closure of schools and all commercial activities potentially dangerous for the spread of the virus. This "containment-oriented" communication goes hand-in-hand with the management of the emergency from a strictly "sanitary" point of view (medical staff involved, health equipment, set up of intensive care unit and hospitals), with the clear intention of reassuring people but also underlining the great commitment of the Region. What is more, the use of institutional tones (even reinforced by the choice to always record his messages or perform his live broadcasts sitting behind the presidential desk) often gives way to peremptory statements, marked by strict calls for individual responsibility (primarily addressed to young people) aimed both at raising awareness and "moralizing" public opinion. His strong personality as a leader expresses itself through direct communication, which leaves no room for personal interpretation, but intentionally leads to the understanding of what is "right" or "wrong" sometimes boasting an ironic tone, which will be a distinctive feature of his communication in the later stages of the lockdown.

Both actors make use of the generic hashtag #coronavirus in their posts. While *Campania* does not make large use of multimedia contents (just standard pictures for several posts), De Luca's Facebook page is instead characterized by the frequent use of videos, especially live broadcasts, but also video excerpts of TV and radio programmes. With specific reference to live broadcasts, they are often re-posted in the following days, divided into several episodes, according to the single issues covered during his interventions, in order to reiterate at various times the key point of his emergency guidelines.

Sicilian Region/Nello Musumeci

The Sicilian Region President, Nello Musumeci, presents a slightly more intense communication flow on Facebook while the Sicilian Region (*Sicily*) prefers Twitter as main information channel.

The political leader leaves space for the voice of the Institution in the communication arena often retweeting the contents shared by the Region. If *Sicily* starts dealing with the coronavirus topic on January 30, Musumeci faces the same topic only since February 22. Two main voices narrate the emergency situation in Sicily (President Musumeci and the Councilor for Health Ruggero Razza), reassuring citizens about the limited number of cases in the Region and the efficiency of the regional health service. Through Musumeci's words, new measures are announced which are established from time to time to slow down the spread of the pandemic on the island. The local pandemic narrative is flanked by two other voices: Palermo Airport and Catania Airport. In fact, by retweeting the two accounts, the measures taken during those days are frequently communicated. This shows that one of the main concerns of Sicily is precisely that the virus would come "from outside", specifically by people traveling from Northern

Regions who are experiencing a more acute emergency stage. The retweets of the messages from the two airports become part of the "reassuring" and at the same time "informative" narrative already initiated by President Musumeci and Councilor Razza. Obviously, the informative narrative about the measures adopted is also supported by sharing Conte's messages when the intention is to talk about measures contained in Government decrees and not those issued by the Sicilian President.

Also in this context, cross-mediality is central. This time, it is the Sicilian Region that shares the posts and tweets launched by President Musumeci following the decision of voluntary isolation after the President of the Lazio Region, Nicola Zingaretti, communicated his positivity to the virus. Starting from March 8, therefore, Musumeci is the protagonist of many television interviews (for example on La7 and Rete4) shared through his social channels and often re-launched by *Sicily*.

FROM SITUATION AWARENESS TO SOCIAL CRISIS AGENDA

The second phase of the analysis focuses on the leading role played by the examined institutional and political actors, with the main aim to point out *if*, *when* and to *what extent* they became aware about the Covid-19 emergency as well as to underline the relation between the degree of *crisis awareness* and the main strategies of "social crisis communication" aimed at setting the agenda and developing citizens' responsibility in their everyday life.

Palazzo Chigi/Giuseppe Conte

For both actors (Palazzo Chigi and Giuseppe Conte), a high degree of crisis awareness is detectable starting from the very early stages of the health emergency in Italy. The social posting about Covid-19 coming from both accounts starts indeed as early as the news about the very first cases of coronavirus infection in Italy is published, on January 30. Although the emergency has not yet exploded on a national level, Italian Government starts to pay close attention to the issue from its very beginning, which is obviously expected to increase along with the pandemic's development. The first containment measures adopted at national level consists of the complete suspension of flights between Italy and China, right from the disclosure about the infected Chinese tourist couple coming from Wuhan, the first world epicentre of the virus.

The social posting of both social network accounts seems to perform the function to "address" citizens' attention towards the issues of more relevant public interest, which matches with the first level of agenda setting, summarized in the attempt to transfer the relevance of some topics from the "macro-agenda" of the Institutions to that of public opinion. That function, however, takes place within a very particular context – such as that of a health emergency – which already in itself is able to impress public opinion, since it embodies the most significant news values typical of a critical event. Palazzo Chigi's agenda setting takes shape through the use of direct and concise messages, mainly on Twitter, reflecting the medium by which they are hosted, that does not leave much room for commentary. On the other side, Conte decides to develop the first level of agenda setting on his Facebook page, which seems to be the "elected place" chosen by the Italian Premier to directly address his crisis communication to Italian people.

Therefore, if both actors contribute to "set the agenda", it is primarily Conte's social communication to provide also the "frames" through which people are called to deal with the most relevant information about coronavirus emergency (and thus to fulfil the second level of the agenda setting). While remaining faithful to his institutional function, Conte tends to customize his social contents. First of all, he is able to secure a steady stream of information, but he does not "hyper-communicate"; the number of overall posts is indeed relatively low, but this seems to reflect a targeted choice rather than a communication deficit.

With specific regard to the above mentioned "framing activity", the choice to share his live broadcasts from his personal Facebook page reflects the need to establish a direct dialogue "from the Prime Minister to the citizens", without the intermediation of any formal Institution. His many live broadcasts are also aimed at embodying the key function of a "ritual", since they are characterized by the sharing of standard contents both before and after the social event, as well as by recurring features (e.g. the "usual" wait between the scheduled and actual start time), up to the symbolic meaning the event itself has for citizens, as unifying moment but also as subject of social debate (in the further days).

Ministry of Health/Roberto Speranza

For the roles and functions performed by these two actors, the awareness of the crisis is immediately evident and with the same immediacy is communicated externally.

The storytelling begins with Roberto Speranza who, on January 30, announces the start of the coronavirus task force and his interventions in the Chamber and in the Senate scheduled for the morning. A few hours later, the Ministry of Health begins its storytelling by assuming a hegemonic narrative role if compared to the narration carried out by its political leader. Two ways of narrating that are close but that retain their peculiarities. Minister Speranza is "more taciturn", while the Ministry affirms and incites more; the first one makes a few statements that he prefers to alternate with the retweets and posts of the other actors and Institutions who support him in the fight against contagions; the second one leads a real campaign of responsibility (#iorestoacasa) but retweets and shares the posts of other bodies, Institutions and accounts that support what has been said up to that moment or help to focus attention on a particular topic while still recognizing the authority of "another source". It is precisely with the urgency of the emergency that both actors trust and recognize the authoritativeness of the other sources in their narration.

For this reason, it can be said that the emergency narration, through which these two accounts set the crisis agenda, is put into action on three levels: the story "through their own voice", that done "through the voice of others" and that done "by others through their voice". The first level consists of tweets and posts launched by the Ministry/Minister. The most numerous are certainly launched by the Ministry which knows how to assume and defend its leadership in the health sector by frequently launching messages that call attention to the right behaviours to be put in place to combat the infection. This first level is directly connected to what we have defined here as "third level", i.e. the consideration of these messages by other actors (in the actors analysed the material disclosed or the messages launched online by the Ministry have always been shared) that recognize the authoritativeness of this institution par excellence and follow its responsibility message, making the message viral and easily recognizable on the whole national territory. The second level (the voice of the others) is frequently imposed and seems to be a mechanism for recalling and simultaneously recognizing the other actors in reference to more informative content. The information and updating, therefore, is narrated with the voice of others, the responsibility is narrated with your own voice waiting for the other voices to join in the chorus.

Therefore, it is the Ministry that sets the agenda, but also guides public opinion toward a proper understanding about the mentioned topics (1° and 2° level). However, if compared to other institutional actors who perform this task making use of a communication strategy almost entirely self-produced, in this case the agenda setting function is widely performed both through "the voice of others" (Conte/

Regions/Civil Protection) and mainly in an autonomous way through the awareness/educational campaigns carried out by the Ministry.

Lombardy Region/Attilio Fontana

The joint analysis based on the social media profiles of the Lombardy Region and its President Attilio Fontana clearly shows that the two "actors" are aimed at performing different but complementary tasks during the coronavirus emergency.

A common point is just detectable in the lack of crisis awareness they show on the occasion of the "time zero" of the pandemic, when the coronavirus emergency is not yet perceived in such terms by either of them. At that time, indeed, the few tweets and posts about the health emergency get lost in wider communication flow of *"no-Covid"* issues while the only mentions about Covid-19 can be found in some tweets reporting the official denials about an alleged case of coronavirus in Monza, reinforced by the use of specific hashtags aimed at reducing the topic to a "Chinese trouble" (#VirusCina). This lacking awareness would later be mentioned on several occasions by the Italian national media, highlighting the superficial approach that leads *Lombardy* to underestimate the health emergency, and thus to make serious mistakes in the management of the crisis at regional level.

A growing awareness starts developing from February 22, at the time when the spread of contagions has already increased within the regional borders and both actors are called to find remedies to contain them as much as possible. Starting from this date *Lombardy*, thanks to its broad social media coverage, clearly performs the task of agenda setting, serving the double purposes of highlighting the main issues "around which" to think around (first level) as well as suggesting "how to think" around them (second level), also showing some distinctive and personalized features. First of all, the use of "episodic" communication as expedient for diluting the information about the new virus in progressively numbered posts: this innovative way to communicate is carried out firstly "borrowing" the institutional campaign carried out by the Ministry of Health and then making its own "regional" campaign, and thus showing an increasing professionalism that goes hand-in-hand with the developments of the pandemic.

Moreover, the advanced communication skills shown by *Lombardy* (regardless of the specific coronavirus emergency), combined with its condition of being "the most affected Region" in Italy, also explain the relative "autonomy" by which this actor manages the news coverage of the health crisis as against the national guidelines. The same news about the national decrees and containment measures are reported merely as ordinary tips about the emergency. The main focus is, obviously, on the regional management of virus containment policies: hence the great visibility given to the regional press conferences and guidelines as well as to the main updates provided by its key representatives (from President Fontana, to the Councilor for Health Giulio Gallera, to the mayors of towns hardest affected by the viral contagions), together with the different timing of upcoming decrees and declarations, which develop in parallel with national measures, closely following the contagion trends within the regional borders.

Another distinctive feature is furthermore detectable in the multiple focus on the key issue of responsibility, which is declined in a series of sub-themes. The call for responsibility is primarily addressed to journalists, with the aim of preventing the spread of fake news (in the "time zero" of the emergency, when the primary goal is to deny the first alleged cases of coronavirus infection, as well as in the midst of the health crisis, when it comes the need to deny the rumours disclosed by CNN about the alleged spread of the national decree, dated March 8, before its official signing). Secondly, "responsibility" is seen in its conventional meaning of appeal to (and praise of) the civic sense of the Lombardy people in dealing with the emergency.

While *Lombardy* is the only actor to "set the agenda", President Fontana contributes to provide his own interpretative frames about the most relevant issues, thus supporting the Region in performing the second level of agenda setting. His social communication is marked indeed by the use of calm and reassuring tones, aimed at avoiding any kind of incumbent alarm but rather to inform and reassure. Even performing the main task to update citizens about the sanitary emergency, his focus often moves from the mere technical data (number of infected, medical equipment, etc.) to the more emotional level expressing his personal closeness and gratitude towards medical and health personnel, with the primary goal of developing a shared sense of gratitude and regional pride. A similar argument can be made for the communication about the containment measures involving schools, workers, families and businesses affected by the emergency: even in this case, the clear intention behind the social posting of Fontana, besides informing, is traceable to the will of reassuring people about the dedication of regional Institutions in dealing with the social and economic implications of the virus.

Veneto Region/Luca Zaia

The joint analysis focused on the social media coverage of *Veneto* and its President Zaia highlights some specific features, related both to *when* and *how* the two actors analysed become aware about the emergency, as well as to the different role of agenda builder they play through their own crisis communication strategies.

Although Veneto would prove to be one of the Regions hardest hit by Covid-19, the crisis communication carried out in the very first phase of the emergency doesn't point out a high level of crisis awareness. Indeed, as already highlighted in the analysis focused on the social posting of the two actors, in the "time zero" of the pandemic it is almost absent any clear reference to Covid-19 emergency, which instead will gain full visibility in the following weeks, going hand-in-hand with the spread of contagions within the regional borders.

However, if the overall awareness of being in a crisis situation lags behind, it is equally evident that the two actors (and especially Zaia) seem to quickly recover the "lost time", by adopting political measures – supported by a strategic use of crisis communication – that contribute to restore full control over an emergency situation that threatened much more serious consequences than its actual course. It is precisely the role played by Zaia to perform both an informative function (usually performed by Regions/Institutional actors), as well as to represent the main "reference point" for the emergency at regional level.

On one hand, *Veneto* indeed limits itself to set the agenda only at the early stages of the crisis, at the time marked by the adoption of the first containment measures, when there is a need to provide citizens with practical instructions and good practices to be followed in order to avoid the spread of infections. This function, however, is mainly performed on Facebook, since the Region's Twitter account acts rather as a "sounding board" for the crisis communication carried out by other institutional actors. On the other hand, President Zaia fully plays the role of agenda builder on both social networks, providing the selection of topics around which his audience is called to develop its own opinion, as well as the "frames of meaning" through which he guides people's interpretation of the events. His framing activity relies mainly on the use of a communication strategy combining traditional media with the new ones, marked by a frequent use of videos through which he shares time-delay interviews given to tv programmes as well as to both national and local news. His wise reliance on crossmediality is also enriched by the use of one of the most traditional media, the radio, allowing his followers to easily recover the podcasts about his interventions directly from the broadcasters' platforms. What is more, on the occasion of his daily updates and press declarations about the developments of the emergency, Zaia is very careful to confine his role at a strictly institutional level (pointing out, at most, the political merits underlying his policies); any further information or clarification of a technical-scientific nature is instead entrusted to Manuela Lanzarin (Health Councillor) and Francesca Russo (Head of the Prevention Department of the Region), thus drawing public's attention to a "univocal" interpretation of data, according to a precise strategy aimed at highlighting the division of roles and responsibilities.

Campania Region/Vincenzo De Luca

Considered as a whole, these two actors play an "original" role in the communication process of coronavirus emergency. More particularly, the high level of attention on the spread of contagions would suggest that the Region was strongly affected by the virus (when that is not the case), not so much because of the amount of social posts, but rather for the kind of topics they deal with as well as the frequency of virus updates. De Luca performs, at regional level, the same "first-hand communication" carried out at national level by Giuseppe Conte, while the social posting of *Campania* is limited to the role of "press office" of the President.

Both actors contribute to set the agenda but it is primarily the social communication of De Luca (similarly to what happens with the Premier) who guides his citizens to "understand" the emergency, providing them some interpretative frames with the aim to educate, raise their awareness and motivate them to follow a shared course of action. De Luca's critical attitude towards the Government containment measures (expressed through the will to constantly integrate them with stricter regional measures), together with his habit of explaining and reiterating "permissions" and "prohibitions", seems to be aimed at serving an educational function even more than an institutional one. That explains his difficult task of empowering and reassuring people but also blaming them, if and when necessary. What is more, De Luca decides to perform this challenging task relying only on his own resources, without calling on celebrity endorsements (as happens, for example, in Lombardy) thus becoming, in a certain way, the "testimonial of himself".

The close attention he paid to manage the emergency, right from the first and limited contagions, testifies his marked propensity for crisis awareness, under a preventive perspective, which results in his repeated effort to make citizens responsible for appropriate behaviours in their daily life. At the first stage of the emergency, De Luca's interest in dialoguing with his regional audience, as a priority, is reflected in the targeted choice to take part in tv and radio programmes aired by local broadcasters. Indeed, he will expand his participation to national networks only later, at the moment when he feels confident to praise the Campania Region as a virtuous example of management of a health crisis, thus aiming at gaining consensus on a large scale.

Sicilian Region/Nello Musumeci

Both actors show a high degree of crisis awareness from the first moment. In fact, the coronavirus is treated immediately in the posts and tweets of the very first days. The peculiarity lies in the fact that, until February 22, the epidemic is seen as an external problem since, despite the episode of a tourist hospitalized in the Region, citizens are invited to avoid alarms and reassured on the fact that the regional health system is well established and ready to face any type of emergency.

It is precisely the theme of security that is dear to the narrative in this context. On one hand, the health system which is often presented as suitable and ready to manage any degree of emergency, supported by the regional measures aimed at integrating the ones issued by the central Government. The intent is to protect the Island in a preventive way. In order to emphasize the security issue, the narrative also brings practical examples. Sharing the messages of the Sicilian airport accounts serves to demonstrate what is actually being done after it has been announced, precisely to ease the insecurity and fear perceived by citizens. On the other hand, they also proceed with an optimistic tale that tries to remove the threat of coronavirus from the minds. The continuous retweets and shares of the social account @visitsicily serve to fuel a reaction and another type of protection: that of the image of Sicily which cannot risk being affected due to the low number of cases and which must therefore be protected in view of a new take-off of the tourism sector when the battle is won.

The narrative begins to take on different tones starting from Conte's press conference (February 22) in which the first measures to contain the virus take place, a sign that Covid is starting to be an Italian problem. *Sicily* also starts communicating the first containment measures which include quarantine for migrants landed in Pozzallo, the suspension of educational and school trips, school checks, airports, etc. However, if at first *Sicily* speaks "with its voice" (a voice amplified also through the sharing of its messages by Musumeci), starting from February 22, it starts speaking "with the voice of others" by implementing a strategy for sharing messages sent by other institutions or personalities such as the Ministry of Health, the airports of Palermo and Catania, the Civil Protection, Palazzo Chigi, etc. The intent is to inform, reassure and empower at the same time in the attempt to keep the Sicilian island isolated from the virus.

This sense of responsibility reaches its peak on March 8, the day on which the voluntary isolation of the President of the Region is announced by the voice as Nello Musumeci, by sharing his video ad that has gone viral. *Sicily* "takes back the word" only sometimes when, together with the sharing of the messages of the other frequently shared actors, it narrates by itself the further measures established by Nello Musumeci or his staff to lead the fight against the pandemic victorious.

DISCUSSION AND CONCLUSION

Among the different emergencies which our society has had to face over the last decades, the Covid-19 health crisis emerges maybe as the worst one at least from two different perspectives: on one hand, it has forced Institutions and Governments to take on serious containment measures aimed at protecting citizenship, but often these decisions have been perceived by the same citizens as "unpopular" due to the sacrifices, often not suddenly understood, they have been requiring; on the other hand, because of its communication management, in which the need to gain people awareness in respect of the crisis seriousness has daily blended with the need to shift from the fear of death/refusal of the danger to a responsible reaction to the global pandemic.

Referring specifically to the crisis communication perspective, the increase in the number of cases of contagion has been accompanied by a multiplication of official press releases and recommendations by the Institutions and health authorities, by an infinite number of services broadcasted on the news, continuous updates by radio broadcasters, online press and all news channels, all this overloaded by the buzz on social networks (Manfredi, 2019). Nonetheless, even though crisis communication has been of paramount importance in facing the global health emergency, it hasn't been blameless, due to lack of consistency and efficacy which has characterized the official sources (public service communication, political communication, journalism) (Scaglioni & Sala, 2020), not to say about the confusion generated by fake news or by the constant chatter on the Net (Mantineo, 2020).

Analysing real-time the media coverage of the Covid-19 health emergency, several scholars (Guigoni & Ferrari, 2020; Scaglioni & Sala, 2020; Morcellini, 2020) have pointed out that one of the biggest mistakes of Italian crisis communication as a whole is that it has not allowed citizens to gain immediately the correct crisis awareness and, as a consequence, people have delayed to take on the right behaviours/conducts. The crisis awareness also represents the main issue on which this research focuses, aiming at understanding *when* and *how* the most relevant Italian PAs involved in crisis management in its first phase became aware about the health emergency, and *to what extent* their crisis awareness affected the leading function of "agenda setting" toward citizens.

However, answering the research question needs a preliminary crossed analysis of the social posting of the above-mentioned actors. To this purpose, it is useful to take into account at least the five meaningful variables which emerge from data analysis, that is: the typology of account (institution/politician), the typology of Institution (national/local), the degree of involvement in the emergency (North/South), the key role played by the analysed social media (Facebook/Twitter), the communication flow over the five topical moments analysed.

What Matters is the Source

The first meaningful variable consists of the typology of account, which can be split distinguishing between *institutional accounts* (i.g.: Palazzo Chigi, Ministry of Health, Regions) and *personal ones* (i.g.: Giuseppe Conte, Roberto Speranza, Attilio Fontana, etc.). This distinction transfers to the analysed case study one of the meaningful structural features of the Italian scenario of PAs communication, which is characterized since its origin both by a strong (theoretical) contrast and a (practical) enduring hybridization between the public services dimension and the political one (Faccioli, 2006; Mancini, 2011; Mazzoleni, 2012; Grandi, 2013).

The "dangerous relationship" between the two mentioned dimensions takes shape also referring to the Covid-19 crisis communication: in fact, while the PAs accounts usually carry out a public service function aimed at creating awareness and suggesting proper behaviours (and, doing so, usually preserving a "safe distance" toward citizens), the personal ones tend to prefer a more "political/persuasive" approach, which produces a higher level of citizens' engagement (aimed both at supporting and opposing the political leader who posts or tweets). This political/persuasive approach emerges especially referring to Zaia and De Luca, who stand out (due to their political charisma and the leadership they are able to express) as Premier Conte's main competitors, and it affects also their communication management, which sets the agenda more explicitly (and maybe efficiently) rather than Musumeci and Fontana. In respect of the last one, it is also interesting to underline that Fontana's personal approach seems to be based on an emotional approach rather than on a political one, and doing so it introduces a third level of agenda setting.

Comparing Competences: Who Decides What?

Focusing on the second variable, on one side, the analysis highlights the relationship between the two main national actors (Presidency of the Council of Ministers and Ministry of Health), pointing out how they tend to perform their role in a complementary and standardised manner. Indeed, the joint social posting of Palazzo Chigi and Conte tends to *set the agenda* about the institutional guidelines to deal with the emergency, while the Ministry of Health, on its turn, limits itself to *share the agenda*, with the exception of the issues related to its specific sphere of competence (especially the ones focusing on the awareness/educational campaigns promoting good practices and behaviours), where instead the opposite happens.

On the other side, the global health emergency tends to re-define the distinct roles between national and local Institutions about health issues, leading the national authority to claim for itself the crisis management as a whole, and thus involving also the crisis communication. Indeed, from a communication point of view, the local PAs tend to interact with the Italian Government according to three different approaches: *opposition* (that is the official position of Veneto), embodied by its President, who firmly criticizes the containment measures adopted at national level, contrasting them with the proposal of autonomous regional interventions); *integration* (embodied by Musumeci and De Luca, who frequently "integrate" national decrees with stricter regional measures, aimed at protecting their own borders from the virus); *autonomy* (which is the one reflecting the position of the Lombardy Region, urged by necessity to develop its own "autonomous" management of the crisis).

"There Is No More Time" vs. "We Are Still in Time"

With regard to the third variable, as already mentioned, the selection of the PAs (and in this case the four Italian Regions) relies on their different degree of involvement in the health emergency in the analysed period, with also a look toward a geographical distribution, reflecting the polarization (North vs. South/Islands) that marked the pandemic's course in Italy.

On one hand, Lombardy and Veneto don't show a high level of crisis awareness at the very early stages of the emergency, but rather "react" to the pandemic when it has already started its diffusion in order to find a remedy. Specifically, they go straight from the starting idea *"There is no crisis"* to the alarming awareness *"There is no more time"*. In doing so, however, the main difference between the two Northern Regions relies in the different outcomes of their crisis management: if the delayed awareness of Lombardy translates into an overall crisis mismanagement, to the point of being charged with "negligent epidemic" (Nava *et al.*, 2020), Veneto manages to make up for lost time and to contain the spread of contagions.

On the other hand, Campania and Sicily, despite being two Regions slightly affected by the crisis, immediately show a high level of crisis awareness which results in the adoption of preventive measures, aimed at containing the spread of contagions as much as possible, strongly relying on the main assumption *"We are still in time"*. In doing so, their virtuous examples of crisis management allow them to point out the relative "safety" of their Regions which becomes, at the same time, a kind of source of pride.

To Tweet, or To Post? That Is the Question

Referring to the fourth variable, talking about the key role played by the two analysed social media suggests to consider two different questions: what are the more effective social tools to communicate consciously? How to combine the various communication needs in an emergency situation and above all on which platform is it more appropriate to convey the main messages? Obviously, there is not a right or a wrong answer to these questions, and the research confirms that. In fact, each of the analysed actors uses the channels that, according to their perspective, allow the message to be better communicated and reach a wider audience by selecting their targets. Here we can produce a broad update of what Marshall McLuhan (1964) said according to which "The medium is the message".

At the end of the analysis, Twitter stands out as the main channel for PAs communication, especially referring to the tweets shared by institutional accounts (Palazzo Chigi, Ministry of Health, Regions), confirming the strategic role it already played in other crisis events (i.e.: terrorism, natural disasters as well as the Arab Springs) (Austin & Jin, 2017; Ferrigni & Spalletta, 2018). The characteristics of this platform, which consist in the shortness of the messages that become viral through the wise use of hashtags and the continuous bounces of the same messages from one account to another thanks to mentions and retweets (Covello, 2006), also in the analysed case help to affirm the centrality of the Institutions in crisis and emergency communication. The PAs social activity, founded on their reputation as authoritative sources, also represents a tool against the spread of fake news (Citarella, 2017).

If the institutional accounts manage the main communication between tweets and retweets, on the contrary they turn to Facebook to deepen and expand the messages, integrating them by sharing videos and pictures which allow to reinforce their meaning and efficacy.

Moving from the institutional accounts to the personal ones, the research shows that political leaders tend to make a less extensive use of social media channels, with the only exceptions of De Luca and Zaia whose role is of paramount importance during the emergency. When political actors communicate, they preferably do it on Facebook, because also in this case the Zuckenberg's social network allows to deepen, explain better, support the message with multimedia tools that complete the sense of communication and, above all, because it allows politician to establish both a face-to-face and a face-to-many relationship with their followers (and, as Zaia clearly suggests, their potential voters). Of course, these are very important characteristics for a political leader.

Finally, it is interesting to underline that, during this emergency, Facebook played another very important role by becoming the social network of live moments. The press conferences held by Conte and any other political leaders were hosted simultaneously in the *halls* of the Institutions as well as on the *walls* of their official social accounts through live videos. Therefore, if Twitter has become the channel of immediacy, Facebook is that of the deepening and live action, making leaders and Institutions central and bringing citizenship closer to politics.

Five Moments for Five Levels of Awareness (and so Many Hashtags)

The five topical moments to which the research refers can be summarized in respect of so many meaningful crisis stages and related hashtags and claims.

The first one is the moment of the *discovery* of the coronavirus in Italy (January 30). Starting from this moment, the virus is no longer a "Chinese trouble", far away from our daily lives, but a threat that exists and begins to worry us. On a communicative level, the hashtag #coronavirus emerges as the keyword that will become increasingly inflated. Although #Covid-19 is among the trendiest hashtags of the moment, the analysed PAs seem (casually?) to prefer #coronavirus which, in the scientific field, is something that "comes first" because it designates the family to which the virus belongs (while Covid-19 is the generated disease).

At that time, however, the virus is still "abstract". In fact, its *materialization* occurs in the second moment (February 22), the very first instant in which the threat really seems to have landed in our house. This "materialization" is expressed also in communication strategies, through the hashtag #conferenzastampa (#pressconference) which refers to the first in a close series of meetings involving Premier Giuseppe Conte and the other institutional actors, followed by direct indications to citizenship and a long series of decrees that will influence life and behaviours of Italians.

The third topical moment is that of *awareness*. The virus starts to be very, too present. The life of each of us begins to change, but not to stop. The closure of all schools and universities in the Country is announced on March 4, but teaching – like all the other activities of daily life – keeps going on, only in different places and ways. #lascuolanonsiferma (#schooldoesnotstop) summarizes the attitude of Italians who don't give up.

Reaction is the attitude that accompanies the Italians on the following days, up to the moment when the red-zone is extending beyond the initially imaginable limits (March 8). This stage is deeply influenced by the local dimension: in fact, each Region reacts referring to its own experience. Therefore, Lombardy praises for resistance through the hashtag #fermiamoloinsieme (#LetStopItTogether), while Veneto expresses its willpower promoting regional pride and cohesion through the claim "I Love Veneto". On the contrary, the Southern Regions show different attitudes: in fact, *Sicily* reacts to the incumbent crisis continuing to promote tourism (Visit Sicily), while *Campania* tends to present itself as a safe and responsible Region through the hashtag #CampaniaSiCura, which means at the same time "safe Campania" and "Campania cares".

As soon as the whole Italy is placed under lockdown (March 9), it is the *responsibility* that "infects" everyone. #iorestoacasa (#stayathome) – that is the claim both of the Ministry of Health's social campaign and of Premier Conte's decree which implement social distancing by staying in one's own home, leaving only for reasons of necessity – means fighting to win. A fight that goes viral.

Setting the Crisis: PAs Communication as the Crisis Agenda Builder

At the end of this work, it is possible to answer the research question which, as above mentioned, aims to understand *if* and *how* there is a relationship between the crisis awareness, which represents the primary goal of crisis communication (Coombs & Holladay, 2010; Svedin, 2012), and the agenda setting effect which PAs communication clearly aims to produce suggesting citizens the topics around which they have to think and how they have to think around them (McCombs & Shaw, 1972; McCombs, 2004). Although the employed methodology, based on a qualitative media content analysis, forbids to detect evidence of absolute value, nevertheless it suggests some interesting highlights which allow to state that the supposed relationship exists at least concerning the analysed case, the selected actors and the considered time period.

First of all, it is interesting to note how, during the time-frame analysed, a gradual transition from risk communication to crisis communication takes shape (Morgan *et al.*, 2002; Reynolds & Seeger 2005; Littlefield & Sellnow, 2015). If at the time zero of the pandemic, the Italian PAs communication activity aimed at providing useful information for the understanding of the Covid-19 emergency, as an unexpected and unknown event, in the following weeks, alongside the increase of contagion and the related requirement for effective containment measures, they start performing a real crisis communication, aimed at motivating the adoption of responsible behaviors.

More specifically, the research shows that the overall crisis communication coming from the analysed PAs in the first stages of the Covid-19 health emergency aims at transferring to citizens that situation awareness which they themselves are gradually gaining through a five steps process which starts from the *discovery* of the crisis (January 30) and ends with a call for *shared responsibility* (March 9). However, in order to do so, all the analysed actors tend to adopt communication strategies which are influenced by their own approach (institutional/political-persuasive), reflect their nature (national actor *vs.* local one), are affected by their own degree of crisis involvement (North/Sud), express their own being differently "social media confident". Going beyond the physiological differences, the overall crisis communication seems to be inspired by a logic aimed at gaining the awareness about what is going (and how the crisis may evolve). This goal represents both the first and the second level of the PAs crisis agenda (McCombs & Shaw, 1972; McCombs, 2004).

But there's more. The social media content analysis shows that the analysed actors, especially through their personal accounts, set the crisis agenda not only suggesting citizens the issues around which they have to think (first level: *we have a problem and it consists of the Covid-19 health emergency*) and how they have to think around them (second level: *we need to react to the global pandemic by adopting a shared responsibility*), but also introducing a third level of agenda (Guo & McCombs, 2011; Guo, 2014), which consists of setting the online network in order to ensure that the containment measures become viral as well as the virus. From this point of view, the most meaningful example in the first stage of the emergency, on which the research focuses, is represented by the crisis communication coming from Attilio Fontana. In fact, if the Lombardy Region communication sets the agenda at its first and second levels, Fontana's social communication performs the important function of "social glue" during the emergency, through an intensive public relations activity which takes shape in his heartfelt and participatory messages addressed to Lombard citizens, as well as his frequent expressions of gratitude and praises toward medical and health personnel engaged in the front line against Covid-19, even enhanced through his personal and institutional support to private fundraising activities. And it is precisely to this task that it is possible to identify some key features of the third level of agenda setting (Guo & McCombs, 2011; Guo, 2014), since he expresses a clear call for public engagement.

Of course, in the shift from the first crisis stage (which ends on March 9) to the second one (which begins in the midst of the crisis and consists of the global lockdown), the third level of the agenda setting process finds in Vincenzo De Luca its undisputed star. But that's matter for another research (De Rosa, 2020).

REFERENCES

Adam, B., Beck, U., & van Loon, J. (Eds.). (2000). *The Risk Society and Beyond: Critical Issues for Social Theory*. Sage. doi:10.4135/9781446219539

Altheide, D. L., & Schneider, C. J. (2013). Qualitative Media Analysis. *Sage (Atlanta, Ga.).*

Austin, L. L., & Jin, Y. (Eds.). (2017). *Social Media and crisis communication*. Routledge. doi:10.4324/9781315749068

Bauman, Z. (1999). *La società dell'incertezza* [The society of uncertainty]. il Mulino.

Beck, U. (1992). Risk Society: Towards a New Modernity. *Sage (Atlanta, Ga.).*

Bruns, A., & Burgess, J. (2014). Crisis Communication in Natural Disasters: The Queensland Floods and Christchurch Earthquakes. In A. Bruns, M. Mahrt, K. Weller, J. Burgess, & C. Puschmann (Eds.), *Twitter and Society* (pp. 373–384). Peter Lang.

Citarella, P. (2017). *Social media e P.A. La comunicazione istituzionale ai tempi di Facebook* [Social media and PAs. Institutional communication in the age of Facebook]. Angeli.

Comunello, F. (2014). *Social media e comunicazione d'emergenza* [Social media and crisis communication]. Guerini e Associati.

Comunello, F., & Mulargia, S. (2017). Tra risposte protocollate e «social sensing». L'uso dei social media per la comunicazione d'emergenza nelle istituzioni locali italiane [Between registered answers and social sensing. The use of social media in PAs crisis communication]. *Sociologia e ricerca sociale, 112*, 111-137.

Coombs, W. T. (2002). Assessing Online Issue Threats: Issue Contagions and Their Effect on Issue Prioritisation. *Journal of Public Affairs, 2*(4), 215–229. doi:10.1002/pa.115

Coombs, W. T. (2004). Impact of Past Crises on Current Crisis Communications: Insights From Situational Crisis Communication Theory. *Journal of Business Communication, 41*(3), 265–289. doi:10.1177/0021943604265607

Coombs, W. T. (2007a). Protecting Organization Reputations During a Crisis: The Development and Application of Situational Crisis Communication Theory. *Corporate Reputation Review, 10*(3), 163–176. doi:10.1057/palgrave.crr.1550049

Coombs, W. T. (2007b). Ongoing Crisis Communication – Planning, Managing and Responding. *Sage (Atlanta, Ga.).*

Coombs, W. T. (2014). Applied Crisis Communication and Crisis Management: Cases and Exercises. *Sage (Atlanta, Ga.).* Advance online publication. doi:10.4135/9781544308531

Coombs, W. T., & Holladay, S. J. (Eds.). (2010). *Handbook of Crisis Communication*. John Wiley & Sons. doi:10.1002/9781444314885

Coombs, W. T., & Holladay, S. J. (2014). How Publics React to Crisis Communication Efforts. *Journal of Communication Management (London), 18*(1), 40–57. doi:10.1108/JCOM-03-2013-0015

Covello, V. T. (2006). Risk communication and message mapping: A new tool for communicating effectively in public health emergencies and disasters. *Journal of Emergency Management (Weston, Mass.), 4*(3), 25–40. doi:10.5055/jem.2006.0030

De Rosa, P. (2020). Da locale a virale. L'emergenza permanente dello "Sceriffo" Vincenzo De Luca. In V. De Luca & M. Spalletta (Eds.), *Pandemie mediali. Comunicazioni, narrazioni, socializzazioni e consumi del MediaVirus* [Media Pandemics. Communication, storytelling, socialization and consumption practices of the MediaVirus]. Aracne.

Ducci, G. (2017). *Relazionalità consapevole: la comunicazione pubblica nella società connessa* [Conscious relationality: public communication in the connected society]. Angeli.

Endsley, M. R. (1995). Toward a Theory of Situation Awareness in Dynamic Systems. *Human Factors Journal, 37*(1), 32–64. doi:10.1518/001872095779049543

Endsley, M. R., & Garland, D. J. (Eds.). (2000). *Situation Awareness Analysis and Measurement*. Lawrence Erlbaum Associates Publishers. doi:10.1201/b12461

Faccioli, F. (2006). *Comunicazione pubblica e cultura del servizio* [Public communication and culture of service]. Carocci.

Ferrigni, N., & Spalletta, M. (2018). La paura viene twittando. Social media, terrorismo e percezione della sicurezza [Fear comes with Twitter. Social media, terrorism and social perception of security]. *Sociológia, 1*, 193–204.

Fraser, J. (1981). Italy: Society in Crisis, Society in Transformation. Routledge & Kegan.

Giddens, A. (1990). *The Consequences of Modernity*. Polity Press.

Grandi, R. (2013). *La comunicazione pubblica* [Public communication]. Carocci.

Grant, J. M., & Mack, D. A. (2004). Preparing for the Battle: Healthy Leadership During Organizational Crisis. *Organizational Dynamics, 33*(4), 409–425. doi:10.1016/j.orgdyn.2004.09.007

Guigoni, A., & Ferrari, R. (2020). *Pandemia 2020. La vita quotidiana in Italia con il Covid-19* [Pandemic 2020. The Italian everyday life with the Covid-19]. M & J Publishing House.

Guo, L. (2014). Toward the Third Level of Agenda Setting Theory: A Networked Agenda Setting Model. In T. Johnson (Ed.), *Agenda Setting in a 2.0 World. New Agendas in Communication*. Routledge.

Guo, L., & McCombs, M. (2011). Networked Agenda Setting: A Third Level of Media Effects. *Annual Conference of the International Communication Association.*

Heide, M., & Simonsson, C. (2019). *Internal Crisis Communication: Crisis Awareness, Leadership and Coworkership*. Routledge. doi:10.4324/9780429425042

Hornmoen, H., & Backholm, K. (Eds.). (2018). *Social Media Use in Crisis and Risk Communication. Emergency, Concern and Awareness*. Emerald. doi:10.1108/9781787562691

Kaufmann, M. (2015). Resilience 2.0: Social Media Use and (Self-)care During the 2011 Norway Attacks. *Media Culture & Society, 37*(7), 972–987. doi:10.1177/0163443715584101 PMID:29708120

Krippendorff, K. (2004). Content Analysis: An Introduction to Its Methodology. *Sage (Atlanta, Ga.).*

Littlefield, R. S., & Sellnow, T. L. (Eds.). (2015). *Risk and Crisis Communication: Navigating the Tensions between Organizations and the Public*. Lexington Books.

Liu, B. F., Jin, Y., & Austin, L. L. (2013). The Tendency To Tell: Understanding Publics' Communicative Responses To Crisis Information Form and Source. *Journal of Public Relations Research, 25*(1), 51–67. doi:10.1080/1062726X.2013.739101

Lovari, A. (2013). *Networked citizens: comunicazione pubblica e amministrazioni digitali* [Networked citizens: public communication and digital administration]. Angeli.

Maireder, A., & Ausserhofer, J. (2014). Political Discourse on Twitter Networking. Topics, Objects and People. In A. Bruns, K. Weller, J. Burgess, C. Puschmann, & M. Mahrt (Eds.), *Twitter and Society* (pp. 305–318). Peter Lang.

Mancini, P. (2011). *La comunicazione pubblica* [Public communication]. Laterza.

Manfredi, G. (2019). *Infodemia: I meccanismi complessi della comunicazione nelle emergenze* [Infodemia. The complex mechanisms of emergency communication]. Guaraldi.

Mantineo, A. (2020). *Fakecrazia. L'informazione e le sfide del coronavirus* [Fakecracy. Coronavirus information and challenges]. Media Books.

Mazzoleni, G. (2012). La comunicazione politica [Mulino.]. *Political Communication*, il.

McCombs, M. E. (2004). *Setting the Agenda: Mass Media and Public Opinion*. Polity Press.

McCombs, M. E., & Shaw, D. L. (1972). The Agenda-Setting Function of Mass Media. *Public Opinion Quarterly*, *36*(2), 176–187. doi:10.1086/267990

McLuhan, H. M. (1964). *Understanding media. The extension of man*. McGraw-Hill.

Morcellini, M. (2020). *Antivirus. Una società senza sistemi immunitari alla sfida del Covid-19* [Antivirus. A society without immune systems to the challenge of Covid-19]. Lit Edizioni.

Morgan, M. G., Fischhoff, B., Bostrom, A., & Atman, C. J. (2002). *Risk communication: A mental models approach*. Cambridge University Press.

Nava, F., Gambino, G., Lucarelli, S., Di Benedetto Montaccini, V., Telese, L., & Revelli, M. (2020). *Epidemia colposa? Le verità nascoste sulla mancata zona rossa nella Val Seriana* [A negligent epidemic? The hidden truths about the missing red zone in Val Seriana]. The Post Internazionale.

Neuendorf, K. A. (2002). The Content Analysis Guidebook. *Sage (Atlanta, Ga.)*.

Perry, R. W. (2007). What Is a Crisis? In H. Rodriguez, E. L. Quarantelli, & R. R. Dynes (Eds.), *Handbook of Disaster Research* (pp. 1–15). Springer. doi:10.1007/978-0-387-32353-4_1

Quarantelli, E. L. (2005). A Social Science Research Agenda for the Disasters of the 21st Century. In R. W. Perry & E. L. Quarantelli (Eds.), *What Is a Disaster? New Answers to Old Questions* (pp. 325–296). Xlibris.

Reynolds, B., & Seeger, M. (2005). Crisis and Emergency Risk Communication as An Integrative Model. *Journal of Health Communication*, *10*, 43–55. doi:10.1080/10810730590904571 PMID:15764443

Scaglioni, M., & Sala, M. (Eds.). (2020). *L'altro virus. Comunicazione e disinformazione al tempo del Covid-19* [The other virus. Communication and disinformation in the age of the Covid-19]. Vita & Pensiero.

Svedin, L. (2012). *Accountability in Crises and Public Trust in Governing Institutions*. Routledge. doi:10.4324/9780203120149

van der Meer, T. G., & Jin, Y. (2020). Seeking Formula for Misinformation Treatment in Public Health Crises: The Effects of Corrective Information Type and Source. *Health Communication*, *35*(5), 560–575. doi:10.1080/10410236.2019.1573295 PMID:30761917

KEY TERMS AND DEFINITIONS

Coronavirus: A family of viruses that cause illness ranging from the common cold to more severe diseases, such as Middle East Respiratory Syndrome (MERS-CoV) and Severe Acute Respiratory Syndrome (SARS-CoV). The novel coronavirus recently discovered has been named SARS-CoV-2 and it causes COVID-19.

COVID-19: The name of the disease caused by the novel coronavirus, SARS-CoV-2, and is short for "Coronavirus Disease 2019".

Crisis: A situation that is perceived as difficult. Its greatest value is that it implies the possibility of an insidious process that cannot be defined in time, and that even spatially can recognize different layers/levels of intensity.

Emergency: An unexpected event which places life and/or property in danger and requires an immediate response through the use of routine community resources and procedures.

Pandemic: An epidemic that has spread over several countries/continents, usually affecting a large number of people.

Quarantine: Separating and restricting the movement of people exposed (or potentially exposed) to a contagious disease.

Situation Awareness: The perception of environmental elements with respect to time and/or space, the comprehension of their meaning, and the projection of their status after some variable has changed, such as time, or some other variable, such as a predetermined event.

Social Distancing: Measures taken to reduce person-to-person contact in a given community, with a goal to stop or slow down the spread of a contagious disease. Measures can include working from home, closing offices and schools, canceling events, and avoiding public transportation.

ENDNOTES

1. The chapter is the joint work of all the authors. In particular, Marica Spalletta wrote paragraph 5, Dario Fanara wrote paragraphs 1 and 3, Paola De Rosa wrote paragraphs 2 and 4.
2. According to the data provided by the Italian Department of Civil Protection, over the examined time period, the Northern Regions of Lombardy and Veneto (together with Piemonte and Emilia Romagna) are the hardest hit by the pandemic, while the Southern Regions, among which Campania and Sicily, represent a much lower percentage of the total cases and recorded deaths in Italy.

Chapter 11
"Keep Calm and Get Informed": Risk Communication and Engagement During the COVID-19 Pandemic in Romania

Camelia Cmeciu
Faculty of Journalism and Communication Studies, University of Bucharest, Romania

ABSTRACT

COVID-19 is a new ongoing pandemic and an alarming public health issue which urges emergency measures at a national and international level. This chapter uses a collaborative message-centered approach and explores how the Department for Emergency Situation (DES) in Romania tailored its verbal and visual messages to this pandemic and how the Romanian online users assessed the DES involvement in the risk and crisis management. The quantitative and qualitative content analyses showed that DES laid an emphasis on the narratives of informational responsibility and responsibility in action, whereas online users focused on the narratives of multilayered rationalization and a multifaceted incompetence.

INTRODUCTION

Serious threats to the health of people worldwide, uncertainty about the transmission or symptoms, anxiety turned into panic, lack of rapid diagnostic tests, media hype are some of the challenges that organizations and publics face throughout a pandemic or an epidemic (Liu, Kim, 2011; Jin et al. 2019; Ali, 2020). The year 2020 added the coronavirus COVID-19 pandemic to the already known health epidemics and pandemics (AIDS, BSE, SARS, or H1N1). Firstly identified in China at the end of 2019, within the first two months of 2020 the virus extended from China to over 140 countries[1].On January 30, 2020 COVID-19 was declared a global health emergency and 20 days later, it was declared a global pandemic[2]. The shift from urgency to action is essential during a pandemic since the lives of people and organizations worldwide are affected. Event prohibition, lockdowns, or shutdowns are just a few measures taken by governments in order to slow down the COVID-19 spread and to avoid "lethal capacity overloads of national healthcare systems" (Wenzel et al., 2020, V8).

DOI: 10.4018/978-1-7998-6705-0.ch011

Romania made no exception to these immediate governmental actions. End of February 2020 was the date when the danger of coronavirus was looming over Romania since 76 persons were reported to be infected with this virus in Italy, in the regions where a lot of Romanians work and live[3]. The first preventive measures were taken by the Romanian government on February 22, one of them being a 14-day quarantine for citizens returning from the affected regions in Italy, a measure that was later on extended to every person coming from abroad. What seemed to be some remote danger, at that moment, turned into something that had a high risk for Romania.

The Department for Emergency Situations (DES) within the Ministry of Internal Affairs is one of the Romanian authorities which started to communicate instructional messages regarding COVID 19. The first DES Facebook post was on February 24, 2020. Two days later, the first case of Coronavirus was reported in Romania[4]. This news brought fear and anxiety among Romanians who started to make provisions for a possible quarantine period. In the following days, this state of panic and hysteria was fueled by news media and visual and verbal social media framings of empty shelves or enormous queues in supermarkets. By beginning of May 2020, the situation of the Coronavirus pandemic in Romania was the following: 11,616 diagnosed persons, 3 cured persons, 21 persons in quarantine, and 12,158 persons home monitored and 726 deaths[5].

Unlike institutional crises, pandemics facilitate more a two-way communication because "interests are more aligned and a coproduction by institutional actors and publics is called for." (Hyvärinen, Vos, 2016, p. 97). Since risk communication is defined as "an interactive process of exchange of information and opinion among individuals, groups, and institutions" and since it "involves multiple and competing messages" (Sellnow et al. 2009, p. 4), in this chapter the author will use a perspective of interacting arguments to the crisis and risk communication of the Romanian Department for Emergency Situations (DES). Going beyond an agency approach, the author will employ what Heath and Palenchar (2016, p. 441) identify as a collaborative perspective. The collaborative message-centered approach will be applied to the online content generated on the Facebook page of the Romanian Department for Emergency Situations (DES) during the COVID-19 pandemic by the Romanian authority and citizens. The focus of this chapter will be on Facebook since Romanians have a higher interest[6] in this social media platform (11.000.000 users) compared to Twitter (374, 923 active users). The main objectives of the chapter are the following: (a) identification of message tailoring elements and the integrating narratives in DES Facebook posts; (b) identification of the levels of interpretation process and the integrating narratives in online users' comments.

BACKGROUND

Social Media, Health Risk and Crisis Communication

Unlike other types of crisis (corporate, environmental etc.), public health emergencies are different since they imply an urgent and unexpected crisis and official communicators should focus on explaining, persuading, and empowering decision making (Freimuth et al., 2008, p. 37S). There are five communication challenges that public health organizations face in health crises (Freimuth, 2006): communicating about uncertainty, selecting credible spokespersons, collaborating with and across organizations, satisfying media demands, and quickly providing needed information.

The research on crisis and risk communication in pandemics focused on two main strands: on the communication channel (traditional, new and social media) and on the communication content.

The Role of Traditional and Social Media in Health Risk and Crisis Communication

The urgency of sense-making and sense-giving in the interaction with citizens (Palttala & Vos, 2012) throughout a health crisis could be achieved by use of (traditional, new or social) media. Although authors (Tursunbayeva et al. 2017; Eckert et al., 2018) found in their studies that government health agencies do not use social media as a routine practice, a plea is made for the usage of social media because they may help in spreading truthful information, in verifying information or in dispelling rumors during disasters or pandemic crises. At the same time, during an epidemic, social media are considered to reduce the extent of spread (Yoo et al., 2016).

Mats Eriksson's (2018) extensive systematic and interdisciplinary review on social media crisis communication revealed five research-based lessons: (1) the importance of online dialogue, informational messages, source and timing; (2) the necessity of blending precrisis communication with the social media logic; (3) the relevance of social media monitoring; (4) a continuation of traditional media usage in crisis situations alongside with social media; (5) an adoption of social media in strategic crisis communication. As observed, Eriksson (2018) makes a plea for a blending of traditional and social media. Studies on media exposure and information seeking during crisis situations show that: television broadcasts, mobile phones and internet are mostly used to obtain disaster information (Xu et al., 2020), traditional media are considered a more credible information source during a crisis and play a significant role in the amplification and diffusion of information about infectious disease threats (Jin et al., 2019).

In their study on organizational framing of the 2009 H1N1 pandemic via social and traditional media, Liu and Kim (2011) found traditional media were more salient when this flu crisis was framed as a disaster, a health crisis, or a general health issue. Government and corporate organizations tended to use equally social media and traditional media when framing the H1N1 flu crisis as a general crisis. These findings were in line with Lin et al. (2014)'s review of the literature on the H1N1 pandemic: despite the dominance of traditional forms of mass media as the primary source of information, Facebook and Twitter were recognized as useful tools of raising awareness and of promoting intervention measures. In a study on MERS-preventive behavioral intentions, Yoo et al. (2016) showed that social media are effective in changing or reinforcing attitudes and behaviors during a health crisis.

Guidry et al. (2017) examined the Twitter and Instagram posts related to the Ebola crisis coming from three international health organizations (CDC, WHO and MSF) and found risk perception variables (information about adverse outcomes) and tips for avoiding infection or health organizational actions were more salient in Instagram posts than in tweets. In their study on the 2016 Zika virus epidemic, Park et al. (2019) highlighted that television news was the most important channel for Zika information. Despite this prevalent importance of traditional media in health crisis communication, Guidry et al. (2017) suggest that effective health crisis and risk communication social media messaging might work best when information is made available across social media platforms. In the same line, DiStasso et al. (2015) consider that social networks allow organizations during crisis situations to monitor the online ecosystem and to correct rumors before they turn into a fact, to post counter statements or provide links providing true information.

Health Crisis Communication – Message Tailoring

The main challenge for public and private organizations is to tailor the messages in order for the public to adopt the suggested behavior. The studies mentioned below focus on four aspects: (a) frame divergence in the messages of governmental and corporate organizations, (b) fake news spread during a health crisis, (c) credibility of sources, and (d) dialogic content.

(a) *Frame divergence.* In their comparative study on government and corporate messages during the 2009 H1N1 pandemic, Liu and Kim (2011) found that government organizations used general health issues and disaster frames whereas corporate organizations tended to frame the pandemic as a general crisis. Having public good as primary goal, government organizations employed alert, sympathy, severity and seasonal indicators, disease prevention and health care services unlike corporate organizations whose messages focused more on confusion. One important finding was the discrepancy in the usage of the uncertainty indicator which was more frequently adopted by government organizations whereas the reassurance and conflict indicators were more salient in messages issued by corporate organizations. Liu and Kim (2011) emphasize previous findings suggesting that addressing uncertainty inherent in (pandemic) crisis implies a limitation of internal and external confusion.

At the level of the content, organizations should counterbalance negative/ problem-based messages with more positive/ solution-based messages and embed more visuals since they stay longer in people's memories during high-stress situations (Guidry et al., 2017). At the same time, the authors highlight that it is important for organizations to acknowledge people's fear and the dangerous aspects of a health crisis.

(b) *Fake news.* Social media lack "traditional gatekeeping processes" (van Zoonen, van der Meer, 2015) and may breed "the rapid dissemination of misinformation" (DiStaso et al., 2015, p. 222) which could hardly be corrected (Lewandowsky et al., 2015), especially when it comes to health issues.

As Veil et al. (2011, p. 118) noted, social media can also be used by stakeholders who want to "create and disseminate their own influence, de-centralizing the dissemination of information and reducing official control, which is often very daunting to organizations." During the Zika virus crisis, social media channels were full of misinformation, conspiracy theories, pseudo-scientific claims and other such unreliable but viral information from non-official health sources (Avery, 2017). During the Ebola crisis, misinformation was again rampant on social media platforms and the official health organizations did not /rarely address(ed) that issue, missing the chance to correct misinformation (Guidry et al., 2017). Thus, combating false health information is important and authors mention that source credibility and dialogic content may be two solutions to stop misinformation.

(c) *Credibility of sources.* An important challenge to pandemic communication is the type of source providing the online information. Although the publics have turned into prosumers of information, organizations should be aware that the user-generated content on the internet "is not exclusively produced by reputable sources" (Faasse et al., 2016, p. 5813).

Park et al. (2019) showed in their study on Zika that it is important to take into account the people's levels of risk perception when addressing the issue of credible sources. Their findings revealed that persons who perceived low Zika risk relied more on medical professionals and friends and family for Zika information whereas people who perceived high Zika risk, thus feeling vulnerable, used television news and Facebook for information. Thus, Park et al. (2019) recommend that communication practitioners should add value to the information by placing medical professionals as sources on TV news. One aspect that should be considered is that trust is socially context dependent. In Romania, doctors should be used moderately as credible sources of information in health crisis since Romanians tend to make generalizations and associate all doctors with the ineffective health care system (Craciun, Bababan, 2012; Cmeciu et al., 2018).

(d) *Dialogic content.* Real-time dialogue and timing are essential in health crisis communication and health organizations should get involved in conversations. Empathetic communication is considered to be important in order to foster trust and to motivate publics to take positive actions (Seeger et al., 2003; Coombs, 2019). Eriksson (2018) highlighted in his review that during emotionally-laden crisis, emotional conversational message tactics on Facebook are very effective. Tirkkonen and Luoma-aho (2011)'s content analysis of the Finnish discussion forums during the 2009-2010 swine flu epidemic highlights that a more proactive authority communication and dialogue between the publics and authorities are necessary if the latter does not want to lose credibility. According to Du Plessis (2018, p. 829), organization-public dialogue "may also boost stakeholder support and encourage relationship building to help move the organization forward after the crisis with dialogic communication".

Health Risk Communication – A Message-Centered Approach

Authors acknowledge that risk communication during pandemic situations urge an interactive process between authorities and publics (Sellnow et al. 2009; Hyvärinen, Vos, 2016). This dialogue between authorities and stakeholders implies multiple and competing messages. Starting from Perelman and Olbrechts-Tyteca's concept of plurality (1969), a message convergence framework (Sellnow et al., 2009; Anthony et al., 2013) is developed, thus admitting that stakeholders play an active role in a risk situation by collecting information from various (credible or biased) sources and trying to make sense of contradictory information. In this study, stakeholders' active role will be associated with engagement, defined by Johnston and Taylor (2018, p. 3) "as an iterative, dynamic process, where participation, experience, and shared action emerge as central components of engagement". Sellnow et al. (2009) encompass a message-centered approach, providing a perspective of interacting arguments to risk communication.

The analyses using a message-centered approach focus on a two-fold stage:

a. An examination of the risk messages issued by the organizations in charge, by using the conceptual framework of the best practices of risk communication (Sellnow et al., 2009, pp. 24-28, p. 57). This chapter will refer to two main best practices: (1) *acknowledge levels of risk tolerance* (instructions provided to the public about the levels of risk to be experienced); (2) *account for uncertainty inherent in risk* (reinforcement of the unknown as an argument when framing risk messages to the public).
b. An examination of the four-level process of interpretation (Sellnow et al., 2009, pp. 8-9) provided by people since they actively seek and recognize points of convergence: (1) *systematic versus heuristic interpretation* (an evaluation based on cognitive evidence or on the source's identity and non-content cues); (2) *high versus low probability* of a risk situation to evolve into a crisis situation; (3) *high versus low self-efficacy* (people's degree of conviction that they will perform the activities mentioned in the risk messages); (4) *high versus low credibility* (the degree of openness, objectiveness, competence that people associate with the organization sending their risk messages).

Within this message-centered approach, the IDEA model for effective risk and crisis message design is developed (Sellnow et al. 2017; Sellnow-Richmond et al. 2018). The model consists of four elements:

- *internalization* is obtained through personal relevance, potential impact, proximity, timeliness.
- *distribution* focuses on the channel or channels the message is sent through.
- *explanation* implies to provide accurate information about what is happening and being done about the event. Internalization and explanation are part of an information strategy and they could be linked to one best practice in risk and crisis communication (Sellnow et al. 2009, 24-28; 57), namely to account for *uncertainty* inherent in risk (reinforcement of the (un)known as an argument when framing messages to the public).
- *action* focuses on the usage of clear messages which should include specific preparation action steps and specific response steps. Action is part of the instructional strategy and it could be linked to acknowledge levels of *risk tolerance* (instructions provided to the public about the levels of risk to be experienced).

In their study on the thematic analysis of the messages issued by local, national and international media during the 2014 Ebola outbreak, Sellnow-Richmond et al. (2018) combined exemplification theory with the IDEA model. Exemplars (evocative words, phrases, images, and sometimes sounds) have a positive or a negative valence and their usage may help people internalize the relevance and potential impact of a pandemic on them and their loved ones, "may serve as effective cognitive shortcuts to increase an understanding of complex information about the virus (explanation) and, ultimately, to foster appropriate self-protective actions". (Sellnow-Richmond et al., 2018, p. 142). The findings of their analysis on the Ebola outbreak revealed that each organization preferred to embed in their messages elements of explanation over internalization and action, as well as negative over positive exemplification. The authors recommend the strategic use of positive exemplification in instructional risk preparedness and crisis response messages because in this way, people will be motivated to attend these messages (*internalization*), to reduce potential misunderstandings (*explanation*), and to take appropriate self-protective actions (*action*).

This literature review showed that health organizations should adopt a proactive and an immediate reactive approach to pandemic crisis and risk communication since people panic and are in search of fast and accurate information. As mentioned above, dialogue, informative message, timing and credible source are four 'ingredients' that guarantee long-term organization-public relationships and individual or/ and community compliance with a suggested health behavior.

Making Sense of the Covid-19 Pandemic Within the Romanian Context

Patrick Brown (2020) provides an insight into the COVID-19 risk taking into account Heyman's risk features (2010): configuration of *probabilistic knowledge, time-framing, categories, and values.*

The most discussed *probabilities* regarding this virus refer to the risk of death and of being infected. Brown (2020) considers that these two probabilities have become germane to the sense-making process and debate over COVID-19. In Romania, besides mortality rates and speed of infection related to the probabilistic knowledge mentioned by Brown (2020), several key sets of numbers are presented by the official authorities: rates of quarantined, of isolated, of tested and of cured.

Risk governance approaches and levels of trust are two issues discussed related to *time-framing* of COVID-19. Brown (2020) considers that the different approaches to governments' risk governance lie in the cultural norms shaping the political imaginaries regarding the interventions of the state into the private sphere. In the case of Romania, the government's risk governance included: a two-month state of emergency (March 11 – May 15), ten military ordinances by end of April 2020, the lockdowns of schools, universities, shopping malls, parks, or of any public gathering, the flight cancellation to and from Italy, Spain, UK, France, USA etc. Within the two-month of state of emergency, the movement of all persons on the territory of Romania was conditioned to the filling-in of the self-declaration for leaving the premises. By end of April 2020, the fines for breaking lockdown regulations amount to 78 mld euros[7]. As mentioned above, trust is important within the discursive time-frames. National or international experts in public health, epidemiology are considered by the UK and Netherlands governments to be trustworthy sources of information during the COVID-19 pandemic (Brown, 2020, p. 5). But trust is a relative contextual issue. The IRES survey[8] shows that Romanians get the information about COVID-19 from news media (77%), Facebook (29%), state institutions (25%), and physicians (19%), and family members (2%). Despite the fact that Romanians do not get their information from doctors, they consider that doctors (15%) and Raed Arafat[9] (27%), a State Secretary at the Ministry of Health and a former physician, have the greatest merit in the COVID-19 crisis management[10]. It is interesting to observe that Raed Arafat is also mentioned in another survey on the levels of trust in the national public personalities. Thus, Romanians trust more Raed Arafat (78%) than the president of Romania (39%), the minister of health (37%), or the Prime Minister (24%)[11]. At an institutional level, the Department for Emergency Situations (DES, 75%) is the second most trusted institution after the Romanian army (84%)[12].

A boundary of age has emerged as a form of *categorization* within the COVID-19 pandemic (Brown, 2020). In Romania, the military ordinances focused on three main categorizations: age (above 65 years old), geographical (Romanians coming from abroad and citizens from the Suceava county which was completely shut down) and professional (beekeepers, commercial fishermen, farmers, cross border workers, or dentists). Following previous research on risk, Brown (2020, p. 7) mentions that "categorisations of people, events and conditions are inextricably bound up with processes of (de)valuing and (de) legitimation". Familiarity and stability are two values related to a risk. COVID-19 has an unfamiliar and fluid nature which will be experienced as problematic and dangerous (Brown, 2020, p. 7). The CURS survey[13] showed that Romanians express the following fears: of getting infected (89%), of infecting others (65%), of salary cuts (57%), or of losing one's job (51%). Thus, within the COVID-19 context Romanians' biggest fears are related to infection, salary and job stability.

MAIN FOCUS OF THE CHAPTER AND METHODOLOGY

Starting from the best practices in risk communication and the IDEA model presented above, this chapter will provide an insight into the polyphony of voices present on the Department for Emergency Situations (DES) Facebook page throughout the #COVID-19 pandemic in Romania. This study seeks to apply a message-centered approach to the online content generated by the Romanian Department for Emergency Situations (DES) and the online users during the first month since the first case of COVID-19 appeared in Romania.

Using a message-centered approach, this chapter addresses the following research questions:

RQ1: What are the most salient IDEA model categories present in the Romanian DES Facebook posts?
RQ2: What narratives could be associated with the IDEA model categories?
RQ3: What elements of the interpretation process are dominant in the online users' comments?
RQ4: What narratives prevail in online users' assessment of DES messages during the COVID-19 pandemic?

Method

The research design used in this study followed four phases: data collection phase, data mining, cross-tabulation and integrating through narrative.

A. *Data collection phase*. QDA miner 5.0.15 – Wordstat 7.1.22 software, a qualitative software for mixed methods (Silver & Lewins, 2014), was used to collect the Facebook posts and comments from the official page of the Romanian Department of Emergency Situations between February 24 – March 24, 2020. The data included 102 DES posts and 11,122 comments. For the analysis, all 102 posts were analysed and the comments were sampled by choosing every fifth applicable comment for analysis and a total of 2,225 comments were included for the analysis.
B. *Data mining phase*. As mentioned above, the QDA miner 5.0.15 facilitated the extraction of the data (posts and comments). Besides this, it was used in the coding of the units of analysis (ranging from words to sentences) using a coding scheme and in generating a frequency analysis.

This study employed both a deductive and an inductive method. Starting from the IDEA model, combined with the exemplification theory and the four-level process of interpretation, the coding scheme was adapted to the discursive specificity of the COVID-19 pandemic. The following categories were included in the coding scheme for each of the IDEA model elements.

The categories for *internalization* (I) were the following:

- *I-proximity* (information regarding the explicit location spread of COVID-19 in Romania);
- *I-personal relevance* (information regarding categories of risk who are affected by COVID-19 – old people over 65 years old, for example; rates of infection; mortality rates);
- *I-timeliness* (information regarding various time frames: time pressure imposed by the military ordinances, time interval of getting infected with COVID-19, time interval of getting exposed to COVID-19)
- *I-exemplification* (information regarding specific exemplars – patients infected, metaphorical representations of the virus or other abstract concepts). The author will identify I-positive exemplars (patients who survived, positive-laden metaphor) versus I-negative exemplars (patients who died, patients who did not comply with the measures, negative-laden metaphors etc.).

Although the IDEA model refers to channels of distribution (traditional, new, social media), in this chapter distribution will refer to the message form since the analysis will be on a single channel, namely Facebook. *Distribution* was coded according to the message form: *D-text, D-image, D-video, D-text+image, D-text+video, D-image+video, D-text+image+video.*

Explanation (E) was associated with lack of uncertainty (LoU) and source credibility (SC) . Lack of uncertainty refers to all the information DES posts about the cause, possible effects, cure related to COVID-19, measures taken by authorities. The categories belonging to explanation were:

- *E-LoU-infection cause, E-LoU-symptoms, E-LoU-cure, E-LoU-official measures*
- *E-SC-national experts, E-SC-international experts, E-SC-RO-official authorities, E-SC-international-official authorities*
- *E-exemplification* (refers to words and/ or images that help making complex information easier). The author will divide between *E-positive exemplars* (for example, infographics) and *E-negative exemplars* (for example, images that are misleading or may foster misunderstanding)

Action (A) was associated with *high level of risk tolerance* (instructions and tasks provided to people for self-protection). The categories will be:

- *A-self-protective action steps* (information about specific actions to follow not to get infected with COVID-19. For example, to use masks, sanitizers etc.)
- *A-information action steps* (information about to get informed from trustworthy and official sources of information)
- *A-community action steps* (information about helping others, about actions to show respect to doctors)
- *A-positive exemplars* (the action steps include visuals referring to specific actions, for example, to wash one's hands, how to wear a mask etc.)
- *A-negative exemplars* (the action steps include visual referring to how not to follow specific actions)

This study will focus on what Johnston and Taylor (2018) identify as mid-level of engagement, namely an individual level of comment analysis. The coding scheme used for the analysis of the Romanians' comments to the DES posts was the following:

- *evaluation* (Ev): Ev-*systemic* (reference to objective evidence related to the competing claims of the organizations involved in the COVID-19 pandemic) vs Ev-*heuristic* (reference to the organizational representatives' identity or other cues not related to the COVID-19 pandemic);
- perceived probability: *high perceived probability* (an event with high probability and high severity would warrant immediate action because it is perceived as dangerous) vs *low perceived probability* (an event would instill little concern).
- efficacy: *high self-efficacy* (comments including people's conviction that they will perform the activities that would reduce their level of risk), *low self-efficacy* (comments including people's doubts about the activities suggested in the DES messages) vs *low other-efficacy* (comments including references to others' improper behavior);
- credibility: *high credibility* (reference to the trust, openness or information accuracy associated with the sources providing information throughout the COVID-19 pandemic) vs *low credibility* (reference to the lack of confidence and trust in the sources).

C. Cross-tabulation phase. Using Wordstat 7.1.22., a computer-program-assisted text analysis based on a text mining program, a cross-tabulation analysis was employed to identify the relationship between keywords or phrases and the IDEA model elements, on the one hand, and the levels of interpretation, on the other hand.

D. Integrating through narrative phase. This phase refers to the integration of qualitative and quantitative data at the interpretation level. A weaving approach (Fetters et al., 2013) was used by involving both quantitative and qualitative findings together on a theme-by-theme basis.

FINDINGS

The Facebook Posts of the Romanian DES

Salience of IDEA Model Categories in DES Posts

To evaluate the use of the elements of the IDEA model in the DES's online communication on Facebook, 102 posts were coded for internalization, distribution, explanation and action. As observed in Table 1, posts embedding text and image (61%, n = 62) were the most frequently used, followed at a significant distance by posts with images (17%, n = 18) and posts with text and video (8%, n = 8).

DES tailored its messages more on explanation (86%), followed by internalization (75%) and action (68%). The three most salient sub-categories were: explanation (E) through positive exemplars (46%), action rendered through information action steps (46%) and internalization through personal relevance (40%).

The three most used IDEA sub-categories for *explanation* were: positive exemplars (46%), Romanian official authorities as source credibility (E-SC-RO-official authorities, 36%), and official measures as signs of lack of uncertainty (E-LoU-official measures, 36%). Personal relevance (40%) and proximity (25%) were the first two elements mostly present in the DES posts focused on the IDEA category of *internalization*. At the level of *action*, DES laid on emphasis more on information action steps (46%) than on self-protective action steps (25%).

DES Posts – Narratives

The qualitative analysis and the cross-tabulation results revealed two main narratives: *informational responsibility* and *responsibility in action.*

The 'Informational Responsibility' Narrative

A greater focus was placed on an informational responsibility through a salience laid on *internalization* through *personal relevance*, *proximity* and on *information action steps*. Personal relevance was clustering with words, such as 'quarantine' (n=35), 'infected' (n=35), 'tests' (n=35), or 'telverde'/'green line'[14] (n=35). Since the first day of the pandemic, the Romanian DES shaped its discourse on categorization, by pinpointing various categories of people upon whom COVID-19 might have an impact, thus creating personal relevance for online users. The DES communication about the pandemic situation in Romania was daily: by March 19, twice a day (in the morning and in the afternoon) and since March 19, DES made a Facebook post around noon. Categorization evolved throughout February 24 – March 24, 2020. At the beginning, an emphasis was laid on rates of people infected, people in quarantine, people monitored at home and on numbers of tests. Two days later, the category of Telverde/ Green Line and emergency calls was added. On March 1, number of healed people as a category was revealed in the daily information. Since March 22, the mortality rate category is embedded in the DES messages. Personal relevance through categorization was combined with geographical proximity through specific mentions of the location of the people infected, of the people in quarantine or of the deceased people. This information on the closeness of the COVID-19 was essential for Romanian citizens since they wanted to know what cities and towns are the most affected by the virus.

Table 1. IDEA model categories – DES Facebook posts (n=102)

IDEA categories	Sub-categories	Percentage
Internalization (I) – 75% (n=77)	I-proximity	25% (n = 26)
	I-personal relevance	40% (n = 41)
	I-timeliness	10% (n = 10)
	I-positive exemplars	0% (n = 0)
	I-negative exemplars	17% (n = 18)
Distribution (D)	D-text	6% (n = 6)
	D-image	18% (n = 18)
	D-video	1% (n = 1)
	D-text+image	61% (n = 62)
	D-text+video	8% (n = 8)
	D-*image+video*	5% (n = 5)
	D-*text+image+video*	2% (n = 2)
Explanation (E) – 86% (n = 88)	E-LoU-infection cause	5% (n = 5)
	E- LoU -symptoms	2% (n = 2)
	E- LoU -official measures	36% (n = 37)
	E- LoU -cure	5% (n = 5)
	E-SC-national experts	2% (n = 2)
	E-SC-international experts	7% (n = 7)
	E-SC-RO-official authorities	36% (n = 37)
	E-SC-international-official authorities	2% (n = 2)
	E-positive exemplars	46% (n = 47)
	E-negative exemplars	6% (n = 6)
Action (A) – 68% (n=82)	A-self-protective action steps	25% (n = 26)
	A-information action steps	46% (n = 47)
	A-community action steps	0.98% (n = 1)
	A-positive exemplars	5% (n = 5)
	A-negative exemplars	3% (n = 3)

Another important salience could be observed at the level of *explanation* through the presence of *official measures* and *national authorities*. The informational responsibility presented above was accompanied by the lack of uncertainty rendered through the enumeration of various official measures taken by the Romanian authorities. These measures were always associated with *sources of credibility* from Romania, the phrase 'strategic communication group' being mentioned in 41 posts. Its power-based authorization was revealed through posts on micro-level measures and macro-level measures. 'Hand washing' (n=21), 'sanitizers' (n=15), 'home' (n=22), 'quarantine' (n=15) or 'ordinance' (n=11) clustered with *E-LoU-official measures*. The micro-level measures focused on information about personal hygiene measures (hand washing etc.) whereas the second type of measures had a threefold nature. Firstly, they referred to measures regarding the quarantine procedures for those coming from abroad, the testing capacity, or the acquisition of protection equipment for doctors. Secondly, DES made posts about the three military ordinances issued within the time span of this analysis. The measures ranged from suspension of all cultural, scientific, artistic, religious, sports, entertainment or gambling, spa treatment and personal care activities up to the progressive suspension of flight from and to Romania or to travel restrictions for persons aged 65 and above. Thirdly, the measures referred to the digital platform (stirioficiale.ro/ officialnews.ro), an initiative of DES, Authority for Digitalization of Romania, Romanian Government and Code for Romania Task Force. This last measure is related to the sub-category (*A-information action steps*), thus showing that DES framed its responsibility as a battle not only against the virus but also against fake news.

The 'Responsibility in Action' Narrative

A second narrative was *responsibility in action*. It is interesting to observe that *A-information action steps* (46%), clustering with 'fake news' (n=11), 'inform yourself' (n=9), or 'credible sources' (n=12), was more salient than *A-self-protective action steps* (25%), thus highlighting the importance laid on being close to citizens and on providing accurate information. DES urged Romanian citizens to use the Telverde/ Green Line phone number in case they have an emergency, but at the same time, to get informed only from official sources.

26 (25%) posts focused on *self-protective action steps*. DES posted messages embedding concrete tasks regarding washing one's hands, face, eyes, nose or mouth, proper distancing, frequent surface disinfection etc.

At the level of *exemplification*, two types of exemplars were dominant: positive exemplars (n= 47, 46%) incorporated in explanation and negative exemplars (n=18, 17%) incorporated in internalization. Informational responsibility and responsibility in action were supported by explanation through positive exemplars (*E-positive exemplars*). The daily information bulletins were accompanied by infographics comprising the categorization presented above. This knowledge-based legitimation of DES was also visually embedded in graphic illustrations as cognitive shortcuts regarding doubtful information about the virus. This visual communication was part of an information campaign run together with the Romanian Health Observatory to answer all the fake news or the uncertainties related to ways of contracting the virus or of curing. Various issues were answered using images which embedded a question and a juxtaposed answer (yes or no) highlighted in a green or red square. For example, "Is it preferable to stay home during this time?"/ "answer: yes", "Is Coronavirus transmitted through sneezing or coughing?" / "answer: yes", or "Does lemon consumption prevent from Coronavirus infection? / "answer: no".

One important element within *internalization* was the blending of personal relevance with negative exemplars (*I-negative exemplars*). Using this negative exemplification, DES made use of loss-framed messages by appealing to fear: the fear of getting infected and eventually the fear of death and the fear of a financial loss. Two main categories of information could be identified: on the one hand, exemplification of people who were infected, by specifying their location and the hospital where they were treated, and on the other hand, exemplification of types of fines applied to the people who did not comply to the quarantine rules and/ or the emergency ordinance regulations.

Online Users' Assessment of the DES Posts

Salience of Levels of Interpretation

The quantitative analysis of the Facebook comments to the posts generated by the Romanian Department of Emergency Situations showed the prevalence of three main levels of the interpretation process (table 2): *evaluation* through *systemic* (39%) and *heuristic* (19%) interpretation, *high perceived probability* (16%), and *low credibility* (14%).

Table 2. Levels of interpretation – online users' comments (n = 2,225)

evaluation	Ev-systemic evaluation	39% (n = 857)
	Ev-heuristic evaluation	19% (n = 416)
perceived probability	high perceived probability	16% (n = 346)
	low perceived probability	1% (n = 18)
efficacy	high self-efficacy	8% (n = 175)
	low self-efficacy	3% (n = 67)
	low other-efficacy	5% (n = 110)
credibility	high credibility	7% (n = 161)
	low credibility	14% (n = 302)

Online Users' Comments – Narratives

The qualitative analysis of the online users' comments showed two narratives associated with these levels of interpretation: *a multilayered rationalization* and *a multifaceted incompetence.*

The 'Multilayered Rationalization' Narrative

Online users tried to make sense of the new crisis situation perceived as highly harmful, by providing a rational (de)legitimation of the DES's actions. *High perceived probability* is clustering with words, such as 'die' (n=25), 'fake news' (n=25), 'panic' (n=23), 'fear' (n=23), 'work' (n=21), 'close' (n=14), 'continue' (n=14), 'immunity' (n=13), or 'money' (n=12). These words highlight a threefold fear: the fear of getting infected ("This is the only way of reducing the number of contaminations. Romanians do whatever they are asked if they are afraid… remember communism", March 20), the fear of misinformation ("publishing fake news may lead to confusion, panic, wrong information", March 20), and fear associated to economic implications ("We stay home but the day will come when we have to put something on the table. … When famine strikes the whole population, nothing will stop it…", March 21). The emotional state of increasing wary made online users to assess the DES's measures from the very beginning the COVID-19 crisis occurred in Romania.

As mentioned above, this assessment focused mainly on a *systemic evaluation* (present in 857 comments) and included two main layers of rationalization: a micro layer and a macro layer. The micro layer of rationalization was emphasized by words such as 'mask' (n=25), 'expensive' (n=25), 'wash' (n=22), or 'home' (n=21). The COVID-19 impact upon one's family was a major theme that emerged within the online users' comments related to this type of rationalization. In their messages, they made reference to the objective aspects related to the high price of masks, to issues of personal hygiene and they evaluated the DES recommendation of staying home as an important social practice.

'Tests' (n=32), 'county' (n=28), 'near' (n=25), 'rights' (n=23), 'situation report' (n=22), 'fine' (n=21), 'measures' (n=20), 'recommendation' (n=20) are the most frequently used words which clustered with *systemic evaluation*, highlighting a macro layer of rationalization. Some Romanian citizens wanted DES to provide more thorough information through categorization. They urged DES to add a category based on a county proximity related to the people infected, in quarantine or dead, arguing that they want to know the real situation in their county. On the other hand, this categorization was perceived as a way of emphasizing the DES helplessness. The commentators did not want recommendations or numbers of people who are infected, or dead, they wanted to know the exact measures taken by the Romanian authority. The assessment of risk governance had a positive valence through acknowledging fines as a necessary measure for those citizens who consider that rights come without obligations ("(…) It is a real pity that we are not aware of the reality, but we just want to be given something without assuming some obligations", March 22). Within this context, some online users considered that the military ordinances were properly institutionalized social practices. The negative valence of risk governance was associated with the reduced number of tests: "WHO recommends to run as many tests as possible. (…) 500 tests is a small number" (March, 17).

The 'Multifaceted Incompetence' Narrative

This narrative focuses on a twofold delegitimation targeted, at a macro level, towards the Romanian official authorities through *low credibility* using *heuristic evaluations* and, at a micro level, towards other online users through *heuristic evaluations*.

The negative presentation through emotionally loaded aggressive words ("Why don't you test, you scumbags? We are in community transmission? Why don't doctors have proper equipment? You, morons! You kill those who can save lives", March 2) clearly shows that this pandemic triggered severe threats to some Romanians' expectations regarding the competence of the DES and other Romanian official institutions. The crosstabulation results showed the dominance of 'imbecility' (n=31), 'incapable' (n=25), 'stupid' (n=22), 'lies' (n=21), or 'promises' (n=14) clustering with the heuristic evaluation regarding the official measures. Some Romanian citizens criticized these measures but without bringing cognitive arguments related to their labeling ("A total imbecility… in two or three weeks, it will be a disaster!", February 27). The low credibility was associated with various social actors: political actors (the incumbent government, the prime minister, the president, the mayors for improper crisis management), economic actors (online retailers for the high prices of masks and sanitizers) and media actors (media outlets for presenting sensational, anxiety and fearful loaded news). One of the most representative comments for this narrative is the following: "The problem is that we live in a country ruled by incompetents and sooner or later we will all be affected by their stupidity" (March, 17).

One important feature of Facebook engagement is user-user conversation. The COVID-19 pandemic brought disputes between some Romanian online users who either agreed or attacked others' points of view. Some online users' identity was attacked, being accused of behaving like experts in infectious diseases or journalists from tabloid newspapers: "Are you a graduate of some medical school? Shut up"; "Are you a reporter? Now I understand why Libertatea/Liberty is a site full of stupid articles. You wrote there." (March, 20).

Considered "a vital element in the public's sense of risk" (Sellnow et al. 2009, p. 9), efficacy suggests the degree of a person's compliance to carry out the behaviors suggested in the messages. As observed in table 2, the three types of efficacy had low frequencies, not all Romanian online users being willing to express their response to the Romanian official authorities' suggested recommendations. Despite this reduced salience, it is important to highlight the juxtaposition between self-presentation and other-presentation in terms of efficacy. On the one hand, some users presented their daily activities during the isolation time, emphasizing the importance of individual prevention. Although *high credibility* is not as much mentioned as *low credibility*, it is also present in the comments of the users who adopted the recommended behavior. Some users showed their gratitude and respect to doctors, others appreciated the open, accurate and complete information provided by DES and WHO to citizens.

On the other hand, some other users considered that all restrictions were a violation of one's rights and they rejected to comply to some exaggerated rules: "Paranoid measures. Death is part of life. I go to work, I don't stay home" (March 19). 'Full' (n=12), 'malls' (n=11), or 'buses' (n=11) are three words clustering with *low other-efficacy*. Some online users were outraged that people did not adopt the suggested behavior of staying home and expressed their worry about a possible worsening of the situation: "But buses are full, old people keep walking in the streets without any stress… I do not know what will happen to us" (March 19).

CONCLUSION

This study sought to use a collaborative perspective on the risk and crisis communication of the Department for Emergency Situations during the first month of the COVID-19 pandemic in Romania. The unfamiliar and fluid nature associated with COVID-19 (Brown, 2020) brought multiple and competing messages coming either from DES or from online users.

The first research question focused on the frequency analysis of the four categories of the IDEA model: internalization, distribution, explanation and action. At the level of distribution, the Romanian authority mainly used text accompanied by images in its posts. This finding goes in line with what previous studies have already highlighted, namely the importance of embedding more visuals since people will remember them during high-stress situations (Guidry et al., 2017).

Unlike the results of Sellnow-Richmond et al.'s study (2018) on Ebola outbreak, where explanation prevailed over internalization and action and negative exemplification over positive exemplification, this chapter showed that during the first month of COVID-19 pandemic in Romania, DES tried to get a balance between explanation, internalization and action and laid more an emphasis on positive exemplars. The findings showed that there is not a high discrepancy in DES posts between explanation and the other two categories of IDEA model, namely internalization and action. In terms of exemplification, positive exemplification was more salient when associated with explanation and negative exemplification was more associated with internalization. Exemplification was scarcely used in messages focused on action. The strategic use of positive exemplification for explanation, internalization and action should be adopted by DES since, according to Sellnow-Richmond et al. (2018), it may trigger attendance of messages, reduction of fake news and adoption of self-protective actions.

The second research question focused on the DES narratives associated with the IDEA model categories. Two main integrating narratives were observed in the DES message tailoring: *informational responsibility* and *responsibility in action*. Information seeking is essential to be addressed by authorities in risk and crisis situations. In line with Liu and Kim (2011)'s suggestion that government organizations should address uncertainty, DES laid an emphasis on internalization through personal relevance and proximity, combined with information action steps. Unlike the official health authorities that did not correct the misinformation during the Ebola crisis (Guidry et al., 2017), the Romanian authority (DES) seemed to be aware of the importance of combating false health information on social media and tailored its daily posts on categorization of infection or mortality rates and brought updates on the information campaign run together with Code for Romania Task Force and other official institutions.

Studies show that source credibility obtained through medical professional and dialogic content play a significant role in stopping misinformation (Tirkkonen & Luoma-aho, 2011; Guidry et al., 2017; du Plessis, 2018; Park et al., 2019). The research showed that DES did not respond to any of the comments and credibility was rendered through a group (the strategic communication group), not through individual experts. Despite this lack of online dialogic involvement and of a pervasive usage of experts, DES tailored its messages on explanation mentioning the measures to be taken. The emphasis on micro-level and macro-level measures revealed a power-based authorization, DES presenting either the personal hygiene steps to be followed or the content of military ordinances.

The second narrative, *responsibility in action*, brought a surprising result: a prevalence of information action steps over self-protective action steps. DES incorporated in its messages tasks to be followed for individual and group protection, but laid a higher emphasis on combating fake news, urging Romanian citizens to get informed from credible sources and it also tried to acknowledge people's fear by providing answers to some rumors (for example, lemon consumption prevents COVID-19 infection). Two main aspects should be highlighted here: the collaborating with Code for Romania Task Force and Romanian Health Observatory in this awareness campaign and the visual tailoring of the messages. These findings show that DES addressed two main challenges mentioned in the literature (Freimuth, 2006; Guidry et al., 2017): collaborating with other organizations and communicating through visual images. The visual tailoring of messages should be associated with positive exemplification (Sellnow-Richmond et al., 2018). Explanation through positive exemplars and internalization through negative exemplars prevailed. DES used infographics and graphic illustrations to provide a knowledge-based visual legitimation, these two visual representations comprising information on categorization and on factual data about the virus.

The efficacy of DES messages was tackled in the analysis of online users' comments. The third research question addressed the dominance of elements of the interpretation process present in the online users' comments. Romanian commentators mostly used evaluation through systemic and heuristic evaluation, followed by high perceived probability and low credibility. Thus, they tried to provide objective evidence related to the DES posts on the COVID-19 pandemic. Although the heuristic evaluation was not as frequent as systemic evaluation, it showed that Romanians commented either on the identity of the official authorities or on that of other commentators.

The last research questions focused on the online users' assessment of DES messages in terms of narrativization. Two narratives emerged from the analysis: a *multilayered rationalization* and a *multifaceted incompetence*. Online users tried to make sense of this new crisis situation by critically analyzing the official measures at a micro and macro level. The high impact of the virus upon one's family and the fear of infection and death may be considered an essential reason why they cognitively evaluated the DES messages. They asked the official authorities for more information on the virus and measures in order to reduce potential misunderstandings. At the same time, they positively evaluated the risk governance related to financial punishments for non-compliance to rules and to military ordinances as institutionalized social practices. The multifaceted incompetence focused on a delegitimation of online users and Romanian official authorities. Unlike previous studies (Tirkkonen & Luoma-aho, 2011; Guidry et al., 2017; du Plessis, 2018; Park et al., 2019) which showed a high degree of public confidence in authorities, the results of this chapter reveal a reluctance towards authorities. This goes in line with the IRES survey (see endnote 11) which showed that there is a high discrepancy in the trust showed towards the individuals associated with the respective institution. Romanians trust more Raed Arafat (78%), a State Secretary at the Ministry of Health, than the president of Romania (39%). Since Romanians display a greater trust in individuals than in institutions, a recommendation for Romanian communication professionals would be to use these persons as sources of credibility. This aspect was also noticed in the analysis of the DES messages, namely scarce mentions of individual experts who may be considered trustworthy by citizens. (De)legitimation of other users is a feature of dialogic communication and it becomes even more salient when the organization does not get involved in the conversation, as it was the case of the Romanian Department for Emergency Situations which did not provide any comment to the online users' messages.

Although DES tailored its messages to balance the elements of explanation with internalization and action and incorporated more positive exemplars associated with exemplification, the Romanian official authority did not adopt a real-time dialogue. This lack of dialogic content between an official authority and commentators was also found in Tirkkonen and Luoma-aho (2011)'s study on Finnish dialogic communication during the swine flu epidemic. The lack of online conversation involvement could be one explanation for the prevalence of low credibility in the online users' comments which did not receive any official answer. Previous studies (Seeger et al., 2003; Eriksson, 2018; Du Plessis, 2018; Coombs, 2019) showed that empathetic conversation between authorities and citizens may foster trust, boost stakeholder support and motivate them to take positive actions.

The IDEA model combined with exemplification and the levels of interpretation process constituted valuable frameworks for the content analysis of the DES posts and Romanian citizens' comments related to the COVID-19 pandemic. On the one hand, when tailoring messages during crisis situations, communication practitioners should combine internalization with exemplification, explanation and action in order to motivate people to attend the respective message, to make them understand and to determine them to take appropriate self-protective actions. On the other hand, communication practitioners should be aware that social media bring a compulsory dialogic communication where citizens can evaluate the organizations' activity, perceive the probability of an event, show their trust and express their self-efficacy.

FUTURE RESEARCH DIRECTIONS

Although this study was limited to the content analysis of COVID-19 pandemic at a national limit, further research may be carried out: a comparative analysis of various departments of emergencies from other countries; a longitudinal study on different pandemics and outbreaks. Another strand of research should focus on an extension of methodology: the content analysis could be combined with interviews focused on official authorities and with surveys applied to online users.

REFERENCES

Ali, I. (2020). *The COVID-19 pandemic: Making sense of rumor and fear.* Academic Press.

Anthony, K. E., Sellnow, T. L., & Millner, A. G. (2013). Message convergence as a message-centered approach to analysing and improving risk communication. *Journal of Applied Communication Research, 41*(4), 346–364. doi:10.1080/00909882.2013.844346

Avery, E. J. (2017). Public information officers' social media monitoring during the Zika virus crisis, a global health threat surrounded by public uncertainty. *Public Relations Review, 43*(3), 468–476. doi:10.1016/j.pubrev.2017.02.018

Brown, P. (2020). Studying COVID-19 in light of critical approaches to risk and uncertainty: Research pathways, conceptual tools, and some magic from Mary Douglas. *Health Risk & Society, 22*(1), 1–14. doi:10.1080/13698575.2020.1745508

Cmeciu, C., Coman, I., & Coman, C. (2018). *Beyond The Ministry Of Health's Vaccination Campaign during the Measles Outbreak in Romania – Public Assessment and interaction.* Paper presented at the ICA conference, Prague.

Coombs, W. T. (2019). *Ongoing crisis communication: Planning, managing, and responding* (5th ed.). Sage.

Craciun, C., & Baban, A. (2012). "Who will take the blame?": Understanding the reasons why Romanian mothers decline HPV vaccination for their daughters. *Vaccine, 30*(48), 6789–6793. doi:10.1016/j.vaccine.2012.09.016 PMID:23017603

DiStaso, M. W., Vafeiadis, M., & Amaral, C. (2015). Managing a health crisis on Facebook: How the response strategies of apology, sympathy, and information influence public relations. *Public Relations Review, 41*(2), 222–231. doi:10.1016/j.pubrev.2014.11.014

Du Plessis, Ch. (2018). Social media crisis communication: Enhancing a discourse of renewal through dialogic content. *Public Relations Review, 44*(5), 829–838. doi:10.1016/j.pubrev.2018.10.003

Eckert, S., Sopory, P., Day, A., Wilkins, L., Padgett, D., Novak, J., Allen, T., Alexander, N., Vanderford, M., & Gamhewage, G. (2018). Health-related disaster communication and social media: Mixed-method systematic review. *Health Communication, 33*(12), 1389–1400. doi:10.1080/10410236.2017.1351278 PMID:28825501

Eriksson, M. (2018). Lessons for crisis communication on social media: A systematic review of what research tells the practice. *International Journal of Strategic Communication, 12*(5), 526–551. doi:10.1080/1553118X.2018.1510405

Faasse, K., Casey, J., Chatman, C. J., Leslie, R., & Martin, L. R. (2016). A comparison of anguage use in pro- and anti-vaccination comments in response to a high profile Facebook post. *Vaccine, 34*(47), 5808–5814. doi:10.1016/j.vaccine.2016.09.029 PMID:27707558

Fetters, M. D., Curry, L. A., & Creswell, J. W. (2013). Achieving integration in mixed methods designs: Principles and practices. *Health Services Research, 48*(6), 2134–2156. doi:10.1111/1475-6773.12117 PMID:24279835

Freimuth, V. S. (2006). Order out of chaos: The self-organization of communication following the Anthrax attacks. *Health Communication, 20*(2), 141–148. doi:10.120715327027hc2002_5 PMID:16965251

Freimuth, V. S., Hilyard, K. M., Barge, J. K., & Sokler, L. A. (2008). Action, not talk: A simulation of risk communication during the first hours of a pandemic. *Health Promotion Practice, 9*(4_suppl), 35S-44S.

Guidry, J. P. D., Jinb, Y., Orr, C. A., Messner, M., & Meganck, S. (2017). Ebola on Instagram and Twitter: How health organizations address the health crisis in their social media engagement. *Public Relations Review, 43*(3), 477–486. doi:10.1016/j.pubrev.2017.04.009

Heath, R. L., & Palenchar, M. J. (2016). Paradigms of risk and crisis communication in the 21st century. In A. Schwarz, M.W. Seeger, & C. Auer (Eds.), The handbook of international crisis communication research (pp. 437-446). Bognor Regis: John Wiley & Sons Ltd.

Heyman, B. (2010). The concept of risk. In B. Heyman, M. Shaw, A. Alaszewski, & M. Titterton (Eds.), *Risk, safety and clinical practice: Healthcare through the lens of risk* (pp. 15–36). Oxford University Press.

Hyvärinen, J., & Vos, M. (2016). Communication concerning disasters and pandemics: coproducing community resilience and crisis response. In A. Schwarz, M. W. Seeger, & C. Auer (Eds.), *The handbook of international crisis communication research* (pp. 96–108). John Wiley & Sons Ltd. doi:10.1002/9781118516812.ch10

Jin, Y., Austin, L., Santosh Vijaykumar, S., Jun, H., & Nowak, G. (2019). Communicating about infectious disease threats: Insights from public health information officers. *Public Relations Review, 45*(1), 167–177. doi:10.1016/j.pubrev.2018.12.003

Johnston, K. A., & Taylor, M. (2018). Engagement as communication pathways, possibilities, and future directions. In The handbook of communication engagement (pp. 1-15). John Wiley & Sons.

Lewandowsky, S., Ecker, U. K. H., Seifert, C. M., Schwarz, N., & Cook, J. (2012). Misinformation and its correction: Continued influence and successful debiasing. *Psychological Science in the Public Interest, 13*(3), 106–131. doi:10.1177/1529100612451018 PMID:26173286

Lin, L., Savoia, E., Agboola, F., & Viswanath, K. (2014). What have we learned about communication inequalities during the H1N1 pandemic: A systematic review of the literature. *BMC Public Health, 21*(1), 1–13. doi:10.1186/1471-2458-14-484 PMID:24884634

Liu, B. F., & Kim, S. (2011). How organizations framed the 2009 H1N1 pandemic via social and traditional media: Implications for U.S. health communicators. *Public Relations Review*, *37*(3), 233–244. doi:10.1016/j.pubrev.2011.03.005

Palttala, P., & Vos, M. (2012). Quality indicators for crisis communication to support emergency management by public authorities. *Journal of Contingencies and Crisis Management*, *20*(1), 39–51. doi:10.1111/j.1468-5973.2011.00654.x

Park, S., Boatwright, B., & Avery, E. J. (2019). Information channel preference in health crisis: Exploring the roles of perceived risk, preparedness, knowledge, and intent to follow directives. *Public Relations Review*, *45*(5), 101794. doi:10.1016/j.pubrev.2019.05.015

Seeger, M. W., Sellnow, T. L., & Ulmer, R. R. (2003). *Communication and organizational crisis*. Praeger.

Sellnow, T. L., Ulmer, R. R., Seeger, M. W., & Littlefield, R. S. (2009). *Effective risk communication. A message-centered approach*. Springer. doi:10.1007/978-0-387-79727-4

Sellnow-Richmond, D. D., George, A. M., & Sellnow, D. D. (2018). An IDEA model analysis of instructional risk communication in the time of Ebola. *Journal of International Crisis and Risk Communication Research*, *1*(1), 135–166. doi:10.30658/jicrcr.1.1.7

Silver, C., & Lewins, A. (2014). *Using software in qualitative research: A step-by-step guide*. SAGE. doi:10.4135/9781473906907

Tirkkonen, P., & Luoma-aho, V. (2011). Online authority communication during an epidemic: A Finnish example. *Public Relations Review*, *37*(2), 172–174. doi:10.1016/j.pubrev.2011.01.004

Tursunbayeva, A., Franco, M., & Pagliari, C. (2017). Use of social media for e-Government in the public health sector: A systematic review of published studies. *Government Information Quarterly*, *34*(2), 270–282. doi:10.1016/j.giq.2017.04.001

Van Zoonen, W., & van der Meer, T. (2015). The importance of source and credibility perception in times of crisis: Crisis communication in a socially mediated era. *Journal of Public Relations Research*, *27*(5), 371–388. doi:10.1080/1062726X.2015.1062382

Veil, S. R., Buehner, T., & Palenchar, M. J. (2011). A work-in-process literature review: Incorporating social media in risk and crisis communication. *Journal of Contingencies and Crisis Management*, *19*(2), 110–122. doi:10.1111/j.1468-5973.2011.00639.x

Wenzel, M., Stanske, S., & Lieberman, M. B. (2020). Strategic responses to crisis. *Strategic Management Journal*, *41*, V7–V18.

Xu, D., Zhuang, L., Deng, X., Qing, C., & Yong, Z. (2020). Media exposure, disaster experience, and risk perception of rural households in earthquake-stricken areas: Evidence from rural China. *International Journal of Environmental Research and Public Health*, *17*(9), 3246. doi:10.3390/ijerph17093246 PMID:32384741

Yoo, W., Choi, D.-H., & Park, K. (2016). The effects of SNS communication: How expressing and receiving information predict MERS-preventive behavioral intentions in South Korea. *Computers in Human Behavior*, *62*, 34–43. doi:10.1016/j.chb.2016.03.058 PMID:32288174

KEY TERMS AND DEFINITIONS

Action: Specific preparation action steps and specific response steps to be included in a message.

Credibility: The degree of openness, objectiveness, competence that people associate with the organization sending their risk messages.

Distribution: The form and the channel the message is sent through.

Explanation: Accurate information about what is happening and being done in a crisis and/or risk situation.

IDEA Model: The acronym of internalization, distribution, explanation and action. Developed by Sellnow et al. (2009).

Internalization: Personal relevance, potential impact, proximity, timeliness that should be embedded in a message.

Self-Efficacy: People's degree of conviction that they will perform the activities mentioned in the risk messages.

ENDNOTES

1 World Health Organization (WHO). " Coronavirus Disease 2019 (COVID-19) Situation Report 55. Geneva, Switzerland: WHO." (2020) Retrieved May 08, 2020, from https://www.who.int/docs/default-source/coronaviruse/situation-reports/20200410-sitrep-55-covid-19.pdf?sfvrsn=ca96eb84_2

2 World Health Organization (WHO). " International Health Regulations (IHR) Emergency Committee regarding the Outbreak of Novel Coronavirus (2019-ncov). Geneva, Switzerland: WHO.", (2020) Retrieved May 08, 2020, from https://www.who.int/ihr/procedures/novel-coronavirus-2019/en/

3 World Health Organization (WHO). "Coronavirus disease (COVID-19) pandemic." (2020) Retrieved May 08, 2020, from https://www.who.int/emergencies/diseases/novel-coronavirus-2019

4 Ministry of Health in Romania, "Press releases." (2020) Retrieved March 12, 2020, from http://www.ms.ro/comunicate/

5 Department for Emergency Situations (May 2020). Retrieved May 30, 2020, from https://www.facebook.com/departamenturgente/

6 Zelist.ro. "Overview social media in Romania" (March 2020) Retrieved April 14, 2020,from https://www.zelist.ro/blog/overview-social-media-in-ro-martie-2020/?fbclid=IwAR0N51IOv7bE2CMKJNHVfiJUIcPNZ6T0hH7CW4vOTLaIP9dqCIYDFG_zgrk

7 Romania-insider.com. "EUR 78 mln worth of fines for breaking lockdown regulations in RO so far" (2020). Retrieved May 05, 2020,from https://www.romania-insider.com/fines-covid-apr-21-2020

8 Romanian Institute for Evaluation and Strategy (IRES). "Romanians' perception on COVID-19 pandemic." (March 2020). Retrieved April 20, 2020, from https://ires.ro/articol/381/percep-iile-romanilor-cu-privire-la-epidemia-de-coronavirus-covid-%E2%80%93-19

9 Raed Arafat is a State Secretary at the Ministry of Health, a former intensive care physician and the founder of the Mobile Emergency Service for Resuscitation and Extrication.

10 Center for Urban and Regional Sociology (CURS). "Survey at national level" (April 2020). Retrieved May 10, 2020, from https://curs.ro/ro/cum-reationeaza-romanii-la-epidemie-aprilie-2020/

11 Romanian Institute for Evaluation and Strategy (IRES). "State and authority. The trust in personalities and institutions. An evaluation of measures." (April 2020). Retrieved May 12, 2020, from https://ires.ro/articol/386/romania-in-pandemie-partea-i

12 Ibidem.

13 Center for Urban and Regional Sociology (CURS). "Survey at national level" (April 2020). Retrieved May 10, 2020, from https://curs.ro/ro/cum-reationeaza-romanii-la-epidemie-aprilie-2020/

14 Telverde (Green Line) is a telephone number used for informing Romanian citizens about various social issue. In this case, it was used for providing information about the coronavirus pandemic.

Chapter 12
Social Media and Digital Information Sources in News Coverage of Crisis, Disaster, and Emergency Situations:
A case study from Spain

Marcos Mayo-Cubero
https://orcid.org/0000-0001-6306-3090
Nebrija University, Spain

ABSTRACT

Journalists play a key role in digital emergency situations. In the midst of the chaos generated by a pandemic like the COVID-19 crisis, the information transmitted by the journalist is crucial to save lives and minimize damage. This chapter explores journalists' reliance on and trust in official and unofficial digital information sources. Specifically, the authors explore the authorities' official information sources involved in crisis management and the unofficial information sources (i.e., victims and those affected). The research suggests a model for journalists' relationship with their digital information sources in covering crises by exploring dependence, type of contact, trust, and purpose. It also synthesizes the main communication errors of the public authorities that manage the response to the crisis.

INTRODUCTION

Journalists and news media play a crucial role in the coverage of the crisis, disaster, and emergency situations. Amid the chaos generated by crises, like the current COVID-19 pandemic, an earthquake, or a chemical escape, information is a vital tool to save lives and mitigate harm. Until the massive penetration of the internet and social media in contemporary societies, the public needed news outlets intermediation to be adequately informed. Nevertheless, in the current digital media ecosystem, mass media could

DOI: 10.4018/978-1-7998-6705-0.ch012

have lost their monopoly between information sources and the public. This chapter explores journalistic coverage of crises, disasters, and emergencies in the digital world through a case study in Spain. The authors use this methodological approach because case studies are a common and effective method of descriptive research in the scientific field of crisis communication (Coombs & Holladay, 2010).

The empirical evidence in designing this chapter has been collected through a mixed methodology that integrates several research techniques with a quantitative (survey conducted through a panel of experts) and a qualitative approach (analysis of the journalistic coverage of the recent crises in Spain). The authors have built the theoretical frame of this research by a selection of emergencies analyzed based on their journalistic, social, and political relevance. They are representative of situations like an epidemic, a public health alarm, a food crisis, or natural and human disasters. This chapter considers that the breadth of the sample analyzed includes all the typology present in the scientific literature on risk communication and, specifically, the field of communication in crisis contexts. The quantitative data come from a nationwide survey applied to the 30 most relevant Spanish news outlets in four media mediums: television, radio, press, and online media. Therefore, the sampling was designed according to the relevance of their audience by taking four information sources: OJD (press), EGM (radio), Kantar Media (television), and ComScore (online media). The authors selected the seven newspapers with the highest circulation (OJD), the nine television outlets with the highest audience (Kantar Media), the four radio outlets with the highest audience (EGM), and the ten digital native media with the highest online consumption (ComScore). The team conducted the research through an online questionnaire answered by the editors-in-chief of the newsrooms. The news outlets selected for the sample meet two criteria: national scope and generalist sense. The global response rate achieved was 76 percent, and the partial response rates in each of the mediums were over 50%. The literature review lets to qualify it as highly satisfactory (Keyton, 2019: 154) (Denscombe, 2017: 186). Specifically, this research focuses on the use of official information sources (those coming from public institutions involved in the communication management of the crisis: politicians, civil servants, officials, police, firemen, health workers, etc.) and non-official sources (victims and those affected and general public) in crisis, disaster and emergencies. The following six research questions guide this chapter:

RQ 1: How is the journalist's reliance on digital information sources when covering crises?
RQ 2: How do journalists contact to digital information sources in crises?
RQ 3: How much trust do journalists have in official and unofficial digital information sources on covering crises?
RQ 4: For what purpose do journalists use information from social media in crises?
RQ 5: How do journalists exercise social responsibility in the digital coverage of crises?
RQ 6: What are the main communication errors made by the authorities in crises?

Background

The journalistic coverage of these crises and emergencies is situated in the theoretical framework established by Beck's Social Risk Theory (2002). The media are involved in dynamics, of particular relevance due to their social function, when they make visible the risks that escape the superficial perception of the public, such as genetic manipulation, nuclear hazard, chemical contamination, climate change, etc. (Mythen, 2004). If the press does not report on this, society's knowledge of the risk is greatly diminished. Therefore, a correlation is established between the social responsibility of the media and their orientation towards generating a public, critical, and reflexive debate on these risks (Mayo-Cubero, Lavín, & Gallardo-Camacho, 2017). In this sense, it is worth reflecting on the social transcendence of journalistic coverage in risk situations generated by crises, emergencies, and disasters. Journalist companies are not mere information transmission belts, but actors who influence the process of building social reality (Quarantelli, 2000). For instance, determining the place of a news item on the front page of the newspaper or its site in the TV news playlist determines the public's perception of risk. A communicative process that, of course, does not end the moment the audience receives the message. Society's interpretations and reactions condition, in turn, the decision-making process of the agents in charge of risk management (Arkin, 1989: 128). As for the informative narrative, the theoretical review allows us to detect standard features synthesized in the idea of the primacy of the spectacular over the informative (Camps, 1999), (Coté & Simpson, 2000), (Hight & Smyth, 2003), (Rodríguez & Odriozola-Farré, 2012)

The global pandemic generated by COVID-19 has been accompanied by an unprecedented infodemic. Thus, a sound reflection on the social responsibility of journalists in crisis situations is more necessary than ever. A digital ecosystem in which information flows and easily overcomes technical, social, and geographical borders, and established a new framework to move through and beyond journalism as it has traditionally been conceptualized and practiced, and understanding of the myriad of practices that make up journalism (Deuze & Witschge, 2018: 165). Therefore, it is crucial to understand the role of digital platforms through a communicational perspective by understanding the journalist's needs and the use of these online social networks in these risky scenarios. According to the norms and values that guide their work, it is essential to understand and respect the perspective of journalists. A more holistic study of how journalists cover crises and disasters concludes that journalists experience coverage of any crisis as journalists, leaders, victims, and community members (Tandoc & Takahashi, 2018: 917). Both personal and professional experiences are inextricably linked and cannot be understood without each other (Mayo-Cubero, 2017). However, very little has been written about best practice approaches to news media coverage of such events (Ewart & McLean, 2019: 1). Above all, it is considering the current media ecosystem that includes social networking platforms in addition to traditional media. In this sense, we cannot escape the impact that misinformation and disinformation have on the credibility of the information. Thus, managing the informative sources from social media platforms is vital for journalists to build honest, truthful, and well-balanced news stories.

In the current context generated by the COVID-19 crisis, researchers have identified the need for more scientific evidence to examine the relationship between trust in information sources and misinformation (Stecula, Kuru, & Hall Jamieson, 2020: 2). A recent cross-national study of six countries (Argentina, Germany, South Korea, Spain, the UK, and the US) has shown that while news organizations are the single most widely identified source of news and information about coronavirus, many say do not rely on them. Other sources, including governments (national and local), health authorities (international and national), and experts, are as reliable as the news (Nielsen, Fletcher, Newman, Brennen, & Howard, 2020: 9). Experts have also expressed a lack of a deeper understanding of the fundamental role played by social media in crises (Reuter, Stieglitz, & Imran, 2020: 4). Concerning platforms such as Twitter, Facebook, LinkedIn, YouTube, TikTok, WhatsApp, etc., it is interesting to explore whether the evidence suggests any connection between the increased public confidence in the media as a source of information in crisis situations. Or if, on the contrary, the population relies more on its "personal" sources of information from social networks (e.g., family and friends) when a global health crisis such as the current one occurs. Specifically, researchers have shown interest in determining whether this evidence shows consistency with the so-called "trust nexus" theory (Hanitzsch, Van Dalen, & Steindl, 2018: 19). In other words, citizens' trust in the media and citizens' trust in political institutions are closely related. Currently, these conclusions would be very significant because they are essential for policymakers when designing practical policies to fight misinformation caused by the COVID-19 crisis. In the specific case of Spain, according to Comscore (2020), during the last week of March 2020, the consumption of information from social media grew by 55%. Current research has detected the absence of early announcements, which caused a very abrupt transition to the next stage of the crisis (Costa-Sánchez & López-García, 2020), (Ibáñez-Peiró, 2020).

The COVID-19 pandemic has once again highlighted the importance of studying the dependence and trust that journalists place in digital sources of information. In this sense, one of the essential sources for information on international health organizations. An observation of the speeches of major world leaders such as President Trump or President Macron reveals that, on occasion, political leaders have publicly discredited their data or directly they have ignored their warnings. For instance: "Trump calls WHO's the global death rate from coronavirus 'a false number'" (The Guardian, 2020). Or "WHO chief warns that countries are not taking coronavirus seriously: 'This is not a drill'" (France24, 2020). For the national media, the victims of the pandemic are human interest stories with names and surnames. A powerful example was the front page of the New York Times: "They weren't just names. They were us" (Figure 1). This chapter also explores, from a journalist's perspective, what the most effective, reliable, and consistent digital sources for engaging the public with accurate, timely, and truthful news.

FINDINGS

Reliance on Information Sources in Digital Crises

As for the reliance on digital news sources (Table 1), findings suggest that more than half of the respondents usually (57%) or always (39%) use the digital news outlets as a source of information in the coverage of the crises. Journalists mostly always choose digital official sources accounts as an information source (26%) or usually (48%). However, digital unofficial sources accounts do not get the same reliance on. Despite the risks related to the reliability of the data provided by unofficial sources, professionals suggest that they always (17%) or usually (35%) rely on them, and nearly half say that they are rarely consulted (48%). It is especially significant that the option of never records 0% in the entire results matrix.

Table 1. Reliance on digital information sources

	Digital news outlets	**Digital official sources accounts**	**Digital unofficial sources accounts**
Always	39%	26%	17%
Usually	57%	48%	35%
Rarely	4%	26%	48%
Never	0%	0%	0%
DK/NA	0%	0%	0%

Source: Adapted from (Mayo-Cubero, 2019a)

Figure 1. The front page of The New York Times for May 24, 2020

The New York Times

U.S. DEATHS NEAR 100,000, AN INCALCULABLE LOSS

Contact With Information Sources in Digital Crises

Outcomes are robust in pointing out that contact with information sources in digital crises must be widespread. Specifically, journalists say that the primary contact is always made (57%) through the press conference, that is, in a face-to-face relationship (Table 2). Respondents stressed that they would always contact the Communications department (48%), the official social media accounts (43%), and obtain the most up-to-date data from the official web (35%). It is especially significant that the option of never records 0% in the entire results matrix.

Table 2. Types of contact with information sources

	Press conferences	Communications department	Official social media accounts	Official web
Always	57%	48%	43%	35%
Usually	35%	39%	39%	48%
Rarely	9%	13%	17%	17%
Never	0%	0%	0%	0%
DK/NA	0%	0%	0%	0%

Source: Adapted from (Mayo-Cubero, 2019a)

Trust in Digital Information Sources in Digital Crises

By analyzing trust in information sources, most journalists trust the official web more than the official social media accounts (Table 3). Practically 70% usually or always rely on the data provided by the official website (61% and 9%, respectively). As for official social media accounts (mainly Twitter and Facebook), they offer reasonably reliable data for more than half of the respondents (57%). However, findings recorded concerning the unofficial social media accounts (victims, affected parties, citizens, etc.) are radically different. More than 80% of journalists claim to rarely or never trust in the unofficial social media accounts as a source of information in the coverage of digital crises (59% and 23%, respectively).

Table 3. Trust in digital information sources

	Official web	Official social media accounts	Unofficial social media accounts
Always	9%	4%	0%
Usually	61%	57%	18%
Rarely	22%	35%	59%
Never	9%	4%	23%
DK/NA	0%	0%	0%

Source: Adapted from (Mayo-Cubero, 2019a)

Trust in Unofficial Social Media Accounts in Digital Crises

Journalists give relatively low credibility to unofficial social media accounts as a source of information in emergency environments (Table 4). 39% of respondents usually trust websites, and 35% trust Twitter as a source of information in crises. The degree of reliability of trust is decreasing in the other platforms analyzed. YouTube records rarely trust for 78% of the respondents. Facebook is the social network that inspires the least confidence in journalists in this type of coverage: almost 80% of journalists have rarely or never confidence (57% plus 22%). Significantly, Instagram generates more trust than Facebook and YouTube: 22% trust Instagram as a source of information, usually (22%), and rarely (65%). WhatsApp, in the category of others, generates very little confidence to journalists as a source of information: 82% have rarely or never trust the platform.

Table 4. Trust in unofficial social media accounts

	Web	Twitter	YouTube	Facebook	Instagram	Others: WhatsApp, Reddit, Pinterest ...
Always	9%	4%	0%	0%	0%	0%
Usually	39%	35%	17%	17%	22%	4%
Rarely	52%	57%	78%	57%	65%	65%
Never	0%	0%	0%	22%	9%	17%
DK/NA	0%	4%	4%	4%	4%	13%

Source: Adapted from (Mayo-Cubero, 2019a)

Goals of Using Digital Information Sources in Digital Crises

In terms of the purpose of use (Table 5), journalists mostly use the information obtained from digital information sources in the journalistic coverage of the crisis to make contacts (65%). Besides, almost half of those surveyed said that the data extracted from digital information sources are used to verify official data (43%). In the same proportion (43%), information from digital information sources is considered to generate news by itself. At this point, the authors want to clarify that the question in the questionnaire to which Table 5 answers was a closed choice. The process followed for the construction of the survey questionnaire was as follows. First, the bibliographic review allowed the design of a first draft that was evaluated by a panel of three experts in the field of research. Then, a second draft was written with significant improvements over the first one, which was applied to a sample of experts of ten journalists in the four mediums (press, TV, radio, and online). The feedback of the experts in this second test served to collect all the nuances in the answers, refine the writing, and improve the clarity to ensure the validity and reliability of the data collection instrument.

Main Communication Errors of the Authorities in Digital Crises

Findings suggest that the main communication error of the authorities in crises is the shoddy work of the spokesperson (Table 6). Journalists say that other mistakes of the authorities in crisis information management are the lack of information coordination between the different actors involved (28%), the dissemination of contradictory messages (26%), the political use (23%), and the spreading of incorrect data (15%).

Table 5. Goals of using digital information sources

Get contacts	65%
Verify information	43%
Information source with the proper entity	43%
DK/NA	4%

Source: Adapted from (Mayo-Cubero, 2019a)

Table 6. Main communications errors of the authorities

Poor spokesperson	35%
Lack of information coordination	28%
Contradictory messages	26%
Politicizing the crisis	23%
Incorrect data	15%

Source: Adapted from (Mayo-Cubero, 2019b)

DISCUSSION

Hoaxes, fake news, misinformation, rumors, and disinformation are multiplying in a crisis context. Nowadays, a new media ecosystem is emerging under the powerful impact of social media platforms. Thus, the authors consider it essential to explore how journalists manage the official information sources of the authorities involved in the crisis management response and unofficial information sources. Our research suggests a model for journalists' relationship with their digital information sources in covering crises by exploring dependence, type of contact, trust, and purpose.

Regarding reliance on digital news sources, findings suggest that more than half of the respondents usually (57%) or always (39%) use the digital news outlets as a source of information in the coverage of the crises. Journalists mostly always choose digital official sources accounts as an information source (26%) or usually (48%). Despite the risks related to the reliability of the data provided by unofficial sources, professionals suggest that they always (17%) or usually (35%) rely on them, and nearly half say that they are rarely consulted (48%). The findings suggest that the crisis narrative is mainly constructed with data from official digital sources. However, journalists do not give up on unofficial digital sources to contrast and verify these official data. Thus, news media flee from the rigor and focus on facilitating audience access to experiencing the instability generated by the disaster. For instance, in the audiovisual narrative, the abuse of repetition of image sequences of destruction, suffering, and looting is reiterative. These dynamics tend to be accelerated by the pressures (media, social, political, and economic) that influence all the actors involved.

The importance of providing contextualized, accurate, and responsible information to assist in disaster recovery and reduce vulnerability to future disasters leads journalists to shy away from unreliable sources. The emergence of social media has forever changed the relationship between information sources and journalists. Therefore, it is particularly significant that more than half of the journalists still consider the press conference as the priority medium of obtaining reliable information (57%). Despite the new communicational paradigm that is taking shape in the digital society, the direct, face-to-face relationship of the journalist with the emergency spokesperson cannot be replaced by an indirect relationship mediated by a social media platform (Mayo-Cubero, 2019a: 51).

Journalists informing in crisis situations handle extremely sensitive data such as the number of victims. Outcomes are robust in pointing out that contact with information sources in digital crises must be widespread and persistent. Specifically, journalists say that the primary contact is always made (57%) through the press conference, that is, in a face-to-face relationship. Respondents stressed that they should always contact the Communications department (48%), the official social media accounts (43%), and obtain the most up-to-date data from the official web (35%). The authors deduce that this could be due to the complex and extreme particularity of the informative scenarios of the crises. Therefore, they are chaotic environments submitted to the impact of all types of pressures (economic, political, social, international) by a variety of informative agents (official, non-official, political, technical, press, NGOs, victims, affected). Thus, it is challenging for the information source to overcome all the barriers by circulating exclusively through formal channels and manage to provide reliable and truthful information.

Regarding trust in digital information sources. 70% of journalists usually or always trust in the data provided by the official website. As for official social media accounts (mainly Twitter and Facebook), they provide relatively reliable data for more than half of the respondents (57%). However, findings recorded about the unofficial social media accounts (victims, affected parties, citizens, etc.) are radically different. More than 80% of journalists claim to rarely or never trust the unofficial social media accounts as a source of information in the coverage of digital crises. Therefore, in risky scenarios, phenomena such as fake news, disinformation, and hoaxes intensify the distrust of journalists.

News media professionals give relatively low credibility to unofficial social media accounts in emergency environments. 39% of respondents usually trust websites, and 35% trust Twitter as a source of information in crises. Trust decreases in the other platforms analyzed. Facebook is the social media platform that inspires the least confidence in journalists in this type of coverage. Significantly, Instagram generates more trust than Facebook and YouTube: 22% trust Instagram as a source of information, usually (22%), and rarely (65%). WhatsApp, in the category of others, generates very little confidence to journalists as an information source: 82% have rarely or never trust the platform.

Regarding platforms such as Twitter, Facebook, LinkedIn, YouTube, TikTok, WhatsApp, etc., the research outcomes suggest that an increase in the trust of social media as an information source correlates with an increase in confidence in family and friends and local government as a news source. It would be particularly interesting as a research strand in the future to explore whether these conclusions show consistency with the so-called 'trust nexus' theory: that citizens' trust in the media and political institutions is closely linked (Hanitzsch et al., 2018: 19).

In other words, Twitter and web portals are the sources that generate the most confidence to search for information in crisis scenarios. Nevertheless, daily journalistic practices in newsrooms reveal a contradiction. It has been observed how Facebook and YouTube are platforms widely used by the media to obtain amateur video images in the recent crisis, emergency, and disaster situations (Mayo-Cubero, 2017: 225). Regarding purpose, most journalists use information from social media in crises to obtain contacts (65%) and verify information (43%). Although it has been shown that the view that the media's role is to cooperate with the government to communicate during emergencies, even if this means sacrificing their watchdog role (Okumura, Hayashi, Igarashi, & Tanaka, 2019). Social responsibility should guide the transmission of the information message to an affected community, usually under severe post-traumatic stress, and in need of accurate information about the survivors and how to manage their resilience.

In this sense, it is significant that previous research has shown that the use of non-official sources by the media takes precedence over official government sources in the news narrative in crisis coverage. It is especially relevant that journalists give the same credibility as an informative source to public institutions as they do to victims (Mayo-Cubero, 2020: 8). In a media ecosystem dominated by the presence of fake news, hoaxes, and misinformation, it is essential to identify the main communication errors in crises to combat them. The authors believe that the better the communication policy of the authorities in charge of managing the crisis, the better the media coverage will be. In other words, better communication is in the interest of all actors: authorities, media, and the public.

Describing what the main communication errors are and what their relevance is the first step to be able to solve them. Findings suggest that the main communication error of the authorities in crises is the poor quality job of the spokesperson. Journalists say that other mistakes of the authorities in crisis information management are the lack of information coordination between the different actors involved (28%), the dissemination of contradictory messages (26%), the political use (23%), and the spreading of incorrect data (15%). In the specific case of Spain, the government's failure to anticipate the negative scenarios towards which the pandemic was heading, and the coincidence with the peak in the number of cases during April, generated a situation of alarmism and fear in the public. It can be inferred that the initial messages of calm in the face of the impact of the pandemic, transmitted by the authorities, overprotected the population and did not adequately prepare it for the worst possible scenario.

These outcomes suggest a possible model to solve the main communication errors made by the authorities in digital crises. It is highly recommended to conduct communications through a single spokesperson who specializes in crisis communication and is trained to work with journalists. Avoiding the formula of multiple spokespersons would also reduce the lack of information coordination, contradictory messages, and the dissemination of incorrect data. Finally, the best way to avoid the politicization of a crisis is that its management is always guided by technical, expert, and professional criteria.

By analyzing the on-going news coverage of the COVID-19 crisis, this chapter observes that a recent cross-national study of six countries (Argentina, Germany, South Korea, Spain, the UK, and the US) has shown that while news organizations are the single most widely identified source of news and information about coronavirus, many say do not rely on them. And other sources, including governments (national and local), health authorities (international and domestic), and experts, are as reliable as the news (Nielsen et al., 2020: 9). In this regard, a question arises: does the sharp increase in daily information consumption that we are witnessing during this pandemic correlate with trust in official information sources?

As the authors pointed out, the COVID-19 crisis has once again highlighted the importance of studying trust in digital sources of information. Specifically, one of the essential information sources is international health organizations. This chapter deduces that the fact that WHO, precisely because of its global nature, is perceived by citizens as being indifferent towards national tragedies. To national media outlets, victims are human interest news stories with names and surnames—for instance, The New York Times' front page (Figure 1). Conversely, the same victims are to WHO just cold statistics. This implication is critical for international health communication officers. Rigorous reflection on content and channels is necessary to ensure that health messages reach the public. Moreover, one question arises: would the WHO's messages reach better if that cold statistics were more human?

CONCLUSION

The chapter helps readers to understand the coverage of crises and disasters by journalists through digital platforms in Spain. Thus, the authors propose a model for journalists covering crises and disasters by exploring different sources/contacts for obtaining crisis information, type of contact, trust in such connections and purpose. Also, despite the on-going changes in sourcing for news in the digital world, journalists still rate press conferences higher than social media sources for obtaining reliable information. Getting crisis information first-hand is a great way to tackle misinformation and fake news.

From the perspective of public communication, the authors suggest a possible model to solve the main communication errors committed by the authorities in digital crises. It is strongly recommended that communications be conducted through a single specialized spokesperson to reduce informative disco-ordination, contradictory messages, and inaccurate data. Finally, the best way to avoid the politicization of a crisis is with communication management guided by technical, expert, and professional criteria.

ACKNOWLEDGMENT

This research received no specific grant from any funding agency in the public, commercial, or not-for-profit sectors.

REFERENCES

Arkin, E. (1989). Translation of Risk Information for the Public: Message Development. In V. Covello, D. McCallum, & M. Pavlova (Eds.), *Effective Risk Communication. Contemporary Issues in Risk Analysis* (pp. 127–135). Springer. doi:10.1007/978-1-4613-1569-8_19

Beck, U. (2002). *La sociedad del Riesgo Global*. Siglo XXI Editores.

Camps, S. (1999). *Periodismo sobre catástrofes*. Ediciones Paulinas.

Coombs, W. T., & Holladay, S. J. (2010). *The Handbook of Crisis Communication*. WILEY Blackwell. doi:10.1002/9781444314885

Coté, W., & Simpson, R. (2000). Covering violence. A guide to ethical reporting about victims and trauma. New Yok: Columbia University Press.

Denscombe, M. (2017). *The Good Research Guide: for small-scale social research projects* (6th ed.). McGraw-Hill Education.

Deuze, M., & Witschge, T. (2018). Beyond journalism: Theorizing the transformation of journalism. *Journalism: Theory. Practice & Criticism*, *19*(2), 165–181. doi:10.1177/1464884916688550 PMID:29417952

Ewart, J., & McLean, H. (2019). Best practice approaches for reporting disasters. *Journalism*, *20*(12), 1573–1592. doi:10.1177/1464884918757130

Hanitzsch, T., Van Dalen, A., & Steindl, N. (2018). Caught in the Nexus: A Comparative and Longitudinal Analysis of Public Trust in the Press. *The International Journal of Press/Politics*, *23*(1), 3–23. doi:10.1177/1940161217740695

Hight, J., & Smyth, F. (2003). *Tragedias & Periodistas. Guía para una cobertura más eficaz*. Dart Center for Journalism and Trauma.

Keyton, J. (2019). *Communication Research: Asking Questions, Finding Answers* (5th ed.). McGraw-Hill Education.

Mayo-Cubero, M. (2017). *Journalistic specialization as a strategic tool in disaster communication*. Complutense University of Madrid. https://eprints.ucm.es/42138/

Mayo-Cubero, M. (2019a). Use of social media in news media coverage of the crisis, disaster, and emergencies in Spain. *Revista Española de Comunicación en Salud, 1*, 43–54. doi:10.20318/recs.2019.4428

Mayo-Cubero, M. (2019b). Assessment of journalistic coverage and public institutional communication of the most recent crises and emergencies in Spain. *University of Seville and Spanish Association of Investigation in Communication Conference,* 1–10. https://www.researchgate.net/publication/334479921_Assessment_of_journalistic_coverage_and_public_institutional_communication_of_the_most_recent_crises_and_emergencies_in_Spain_Evaluacion_de_la_cobertura_periodistica_y_de_la_comunicacion_institucional

Mayo-Cubero, M. (2020). News sections, journalists and information sources in the journalistic coverage of crises and emergencies in Spain. *El Profesional de la Información, 29*(2), 1–12. doi:10.3145/epi.2020.mar.11

Mayo-Cubero, M., Lavín, E., & Gallardo-Camacho, J. (2017). The news coverage of the Lorca Earthquake in Spain: the journalist's social responsibility. *Ámbitos. Revista Internacional de Comunicación, 35*, 1–17. https://idus.us.es/handle/11441/66415

Mythen, G. (2004). *Ulrich Beck: A Critical Introduction to the Risk Society*. Pluto Press.

Nielsen, R. K., Fletcher, R., Newman, N., Brennen, J. S., & Howard, P. N. (2020). *Navigating the "Infodemic": How People in Six Countries Access and Rate News and Information about Coronavirus.* https://reutersinstitute.politics.ox.ac.uk/infodemic-how-people-six-countries-access-and-rate-news-and-information-about-coronavirus

Okumura, N., Hayashi, K., Igarashi, K., & Tanaka, A. (2019). Japan's media fails its watchdog role: Lessons learned and unlearned from the 2011 earthquake and the Fukushima disaster. *Journalism*, (1), 2–6. doi:10.1177/1464884919891270

Quarantelli, E. L. (2000). *Emergencies, Disaster and Catastrophes are different phenomena.* DRC Preliminary Paper.

Reuter, C., Stieglitz, S., & Imran, M. (2020). Social media in conflicts and crises. *Behaviour & Information Technology, 39*(3), 241–251. doi:10.1080/0144929X.2019.1629025

Rodríguez, P., & Odriozola-Farré, B. (2012). Catástrofes y periodismo: El relato, los escenarios, las interacciones y las necesidades prácticas y psicológicas de todos los implicados. *Estudios sobre el Mensaje Periodístico, 18*(2), 577–594. doi:10.5209/rev_ESMP.2012.v18.n2.41033

Stecula, D. A., Kuru, O., & Hall Jamieson, K. (2020). How Trust in Experts and Media Use Affect Acceptance of Common Anti-Vaccination Claims. *The Harvard Kennedy School (HKS). Misinformation Review, 1*(1), 1–11. doi:10.37016/mr-2020-007

Tandoc, E. C. Jr, & Takahashi, B. (2018). Journalists are humans, too: A phenomenology of covering the strongest storm on earth. *Journalism, 19*(7), 917–933. doi:10.1177/1464884916657518

Trump calls WHO's global death rate from coronavirus 'a false number.' (2020, March 5). *The Guardian.* Retrieved from https://bit.ly/2yNALff

WHO chief warns that countries are not taking coronavirus seriously: 'This is not a drill.' (2020, March 6). *France24*. Retrieved from https://bit.ly/2S98uXH

ADDITIONAL READING

Bavel, J. J. V., Baicker, K., Boggio, P. S., Capraro, V., Cichocka, A., Cikara, M., Crockett, M. J., Crum, A. J., Douglas, K. M., Druckman, J. N., Drury, J., Dube, O., Ellemers, N., Finkel, E. J., Fowler, J. H., Gelfand, M., Han, S., Haslam, S. A., Jetten, J., ... Willer, R. (2020). Using social and behavioural science to support COVID-19 pandemic response. *Nature Human Behaviour*, *4*(May), 460–471. Advance online publication. doi:10.103841562-020-0884-z PMID:32355299

Bode, L., & Vraga, E. K. (2018). See Something, Say Something: Correction of Global Health Misinformation on Social Media. *Health Communication*, *33*(9), 1131–1140. doi:10.1080/10410236.2017.1331312 PMID:28622038

Casero-Ripollés, A. (2020). Impact of Covid-19 on the media system. Communicative and democratic consequences of news consumption during the outbreak. *El profesional de la información*, 29, 2, e290223. doi:10.3145/epi.2020.mar.23

Cottle, S. (2009). *Global Crisis Reporting: Journalism in the Global Age*. Open University Press.

Donovan, J. (2020). Social-media companies must flatten the curve of misinformation. *Nature*. Advance online publication. doi:10.1038/d41586-020-01107-z PMID:32291410

Gallotti, R., Castaldo, N., Valle, F., Sacco, P., & De-Domenico, M. (2020). COVID19 Infodemics Observatory. doi:10.17605/OSF.IO/N6UPX

Glik, D. C. (2007). Risk Communication for Public Health Emergencies. *Annual Review of Public Health*, *28*(1), 33–54. doi:10.1146/annurev.publhealth.28.021406.144123 PMID:17222081

Lazer, D. M. J., Baum, M. A., Benkler, Y., Berinsky, A. J., Greenhill, K. M., Menczer, F., Metzger, M. J., Nyhan, B., Pennycook, G., Rothschild, D., Schudson, M., Sloman, S. A., Sunstein, C. R., Thorson, E. A., Watts, D. J., & Zittrain, J. L. (2018). The science of fake news. *Science*, *359*(6380), 1094–1096. doi:10.1126cience.aao2998 PMID:29590025

Lee, S. T., & Basnyat, I. (2013). From Press Release to News: Mapping the Framing of the 2009 H1N1 A Influenza Pandemic. *Health Communication*, *28*(2), 119–132. doi:10.1080/10410236.2012.658550 PMID:22439616

Reuter, C., Stieglitz, S., & Imran, M. (2020). Social media in conflicts and crises. *Behaviour & Information Technology*, *39*(3), 241–251. doi:10.1080/0144929X.2019.1629025

Vosoughi, S., Roy, D., & Aral, S. (2018). The spread of true and false news online. *Science*, *359*(6380), 1146–1151. doi:10.1126cience.aap9559 PMID:29590045

Zarocostas, J. (2020). How to fight an infodemic. *Lancet*, *395*(10225), 676. doi:10.1016/S0140-6736(20)30461-X PMID:32113495

KEY TERMS AND DEFINITIONS

Authorities: Generally, public organizations involved in crisis response management (political, police, firefighters, health, etc.) They may have a regional, national, or international dimension.

Crisis: In the context of risk communication, this dynamic situation of change can be generated by the impact of a natural phenomenon (hurricane, earthquake, epidemic, etc.) or a human one (industrial accident, terrorist attack, etc.).

Digital Official Information Source: Digital information source belonging to the organizations involved in response to the crisis (government, police, firefighters, etc.) This source will usually transmit the data through social networking platforms and their official accounts (Twitter, Facebook, YouTube, etc.)

Digital Unofficial Information Source: Digital information source from victims and those affected by the crisis. This source will generally transmit the data through social network platforms and their unofficial accounts (Twitter, Facebook, YouTube, etc.)

Reliance: Dependence on a source of information. Accurately, it measures the frequency of use of the information source.

Spokesperson: Person(s) in charge of transmitting to journalists and the public the organization's official position.

Trust: Credibility entitled to the data obtained from an information source.

Chapter 13
Communication During a Pandemic:
An Analysis Through the Lenses of Brand Management Strategy

Cássia Liandra Carvalho
University of Aveiro, Portugal

Belem Barbosa
https://orcid.org/0000-0002-4057-360X
University of Aveiro, Portugal

ABSTRACT

Although the literature on crisis communication is quite vast, business communication related to global crises (e.g., natural disasters) is largely unexplored. This chapter aims to fill this gap and shed light on brand communication strategies during a pandemic. A netnographic study was carried out with the purpose of identifying brand positioning and communication strategies during the COVID-19 pandemic outbreak and of understanding the engagement of brands' followers during that period. The study included four brands of large Brazilian companies and comprised the analysis of brands' feed on Instagram during the first five weeks of the outbreak in Brazil. Findings enable to identify two distinct profiles: unprepared brands and leading brands. The chapter provides valuable clues for both managers and researchers dealing with crisis communication.

INTRODUCTION

In order to become consumers' choice, a brand must hold important and consolidated attributes. It is essential for a successful brand to have a consolidated positioning in the minds of the customers that reflects its values, its vision for the future, and its essence. Brand positioning refers to the unique space that a brand occupies in customers' minds (Aaker, 1997; Pogorzelski, 2018; Rajagopal, 2019; Ries & Trout, 1981; Wang, 2015). It enables customers to see a specific brand in a unique way, to associate

DOI: 10.4018/978-1-7998-6705-0.ch013

the brand with emotions, characteristics, and feelings, and consequently to distinguish the brand from its competitors. Another essential brand attribute is the recognition by its target markets. Brand awareness identifies how brands are perceived by their customers (Aaker, 1997; Arli, 2017; Cheung, Pires, & Rosenberger, 2019; Romaniuk, Wight, & Faulkner, 2017) and portrays how each company is associated with its products. But both brand attributes are volatile, particularly during crises. In fact, crises can directly impact a brand, changing the course of its entire marketing planning.

Although the literature on crisis communication is quite vast, it appears that most of it is related to internal crises, that is to say, those that are for instance generated by rumours and events and that are specific to a company or brand (Bundy, Pfarrer, Short, & Coombs, 2016; Coombs, 2006). Thus, business communication related to global crises (e.g., natural disasters) is largely unexplored. There are relevant contributions regarding communication crises in social network sites and also on the multiplication of messages during Hurricane Katrina (Garnett & Kouzmin, 2009; Macias, Hilyard, & Freimuth, 2009; Veil & Husted, 2012) and the epidemic H1N1 (Freberg, Palenchar, & Veil, 2013; Lin, Jung, McCloud, & Viswanath, 2014; Lin, Savoia, Agboola, & Viswanath, 2014). Still, little is known about brand positioning strategies during natural disasters and pandemics.

Since the end of 2019, news about COVID 19 or Coronavirus have gained worldwide attention and spread throughout the world. The first case was reported in Wuhan city, in the Chinese province of Hubei, and in March 11th the World Health Organization declared the outbreak as a pandemic. As a consequence, several countries were quarantined for a period of time, impacting global supply chains (Ajami, 2020; Raffaetà, 2020) and the world economy. Brazil had its first case recorded on February 26th (Aquino & Monteiro, 2020) and its first death on March 17th (BBC News, 2020). From its start, the financial markets have been seriously impacted by the pandemic, making it not only an important health and social issue, but also an economic and financial issue that worried companies all over the world.

This chapter main purpose is to understand brand communication strategies and the consequences on brand followers' engagement during a pandemic period. Exploratory netnographic research was conducted considering four brands from large Brazilian companies in different sectors. This research comprised the analysis of brands' feed on Instagram. Data was collected for five weeks, between February (when the first case of COVID-19 was registered in Brazil) and April 1st 2020, and analyzes 155 posts from 4 different brands and a total of 33,168 comments.

This study makes important contributions to the literature. It approaches the unexplored topic of brand communication during global crises, which despite the relevance given by practitioners has been so far disregarded by researchers. It provides anecdotal evidence of several strategies that brands adopt in the context of a global crisis such as the one related to the Covid-19 pandemic. By analysing communication strategies and further relating it to effectiveness indicators of social media communication, this study provides relevant cues for both academics and practitioners, namely by identifying alternative strategies and proposing their effectivess. Indeed, these outputs provide interesting suggestions for both future researchers and strategic communication managers of companies and brands from diverse sectors, namely the ones mostly affected by the pandemic.

The chapter is organized as follows. Next section presents a literature review which covers important concepts such as brand positioning, awareness, crisis management and communication strategies. The method section describes the study carried out and delves into the netnography approach. The results section presents the findings. Finally, the final conclusions and limitations encountered during the process are presented in the conclusion section.

LITERATURE REVIEW

Brand Positioning and Awareness

Brand positioning refers to the set of strategies in order to gain and manage a favorable position in customers' minds (Aaker, 1997; Hartmann, Apaolaza-Ibáñez, & Forcada-Sainz, 2005; Jalkala & Keränen, 2014; Kotler & Keller, 2012; Lee, Kim, & Won, 2018; Ries & Trout, 1981; Wang, 2015), that is aligned with business values and purpose. The general premise of brand positioning can be described as the process of integrating marketing communication to align the perceptions of target consumers with the brand identity (Park, Rajagopal, Dillon, Chaiy, & DeSarbo, 2017; Park, Lee, Kim, & Park, 2018) and to provide psychological meaning to the brand.

According to Pogorzelski (2018), successful positioning combines marketing actions that assign the brand to a privileged perceptual space in customers' minds. Indeed, brand positioning should help the company stand out from its competitors and highlight brand's differences. It is a strategic tool that has important consequences on the company's performance (Heinberg, Ozkaya, & Taube, 2017) and that requires a significant amount of investment. This is because brands provide experience for consumers, and from that it becomes more or less associated with a product. A successful brand positioning needs to represent brand attributes, use value and competitive differentiation into this experience (Adlakha & Sharma, 2019; Hassan & Craft, 2012; Rajagopal, 2019), generating a real connection with customers. As such, brand positioning creates competitive advantage (Pogorzelski, 2018) and attracts customers (Heinberg et al., 2017). In line with these contributions, the first research question proposed for this chapter is:

RQ1: What are the positioning strategies adopted by brands on social media during a pandemic?

Another relevant aspect to strengthen customer relationship is brand awareness, which represents the influence that the brand has on customer's mind (Aaker, 1997; Ahmed, Vveinhardt, & Streimikiene, 2017; Rahman, 2018). This is an important indicator for brands (Romaniuk et al., 2017), acknowledged as one of the main influencers in the purchase decision-making process (Cheung et al., 2019; R. Huang & Saragollu, 2014; Lou & Yuan, 2019), directly impacting the customer and brand relationship. A well-known brand has a higher chance of being chosen by consumers (Arli, 2017; Kucuk, 2011), because brand recognition is associated with popularity (Rahman, 2018).

Brand awareness involves remembering not only its name, but also the general feelings associated with it, information about its products and services, and other experiential details. Recognition refers to consumer memory structures that are associated with the brand (Arli, 2017; Thaichon & Quach, 2015) and are strengthened by marketing actions. Furthermore, it is potentially the main reason behind repeated consumer buying decisions (Dedeoğlu, van Niekerk, Küçükergin, De Martino, & Okumuş, 2019; Kucuk, 2011), as once the consumers are linked to a brand they are more likely to make repeated purchases due to the development of emotions trust (Thaichon & Quach, 2015), positive image (Cheung et al., 2019), and loyalty (Molinillo, Japutra, Nguyen, & Chen, 2017). Hence, the second research question proposed for this chapter is:

RQ2: How do brand awareness levels on social media vary during the pandemic period?

Crisis Management

Brand positioning and awareness are among the factors that most vary when the company faces a crisis (Rea, Wang, & Stoner, 2014; Seo & Jang, 2013), because consumers and stakeholders' perceptions are influenced by the facts and events that constitute the crisis. A crisis is defined here as a significant threat to brand operations that can have negative consequences if not handled properly (Bundy et al., 2016; Coombs, 2004; Coombs & Laufer, 2018). There are different types that crises that directly impact on brands: internal crises, crises generated by rumors and events produced by the company (Bundy et al., 2016; Coombs, 2009; Coombs & Holladay, 2014), and crises resulting from unexpected and global events (Coombs & Laufer, 2018; Johansson, Dimofte, & Mazvancheryl, 2012). Frequently, crises result in financial damage, negative impacts on brand reputation. In some cases it can also cause company bankruptcy. Garnett and Kouzmin (2009) explain that crisis management acts mainly in the supervision of the crisis event, mobilizing response resources, seeking advice from professionals, and calming the public fears. Crisis communication is defined as the collection and processing of relevant information in a crisis situation (Bazago, Guardia, & García, 2020; Cheng, 2018), comprising both emergency planning and preventive actions to be taken in face of the facts. During tense or critical periods of a crisis, public perceptions play a key role in crisis management (Stieglitz, Mirbabaie, & Potthoff, 2018; Veil & Husted, 2012), so they are an important focus of managers' actions.

Natural disasters have been important triggers of communication, as it was the case of hurricane Katrina (Garnett & Kouzmin, 2009; Garnett & Kouzmin, 2007; Macias et al., 2009), Japanese tsunami (Acar & Muraki, 2011; Seong Eun Cho, 2013) and the earthquake in Haiti (Muralidharan, Rasmussen, Patterson, & Shin, 2011). Communication crises were also caused by epidemics such as the H1N1 flu (Freberg et al., 2013; Lin, Savoia, et al., 2014), Ebola (Guidry, Jin, Orr, Messner, & Meganck, 2017; Ratzan & Moritsugu, 2014) and Zika (Vafeiadis, Bortree, Buckley, Diddi, & Xiao, 2019; Wardle & Derakhshan, 2017). These major global crises impact several sectors, including economic, health, education, technology, and innovation (Dzarasov, 2016; Gills & Gray, 2012; Kirchgässner, 2009; Stiglitz, 2009), having strong impacts on the markets. During a global disaster, communication processes need to be fast and close to the public, especially to avoid fake news and global panic (Vafeiadis et al., 2019).

There are different crisis communication strategies, which are used according to the needs and profile of the public. During Hurricane Katrina, communication was dynamic and clear. The main strategy of the American government was to act directly on the spot and fight disinformation in a fast and credible way (Garnett & Kouzmin, 2009; Macias et al., 2009) through formal and online media (such as websites and blogs). During the H1N1 pandemic, there was a huge lack of information, as part of the population did not know how to react to the disease (Lin, Savoia, et al., 2014) and the strategy was maingly informative. The objective was to inform communities about the severity of the disease and the measures that should be taken (Lin, Jung, et al., 2014). The same occurred during the spread of Ebola, when communication was predominantly clear and objective information about the forms of contagion (Guidry et al., 2017). During the SARS pandemic, communication sought engagement between communities and health agencies (Hayllar, 2007; Tiong, 2004).

Although extant literature provides some cues on crisis communication during natural desasters and epidemics, it should be noted that the scope and nature of coronavirus pandemic is very distinct from past situations. Indeed, none of above mentioned crises studied by extant literature compares to COVID-19 neither in their spread nor in the communication context they happened, particularly considering the use of social network sites in interpersonal and institutional communication and even recarding problems of misinformation that arguably are more common nowadays. Unlike disasters that happen in a short period, pandemics remain for a long period of time, and communication strategies are focused on keeping the population healthy, so they are always clear, objective and quick. Commonly brands participate in these initiatives, namely on social media, by posting about the topic, and eventually making recommendations and showing solidarity with the victims.

The Role of Corporate Social Responsibility During Crisis Communication

Overall, companies have an important role in crisis management. Through Corporate Social Responsibility (CSR) they create actions that intend to improve social or environmental conditions in the world (Blomgren, 2011; Carroll, 1999; Garnelo-Gomez & Saraeva, 2019; Jamali & Mirshak, 2007; Schmeltz, 2012; Tetrault Sirsly & Lvina, 2019; Ziek, 2009). Indeed, companies have been striving to create actions that have relevant impacts on local communities or even on a large scale. Actions focused on sustainability (Babiak, 2010; Kucharska & Kowalczyk, 2019; Sardana, Gupta, Kumar, & Terziovski, 2020), in gender issues (Grosser & Moon, 2005, 2019; Schaefer, Terlutter, & Diehl, 2020) and on ways to improve public health (Droppert & Bennett, 2015; Pulker, Trapp, Scott, & Pollard, 2018), just to name a few. CSR actions benefit the community but also provide advantages to the companies that implement them. There is a great interest in products and services that have ethical and philanthropic initiatives (Chowdhury, 2019; Vitell et al., 2016). Consumers appreciate companies care about social issues (Aydın, 2019; Brønn & Vrioni, 2001), particularly the organizations that have an ethical commitment and are attentive to the social and environmental impacts they generate. In addition to being within an ethical profile, companies benefit directly from CSR: increased brand value (Annan-Diab & Jensen, 2017; Hur, Kim, & Woo, 2014), da favorable reputation (Fatma, Rahman, & Khan, 2015; Lee, Chang, & Lee, 2017; Tetrault Sirsly & Lvina, 2019) and profitability (Jain, Vyas, & Chalasani, 2016; Wagner, Lutz, & Weitz, 2009).

Summing up these contributions in the literature that point to diversified strategies to deal with crises, and considering that brand communication during global disasters and pandemics have been disregarded by research the third research question is as follows:

RQ3: What is the relationship between companies' CSR strategies and their crisis communication during a pandemic?

METHOD

Guided by the objectives defined for this research and in order to tackle the research questions formulated in the previous section, a netnographic approach was carried out, comprising both online communication strategy and consumers' reactions (Sharma, Ahuja, & Alavi, 2018). Netnography is an adaptation of ethnographic research techniques to study online communities and computer-mediated social interactions (Kozinets, 2019). It provides a comprehensive understanding of the users' thinking process and behaviors on different digital platforms (Wallace, Costello, & Devine, 2018; Y. S. Wang, 2019b, 2019a; Xharavina, Kapoulas, & Miaoulis, 2019), by mapping and interpreting comments and other interactions on social media.

Netnographers use publicly available information in online forums to understand the needs of relevant online groups and to create scenarios of digital culture (Kozinets, 2002, 2013). Applied to consumer research, netnography is particularly useful to capture the complexity of consumers' interactions on digital media (Loureiro, Serra, & Guerreiro, 2019; Lugosi & Quinton, 2018) and understand the factors that influence decision making and consumer profiles (Kozinets, 2012; Sharma et al., 2018).

This chapter adopted a theory-driven approach to analyze and interpret the data. Table 1 summarizes the main contributions in the literature on brand management that guided the analysis in this chapter. As suggested by Reid and Duffy (2018), in netnography studies coding may be both inductive and deductive, particularly when, as it was the case of this study, research questions were formulated based on contributions in the literature. As such, and considering the scant literature particularly on the specificities of communication under a pandemic, the deductive coding based on the contributions in the literature was followed by an interpretivist approach, by immersing into the data and further understanding users' points of view and contexts of conversation, hence maximizing the netnographic understanding (Reid & Duffy, 2018).

Table 1. Contributions in the literature that guided the analysis

Domains	Contributions in the Literature
Brand Positioning	• Attracts customers (Hartmann et al., 2005; Heinberg et al., 2017). • Acknowledged as a competitive advantage (Jalkala & Keränen, 2014; Park et al., 2017; Pogorzelski, 2018). • Features brand's essence and values (Lee et al., 2018). • Creates a direct connection with the customers (Hartmann et al., 2005; Hassan & Craft, 2012).
Brand Awareness	• Boosts purchase decision making (Arli, 2017; Dedeoğlu et al., 2019; Kucuk, 2011) • Has a positive impact on customer trust (Thaichon & Quach, 2015), brand image (Cheung et al., 2019), and loyalty (Molinillo et al., 2017). • Generates popularity (Rahman, 2018). • Measures customer oppinions (Romaniuk et al., 2017). • Plays a decisive role in consumer behavior (Lou & Yuan, 2019).
Crisis Communication Management	• Fast and credible (Garnett & Kouzmin, 2009; Garnett & Kouzmin, 2007; Macias et al., 2009). • Objective and clear (Guidry et al., 2017; Lin, Jung, et al., 2014; Lin, Savoia, et al., 2014). • Customer engaged (Hayllar, 2007; Tiong, 2004).

Data colletion took place during 35 days, starting on the day of the notification of the first casualty of Covid-19 in Brazil, on February 26th (Aquino & Monteiro, 2020) until April 1, 2020. All posts (N = 155) published on Instagram by four Brazilian brands were analyzed, along with the comments made by Instagram users, totaling 33,168 comments. Instagram stories were not included. Instagram was chosen due to the high number of users in the country, estimated to be 91 million. Currently, Brazil is the third country in the world in terms of Instagram users (Statista, 2020). Para selecionar as marcas a serem estudadas, começou-se por identificar setores de atividade. During the period under review, few studies had been published with data on the economic and business impacts of the pandemic. It was understood that sectors such as airlines, commerce with direct sales and e-commerce, and food and beverages were amongst the ones that would be most impacted by the restrictive measures that were being implemented. After choosing the business sectors, Brazilian brands that had a digital presence, a history of long-term marketing actions, and a considerable number of followers on Instagram (e.g., more than 200 thousand followers) were identified and compared. As a result, four brands from distinct sectors were selected for this study:

- **Ambev**: Beverage industry company present in 19 countries with more than 35,000 employees, has 32 breweries, 2 malt stores, and delivers 30 beverage brands (Ambev, 2020). The most valuable company in Latin America (Economatica, 2018). During the analysis period, its Instagram page (ambev) had 233 thousand followers and 160 publications.
- **Gol:** Airline company with more than 15,000 employees and 715 daily flights through Brazilian and international territories (Gol, 2020). The 40th most valuable brand in Brazil (Barbosa, 2019; Kantar Consulting, 2019). During the analysis period, its Instagram page (voegoloficial) had 866 thousand followers and 1,856 publications.
- **Ifood:** Internet food delivery app. It operates in 912 cities, has 131.3 thousand registered restaurants, 340 thousand delivery people, and places 6.5 million orders per day (IFood, 2020). 2nd food start up with the largest monetary contribution in the world (Brigatto, 2018). During the analysis period, its Instagram page (ifoodbrasil) had 791 thousand followers and 674 publications.
- **Renner**: Fashion retail chain with more than 500 stores, international operations, and around 10,000 employees (Renner, 2020). The 9th most valuable brand in Brazil (Barbosa, 2019; Kantar Consulting, 2019). During the analysis period, its Instagram page (Lojasrenner) had 6.3 million followers and 7,637 publications.

Thus, the study included companies from different sectors and with different characteristics to allow greater diversity and the comparative analysis of different cases. The challenges posed by the pandemic were different for each of them, too. Aircraft companies were amongst the most affected by the pandemic worldwide, retail stores were closed for several weeks in many countries, while the pandemic provided opportunities for home deliversy services. As it will be shown in the next section, they also adapted their communication strategy during the pandemic in distinct ways. Through Netnography it was possible to analyze and narrate the conceptions that are generated within digital environments (Kozinets, 2016, 2019) and to understand consumer perceptions in different situations such as the reactions of consumers and brands during the crisis generated by Covid-19.

The data was extracted manually from the Instagram profile of each of the selected companies. publications in the period under analysis were identified. Text, image or video, and all comments for each publication were considered, and all data were categorized in a spreadsheet. No additional software was used. The analysis process was conducted manually, by analysing the content of each post in detail, including the comments attached to the posts. Symbols and expressions used by users in the comments were independently interpreted and coded by two researchers. After the analyzes, it was possible to build a scenario that translated the users' behavior and their feelings. In addition to the publications made by these companies on Instagram, users' interaction (e.g., likes and comments) were also considered in the data analysis, thus creating a timeline of how brands adapted to the pandemic and how consumers reacted.

RESULTS

On Fevereiro 26, the date of the first death referring to Covid-19 in Brazil, no brand made any mention of the problem. All companies posted content similar to their publishing strategies in the previous weeks. Ambev and IFood posted content related to the carnival, since the 26th was Ash Wednesday. These posts are summarized in Table 2.

Table 2. Examples of posts on the first week of the analysis

Date	Post	Views	Comments	Comment examples
2/26/2020	@ambev: Were you on the block? Graice was there too. And in this video @catracalivre heard a little bit about how that experience was. 🤍#JoyThatTransforms #ambev #BeyondLabels	11,500	54	@ gezagallo> Incridible 👏👏👏
2/26/2020	@ifoodB=brasil: Who said that carnival is over? Arrive at the block with mouth on snack! #OrderUmiFood and enjoy foods revelry. 🍔🍔🍕🥖	8.185	64	@brunascalzo: I made a purchase on February 16th, it was canceled but no refund. I'm texting every day and they just tell me to wait. Total lack of respect for the client, lack of professionalism.
2/26/2020	@lojasrenner: That perfect summer body: yours! With folds, marks and history, your body is part of who you are. Appreciating your details is an exercise to be practiced every day. Our tip is to bet on lingerie that has everything to do with you. Put this part of the production so you can admire your perfect body. Buy online and receive it at home, or pick it up at your favorite store: bra 548953835 + panties 548953051 #renner	15,400	400	@anninhaa123: What a beautiful campaign! I almost didn't notice the lingerie lol I even wanted to do a photo shoot with all my supposed imperfections.
02/27/2020	@voegol: You're at work just wondering when the next vacation will be, right? Then start earning miles with @smiles.oficial, plan to meet and have fun in Mendoza, Argentina.	6,791 likes	75	@vanianpfrocha: Mendonza..... we went there and liked it a lot. I recommend it.

In this period Covid-19 was still having a very low repercussion in the media in Brazil, and accordingly these brands did not make any mention on the subject. Table 2 shows that the brands talked about carnival, launch of new collections, and general advertising campaigns about their products. So, from February 26 to March 9, none of the analyzed brands posted any content related to Covid-19. On the 10th there was one mention the ailine company Gol, as shown on Table 3. The post included a video with the company's President, Paulo Kakinoff, presented the health protocols for travelling, from aircraft cleaning to team safety procedures, which were praised by the followers. Positioning is totally effective when it manages to translate the essence, its company values (Lee et al., 2018) and its emergency actions to the cause. As a consequence, the number of views reflects the brand awareness and its attributes such as trust, loyalty and positive image (Cheung et al., 2019; Molinillo et al., 2017; Thaichon & Quach, 2015).

On the same day, the other companies analyzed did not mention the pandemic, and their posts had a marked advertising nature, promoting their products. Gol returned to the topic on March 16, stressing their initiatives to maintain customers' safety. By that time Covid-19 was gaining prominence in the media and in the comments followers question about travel processes, cancellations, and even if the company would continue to operate as usual.

The Brazilian Ministry of Health and state governments issued confinement, prohibitions, and activity closings from March 17 (Silva, Borda, & Tasca, 2020). On that week the brands changed their digital strategy and all posted content about their action plans to deal with the pandemic (Table 4).

Aligned with the social and health context of those days, one of the brands (Gol) kept posting about their contingency plans, and the other three brands changed their marketing strategies and started issuing communications with reference to their actions during the pandemic. The retail chain Renner announced closing all brick-and-mortar stores and the beverage industry Ambev shared its social responsibility ac-

Table 3. First posts about Covid-19 by aircraft company Gol.

Date	Post	Views	Comments	Comment examples
03/10/2020	@voegol: Coronavirus. How GOL is dealing with the Coronavirus	192,000	1,094	@Guerreirodasmilhas: Excellent president! 💜 @alisonluiz1306: Congratulations Gol or the transparency and all the actions taken within this scenario! Thank you very much for the information Mr. Kakinoff!! 👏👏👏👏👏👏
03/16/2020	@voegol: GOL commitment. If you need to travel, our GOL Commitment is to put your safety and that of our Team as the number one priority. Find out what are GOL initiatives our aircrafts, in order to better face the epidemic caused by the Coronavirus.	55,000	911	@Giovanna.pontes.12: Only you think that it is safe to travel during this period it is totally discarded we're not crazy spare me 😒 @larissamartinsb: All phones unavailable at the moment!!! How do I do to talk with you????

tion of producing alcohol hand sanitizer gel for public hospitals. Both generated high levels of engagement and followers' approval. As suggested by extant literature, understanding the social context and demonstrating empathy with public needs make brands closer to their consumers (Hartmann et al., 2005; Rahman, 2018), having positive impacts on brand awareness and image. The delivery app Ifood chose a different strategy, built on the fact that confinement actually presented huge opportunities for increase sales. As such, its communication stressed the contingency measures to protect customers. Interestingly, both Ifood and Gol posts generated negative comments. In the case of Gol the most notorious cause of negative feedback was the difficulties of customer service to deal with the high number of flight cancelations. In the case of Ifood some negative comments were made regarding the contingency measures that some considered insufficient. As pointed out in the literature, during a crisis, the consumer has a decisive role in the marketing strategy (Stieglitz et al., 2018). In the four cases analysed, it is clear that communication strategies on Instagram were clearly influenced by customers' interactions.

After March 19, brands tried to reconcile the context of the pandemic with their objectives. Some have succeeded and others have not. Dissociation from brands' essence and values and a focus on selling products during the pandemic resulted in customers' allienation in line with suggestions in the literature (Arli, 2017; Dedeoğlu et al., 2019; Kucuk, 2011), confirming that the lack of a clear positioning strategy to deal with the crisis breaks the connection with the followers (Romaniuk et al., 2017). Figure 1 presents a timeline with the two main events related to Covid-19 in Brazil in the period under analysis (the first casualty and the decree to close establisments) and the first posts by each brand related to the pandemic.

Both the strategies adopted and its outcomes are particularly evident in the last week analysed for this study. Table 5 provides examples of the posts made by each brand. Gol communication strategy was still about keeping its business active, communicating the safety measures implemented in its flights. As shown in the example, followers were not happy, and in fact some of the comments were particularly negative and resentful. By this time several countries in Europe and America had forbidden flights, and in Brazil most flights were canceled by then.

Incongruous with the problem that society was struggling with, Renner made a post with clear selling objectives. A different strategy was conducted by Ambev and Ifood, which chose to demonstrate their solidarity with society, emphasizing their willingness to contribute to common good and to offer help.

Table 4. Brands' posts after confinement restrictions in some Brazilian states

Date	Post	Views	Comments	Comment examples
3/17/2020	@voegol: GOL website. Stay tuned to step by step to Reschedule or Cancel your flight with the ease and convenience of your computer. It is our GOL commitment to you.	53,600	864	@bruno.souzabcs Good morning, flight rescheduled how to proceed, as calls unanswered during opening hours. I need another flight for 03/19 and your service channel does not answer
3/17/2020	@ambev: For Ambev people, going beyond the labels is also taking care of each other. Therefore, we will transform the alcohol in our breweries into hand sanitizer filled in our own packaging. The entire production of half a million units will be donated to public hospitals in the cities most hit by COVID-19, so that no one is lacking it. This was one of the ways we found to help at a time when all solidarity makes a difference. #BeyondLabels.	737,000	39,000	@luizfernando_lfs> Congratulations on your initiative 👏👏👏 jucelerequinte> 👏👏👏👏🙏🙏🙏 Congratulations on thinking about others!!!!
03/19/2020	@lojasrenner: It is time to take care of each other, because, for us, people come first. Therefore, we decided to close all of our physical stores as of March 20 and for an indefinite period. Here's how you can continue to connect with us during this time. Together, even apart, we will get over this phase soon. Stay at home and take care of each other ♡ #wekeeptogether #seeyousoon	78,300	6,563	@geocsoares> Congratulations!!!! For the example!!!!!! Lojasrenner> @geocsoares ♡
3/22/2020	@ifoodbrasil: Times like these are made of delivery. And this will be #OurDelivery. Check out all the measures being taken by iFood in relation to Coronavirus (COVID-19) in the bio link.	22,800	275	@ninapitliuk> you guys had to take more care of your couriers! distribute hand sanitizer, think about their food especially now that all establishments are closed... the campaign is beautiful in theory, but in practice there is a lack of humanity with the workers

Figure 1. Timeline: Covid-19 main events in Brazil and bands' posting on Instagram
Source: The Authors

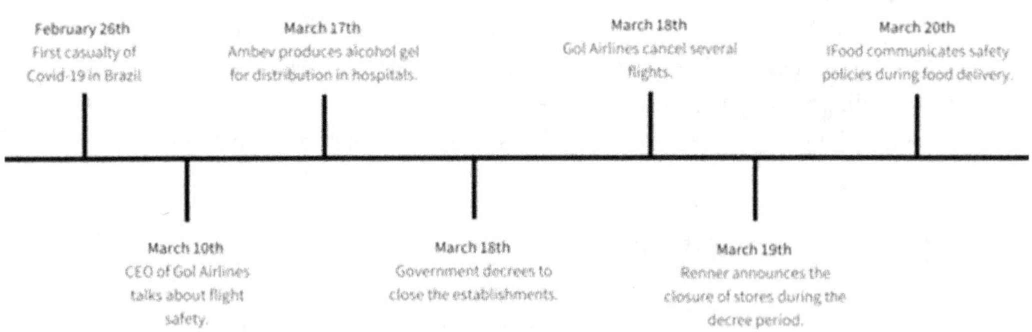

Table 5. Brands' posts while the global public health crisis starts escalating

Date	Post	Views	Comments	Comment example
3/26/2020	@voegol: Our GOL Commitment is to put your safety as our priority so that, if you need to travel, you can do it in comfort and tranquility. For this, we have adopted some hygiene initiatives in our onboard system and cleaning and disinfection of all our aircraft and equipment.	6,028	503	@iisabelavieira_> AND WAIT FOR THE LAWSUITS THAT WILL FOLLOW! COMPLY WITH AT LEAST WHAT YOU PROMISED, BEYOND INFRINGING THE CONSUMER REGULATIONS YOU ARE MAKING MISLEADING ADVERTISEMENT!!!
3/26/2020	@lojasrenner: Why choose between comfort and style? Loungewear are those comfortable clothes that look a big hug! Usually they are made with light fabrics, such as cotton, that leave you free to enjoy your day at home. From self-care or home office: this style goes well with everything! <3	3,522	161	@flaviacristinab203> hello good morning. How we will make payments. the store is closed.
3/27/2020	@ambev: In this moment of unity and solidarity, we are proud to tell you that today we started delivering our alcohol hand sanitizer to public hospitals in the cities of SP, RJ and Brasília. Today, Santa Casa de SP received the first 5,000 bottles and this is just the beginning. Let's go together 😉💚 #BeyondLabels	63,900	4,044	@thiagosansone95> Congratulations, for this beautiful action at this difficult time
3/27/2020	@ifoodbrasil: Stay at home: this is the recommended behavior at this time. So, if you need a market, count on #OurDelivery. Just open the iFood app on your phone and make your purchases. You can also choose Contactless Delivery and arrange with the courier via chat the best place for him to leave your order. Check out all the measures that are being taken by iFood to guarantee the health and safety of all the people that are part of its ecosystem in the bio link	21,300	600	@draericasampaioadv > 👏👏👏💚

Overall, brands that became protagonists during the pandemic and aligned their purpose to the social context resulted in excellent numbers of reach and engagement. The positioning for health, even if unconnected with the brands' products (e.g., Ambev), was aligned with customers' desires (Jalkala & Keränen, 2014; J. Park et al., 2017; Pogorzelski, 2018). The adaptation of brands to the social crisis context and their willingness to contribute to a better world was essential to the success of the communication strategy and the gains in terms of brand awareness and customer involvement.

SOLUTIONS AND RECOMMENDATIONS

Gol's post on March 10 is a preview of its strategy to deal with the public health threat, as the president's speech points out protective measures to diminuish the contagion probability. The week of March 18 marked a no turning point for companies' marketing strategy. Confinement imposition across the country (Silva, Borda, & Tasca, 2020) generated a crisis for most companies, which were not prepared to deal with the quarantine. Crises are always a threat to an organization's reputation (Coombs, 2006; Coombs

& Laufer, 2018; Rea et al., 2014). The main purpose of crisis communication is, therefore, to restore the organization's reputation and customer confidence. This study demonstrates that these brands had distinct strategy to deal with the crisis, directly impacting its awareness and its relationship with customers.

Another interesting fact is the number of posts during the crisis (Table 6), reflecting clear changes in the communication strategy. The sharp decrease in the number of posts shows the unpreparedness of brands in marketing actions to maintain a continuous feed of publications in social media during the crisis. Looking at followers' interaction, the average number of comments increased for all brands, but what changed was the nature of the comments.

Table 6. Posting strategy and customer interaction on Instagram before and after confinement

Brand	Renner	Gol	Ambev	Ifood
Between 02/26/2020 and 03/19/2020 (before confinement)				
Number of posts	67	18	22	21
Likes (average)	12,758	61,115	63,199	3,492
Comments (average)	178	466	208	185
Between 3/20/2020 and 4/1/2020 (during confinement)				
Number of posts	8	9	4	5
Likes (average)	11,030	11,293	38,833	9,968
Comments (average)	470	707	2586	429
Likes variation	-14%	-82%	-39%	185%
Comments variation	164%	52%	1143%	132%

So, brands that were unable to react quickly to the situation (i.e., Gol and Renner) received criticism and queries about their contingency plans for providing their services. The companies that presented solidarity solutions (i.e., Ambev) had positive comments and public support from their customers. Ambev had a significant increase in comments during confinement because the content posted is related to social responsibility actions: free production of alcohol hand sanitizer gel for public hospitals. The 3 posts on hand sanitizer had an unprecedent 842 thousand views and 45,167 comments, marked by customers' acknowledgement and approval of such actions. The company that had the biggest decrease in likes was Gol. Confinement had a direct impact on its business, as thousands of customers had their flights cancelled and rescheduled.

Ifood is the only company in this study for which the crisis created increased business opportunities, and confinement offered a potential of increasing home delivery services for food. Still, it is interesting to note that the communication strategy adopted by this brand included also an emphasis on socially relevant actions. The number of likes and positive comments considerably increased, and so did its business. But what generated the increase in followers' interaction was the fact that the company enabled unemployed people to generate some income as deliverers, and had a positive impact on local stores, as it helped supermarkets and restaurants to keep in activity by partnering with Ifood.

Based on the different strategies carried out, and its outcomes, two distinct profiles stand out from the data: unprepared and leading brands, which are summarized in Table 7.

Table 7. Leading and unprepared brands' strategies to tackle the crisis

Strategies	Leading Brands	Unprepared Brands
Communication Strategy	**Effective strategy** An effective plan is agile, transparent, and connected to the social context, and thus has positive repercussions.	**Weak strategy** Actions without strategy and apparently insensitive to the social crisis evidence lack of preparation for a crisis.
Brand Positioning	**Empathetic Strategy** Pupportive and empathetic positioning during the crisis were perceived as favorable and credible in the minds of consumers. Clear alignment between crisis management strategy and brand values.	**Limited strategy** Focusing on sales and keeping the business contradicts brand's values and becomes irrelevant to followers.
Brand Awareness and Involvement	**Consistent strategy** Positive comments, praise, and recognition for brand's socially relevant actions reflect the consistency of the awareness acquired. Brand involvement is strengthened through positive emotions and interaction.	**Superficial strategy** Negative comments, complaints, and disapproval directly impact brand's awaremess and involvement. The divergence between actions and values negatively affects brand trust.

Unprepared Brands

Some service companies that were directly affected by confinement, such as Renner and Gol, have completely changed their social media communication narratives and actions. Renner closed all of its physical stores and restricted its activity to e-commerce, Gol reduced the number of flights throughout the country. The lack of a clear crisis management strategy and a contingency communication plan resulted in problems for both brands. The number of comments regarding complaints has increased for both. Consumers asked about problems, payments, and returns. These companies were seldom able to timely responses, it was clear that they struggled to manage the crisis and mainly reacted to its immediate consequences, such as complaints.

It was also evident that these brands kept focusing on sales and were seen as insensible and indifferent to the public health issue that was about to escalate in the country. For instance, while the world news was about the increase in the number of deaths, Renner posted about its products, encouraging the sale. The lack of identification with its community and inability to have an active role during the pandemic resulted in low consumer engagement with the brand. Consumer involvement in brand communication during a crisis depends on consumer perceptions of its adaptation process and its ability to change (Rajagopal, 2019). In point of fact, Gol tried to approach the global problem – it was the first of the analyzed brands to post about the pandemic. But its strategy was to avoid shutting down its flight services, and soon the content posted was totally irrelevant to its customers, and in fact contrary to the brands' values (e.g., customer centricity). Messages that do not translate any of the company's values result in customer dissatisfaction and alienate them from the brand (Rahman, 2018; Romaniuk et al., 2017). In line with this, customers stressed the lack of understanding about the seriousness of the problem, which lead to negative comments and dissatisfaction on brands' Instagram page. As stressed in the literature, during a crisis brand communication has to be fast, clear, and objective (Lin, Savoia, et al., 2014; Ratzan & Moritsugu, 2014), in order to become relevant to brand followers and customers.

Hence, unprepared brands lack an adequate strategy to deal with crisis. Understanding what happens in the world and being flexible to adapt to the changes is an important attribute in the positioning of a brand (Aaker, 1997; Heinberg et al., 2017; Pogorzelski, 2018). On the contrary, the ones that are not able to understand the crisis will upset their followers (Cheung et al., 2019; Lou & Yuan, 2019), damage the brand, and eventually loose customers. Indeed, consumers reacted to brands' lack of empathy by not relating the posts to the ongoing health and social crisis, reducing the interaction and keeping the comments mainly to post service complaints.

Leading Brands

Companies such as IFood and Ambev knew how to position themselves when the crisis started and acted with prominence. Ambev decided to help fight the pandemic by producing alcohol hand sanitizer which was delivered to public hospitals free of charge. The company also made a donation to increase the number of beds in a public hospital in São Paulo, which was not communicated in Istagram but disseminated in the news, and reflected in some of the followers' comments on the brand's post on Instagram. The beer production continued normally, but social media channels such as Instagram were used to communicate its social responsibility actions. IFood enabled professionals who were unemployed to become deliverers and earn an extra income and emphasized their willingness to help local businesses to keep selling. Brands that strengthen the common good create a lasting relationship with the consumer, generating high levels of notoriety and social engagement (Arli, 2017; R. Huang & Saragollu, 2014; Molinillo et al., 2017). IFood was also instrumental in securing deliveries to all the people who stayed at home. It acted as a facilitator in the routine of families to access food quickly and safely without leaving home.

Both companies were agile in managing the crisis, creating clear and transparent communication with their consumers (Guidry et al., 2017; Lin, Jung, et al., 2014; Lin, Savoia, et al., 2014). The leading brands showed rapid adaptation to the crisis scenario and became partners with their consumers, showing empathy and solidarity about the situation, The positioning was positive and aligned the essence of companies (Hassan & Craft, 2012; Heinberg et al., 2017; Lee et al., 2018) resulted in high levels of awareness, increase the engagement, and arguably increasing brand equity.

FUTURE RESEARCH DIRECTIONS

This chapter demonstrates that analyzing brands' strategies in dealing with crisis provides valuable insights for both practitioners and academics working with band communication. Despite being limited to four brands and their posts on Instagram during a five-week period, the findings offer several cues that could be the starting point for future research.

It is important to note that the findings need more time to be validated, particularly in a country like Brazil, which struggled to deal with the pandemic for many months after the period under analysis. Eventually, the real impact of the pandemic on the economy, companies included, can only be fully addressed months or maybe even years after these first five weeks. It would be of particular interest to analyze how the crisis communication evolved in a longer period, namely to unveil possible learning processes and changes in communication, either reactive or strategic.

Other interesting topics for future research include but are not limited to the following. In order to validate the findings, it would be interesting that the approach adopted in this chapter could be replicated using other brands. One interesting alternative would be to compare companies from similar activity sectors, eventualy from different socio-cultural contexts. Longitudinal analyses would also be interesting to further understand the learning process that is subjacent to a brand's communication strategy over the duration of a crisis and even reincident crises. Evidently, such analyses may comprise different communication channels, both online and offline.

Different methodological approaches can also provide further understanding of this phenomenon. Indeed, although netnography is a particularly valuable approach to digital communication strategies and their impacts (e.g., customer interaction), other methods such as interviews to managers and surveys to followers are also recommended for future research.

CONCLUSION

Overall, this study demonstrates that brands that during crises have an adequate positioning considering the social context and brand values strengthen their positive image and increase customers' trust and loyalty. In fact, global crises such an pandemic provide some relevant opportunities for companies to create value and demonstrate empathy, and consequently to build strong relationships with their audiences.

Considering the findings, it is clear that companies adopt distinct strategies on social media during a pandemic (RQ1). While leading brands show structured strategies, often build on social responsibility actions, unprepared brands apparently try to keep a focus on selling objectives and instead of prevent customer negative comments, end up reacting to at times huge unsatisfaction and unger reactions, clearly damaging brand image. The results also demonstrate that brand awareness may vary a lot during the pandemic period (RQ2). Social crises such as Covid-19 are particularly intense in terms of emotions, and without surprise followers comments increase and are more extreme in terms of positive and negative emotions. That should be considered, and brands need to plan in advance how their strategy will foster positive involvement and emotions on their target publics – or the company will end up simply reacting to the unhappy crowd. Having a clear CSR strategy alligned with social and environmental aspects affected by the pandemic may guide the communication effort during the crisis (RQ3). As point of fact, this study demonstrates that the pandemic can put the effectiveness of the CRS strategy to the test and accelerate the development of actions that favor publich health and social good.

All summed up, the potential impact of brand communication actions during a pandemic offer an very relevant potential to affect, either positively or negatively, brand positioning, image, and awareness. For that reason, brands should prepare adequate crisis strategies beforehand, in order not only to avoid threats to brand management, but in fact to identify and tackle the opportunities that can arise for brand positioning and customer involvement.

REFERENCES

Aaker, J. (1997). Dimensions of Brand Image. *JMR, Journal of Marketing Research*, *34*(3), 347–356. doi:10.1177/002224379703400304

Acar, A., & Muraki, Y. (2011). Twitter for crisis communication: Lessons learned from Japan's tsunami disaster. *International Journal of Web Based Communities*, *7*(3), 392–402. doi:10.1504/IJWBC.2011.041206

Adlakha, K., & Sharma, S. (2019). Brand positioning using multidimensional scaling technique: An application to herbal healthcare brands in Indian market. *Vision (Basel)*, *1*(11). Advance online publication. doi:10.1177/0972262919850930

Ahmed, R., Vveinhardt, J., & Streimikiene, D. (2017). Interactive digital media and impact of customer attitude and technology on brand awareness: Evidence from the South Asian countries. *Journal of Business Economics and Management*, *18*(6), 1115–1134. doi:10.3846/16111699.2017.1400460

Ajami, R. (2020). Globalization, the Challenge of COVID-19 and Oil Price Uncertainty. *Journal of Asia-Pacific Business*, *21*(02), 1–3. doi:10.1080/10599231.2020.1745046

Annan-Diab, F., & Jensen, B. (2017). Customer Value Perception: Understanding Customer Response to Company's CSR Initiatives. In M. Aluchna & S. Idowu (Eds.), *CSR, Sustainability, Ethics & Governance* (pp. 53–70). Springer., doi:10.1007/978-3-319-39089-5

Aquino, V., & Monteiro, N. (2020). *CORONAVÍRUS | Brasil confirma primeiro caso da doença*. Retrieved from https://www.saude.gov.br/noticias/agencia-saude/46435-brasil-confirma-primeiro-caso-de-novo-coronavirus

Arli, D. (2017). Does social media matter? Investigating the effect of social media features on consumer attitudes. *Journal of Promotion Management*, *23*(4), 521–539. doi:10.1080/10496491.2017.1297974

Aydın, H. (2019). Consumer perceptions and responsiveness toward csr activities: A sectoral outlook. In I. Altinbasak-Farina & S. Burnaz (Eds.), *Ethics, Social Responsibility and Sustainability in Marketing, Accounting, Finance, Sustainability, Governance & Fraud: Theory and Application* (pp. 45–62). Springer. doi:10.1007/978-981-13-7924-6_3

Babiak, K. (2010). CSR and environmental responsibility: Motives and pressures to adopt green management practices. *Corporate Social Responsibility and Environmental Management*, *24*(March), 11–24. doi:10.1002/csr

Bazago, F., Guardia, M., & García, J. (2020). Environmental discourse in natural disaster scenarios. *Ekonomska Istrazivanja*, *33*(1), 3093–3107. doi:10.1080/1331677X.2019.1694425

BBC News. (2020). *São Paulo registra primeira morte por coronavirus e investiga outras quatro*. Retrieved from https://www.bbc.com/portuguese/brasil-51931934

Blomgren, A. (2011). Is the CSR craze good for society? The welfare economic approach to corporate social responsibility. *Review of Social Economy*, *69*(4), 495–515. doi:10.1080/00346764.2011.592329

Brønn, P., & Vrioni, A. (2001). Corporate social responsibility and cause-related marketing: An overview. *International Journal of Advertising*, *20*(2), 207–222. doi:10.1080/02650487.2001.11104887

Bundy, J., Pfarrer, M., Short, C., & Coombs, W. (2016). Crises and Crisis Management: Integration, Interpretation, and Research Development. *Journal of Management*, *43*(6), 1661–1692. doi:10.1177/0149206316680030

Carroll, A. (1999). Corporate social responsibility: Evolution of a definitional construct. *Business & Society*, *38*(3), 268–295. doi:10.1177/000765039903800303

Cheung, M., Pires, G., & Rosenberger, P. (2019). Developing a conceptual model for examining social media marketing effects on brand awareness and brand image. In International Journal of Economics and Business Research (pp. 243–261). doi:10.1504/IJEBR.2019.098874

Cho, S., Jung, K., & Park, H. (2013). Social media use during Japan's 2011. *Media International Australia*, *149*(1), 28–40. doi:10.1177/1329878X1314900105

Chowdhury, R. M. M. I. (2019). The moral foundations of consumer ethics. *Journal of Business Ethics*, *158*(3), 585–601. doi:10.100710551-017-3676-2

Coombs, W. T. (2004). Impact of past crises on current crisis communication: Insights from situational crisis communication theory. *Journal of Business Communication*, *41*(3), 265–289. doi:10.1177/0021943604265607

Coombs, W. T., & Laufer, D. (2018). Global Crisis Management – Current Research and Future Directions. *Journal of International Management*, *24*(3), 199–203. doi:10.1016/j.intman.2017.12.003

Dedeoğlu, B., van Niekerk, M., Küçükergin, K., De Martino, M., & Okumuş, F. (2019). Effect of social media sharing on destination brand awareness and destination quality. *Journal of Vacation Marketing*, *24*(1), 33–56. doi:10.1177/1356766719858644

Droppert, H., & Bennett, S. (2015). Corporate social responsibility in global health: An exploratory study of multinational pharmaceutical firms. *Globalization and Health*, *11*(1), 1–8. doi:10.118612992-015-0100-5 PMID:25886175

Dzarasov, R. (2016). The global crisis and its impact on the Eurasian Economic Union. *European Political Science*, *17*(May), 23–34. doi:10.1080/23745118.2016.1171272

Fatma, M., Rahman, Z., & Khan, I. (2015). Building company reputation and brand equity through CSR: The mediating role of trust. *International Journal of Bank Marketing*, *33*(6), 840–856. doi:10.1108/IJBM-11-2014-0166

Freberg, K., Palenchar, M. J., & Veil, S. R. (2013). Managing and sharing H1N1 crisis information using social media bookmarking services. *Public Relations Review*, *39*(3), 178–184. doi:10.1016/j.pubrev.2013.02.007

Garnelo-Gomez, I., & Saraeva, A. (2019). Yes, we can! Encouraging responsible management through effective CSR communication. In F. Farache, G. Grigore, A. Stancu, & D. McQueen (Eds.), *Responsible People, Palgrave Studies in Governance, Leadership and Responsibility* (pp. 115–134). Springer. doi:10.1007/978-3-030-10740-6_6

Garnett, J., & Kouzmin, A. (2007). Communicating throughout Katrina: Competing and complementary conceptual lenses on crisis communication. *Public Administration Review*, *67*(1), 171–188. doi:10.1111/j.1540-6210.2007.00826.x

Garnett, J., & Kouzmin, A. (2009). Crisis communication post-Katrina: What are we learning? *Public Organization Review*, *9*(4), 385–398. doi:10.100711115-009-0096-5

Gills, B., & Gray, K. (2012). People power in the era of global crisis: Rebellion, resistance, and liberation. *Third World Quarterly*, *33*(2), 205–224. doi:10.1080/01436597.2012.664897

Grosser, K., & Moon, J. (2005). The role of corporate social responsibility in gender mainstreaming. *International Feminist Journal of Politics*, *7*(4), 532–554. doi:10.1080/14616740500284524

Grosser, K., & Moon, J. (2019). CSR and feminist organization studies: Towards an integrated theorization for the analysis of gender issues. *Journal of Business Ethics*, *155*(2), 321–342. doi:10.100710551-017-3510-x

Guidry, J., Jin, Y., Orr, C., Messner, M., & Meganck, S. (2017). Ebola on Instagram and Twitter: How health organizations address the health crisis in their social media engagement. *Public Relations Review*, *43*(3), 477–486. doi:10.1016/j.pubrev.2017.04.009

Hartmann, P., Apaolaza-Ibáñez, V., & Forcada-Sainz, F. J. (2005). Green branding effects on attitude: Functional versus emotional positioning strategies. *Marketing Intelligence & Planning*, *23*(1), 9–29. doi:10.1108/02634500510577447

Hassan, S., & Craft, S. (2012). Examining world market segmentation and brand positioning strategies. *Journal of Consumer Marketing*, *29*(5), 344–356. doi:10.1108/07363761211247460

Heinberg, M., Ozkaya, H., & Taube, M. (2017). The influence of global and local iconic brand positioning on advertising persuasion in an emerging market setting. *Journal of International Business Studies*, *48*(8), 1009–1022. doi:10.105741267-017-0071-2

Huang, R., & Saragollu, E. (2014). How Brand Awareness Relates to Market Outcome, Brand Equity, and the Marketing Mix. In T.-M. Choi (Ed.), *Fashion Branding and Consumer Behaviors* (pp. 113–132). Springer., doi:10.1007/978-1-4939-0277-4_8

Hur, W. M., Kim, H., & Woo, J. (2014). How CSR leads to corporate brand equity: Mediating mechanisms of corporate brand credibility and reputation. *Journal of Business Ethics*, *125*(1), 75–86. doi:10.100710551-013-1910-0

Jain, P., Vyas, V., & Chalasani, D. P. (2016). Corporate social responsibility and financial performance in SMEs: A structural equation modelling approach. *Global Business Review*, *17*(3), 630–653. doi:10.1177/0972150916630827

Jalkala, A., & Keränen, J. (2014). Brand positioning strategies for industrial firms providing customer solutions. *Journal of Business and Industrial Marketing*, *29*(3), 253–264. doi:10.1108/JBIM-10-2011-0138

Jamali, D., & Mirshak, R. (2007). Corporate Social Responsibility (CSR): Theory and practice in a developing country context. *Journal of Business Ethics*, *72*(3), 243–262. doi:10.100710551-006-9168-4

Johansson, J., Dimofte, C., & Mazvancheryl, S. (2012). The performance of global brands in the 2008 financial crisis: A test of two brand value measures. *International Journal of Research in Marketing*, *29*(3), 235–245. doi:10.1016/j.ijresmar.2012.01.002

Kirchgässner, G. (2009). The Global Crisis and the Answer of Economics. *Schweizerische Zeitschrift für Volkswirtschaft und Statistik*, *145*(4), 381–385. doi:10.1007/BF03399285

Kotler, P., & Keller, K. (2012). *Marketing Management* (14th ed.). Prentice Hall., doi:10.1080/08911760903022556

Kozinets, R. (2002). The Field Behind the Screen: Using Netnography For Marketing Research in Online Communities. *JMR, Journal of Marketing Research, 39*(1), 61–72. doi:10.1509/jmkr.39.1.61.18935

Kozinets, R. (2012). Marketing Netnography: Prom/ot(Ulgat)ing a New Research Method. *Methodological Innovations Online, 7*(1), 37–45. doi:10.4256/mio.2012.004

Kozinets, R. (2013). *Netnography: Redefined. Netnography: Redefined.* doi:10.1002/9781118767771.wbiedcs067

Kozinets, R. (2016). Netnography. The Blackwell Encyclopedia of Sociology, 1–2. doi:10.1002/9781405165518.wbeos0782

Kozinets, R. (2019). *Netnography: The Essential Guide to Qualitative Social Media Research.* Sage.

Kucharska, W., & Kowalczyk, R. (2019). How to achieve sustainability? Employee's point of view on company's culture and CSR practice. *Corporate Social Responsibility and Environmental Management, 26*(2), 453–467. doi:10.1002/csr.1696

Kucuk, S. (2011). Push-based brand awareness: The role of product availability and in-store merchandising. *International Review of Retail, Distribution and Consumer Research, 21*(3), 201–213. doi:10.1080/09593969.2011.578793

Lee, J., Kim, Y., & Won, J. (2018). Sports brand positioning: Positioning congruence and consumer perceptions toward brands. *International Journal of Sports Marketing & Sponsorship, 19*(4), 450–471. doi:10.1108/IJSMS-03-2017-0018

Loureiro, S., Serra, J., & Guerreiro, J. (2019). How Fashion Brands Engage on Social Media: A Netnography Approach. *Journal of Promotion Management, 25*(3), 367–378. doi:10.1080/10496491.2019.1557815

Lugosi, P., & Quinton, S. (2018). More-than-human netnography. *Journal of Marketing Management, 34*(3/4), 287–313. doi:10.1080/0267257X.2018.1431303

Macias, W., Hilyard, K., & Freimuth, V. (2009). Blog functions as risk and crisis communication during hurricane Katrina. *Journal of Computer-Mediated Communication, 15*(1), 1–31. doi:10.1111/j.1083-6101.2009.01490.x

Molyneux, L., Holton, A., & Lewis, S. C. (2018). How journalists engage in branding on Twitter: Individual, organizational, and institutional levels. *Information Communication and Society, 21*(10), 1386–1401. doi:10.1080/1369118X.2017.1314532

Muralidharan, S., Rasmussen, L., Patterson, D., & Shin, J. H. (2011). Hope for Haiti: An analysis of Facebook and Twitter usage during the earthquake relief efforts. *Public Relations Review, 37*(2), 175–177. doi:10.1016/j.pubrev.2011.01.010

Park, J., Rajagopal, P., Dillon, W., Chaiy, S., & DeSarbo, W. (2017). A new bayesian spatial model for brand positioning. *Journal of Modelling in Management, 12*(3), 404–431. doi:10.1108/JM2-12-2015-0100

Pogorzelski, J. (2018). Perceptual Branding. In Managing Brands in 4D (p. 180). doi:10.1108/978-1-78756-102-120181003

Pulker, C., Trapp, G., Scott, J., & Pollard, C. (2018). Global supermarkets' corporate social responsibility commitments to public health: A content analysis. *Globalization and Health*, *14*(1), 121. doi:10.118612992-018-0440-z PMID:30497500

Raffaetà, R. (2020). Another Day in Dystopia. Italy in the Time of COVID-19. *Medical Anthropology*, *39*(5), 1–3. doi:10.1080/01459740.2020.1746300 PMID:32212942

Rahman, R. (2018). Building brand awareness: The role of celebrity endorsement in advertisements. *Journal of Global Scholars of Marketing Science*, *28*(4), 363–384. doi:10.1080/21639159.2018.1509366

Rajagopal. (2019). *Competitive Branding Strategies: Maning Performance in Emerging Markets*. Mexico City: Palgrave Macmillan. doi:10.1007/978-3-030-24933-5

Ratzan, S., & Moritsugu, K. (2014). Ebola crisis - Communication chaos we can avoid. *Journal of Health Communication*, *19*(11), 1213–1215. doi:10.1080/10810730.2014.977680 PMID:25356719

Rea, B., Wang, Y., & Stoner, J. (2014). When a brand caught fire: The role of brand equity in product-harm crisis. *Journal of Product and Brand Management*, *23*(7), 532–542. doi:10.1108/JPBM-01-2014-0477

Reid, E., & Duffy, K. (2018). A netnographic sensibility: Developing the netnographic/social listening boundaries. *Journal of Marketing Management*, *34*(3-4), 263–286. doi:10.1080/0267257X.2018.1450282

Ries, A., & Trout, J. (1981). *Positioning: The Battle for Your Mind: How to Be Seen and Heard in the Overcrowded Marketplac*. McGraw-Hill.

Romaniuk, J., Wight, S., & Faulkner, M. (2017). Brand awareness: Revisiting an old metric for a new world. *Journal of Product and Brand Management*, *26*(5), 469–476. doi:10.1108/JPBM-06-2016-1242

Sardana, D., Gupta, N., Kumar, V., & Terziovski, M. (2020). CSR 'sustainability' practices and firm performance in an emerging economy. *Journal of Cleaner Production*, *258*, 120766. doi:10.1016/j.jclepro.2020.120766

Schaefer, S., Terlutter, R., & Diehl, S. (2020). Talking about CSR matters: Employees' perception of and reaction to their company's CSR communication in four different CSR domains. *International Journal of Advertising*, *38*(2), 191–212. doi:10.1080/02650487.2019.1593736

Schmeltz, L. (2012). Consumer-oriented CSR communication: Focusing on ability or morality? *Corporate Communications*, *17*(1), 29–49. doi:10.1108/13563281211196344

Seo, S., & Jang, S. (2013). The roles of brand equity and branding strategy: A study of restaurant food crises. *International Journal of Hospitality Management*, *34*(1), 192–201. doi:10.1016/j.ijhm.2013.02.014

Sharma, R., Ahuja, V., & Alavi, S. (2018). The Future Scope of Netnography and Social Network Analysis in the Field of Marketing. *Journal of Internet Commerce*, *17*(1), 26–45. doi:10.1080/15332861.2017.1423533

Stieglitz, S., Mirbabaie, M., & Potthoff, T. (2018). Crisis Communication on Twitter during a Global Crisis of Volkswagen – The Case of "Dieselgate." In *Proceedings of the 51st Hawaii International Conference on System Sciences* (pp. 513–522). 10.24251/HICSS.2018.066

Stiglitz, J. (2009). The global crisis, social protection and jobs. *International Labour Review*, *148*(1–2), 1–13. doi:10.1111/j.1564-913X.2009.00046.x

Tetrault Sirsly, C. A., & Lvina, E. (2019). From doing good to looking even better: The dynamics of CSR and reputation. *Business & Society*, *58*(6), 1234–1266. doi:10.1177/0007650315627996

Thaichon, P., & Quach, T. N. (2015). From Marketing Communications to Brand Management: Factors Influencing Relationship Quality and Customer Retention. *Journal of Relationship Marketing*, *14*(3), 197–219. doi:10.1080/15332667.2015.1069523

Vafeiadis, M., Bortree, D., Buckley, C., Diddi, P., & Xiao, A. (2018, December). (2019). Refuting fake news on social media: Nonprofits, crisis response strategies and issue involvement. *Journal of Product and Brand Management*. Advance online publication. doi:10.1108/jpbm-12-2018-2146

Vitell, S., King, R., Howie, K., Toti, J., Albert, L., Hidalgo, E., & Yacout, O. (2016). Spirituality, moral identity, and consumer ethics: A multi-cultural study. *Journal of Business Ethics*, *139*(1), 147–160. doi:10.100710551-015-2626-0

Wagner, T., Lutz, R., & Weitz, B. (2009). Corporate hypocrisy: Overcoming the threat of inconsistent corporate. *Journal of Marketing*, *73*(November), 77–91. doi:10.1509/jmkg.73.6.77

Wallace, R., Costello, L., & Devine, A. (2018). Netnographic Slog: Creative Elicitation Strategies to Encourage Participation in an Online Community of Practice for Early Education and Care. *International Journal of Qualitative Methods*, *17*(1), 1–13. doi:10.1177/1609406918797796

Wang, H. (2015). A new approach to network analysis for brand positioning. *International Journal of Market Research*, *57*(5), 727–742. doi:10.2501/IJMR-2015-060

Wang, Y. (2019a). User experiences in live video streaming: A netnography analysis. *Internet Research*, *29*(4), 638–658. doi:10.1108/IntR-01-2018-0029

Wang, Y. (2019b). Virtual cohabitation in online dating sites: A netnography analysis. *Online Information Review*, *43*(4), 513–530. doi:10.1108/OIR-11-2016-0338

Xharavina, N., Kapoulas, A., & Miaoulis, G. (2019). Netnography as a marketing research tool in the fashion industry in Southeast Europe. *International Journal of Market Research*, *147078531985921*. Advance online publication. doi:10.1177/1470785319859210

Ziek, P. (2009). Making sense of CSR communication. *Corporate Social Responsibility and Environmental Management*, *16*(3), 137–145. doi:10.1002/csr.183

ADDITIONAL READING

Deloitt (2020). Management checklist for the COVID-19 crisis. Retrieved from https://www2.deloitte.com/gr/en/pages/about-deloitte/articles/COVID-19-Crisis-Management-Checklist.html

Forbes (2020). Crisis Leadership In Real Time: 8 Pandemic Best Practices. Retrieved from https://www.forbes.com/sites/daviatemin/2020/03/04/crisis-leadership-in-real-time-8-pandemic-best-practices/#8c6e577797ee

Harvard Business Review (2020). Brand Marketing Through the Coronavirus Crisis. Retrieved from https://hbr.org/2020/04/brand-marketing-through-the-coronavirus-crisis

MarketingWeek. (2020). How Corona built brand equity in spite of the pandemic. Retrieved from https://www.marketingweek.com/corona-built-brand-equity-pandemic/

McMahon, J. (2020). *Post-Pandemic: 12 Lessons in Crisis Management*. Liffey Press.

NeoMarketing Podcast. (2020). Adapt, Adopt or Pivot in Corona Economy. Retrieved from https://open.spotify.com/episode/3b6HHJFS1U6f1uewkE41ug?si=O1D8ZKA2RQqR225MUgaaBg

Wall Street Journal. (2020). How Brands Are Advertising During the Pandemic. Retrieved from https://www.youtube.com/watch?v=Q4DpxASLnXo

KEY TERMS AND DEFINITIONS

Brand Awareness: It measures customer's ability to recognize a brand and associate it with a specific company, product, or service.

Brand Positioning: One of the main components of brand strategy. The way a brand differentiates itself from its competitors in order to occupy a distinctive position in the mind of its custumers.

Corporate Social Responsability: A business management model that integrates social and environmental concerns in its business operations, and hence voluntarily collaborate with social goals, values, and development.

Crisis Communication: Public relations' reactive activity that aims to protect a company whenever its reputation is challenged and to effectively deal with all its stakeholders during an emergency.

Crisis Management and Planning: An action that seeks to minimize negative consequences and prepare the company to deal with unexpected and repercussion events.

Nenography: A research method based on the interpretive investigation of social and cultural behavior via the web. It applies practices and principles of ethnography to the online context.

Pandemic: An infectious desease that spread globaly and reaches a high number of people in different countries and continents. The difference between pandemic and epidemic is the number of people affected.

Chapter 14
When Helping Is Dangerous:
Benefits and Risks to Providers Delivering Digital Crisis Intervention

Dana C. Branson
Southeast Missouri State University, USA

ABSTRACT

The purpose of this chapter is to discuss the growth in social services using digital platforms to provide crisis interventions for clients, families, and communities. Digitally provided services afford numerous benefits to clients, such as increased accessibility, timeliness of services, and increase of autonomy. However, there are also new ethical concerns of using digital platforms for social services, as well as dangers to providers—specifically, secondary traumatic stress, vicarious trauma, and manifestations of these phenomena in providers' professional and personal lives. The chapter will discuss these potential concerns and the need for prevention and self-care for providers. Finally, the chapter will review the necessity for ongoing trainings and additional research concerning digitally provided crisis intervention services and occupational hazards for social service providers.

INTRODUCTION:

When individuals and communities experience crisis, fear and chaos erupt. Fortunately, the best of people also emerges in a sudden collective feeling of unity, gratitude, renewed patriotism, and altruism (Herman, 1997). Individuals, families, and groups share their time, skills, resources, and finances to assist others. At times, some even sacrifice their lives for others, often strangers who are in great need. The professionals who take on human carnage and crisis for a living are also on-hand, providing life-saving services, guidance to the masses, and leadership. These are the ones who have gone to school and trained so they can effectively respond to traumas at all levels. Nevertheless, no amount of preparation ever fully prepares a person to deal with trauma, its subjective effect on victims, and its very complex aftermath. Professionals are trained to debrief after especially difficult events, encouraged to seek supervision and the consultation of co-workers in strenuous times, and/or attend in-service trainings to remind staff of

DOI: 10.4018/978-1-7998-6705-0.ch014

the importance of personal safety and self-care (DelTosta et al., 2019; Friedman, 2017; Grundlingh et al., 2017). But what happens when professionals are providing these services in isolation? When client disclosures create intrusive imagery and overwhelm a provider's ability to help, but there is no colleague down the hallway to vent to and self-help is not encouraged because the provider is in control of their own schedule? With the increase in digitally provided crisis interventions and telehealth services, providers are taking on significant amounts of secondary trauma with few supports or reliefs. This places providers in danger of developing secondary traumatic stress and/or vicarious trauma (Tehrani, 2016; Zhou et al., 2020).

The purpose of this chapter is to discuss the increased use of digital platforms as a means of improving access to social services for clients, including crisis intervention during times of community disasters and individual traumatic events. This chapter is a review of contemporary literature surrounding potential dangers to those providing digital social services, as well as a look at seminal sources on the development of terminology concerning ancillary traumatic phenomenon. The theoretical framework that provides an understanding of how trauma is an individual event will be discussed, as well as the secondary effects experienced by providers. Resulting potential ethical issues and practice considerations of digitally provided services will be examined. The chapter will also discuss potential dangers to providers using digital platforms, specifically secondary traumatic stress (STS) and vicarious trauma (VT). The significant differences between the two terms, common symptomology, and risk/protective factors will be reviewed. Finally, the chapter will discuss prevention methods for STS and VT and how social service agencies can work to support providers who deliver crisis services through digital platforms. The need for social service providers has increased and is forecasted to grow an additional 18% from 2018 to 2028. This is faster and higher than the average of other occupations (Bureau of Labor Statistics, 2020). Additionally, as clients, providers, and agencies become more comfortable with offering assistance digitally, more social services will be delivered from professional silos. This increases the potential for negative effects to the provider, as they may be working in isolation and have decreased accessibility to supervision and co-workers for processing of difficult client material. It is imperative that social services be delivered in a manner that increases availability and decreases barriers to needed services. Digital platforms can assist in this endeavor. However, providers of digital services need to be supported and protected from the difficulties of their work so they can continue to provide the professional and effective care giving that clients need.

BACKGROUND

Theoretical Framework

In order to understand STS and VT, it is important to have a working knowledge of the theoretical framework that was used to develop these two terms. The constructivist self-development theory (CSDT) is a heuristic and developmental model that provides an appreciation for the unique and individual reactions one has to traumatic events. The CSDT takes into account an individual's life experiences, worldview, previous exposures to trauma, supply or dearth of social capital, life domains (physical, emotional, mental, cognitive, and spiritual), and other distinct dynamics to elucidate how one is affected by a traumatic event and the aftermath that follows (McCann & Pearlman, 1990). For example, best friends Sarah and Phoebe are both in the same dorm fire at college. They both escape with their lives, however, several of their classmates perished in the fire. While both are emotionally distraught and struggle with grief for their friends, Sarah is able to move into a new dorm room and resume her studies, while Phoebe suddenly struggles with crippling depression and survivor guilt, causing her to be unable to leave her dorm room to attend classes and eventually fail the semester. When looking at the girls' backgrounds, there are distinct differences between the two that might explain the contrast in traumatic aftermath. Sarah was raised by a type-A personality father who encouraged his daughter to never look backward and to focus on the future. While sad, Sarah understands that she could do nothing for those who have died, and she could use the distraction of classwork to positively cope with the unsettling events. Phoebe, however, was from a family of firefighters and struggles with the idea that she could have done more during the fire to help people out of the building safely. She presumes that her family is disappointed in her, which causes her feel that she cannot talk about the situation to anyone, increasing her self-isolation and negative thoughts. The CSDT provide a means of understanding the sequela of events, behaviors, and cognitions of each girl, and how they quickly find themselves on very different paths of coping with the same event.

Although the CSDT explains how victims process and react to trauma, it can also be used to focus on individual adaptations and coping techniques. Even if coping skills are maladaptive, the CSDT sees these behaviors as survivalist and helpful in some manner. Through recognition, purposeful cognitive re-structuring, and intentional behavior changes, people can positively influence how to deal with trauma and experience post-traumatic growth (Ben-Porat, 2015). The goal of trauma-informed care is to help those struggling with traumatic aftermath to find purpose in their life events and to move past dysfunction and stigma to find healing and gratitude (Newsom & Myers-Bowman, 2017; Quitangon, 2019). McCann and Pearlman (1990) took these theoretical underpinnings to develop the term *vicarious trauma* and to develop recommendations and techniques to help social services providers not to be negatively changed over time by the work they do with traumatized clients. Figley (1995) used similar theoretical frameworks to develop the term *secondary traumatic stress* to denote those who are emotionally and cognitively overwhelmed by the amount of service that needs to be provided in a crisis situation, and their inability to adequately meet the needs.

Understanding the theoretical underpinnings of VT and STS provides a logical connection to how delivering social services digitally to those struggling with trauma and crisis situations might create emotional hazards for providers. Those who train, educationally and experientially, to provide social services customarily are driven by a desire to genuinely assist others who are hurting and in need of help. Therefore, when providers open themselves up emotionally, cognitively, and empathically to join a client on a journey into traumatic material, they grasp what the aversive events mean to a client. Additionally, as the provider listens to the client's traumatic details, tactile descriptions, and play by play recall, they create visual pictures of the events. It is not difficult to see the potential for spillover of client-based material into a provider's own worldview (Branson, 2019). While VT and STS are well-established hazards for social service providers in face to face settings, less is known about digitally provided services. However, due to the substantial increase in digitally delivered social services, research is needed to assist agencies and providers to maintain safety, professionalism, and efficacy of services for both the client and the provider.

Methodology

This chapter originated from a narrative literature review, involving theoretical, seminal, qualitative, quantitative, and mixed methods research. Peer-reviewed research was collected from numerous databases, specifically EBSCOhost, ProQuest, Google Scholar, and Omnifile. Additionally, seminal materials were used to establish the theoretical framework of the chapter. Scholarly sources were predominantly used to provide empirical references for information provided. Journals and other resources used provided information on a global level, as a variety of countries and people groups were represented in the studies reviewed. Except for seminal material, contemporary research was used to have accurate and up-to-date research for the chapter. The chapter utilized 69 peer-reviewed sources, with 62 being published within the last five years (2014 to present), which gives the chapter an 89.9% current research rate for the literature review. Additional sources of information were web sites, such as the American Academy of Social Work and Social Welfare, Center for Medicare and Medicaid Services, National Institute of Mental Health, Substance Abuse and Mental Health Services Administration, and The Next Web that provided the most accurate statistics to support discussions within the chapter.

DIGITALLY PROVIDED SOCIAL SERVICES

Demand for Digital Social Services

Those needing social services often struggle with logistical and personal barriers to accessing assistance, such as lack of reliable transportation; social anxiety and/or other mental health issues that restrict their ability to attend offices for services; an inability to comply with agency guidelines, such as being on time, not missing appointments, dress codes, language barriers, restrictions of weapons and substance use; stigma; and other agency regulations (Nguyen et al., 2019; Wray et al., 2016). The use of digital service provision can eliminate these barriers, meet clients where they are (physically and figuratively), and result in better client buy-in for enhanced outcomes (Goldingay & Boddy, 2017). For the purposes of this chapter, digital services refer to social services provided through telehealth, e-therapy, and hotlines. Telehealth is a general term that encompasses health care services provided over a telephone, video chat, or other platforms where the provider may or may not be able to see the patient (Carmona, 2020). An example might be a patient who recently suffered the sudden loss of a loved one who is experiencing physical symptoms of anxiety but is concerned it is a physical aliment. She contacts a doctor through her cell phone FaceTime feature to discuss her symptoms to determine if she needs further medical attention.

E-therapy is a more specific telehealth term that denotes behavioral health care. E-therapy is ideally provided through a digital platform that allows a provider and client to see each other; however, verbal-only exchange and text messaging are also utilized at times to assist clients in need (Carmona, 2020). An example might be a client who struggles with post-traumatic stress disorder calls her therapist during a panic attack that has been triggered by an environmental stimulus. The therapist can provide the client with mindfulness activities and deep breathing techniques to assist in decreasing the biological process of panic and symptoms until the client feels like she is back in control.

Hotlines are established person to person communication links for specific services in a time of crisis (Segal et al., 2019). An example might involve an older adult who utilizes a suicide prevention hotline because he is struggling with the desire to end his life. The older adult will instantly receive an operator who can provide therapeutic responses to his situation and assistance with cognitive re-evaluation. Additionally, if the older adult is unable to find relief for his suicidal ideation, the operator can arrange for appropriate services, such as a wellness check by a police officer, assistance with hospitalization, and/or a referral for counseling. All three digital services involve specifically trained providers and have the goal of delivering social services to those who are unable to access assistance face to face, or prefer a digital format (Carmona, 2020).

The past decade has seen a substantial increase in use of digital platforms in crisis situations (de Albuquerque et al., 2015). With 5.11 billion mobile users, 4.39 billion Internet users, and 3.48 billion social media users in the world (Kemp, 2019), digital platforms are viable mechanisms to be harnessed by social services (Berzin et al., 2015). There are a number of advantages to using digital platforms to deliver social services to clients, such as convenience of location; financial savings of provided services and lack of cost to access the service; new avenues of support—especially for clients who are isolated by location, struggle with social anxiety, and/or struggle with an issue that they are not ready to go public with yet; attractiveness to client; and increase in emotional comfort and level of disclosure due to veiled anonymity (National Institute of Mental Health, 2019). Additionally, in times of crisis, digital platforms can reach more people and more rapidly, keep clients updated with the most accurate information that is highly valued in times of crisis, and allow social service providers to deliver services in real time while maintaining personal safety (Jasmontaite & Dimitrova, 2017; Roberts & Marchais, 2018; Torous et al., 2020).

In March 2020, as the United States prepared for COVID-19, the use of digital platforms for social services came on-line, despite the comfort or readiness of clients and/or providers. The need was so great that for the first time ever, Centers for Medicare and Medicaid Services (CMS; 2020) expanded the services that could be provided by qualified mental health providers, both through telehealth and without state-by-state licensure requirements. Digital services have become a lifeline for clients struggling with mental health issues—specifically, mood disorders, anxiety disorders, Post-Traumatic Stress Disorder, and persistent mental illnesses where daily observation and check-in are vital for overall function (Kauer et al., 2014; Kerst et al., 2020; Rees & Maclaine, 2015; Torniainen-Holm et al., 2016; Turgoose et al., 2018). Digital services are also used to assist family members struggling to care for loved ones with significant mental health issues and/or medical issues, such as dementia, terminal illnesses, or other conditions where ongoing supports are needed to maintain well-being while caregiving (Malhotra et al., 2019). During the COVID-19 pandemic, the ability to access social services digitally greatly reduced the mental health burden communities had to address, and kept numerous patients needing mental health assessments and crisis intervention services out of emergency rooms (Zhou et al., 2020). Due to the worldwide COVID-19 pandemic, the need to embrace the use of technology and provide social services digitally has certainly been demonstrated (Torous et al., 2020); however, there is a great deal of catch-up the field of social services needs to do to be able to provide digital services effectively, comfortably, confidently, and ethically (Oliver et al., 2015).

Before COVID-19, when looking at the services being provided in times of community disasters, the bulk of digital assistance involved information for risk prevention, updated news for real time situations, organization efforts, and other communications concerning community-level disasters (Veil et al., 2011; Wukich, 2016). More recently, social media and other digital platforms are being harnessed by social services to provide crisis intervention to individuals (Martinengo et al., 2019). This use of technology bridges a gap for some clients and decreases barriers to services, but along with these new methods of service provision comes the need to review ethical considerations for providers, clients, and communities (Barsky, 2017; McAuliffe & Nipperess, 2017; Reamer, 2017). Because social service agencies operate under ethical guidelines and are accountable to individual consumers, government funding sources, and public constituents, ensuring quality care to clients is a priority. Social service agencies put in a great deal of time, assessment, and evaluation to ensure quality care (Netting et al., 2017). However, in the pursuit to provide excellent services and care to clients, potential risks to providers are often overlooked. Increasing amounts of research have established positive correlations between trauma work and potential hazards for providers (Baum et al., 2014; DelTosta et al., 2019; Finklestein et al., 2015; Giordano et al., 2016; Greinacher et al., 2019; Hensel et al., 2015; Hopwood et al., 2019; Hricova & Lovasova, 2018; Iqbal, 2015; Quitangon, 2019; Roden-Foreman et al., 2017; Sansbury et al., 2015; Thieleman & Cacciatore, 2014; Waegemakers-Schiff & Lane, 2019).

Technology, Social Services, and Ethics

Social services providers are one of the last groups of professionals to harness digital opportunities for clients, mostly due to concerns of quality care, confidentiality, ethical considerations, and lack of skills on the part of social service professionals (Groshong & Phillips, 2015; Oliver et al., 2015). However, social service providers have become keenly aware of the benefits to clients and are no longer allowing logistical issues to interfere with an increased provision of care to oppressed and vulnerable populations (Branson & Miller, 2020). In 2016, the American Academy of Social Work and Social Welfare (2019) developed the 12 Grand Challenges of Social Work. This was an initiative developed to concentrate efforts moving forward in practice, scholarly research, curriculum development, and policy development for the discipline of social work. One of the 12 challenges is harnessing technology for the social good (Berzin et al., 2015). As a result, social service providers have started to capitalize on the opportunities that digital platforms provide to reach more people in need. However, with any significant change, there is also resistance (Reamer, 2015; Torous et al., 2020). Social service providers are guided in their practice by professional values and ethics. These standards are set in place to protect the client, as well as the provider (Segal et al., 2019). Dynamics most affected by digital platforms being used to provide social services are confidentiality, client privacy, informed consent, the professional relationship between client and provider, and boundaries (Reamer, 2017).

Social service providers are mindful of the need for confidentiality with clients. The ability to create rapport with a client assumes that the client can trust the provider not to share what is discussed in sessions (Egan et al., 2017). Because of this foundational belief, providers are careful to explain the levels of confidentiality to clients before sessions begin for full transparency (Cooper & Lesser, 2014). With the use of digital platforms, social service providers must ensure that a client's confidentiality is protected through electronic communications and record keeping (National Association of Social Workers [NASW] et al., 2017). Similar issues exist with client privacy and informed consent. Digital technology makes locating personal health information and other private material easier and less time consuming.

This also creates windows of opportunity for those with ill intentions to obtain information that they may not need to view. Within social service providers, there is a debate on the use of public information on clients for clinical use (Reamer, 2017). Does a provider have the right to look up a client on social media as part of developing a social history or to gather additional clinical information? Because legal issues are a matter of public record, an Internet search can easily produce a client's arrest and conviction record, previous addresses, mug shots, and newspaper accounts of past life events. The NASW (2017) says that social workers should only use search engines to look up information on clients for "compelling professional reasons" (p. 38). If a client is at serious risk of harm or there is simply no other way to obtain needed information in a time-critical situation, an Internet or social media search might be appropriate. To look up the information to check the accuracy of a client's story or to try and find more information to fill in the gaps of a client's disclosures, however, would be considered inappropriate.

Nevertheless, this is a common practice among social service providers, as they are tasked with collecting a comprehensive assessment to serve as the golden thread for treatment goals, interventions, and discharge/aftercare planning.

In an ideal setting, if Internet and social media searches are routinely utilized to locate client information, clients should be abreast of this on their initial visit through the agency's social media and technology policy and clients should provide informed consent (Reamer, 2017). It is the collective belief of social service providers that clients have the right to know what they are signing up for, how their personal information will be handled, who will view it, how it is being protected, and what they can expect from the services they are being provided (Knowles & Cooner, 2016). Clients should never be caught unawares. Therefore, to forge ahead with routinely collecting information through digital platforms without client consent would be considered highly unethical and inappropriate (Cummings, 2016; Groshong & Phillips, 2015). Even in times of crisis and disaster, the values and ethics that guide quality practice cannot be abandoned.

Professional and personal boundaries are additional issues that are affected by providing social services through digital platforms (Barsky, 2017; Reamer, 2017). Boundaries are overt and/or covert limits that exist to encourage a healthy relationship and serve as guidelines for appropriate client/provider behavior (Cooper & Lesser, 2014). Social service providers are taught about the intersection of personal and professional boundaries in scholastic preparation for their careers and through ongoing trainings and supervision, yet boundary violations remain one of the most common reasons providers lose their licenses (Reamer, 2014). Without special attention to boundaries, social service providers run a risk of developing an inappropriate relationship with a client, physically, emotionally, socially, financially, and/or intimately. Providers understand that proper personal boundaries are foundational to proper professional boundaries. However, with the use of digital social services, there are new and uncharted risks to boundaries that must be navigated.

When working in an agency office, there is the presumed notion of office hours. With digitally provided services, this concept can quickly evaporate, especially if providers are delivering crisis services (Knowles & Cooner, 2016). Crisis workers must be available to deliver services at any point in the day and night. If providers do not have well-established boundaries, they may find themselves working nonstop, on call 24 hours a day, seven days a week. Over time, this can set up dangerous practices to ensure availability, such as the provider giving clients a personal cell phone number for calls and/or texts, an expectation of services on demand, and feelings of abandonment on the part of clients if the provider is not immediately available (Barsky & Reamer, 2018). Additionally, providers put themselves at risk for physical and mental exhaustion, poor decision making, burnout, decreased quality of services, increased strain on relationships with loved ones, and vulnerability to secondary traumatic stress and vicarious trauma (Branson & Miller, 2020; Elbarazi et al., 2017; Finley & Sheppard, 2017; Hopwood et al., 2019; Jarrad & Hammad, 2020; Quitangon, 2019; Willis & Molina, 2019). Good boundaries mean developing formative and envisaged lines to create a separation of the client and the provider. Digitally provided social services, especially in times of crisis, can cause the line to become quite blurry, increasing risk to clients, the services they receive, and the professional livelihood of the provider (Reamer, 2015).

When clients initially come for services, it is common for their disclosures to be guarded, inaccurate, or incomplete for numerous reasons. Clients do not trust social service providers right away—this takes time and rapport building. Additionally, sometimes clients are physically or mentally compromised due to substance use, unresolved trauma, and/or mental health issues, causing them to be poor historians. Clients may be fearful of providing complete honesty in assessments because they fear getting themselves into legal trouble, having their children removed from their care, having their disclosures used against them in a custody battle or with a probation/parole officer, and/or loved ones finding out damning information that could destroy relationships (Cooper & Lesser. 2014). Services provided through digital platforms can be an additional barrier to developing trust and an effective working relationship (Groshong & Phillips, 2015). Although providers may appreciate reasons why clients are not completely honest and forthcoming with information initially, they still have a job to do in a short period of time. Social service providers are often members of multidisciplinary teams and it is their responsibility to craft the social history of a client for the team. Therefore, they may not have the time needed to build rapport and develop the level of relationship where a client will be forthcoming with information. It is understandable why a provider might look to the Internet to find needed information about a client, and why this might become a commonality in one's practice. Yet, this is an ethical violation of a client's privacy and could work negatively against the development of the client/provider relationship, as the information provided by the Internet and social media is one-sided, biased, and often sensationalized (Barsky & Reamer, 2018; Roberts & Marchais, 2018). This can negatively influence the provider's view of the client, which has the potential to create a sequela of poor clinical outcomes.

Protecting client confidentiality, client privacy, informed consent, the professional relationship between client and provider, and boundaries is a difficult task. Digital platforms and social media create additional complications (Barsky & Reamer, 2018; Reamer, 2015). This is especially difficult for providers who themselves are uncomfortable with technology (Torous et al., 2020). It is important to remember that a large portion of social service providers attended secondary education and training programs to work in social services before the Internet and social media. Therefore, each new technology creates a learning curve. Additionally, when social media began, social service providers were cautioned to protect their digital footprint and to avoid digital technology as a means of delivering services because of risks to client confidentiality and quality clinical service (Branson & Miller, 2020). The present-day movement to embrace technology clashes with old ideas, and naturally creates concern and resistance. However, the need for digitally provided social services greatly outweighs the apprehension (Goldingay & Boddy, 2017). Providers need to be urged to offer clinical services and crisis interventions through digital platforms, while being guided by best practices, supported by administration, and encouraged through frustrations when trying something new. Providers also need to be protected from occupational hazards common to social services, such as secondary traumatic stress and vicarious trauma.

Secondary Traumatic Stress and Vicarious Trauma

Secondary traumatic stress (STS) and vicarious trauma (VT) are occupational hazards for those in the helping profession (Branson, 2019; Quitangon, 2019), especially for those who witness, visually or audibly, client-related traumas as part of their occupation (Hricova & Lovasova, 2018; Russell & Cowan, 2018). The use of digitally provided social services has increased faster than research concerning STS and VT occurrence in providers delivering services outside of a traditional methods. However, the CSDT provides a theoretical framework for a logical assumption that STS and VT are potential hazards to digital services providers as well. Although these terms are similar, they are not synonymous, and there are important differences to appreciate. STS is a natural consequence of bearing witness to others' trauma and suffering. This results from a sincere desire to assist others, yet, the helper is overwhelmed emotionally due to the lack the power and/or resources to aid as desired (Bride, 2007). Figure 1 provides an illustration of the elements involved in the development of STS.

Figure 1. Developmental sequence of Secondary Traumatic Stress

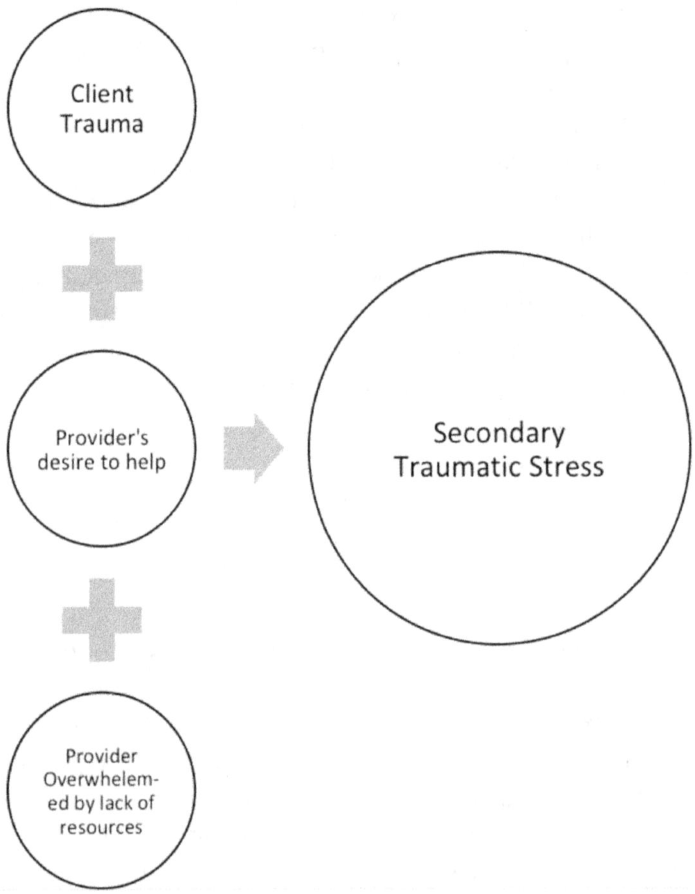

STS can happen to anyone, from community volunteers to trained emergency response personnel who see trauma and crisis as part of their day-to-day jobs. STS is common in police officers, first responders, nurses, firefighters, and other populations that deal with emotional, physical, and sexual traumas on a regular basis (Brown, 2018; Elbarazi et al., 2017; Greinacher et al., 2019; Grundlingh et al., 2017; Jarrad & Hammad, 2020; Kunst et al., 2017, Roden-Foreman et al., 2017; Waegemakers-Schiff & Lane, 2019).

An example of STS might be a veteran firefighter who just became a grandmother for the first time. As a firefighter, she has worked hundreds of fires, car accidents, and community disasters. However, one night when her unit is called to report to a fire where an infant died of smoke inhalation and the mother is inconsolable, the firefighter may find herself overcome with emotions. She can feel the grief of the mother, but also sympathy for the child; ideas of her own daughter having to deal with such a loss and not being able to watch her grandchild grow-up are too much for the firefighter emotionally. Long after the job is over the images still linger. The firefighter finds herself unable to sleep, food is unappealing, she continues to think about the mother and her situation, and she struggles with nightmares. During her next shift, when a call comes in concerning a car accident involving a family with small children, the firefighter reports a fictitious illness to avoid going on the call, as the idea of seeing another distressed mother or injured baby is simply too much to bear.

Compassion fatigue is another term and is synonymous with STS (Figley, 1995). Compassion fatigue typically denotes laypersons who are distressed by their helping duties (Thieleman & Cacciatore, 2014), whereas STS is reserved for professionals who struggle with witnessed trauma as part of their occupations (Quitangon, 2019). Short-term volunteers that assist after a community crisis, such as a tornado, or those who travel for short-term mission/humanitarian work may find themselves struggling with compassion fatigue. An example might be a group of women who travel to a developing country to assist in an orphanage for a week after a devastating mudslide. The excitement of going and helping victims of the disaster is quickly replaced with feelings of insignificance. The group's vision of helping children and being greeted with appreciation for their kindness is quickly replaced with masses of desperate victims and need so great, they do not know where to start. The supplies they brought are only a drop in the bucket of what is needed, and they feel ill-equipped. These feelings of being unprepared and unable to make a difference in the face of tremendous need create negative emotions and unpleasant physical reactions. Volunteers may leave their posts unable to continue, or worse, add to the need with their own requests for services.

VT is a similar term to STS and compassion fatigue, but also has important differences. VT is a term reserved for social service providers who engage in an empathetic relationship with clients and experience negative psychological changes due to an accumulation of client-disclosed traumatic stories (Pearlman & Saakvitne, 1995). To struggle with VT, there must be an empathic relationship present (Quitangon, 2019; Russell & Coran, 2018). Figure 2 providers an illustration of the elements involved in the development of VT.

Figure 2. Developmental sequence for Vicarious Trauma

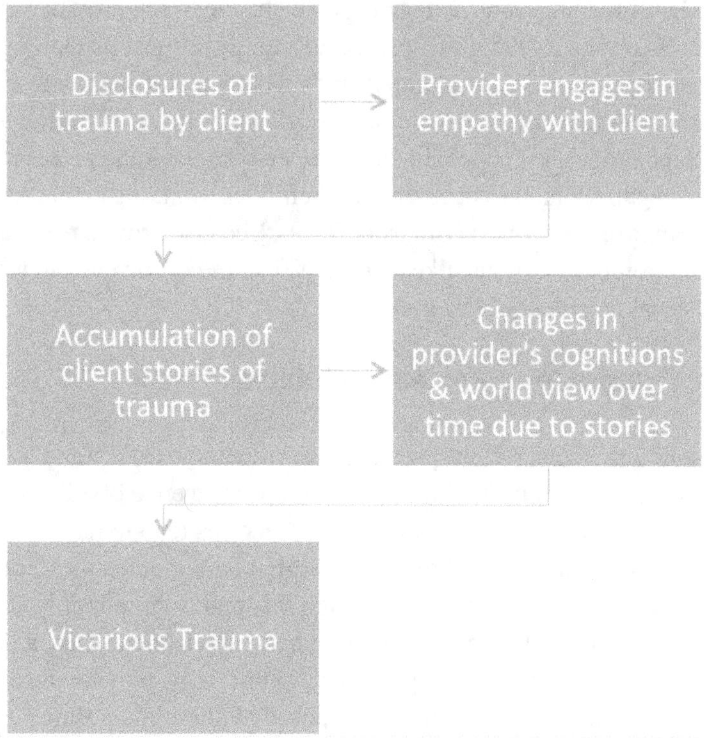

Consequently, although first responders, nurses, and other medical providers might struggle with patient-related traumas, due to the lack of a developed and purposeful empathic relationship, the term VT is inappropriate for their symptomology (Branson, 2019).

To understand VT, one must understand what empathy really is. People commonly report empathy as a personality strength. However, true empathy is a skill that is developed. Although some people, social service providers especially, may have a natural propensity toward empathy, to be genuinely empathetic, one must engage in purposeful behaviors, such as self-awareness, affective sharing, perspective taking, and emotional regulation (Segal et al., 2019). Providers delivering digital social services are trained how to provide empathy, non-judgmental acceptance, and a sense of community through special skills unique to digital platforms (McAuliffe & Nipperess, 2017).

When being empathic with a client, social service providers work diligently to engage in the following behaviors. First, the provider engages in self-awareness; to suspend their worldview, opinions, experiences, and values and adopt the client's. This allows the provider to better appreciate the emotion from the client's point of view. If a provider engages in empathy with a client who discloses a history of sexual trauma, the provider works diligently to understand the events from the client's worldview, values, and personal experiences to deeply grasp how the trauma affects the client. As this is happening, the provider is also creating mental images of the event and the emotional regulation of the client, while simultaneously ensuring their own emotional regulation for proper mental health and professional boundaries.

Empathy is a complex phenomenon that takes a great deal of intentionality on the part of the provider. It is in this connection between the client and the provider that vicarious trauma can develop (Quitangon, 2019). When providers empathize with a client, they are emptying themselves out and adding the events, emotions, cognitions, and opinions of the client into themselves. Because this is a developed and methodical skill of the provider, this can occur in traditional social service settings, as well as digitally provided platforms. Over time, this can create a negative change to the provider's worldview and sense of safety (Pearlman & Saakvitne, 1995). STS can occur rapidly, almost instantly in times of crisis, and because it comes on so quickly, the person understands the source of the distress. However, VT is the result of an accumulation of difficult client disclosures. VT takes time to develop; therefore, sometimes providers do not comprehend what is happening to their personal lives and professional abilities. In their attempt to analyze the negative changes occurring in their lives, thinking errors are common, and silent suffering causes the symptoms to worsen (Branson, 2019). Because providers engaging in digital social services can do so from remote locations, like a home office, the occurrence in isolation is more likely and potential for VT development could be greater.

Symptoms, Risk Factors, and Protective Factors

There are several reasons why social service providers are vulnerable to STS and VT. The set of skills that providers cultivate to be able to connect with clients is the very avenue of STS and/or VT development (Quitangon, 2019). Those who find themselves in social services tend to have common characteristics, such as being empathetic, nonjudgmental, gregarious, vigilant about injustices, supportive of the downtrodden, and self-sacrificing (Branson, 2019). Although this makes for a highly effective social service provider, it also creates a susceptibility to working long hours to ensure clients get the help and services they need, often at the personal expense of the provider. Additionally, providers are willing to join clients on their journeys through traumatic events, and are exposed to client-disclosures with graphic details of the play by play events, emotions connected to the trauma, and tactile sensations that allow the provider to make mental pictures of what happened to clients. These elements are encoded by the provider into their own cognitive schema and there is potential to negatively change their worldview (Pearlman & Saakvitne, 1995; Russell & Cowan, 2018). Reactions of providers to client disclosures are not signs of weakness or incompetence, but side effects of a genuine desire to connect and offer guidance to clients (Branson, 2019). Additionally, resulting negative manifestations can occur in providers' professional and personal lives, creating a potentially negative cascade effect on service provision, clients, social service agencies, communities, and loved ones (Craun et al., 2015; Landers et al., 2020). Common professional symptoms of STS and VT include:

- Cynicism toward clients and client situations
- Misdirected anger at clients and/or colleagues
- Avoidance of certain clients
- Deliberately avoiding assessment material concerning trauma
- Missing work and/or somatic illness
- New anxiety, specific to client disclosures
- Decreased productivity
- Leaving the field

Common personal symptoms include:

- Hypervigilance of safety—for self and others
- Feelings of sadness
- Depression
- Intrusive imagery
- Nightmares
- Withdrawal from loved ones and social group
- Avoidance of physical intimacy
- Pessimistic worldview
- Negative coping skills, such as substance use or other compulsive behaviors (Branson, 2019)

When it comes to trauma, no two people have the same experience or aftermath, further complicating how to understand and treat trauma. The factors that determine how one might be affected by trauma are numerous and each one has the potential to influence trauma recovery differently (Substance Abuse and Mental Health Services Administration, 2014). This is reinforced by the CSDT, which considers the endless personal and environmental variables that create individual differences in traumatic aftermath. Additionally, this process extends to providers (McCann & Pearman, 1990), both traditional and those providing services digitally. Development of STS and VT are similar, as what is a risk factor for one person is a protective factor for another. Literature discusses four broad categories of risk/protective factors: education, experience, history of personal trauma, and gender.

Education

Education has been shown to be both a risk and protective factor for the development of STS and VT. Academic education and trainings are important component of STS and VT awareness and prevention efforts, but it can also create issues for providers that once were not there (Knight, 2010). The concepts and theoretical underpinnings of STS and VT were not developed until the 1990s with Pearlman, Saakvitne, McCann, and Figley (Branson, 2019). Therefore, many social service providers attended preparatory programs before these concepts were discussed as occupational hazards. Today, students are learning about STS and VT in their advanced educations, and they are more likely to identify symptoms as they start to develop. However, there is a learning curve that goes with being an inexperienced social service provider. No matter how good an academic program is, it cannot fully prepare someone for the rigors of the work that comes with helping those who are hurting (Pryce et al., 2007). Adapting to a new job and the reality shock of what the work truly consists of is a common process most providers go through as a rookie. It is important to be able to distinguish between the normal adaption processes and the temporary discomfort of being new, and the development of genuine STS and VT. This is one reason why routine and frequent supervision for new social services providers is vital (DelTosta et al., 2019). Regardless of how social services are being delivered, whether face to face or through a digital platform, the risk factors are equal opportunists as they are based in the provider, not the client or delivery method.

Experience

As with education, experience seems to be a double-edged sword. Research indicates that social service providers with the least amount of experience are most vulnerable to STS (DelTosta et al., 2019; Finley & Sheppard, 2017; Pryce et al., 2007), and those with the most experience are vulnerable to VT (Fahy, 2007; Hensel et al., 2015). New social service providers are often hired into high burnout positions, such as child welfare professionals, case workers, or other front-line employees that are bombarded with client needs. These positions have a number of commonalities, such as high client caseloads, staff shortages, clients that might not like the provider, lack of client and agency appreciation, low pay, billing quotas, constantly changing policies, and a workload that is never done. Additionally, the only time a provider is given attention is usually due to a mistake or negative client outcome, which can sometimes receive media attention and sensationalized judgement from others who do not understand the rigors of the work (Pryce et al., 2007; Roberts & Marchais, 2018). This can cause a provider to leave a position after a short period of time, or the profession altogether before they have time to learn the job and develop the protective "thicker skin" that comes with working in social services.

On the other side of the continuum are those who are veterans of social services—those who have been in the field for years and have the war stories that can only come with experience. Because they have been around, they have often been promoted from the front-line position to managers and supervisors. Because of their experiences and skills, they are often given the clients with the most challenging presentations and consult on others' difficult cases. Therefore, they develop a repertoire of the worst of the worst, equating to numerous client-based traumas that can build up over time and lead to VT (Branson, 2019; Thompson et al., 2014). Experienced providers are also less likely to be under supervision because they are the supervisor, having fewer opportunities to process events that might be causing negative personal and professional issues, putting them at greater vulnerability to VT development.

Personal Trauma

Personal trauma as a factor for STS and VT development also has a double-sided risk and protective factor quality. Some research indicates that providers with a history of personal trauma are more vulnerable to STS and/or VT development (Kulkarni et al., 2013; Sodeke-Gregson et al., 2013), especially when the trauma of the provider is similar to the client's (Hensel et al., 2015). Other research indicates that it can be a protective factor (Kunst et al., 2017; Voss et al., 2011). A deeper investigation into the that sets the stage for STS and VT development. Those providers who have experienced trauma in the past, but now use their story as a means of helping others, have the protectiveness of post-traumatic growth (Ben-Porat, 2015). The movement from victim to survivor to champion not only provides healing for the person, but a platform for delivering assistance to others (Newsom & Myers-Bowman, 2017). Research indicates that many people drawn to the helping professions have their own history of trauma, and this is the motivation behind their desire to be a social service provider (Branson et al., 2019; Shannon et al, 2014). These dynamics create implications for providers and agencies that need to be considered to increase the longevity and efficacy of employees. Additionally, the relationship between personal trauma and post-traumatic growth serves as a springboard for STS and VT prevention.

Gender

A final broad category of risk/protective factor is gender. Research has indicated that being female is a potential risk factor for STS and VT development (Callister & Plante, 2017; Greinacher et al., 2019; Hensel et al., 2015); however, there are important caveats that needs to be considered. First, most social service providers identify as female. The U.S. Bureau of Labor Statistics (2019) reports that 82% of social workers, 76% of counselors, and 79% of social/human service workers are female. Therefore, when conducting research concerning STS and VT with social service providers, women are likely to disproportionately represented. Second, women still provide most of the dependent care, housework, shopping for households, and caregiving to older/ill family members outside of their own children (Schaefer, 2017). This means that female providers have less time to enjoy leisure, decompress, engage in self-care, and are simply more tired and rundown than their male counterparts. Multiple roles place them at more risk for stress-induced illness and phenomena (Ju et al., 2018), such as STS and/or VT. Third, women are more comfortable communicating and sharing concerns with others as a means of working out solutions and processing difficult issues, whereas men are more likely to keep issues to themselves out of fear of looking weak (Adler et al., 2017). Therefore, it is logical to assume that men may deny symptomology of STS and VT because they fear looking incompetent and frail. Regardless, since most social service providers are female and there may be plausible reasons for an increased vulnerability to STS and VT, it is important to find ways to effectively prevent the development of these issues from personal and organizational levels.

SOLUTIONS AND RECOMMENDATIONS

Research indicates that self-care is one of the best means of preventing STS and VT (Butler et al., 2019; Friedman, 2017). Social service providers are often keenly aware of their need to participate in self-care, but because providers are hard workers, desire to help others, and are self-sacrificing, it feels counter-intuitive to take time away from serving others for themselves (Callister & Plante, 2017). As a result, engagement in self-care activities tends to be low (Friedman, 2017; Sansbury et al., 2015). Additional compounding dynamics are a lack of understanding about STS and VT, lack of agency-level support for providers, and lack of academic preparation and attention to STS and VT in supervision (Branson et al., 2019; Brown, 2018). Therefore, for self-care to be effective, it should be intentional at all levels (Bent-Goodley, 2018). Agency administrators, supervisors, and co-workers should encourage self-care opportunities, while also deconstructing thinking errors that serve as a barrier to involvement in activities that provide a sense of renewal. Additionally, providers need to give themselves permission to engage in self-care, even if it feels uncomfortable at first.

Social service agencies that engage in digital services may be at an even greater risk of overlooking the need for providers to use self-care due to misperceptions that digital service providers are not in danger of developing STS and VT. Secondary effects of trauma and crisis have the potential to create problems, regardless of distance from the actual event (Hopwood et al., 2019). It is not the event that creates STS and VT, but the physical, emotional, mental, social, and spiritual changes it creates in providers as they attempt to assist survivors in mitigating negative aftermath of traumatic events (Iqbal, 2015; Pearlman & Saakvitne 1995). Therefore, it is logical to assume that providers working in digital settings are just as vulnerable to STS and VT as providers in more traditional face to face settings. Perhaps, they are even more at risk due to the possibility of providers delivering digital services in professional silos. With purposeful and intentional actions, providers delivering digital crisis services can experience post-traumatic growth, increasing their professional and personal wellbeing, which spills over into increased quality of client care (Ben-Porat, 2015).

FUTURE RESEARCH AND DIRECTIONS

With the increase in trauma-informed care and recognizing STS and VT as occupational risks for numerous professions in human services, the need to support providers has also grown. Additional research is needed on most effective methods for prevention and ways to move providers from having an academic understanding of STS and VT to engagement in intentional behaviors for prevention. One substantial arena that needs further study is the development of STS and VT with digitally provided emergency and crisis intervention services. This area of service delivery will most likely continue to grow and expand, especially in the post COIVD-19 pandemic world.

It is this willingness of providers to assist those who are hurting that is the avenue for STS and VT development. As a result of these potential dangers, awareness of STS and VT starting in preparatory programs and continuing education opportunities is needed. Additionally, ongoing trainings and boosters, prevention methods, and research are necessary to ensure safety of social service providers and continued harnessing of digital platforms for crisis situations.

CONCLUSIONS

STS and VT are occupational hazards for those who work in social services. In times of community disasters, providers are called upon to provide their expertise in human relationships to help people struggling with overwhelming stress. With the use of digital platforms, providers can deliver services to more individuals from a safe physical environment, but the risk to providers' emotional and psychological world is contemporaneous. Social service providers using digital platforms are more likely to be in professional silos, causing them to be distanced from supervisors or other co-workers who can provide much needed processing and consultation. Because providers are known for working extra hours, empathizing with those who are hurting, and forgoing self-care to assist others, they are especially vulnerable to STS and VT. Because digital services may be called upon during non-traditional office hours, those providing care could find themselves working even longer days than those providing face to face services. Ironically, the same characteristics that make effective providers also make them susceptible to secondary effects of trauma and serve as barriers to self-care.

Technology and digital platforms are great assets to social service agencies and can provide valuable services to hard-to-reach clients. In times of crisis, this need is even more urgent, emotionally charged, and graphic. With the commonality of personal technology, the demand for digitally based crisis services is certain to grow. Therefore, it is important to have conversations about how to support providers delivering digital services to protect them from the development of STS and VT. Fortunately, the increased attention on digitally provided crisis and disaster services can be used to prepare a new generation of social service providers to engage in self-care and prevent the development of STS and VT. Additionally, awareness of potential hazards will assist agencies with the adoption of new supportive policies, increase training opportunities, and encourage future research for ongoing efficacy in crisis services. These factors should serve as a call to action for targeted research efforts into the potential dangers of secondary trauma to providers, as well as the post-traumatic growth that can be harnessed to assist with career longevity. While this was already a contemporary conversation for social service providers due to the growth of technology as a platform for client services, the COVID-19 pandemic has forced many into the digital world whether they were ready or not. Therefore, understanding the development of STS and VT in those who provide services to traumatized clients and resulting prevention efforts is vitally important to ensure quality care for clients, provider longevity, and agency support of excellence in service delivery.

ACKNOWLEDGMENT

Special thank you to Michael Heath for his assistance with final preparation of manuscript. His talents are invaluable to this process.

This research received no specific grant from any funding agency in the public, commercial, or not-for-profit sectors.

REFERENCES

Adler, R. B., Rosenfield, L. B., & Proctor, R. F. (2017). *Interplay: The process of interpersonal communication* (14th ed.). Oxford University Press.

American Academy of Social Work and Social Welfare. (2019). *The 12 challenges*. Grand Challenges of Social Work. https://grandchallengesforsocialwork.org/grand-challenges-initiative/12-challenges/

Barsky, A. E. (2017). Social work practice and technology: Ethical issues and policy responses. *Journal of Technology in Human Services*, *35*(1), 1–12. doi:10.1080/15228835.2017.1277906

Barsky, A. E., & Reamer, F. G. (2018). *New practice standards on social work and technology: A situation-based discussion on best practices*. National Association of Social Workers, NASW Special Policy Services.

Baum, N., Rahav, G., & Sharon, M. (2014). Heightened susceptibility to secondary traumatization: A meta-analysis of gender differences. *The American Journal of Orthopsychiatry*, *84*(2), 111–122. doi:10.1037/h0099383 PMID:24826927

Ben-Porat, A. (2015). Vicarious post-traumatic growth: Domestic violence therapists versus social service department therapists in Israel. *Journal of Family Violence*, *30*(7), 923–933. doi:10.100710896-015-9714-x

Bent-Goodley, T. B. (2018). Being intentional about self-care for social workers. *Social Work*, *63*(1), 5–6. doi:10.1093wwx058 PMID:29177487

Berzin, S. C., Singer, J., & Chan, C. (2015). *Practice innovation through technology in the digital age: A grand challenge for social work* (Grand Challenges for Social Work Initiative Working Paper No. 12). American Academy of Social Work and Social Welfare.

Branson, D. C. (2019). Vicarious trauma, themes in research, and terminology: A review of literature. *Traumatology*, *25*(1), 2–10. doi:10.1037/trm0000161

Branson, D. C., & Miller, K. A. (in press). Social work ethics: Harnessing technology for the good of the discipline. *Journal of Sociology and Social Work*.

Branson, D. C., Radu, M. B., & Loving, J. D. (2019). Adverse Childhood Experiences (ACE) scores: When social work students and trauma mix. *The Journal of Baccalaureate Social Work*, *24*(1), 339–360. doi:10.18084/1084-7219.24.1.339

Bride, B. E. (2007). Prevalence of secondary traumatic stress among social workers. *Social Work*, *52*(1), 63–70. doi:10.1093w/52.1.63 PMID:17388084

Brown, S. (2018). The impact of resiliency on nurse burnout: An integrative literature review. *Medsurg Nursing*, *27*(6), 349–378.

Bureau of Labor Statistics. (2020, July). *Occupational outlook handbook, social workers*. U. S. Department of Labor. https://www.bls.gov/ooh/community-and-social-service/social-workers.htm

Butler, L. D., Mercer, K. A., McClain-Meeder, K., Horne, D. M., & Dudley, M. (2019). Six domains of self-care: Attending to the whole person. *Journal of Human Behavior in the Social Environment*, *29*(1), 107–124. doi:10.1080/10911359.2018.1482483

Callister, E., & Plante, T. (2017). Compassion predictors in undergraduates: A Catholic college example. *Pastoral Psychology*, *66*(1), 1–11. doi:10.100711089-016-0729-x

Carmona, M. (2020, May 13). *What's the difference between telehealth and teletherapy?* The Recovery Village. https://www.therecoveryvillage.com/treatment-program/online-counseling/faq/difference-between-telehealth-teletherapy/

Center for Medicare and Medicaid Services. (2020, March 17). *Medicare telemedicine health care provider facts sheet.* https://www.cms.gov/newsroom/fact-sheets/medicare-telemedicine-health-care-provider-fact-sheet

Cooper, M., & Lesser, J. G. (2014). *Clinical social work practice: An integrated approach* (5th ed.). Pearson Education, Inc.

Craun, S., Bourke, M., & Coulson, F. (2015). The Impact of Internet crimes against children work on relationships with families and friends: An exploratory study. *Journal of Family Violence, 30*(3), 393–402. doi:10.100710896-015-9680-3

Cummings, S. (2016). *Social work ethics and "everyday" technology.* National Association of Social Workers. NASW Press.

de Albuquerque, J. P., Herfort, B., Brenning, A., & Zipf, A. (2015). A geographic approach for combining social media and authoritative data towards identifying useful information for disaster management. *International Journal of Geographical Information Science, 29*(4), 667–689. doi:10.1080/13658816.2014.996567

DelTosta, J. E., Ellis, M. V., & McNamara, M. L. (2019). Trainee vicarious traumatization: Examining supervisory working alliance and trainee empathy. *Training and Education in Professional Psychology, 13*(4), 300–306. doi:10.1037/tep0000232

Egan, R., Maidment, J., & Connolly, M. (2017). Trust, power and safety in the social work supervisory relationship: Results from Australian research. *Journal of Social Work Practice, 31*(3), 307–321. doi: 10.1080/02650533.2016.1261279

Elbarazi, I., Loney, T., Yousef, S., & Elias, A. (2017). Prevalence of and factors associated with burnout among health care professionals in Arab countries: A systematic review. *BMC Health Services Research, 17*(1), 1–10. doi:10.118612913-017-2319-8 PMID:28716142

Fahy, A. (2007). The unbearable fatigue of compassion: Notes from a substance abuse counselor who dreams of working at Starbuck's. *Clinical Social Work Journal, 35*(3), 199–205. doi:10.100710615-007-0094-4

Figley, C. R. (Ed.). (1995). *Compassion fatigue: Coping with secondary traumatic stress disorder in those who treat the traumatized.* Routledge.

Finklestein, M., Stein, E., Greene, T., Bronstein, I., & Solomon, Z. (2015). Posttraumatic stress disorder and vicarious trauma in mental health professionals. *Health & Social Work, 40*(2), 25–31. doi:10.1093/hsw/hlv026

Finley, B. A., & Sheppard, K. G. (2017). Compassion fatigue: Exploring early-career oncology nurses' experiences. *Clinical Journal of Oncology Nursing, 21*(3), 61–66. doi:10.1188/17.CJON.E61-E66 PMID:28524893

Friedman, K. (2017). Counselor self-care and mindfulness. *Contemporary Buddhism, 18*(2), 321–330. doi:10.1080/14639947.2017.1373437

Giordano, A. L., Prosek, E. A., Stamman, J., Callahan, M. M., Loseu, S., Bevly, C. M., Cross, K., Woehler, E. S., Calzada, R. R., & Chadwell, K. (2016). Addressing trauma in substance abuse treatment. *Journal of Alcohol and Drug Education, 60*(2), 55–71.

Goldingay, S., & Boddy, J. (2017). Preparing social work graduates for digital practice: Ethical pedagogies for effective learning. *Australian Social Work, 70*(2), 209–220. doi:10.1080/0312407X.2016.1257036

Greinacher, A., Derezza-Greeven, C., Herzog, W., & Nikendei, C. (2019). Secondary traumatization in first responders: A systematic review. *European Journal of Psychotraumatology, 10*(1), 1–11. doi:10.1080/20008198.2018.1562840 PMID:30719236

Groshong, L., & Phillips, D. (2015). The impact of electronic communication on confidentiality in clinical social work practice. *Clinical Social Work Journal, 43*(2), 142–150. doi:10.100710615-015-0527-4

Grundlingh, H., Knight, L., Naker, D., & Devries, K. (2017). Secondary distress in violence researchers: A randomized trial of the effectiveness of group debriefings. *BMC Psychiatry, 17*(1), 1–14. doi:10.118612888-017-1327-x PMID:28578682

Hensel, J. M., Ruiz, C., Finney, C., & Dewa, C. S. (2015). Meta-analysis of risk factors for secondary traumatic stress in therapeutic work with trauma victims. *Journal of Traumatic Stress, 28*(2), 83–91. doi:10.1002/jts.21998 PMID:25864503

Herman, J. (1997). *Trauma and recovery: The aftermath of violence-from domestic abuse to political terror*. Basic Books.

Hopwood, T. L., Schutte, N. S., & Loi, N. M. (2019). Stress responses to secondary trauma: Compassion fatigue and anticipatory traumatic reaction among youth workers. *The Social Science Journal, 56*(3), 337–348. doi:10.1016/j.soscij.2018.08.008

Hricova, M., & Lovasova, S. (2018). Stress, secondary trauma and burnout: Risk characteristics in helping professions. *Ad Alta: Journal of Interdisciplinary Research, 8*(2), 161–165.

Iqbal, A. (2015). The ethical considerations of counselling psychologists working with trauma: Is there a risk of vicarious traumatisation? *Counselling Psychology Review, 30*(1), 44–51.

Jarrad, R. A., & Hammad, S. (2020). Oncology nurses' compassion fatigue, burnout and compassion satisfaction. *Annals of General Psychiatry, 19*(1), 1–8. doi:10.118612991-020-00272-9 PMID:32265998

Jasmontaite, L., & Dimitrova, D. (2017). Online disaster management: Applicability of the European data protection framework and its key principles. *Journal of Contingencies and Crisis Management, 25*(1), 23–30. doi:10.1111/1468-5973.12142

Ju, Y. J., Park, E. C., Ju, H. J., Lee, S. A., Lee, J. E., Kim, W., Chun, S. Y., & Kim, T. H. (2018). The influence of family stress and conflict on depressive symptoms among working married women: A longitudinal study. *Health Care for Women International, 39*(3), 275–288. doi:10.1080/07399332.2017.1397672 PMID:29095122

Kauer, S. D., Mangan, C., & Sanci, L. (2014). Do online mental health services improve help-seeking for young people? A systematic review. *Journal of Medical Internet Research, 16*(3), e66. doi:10.2196/jmir.3103 PMID:24594922

Kemp, S. (2019, January 30). *Digital trends 2019: Every single stat you need to know about the Internet.* The Next Web. https://thenextweb.com/contributors/2019/01/30/digital-trends-2019-every-single-stat-you-need-to-know-about-the-internet/

Kerst, A., Zielasek, J., & Gaebel, W. (2020). Smartphone applications for depression: A systematic literature review and a survey of health care professionals' attitudes towards their use in clinical practice. *European Archives of Psychiatry and Clinical Neuroscience, 270*(2), 139–152. doi:10.100700406-018-0974-3 PMID:30607530

Knight. (2010). Indirect trauma in the field practicum: Secondary traumatic stress, vicarious trauma, and compassion fatigue among social work students and their field instructors. *The Journal of Baccalaureate Social Work, 15*, 32-57.

Knowles, A. J., & Cooner, T. S. (2016). International collaborative learning using social media to learn about social work ethics and social media. *Social Work Education, 35*(3), 260–270. doi:10.1080/02615479.2016.1154662

Kulkarni, S., Bell, H., Hartman, J. L., & Herman-Smith, R. L. (2013). Exploring individual and organizational factors contributing to compassion satisfaction, secondary traumatic stress, and burnout in domestic violence service providers. *Journal of the Society for Social Work and Research, 4*(1), 114-130. https://doi. doi:10.5243/jsswr.2013.8

Kunst, M. J. J., Saan, M. C., Bollen, L. J. A., & Kuijpers, K. F. (2017). Secondary traumatic stress and secondary posttraumatic growth in a sample of Dutch police family liaison officers. *Stress and Health, 33*(5), 570–577. doi:10.1002mi.2741 PMID:28127898

Landers, A. L., Dimitropoulos, G., Mendenhall, T. J., Kennedy, A., & Zemanek, L. (2020). Backing the blue: Trauma in law enforcement spouses and couples. *Family Relations, 69*(2), 308–319. doi:10.1111/fare.12393

Malhotra, S., Chakrabarti, S., & Shah, R. (2019). A model for digital mental healthcare: Its usefulness and potential for service delivery in low- and middle-income countries. *Indian Journal of Psychiatry, 61*(1), 27–36. doi:10.4103/psychiatry.IndianJPsychiatry_350_18 PMID:30745651

Martinengo, L., Van Galen, L., Lum, E., Kowalski, M., Subramaniam, M., & Car, J. (2019). Suicide prevention and depression apps' suicide risk assessment and management: A systematic assessment of adherence to clinical guidelines. *BMC Medicine, 17*(1), 1–12. doi:10.118612916-019-1461-z PMID:31852455

McAuliffe, D., & Nipperess, S. (2017). e-Professionalism and the ethical use of technology in social work. *Australian Social Work, 70*(2), 131–134. doi:10.1080/0312407X.2016.1221790

McCann, I. L., & Pearlman, L. A. (1990). *Psychological trauma and the adult survivor: Theory, therapy, and transformation.* Brunner and Mazel.

National Association of Social Workers. Council on Social Work Education, Association of Social Work Boards, & Clinical Social Work Association. (2017). Practice standards on social work and technology. NASW Press.

National Institute of Mental Health. (2019, September). *Technology and the future of mental health treatment*. https://www.nimh.nih.gov/health/topics/technology-and-the-future-of-mental-health-treatment/index.shtml

Netting, F. E., Kettner, P. M., McMurtr, S. L., & Thomas, M. L. (2017). *Social work macro practice*. Pearson Education, Inc.

Newsom, K., & Myers-Bowman, K. (2017). "I am not a victim. I am a survivor": Resilience as a journey for female survivors of child sexual abuse. *Journal of Child Sexual Abuse*, *26*(8), 927–947. doi:10.1080/10538712.2017.1360425 PMID:28857725

Nguyen, M. X., Go, V. F., Bui, Q. X., Gaynes, B. N., & Pence, B. W. (2019). Perceived need, barriers to and facilitators of mental health care among HIV-infected PWID in Hanoi, Vietnam: A qualitative study. *Harm Reduction Journal*, *16*(1), 1–9. doi:10.118612954-019-0349-8 PMID:31878934

Oliver, D. P., Washington, K., Wittenberg-Lyles, E., Gage, A., Mooney, M., & Demiris, G. (2015). Lessons learned from a secret Facebook support group. *Health & Social Work*, *40*(2), 125–133. doi:10.1093/hsw/hlv007 PMID:26027421

Pearlman, L. A., & Saakvitne, K. W. (1995). *Trauma and the therapist: Countertransference and vicarious traumatization in psychotherapy with incest survivors*. Norton.

Pryce, J. G., Shackelford, K. K., & Pryce, D. H. (2007). *Secondary traumatic stress and the child welfare professional*. Lyceum Books.

Quitangon, G. (2019). Vicarious trauma in clinicians: Fostering resilience and preventing burnout. *The Psychiatric Times*, *36*(7), 18–19.

Reamer, F. G. (2014, July). Eye on ethics: Elastic boundaries in social work-proceed with caution. *Social Work Today*. https://www.socialworktoday.com/news/eoe_070214.shtml

Reamer, F. G. (2015). Clinical social work in a digital environment: Ethical and risk-management challenges. *Clinical Social Work Journal*, *43*(2), 120–132. doi:10.100710615-014-0495-0

Reamer, F. G. (2017, August). Eye on ethics: New NASW code of ethics standards for the digital age. *Social Work Today*. https://www.socialworktoday.com/archive/081617.shtml

Rees, C. S., & Maclaine, E. (2015). A systematic review of videoconference-delivered psychological treatment for anxiety disorders. *Australian Psychologist*, *50*(4), 259–264. doi:10.1111/ap.12122

Roberts, T., & Marchais, G. (2018). Assessing the role of social media and digital technology in violence reporting. *Contemporary Readings in Law and Social Justice*, *10*(2), 9–42. doi:10.22381/CRLSJ10220181

Roden-Foreman, J. W., Bennett, M. M., Rainey, E. E., Garrett, J. S., Powers, M. B., & Warren, A. M. (2017). Secondary traumatic stress in emergency medicine clinicians. *Cognitive Behaviour Therapy*, *46*(6), 522–532. doi:10.1080/16506073.2017.1315612 PMID:28452256

Russell, M., & Cowan, J. (2018). The making of compassion stress injury: A review of historical and etiological models toward a de-stigmatizing neurobehavioral conceptualization. *Challenges*, *9*(1), 7. doi:10.3390/challe9010007

Sansbury, B. S., Graves, K., & Scott, W. (2015). Managing traumatic stress responses among clinicians: Individual and organizational tools for self-care. *Trauma, 17*(2), 114–122. doi:10.1177/1460408614551978

Schaefer, R. T. (2017). *Sociology: A brief introduction* (12th ed.). McGraw Hill Education.

Segal, E. A., Gerdes, K. E., & Steiner, S. (2019). *An introduction to the profession of social work: Becoming a change agent* (6th ed.). Cengage.

Shannon, P. J., Simmelink-McCleary, J., Im, H., Becher, E., & Crook-Lyon, R. E. (2014). Experiences of stress in a trauma treatment course. *Journal of Social Work Education, 50*(4), 678–693. doi:10.1080/10437797.2014.947901

Sodeke-Gregson, E. A., Holttum, S., & Billings, J. (2013). Compassion satisfaction, burnout, and secondary traumatic stress in UK therapists who work with adult trauma clients. *European Journal of Psychotraumatology, 4*(1), 1–10. doi:10.3402/ejpt.v4i0.21869 PMID:24386550

Substance Abuse and Mental Health Services Administration. (2014). Trauma-informed care in behavioral health services. Treatment Improvement Protocol (TIP) Series 57. Author.

Tehrani, N. (2016). Extraversion, neuroticism and secondary trauma in Internet child abuse investigators. *Occupational Medicine, 66*(5), 403–407. doi:10.1093/occmed/kqw004 PMID:26928859

Thieleman, K., & Cacciatore, J. (2014). Witness to suffering: Mindfulness and compassion fatigue among traumatic bereavement volunteers and professionals. *Social Work, 59*(1), 34–41. doi:10.1093wwt044 PMID:24640229

Thompson, I. A., Amatea, E. S., & Thompson, E. S. (2014). Personal and contextual predictors of mental health counselors' compassion fatigue and burnout. *Journal of Mental Health Counseling, 36*(1), 58–77. doi:10.17744/mehc.36.1.p61m73373m4617r3

Torniainen-Holm, M., Pankakoski, M., Lehto, T., Saarelma, O., Mustonen, P., Joutsenniemi, K., & Suviaari, J. (2016). The effectiveness of email-based exercises in promoting psychological wellbeing and healthy lifestyle: A two-year follow-up study. *BMC Psychology, 4*(21), 1–12. https://doi.org.doi:10.118640359-016-0125-4 PMID:27184251

Torous, J., Jän Myrick, K., Rauseo-Ricupero, N., & Firth, J. (2020). Digital mental health and COVID-19: Using technology today to accelerate the curve on access and quality tomorrow. *JMIR Mental Health, 7*(3), e18848. doi:10.2196/18848 PMID:32213476

Turgoose, D., Ashwick, R., & Murphy, D. (2018). Systematic review of lessons learned from delivering tele-therapy to veterans with post-traumatic stress disorder. *Journal of Telemedicine and Telecare, 24*(9), 575–585. doi:10.1177/1357633X17730443 PMID:28958211

United State Bureau of Labor Statistics. (2019). *Labor force statics from the current population survey.* https://www.bls.gov/cps/cpsaat11.htm

Veil, S. R., Buehner, T., & Palenchar, M. J. (2011). A work-in-process literature review: Incorporating social media in risk and crisis communication. *Journal of Contingencies and Crisis Management, 19*(2), 110–122. doi:10.1111/j.1468-5973.2011.00639.x

Voss Horrell, S. C., Holohan, D. R., Didion, L. M., & Vance, G. T. (2011). Treating traumatized OEF/OIF veterans: How does trauma treatment affect the clinician? *Professional Psychology, Research and Practice*, *42*(1), 79–86. doi:10.1037/a0022297

Waegemakers-Schiff, J., & Lane, A. M. (2019). PTSD symptoms, vicarious traumatization, and burn-out in front line workers in the homeless sector. *Community Mental Health Journal*, *55*(3), 454–462. doi:10.100710597-018-00364-7 PMID:30684127

Willis, N. G., & Molina, V. (2019). Self-care and the social worker: Taking our place in the code. *Social Work*, *64*(1), 83–86. doi:10.1093wwy049 PMID:30365017

Wray, L. O., Pikoff, E., King, P. R., Hutchison, D., Beehler, G. P., & Maisto, S. A. (2016). Veterans' mental health beliefs: Facilitators and barriers to primary care-mental health use. *Families, Systems & Health*, *34*(4), 404–413. doi:10.1037/fsh0000231 PMID:27819439

Wukich, C. (2016). Government social media messages across disaster phases. *Journal of Contingencies and Crisis Management*, *24*(4), 230–243. doi:10.1111/1468-5973.12119

Zhou, X., Snoswell, L. E., Harding, L. E., Bambling, M., Edrippulige, S., Bai, X., & Smith, A. C. (2020). The role of telehealth in reducing the mental health burden from COVID-19. *Telemedicine Journal and e-Health*, *26*(4), 1–3. doi:10.1089/tmj.2020.0068 PMID:32202977

ADDITIONAL READING

Campbell, M. (2018). An engaged electronic medical record for social good. *Health & Social Work*, *43*(2), 131–133. doi:10.1093/hsw/hly001 PMID:29462478

Foreman, T. (2018). Wellness, exposure to trauma, and vicarious traumatization: A pilot study. *Journal of Mental Health Counseling*, *40*(2), 142–155. doi:10.17744/mehc.40.2.04

Gobin, R. L. (2019). *The self-care prescription: Powerful solutions to manage stress, reduce anxiety, and increase well-being*. Althea Press.

Goldkind, L., & Wolf, L. (2015). A digital environment approach: Four technologies that will disrupt social work practice. *Social Work*, *60*(1), 85–87. doi:10.1093wwu045 PMID:25643579

Grise-Owens, E., Miller, J., & Eaves, M. (2016). *The A-to-Z self-care handbook for social workers and other helping providers*. The New Social Worker Press.

Phillips, J. (2018). Risk in a digital age: Understanding risk in virtual networks through digital response networks (DRNs). *International Development Planning Review*, *40*(3), 239–272. doi:10.3828/idpr.2018.18

Ries, E. (2019). Beating burnout. *PT in Motion*, *11*(1), 28–39.

Smith, J. (2015). *Master the media: How teaching media literacy can save our plugged-in world*. Dave Burgess Consulting, Inc.

Somoray, K., Shakespeare, F. J., & Armstrong, D. (2017). The impact of personality and workplace belongingness on mental health workers' professional quality of life. *Australian Psychologist*, *52*(1), 52–60. doi:10.1111/ap.12182

Taylor, A. (2017). Social work and digitalisation: Bridging the knowledge gaps. *Social Work Education*, *36*(8), 869–879. doi:10.1080/02615479.2017.1361924

Wood, A., Prins, A., Bush, N., Hsia, J., Bourn, L., Earley, M., Walser, R., & Ruzek, J. (2017). Reduction of burnout in mental health care providers using the Provider Resilience mobile application. *Community Mental Health Journal*, *53*(4), 452–459. doi:10.100710597-016-0076-5 PMID:28070775

KEY TERMS AND DEFINITIONS

Boundaries: Overt and/or covert limits that exist to encourage a healthy and appropriate relationship and serve as guidelines for behavior.

Ethics: Professional norms that are developed from an agreed upon set of values that guide practice behavior and decision making.

Post-Traumatic Growth: Positive emotions, cognitions, and outcomes from trauma that come about through strategically selective reflections. Purposefully focusing on positives of traumatic aftermath, adopting an attitude of gratitude, and searching for ways to be empowered by lessons learned from the past events.

Self-Care: Discretionary and enjoyable activities that provide relaxation, stress-relief, and renewal of body, mind, and spirit. These activities should be leisure in nature, not chores or other "to do" list items that might provide relief because a task has been completed, but do not provide a sense of renewal.

Social Media: Electronic platforms that allow for the sharing of information over the Internet.

Social Service Provider: A professional who has completed the required curriculum and training to provide mental health, crisis, trauma-informed care, and other social service duties to assist those struggling, oppressed, and/or vulnerable in society.

Technology: Evolving devices and/or tools that increase efficiency in work. For the purposes of this chapter, technology is the collection of devices and/or tools that provide new means of communication and service provisions, specifically over electronic pathways.

Chapter 15
Information as Humanitarian Aid:
Delivering Digital Services to Empower Disaster-Affected Communities

Kenny Meesters
https://orcid.org/0000-0002-8324-956X
Delft University of Technology, The Netherlands

Yan Wang
https://orcid.org/0000-0002-6317-7546
Delft University of Technology, The Netherlands

ABSTRACT

For a plethora of decisions we make on a daily basis, we can rely on timely, tailored information delivered via digital services. The technologies and the knowledge needed for the design, development and delivery of digital services have become increasingly accessible. These developments have also made their way to the crisis response domain and resulted in a variety of digital services to deliver information, supporting emergency responders in their decision-making processes. At the same time, affected communities by disasters are also in need of information during such critical events. Timely, relevant information helps affected citizens to understand the situation, make informed decisions, and gain access to life-saving aid. However, designing, creating, and delivering digital services to deliver vital information to communities warrant specific considerations. This chapter per the authors explores the concept of digital services in regard to empowering communities affected by critical events.

INTRODUCTION

Every day, each and every one of us makes decisions. Decisions related to travel or transport, decision related to consumption, how we choose to spend our time. While the decisions we make and the freedom to make choices vary greatly, for every decision we rely on information to support us in the decision-

DOI: 10.4018/978-1-7998-6705-0.ch015

making process (Malhotra, 1982). The availability, price and time, along with the weather may play role in our decision to use certain modes of transport. The availability and prices of food, or inspiration from recipes may influence our choices of food consumption. And we can rely on information from friends, communities or even strangers online to determine how we spend our time. (Lelis & Howes, 2011).

Today, advances in information and communication technology (ICT) provide various options to deliver the right and relevant information at the right time, at the right place and in a understandable manner (Elsey et al., 2002) to us, supported by an increased availability and diversity of data. Weather apps provide localized weather forecasts based on the GPS module of a phone, websites allow you to find specific recipes for a certain budget, and rating & reviews of TV-show are displayed right along the list of available shows. For almost every situation, information need, or decision faced, a digital service exists to deliver information to the user. Or as Apple advertised their new iPhone in 2009: "There is an app for that"(Apple, 2009).

However, for some crucial decisions people accessing this information is not trivial matter, in some situations people may not have acces to technologies, the platforms may not contain the necessary information, or the digital services are not available. (Weber, 1987). Despite the reliable technology, connected platforms and services, we still do not always find all the information needed. Especially in times of crisis, when people are forced to make important decisions, this critical information can be absent (Comes et al., 2019). However, it is exactly at these times that delivering the right information, at the right time to the right person can make a crucial difference and transform communities from victims to resilient and empowered actors (Baharmand et al., 2016).

This chapter examines the critical role of information for disaster-affected communities, the challenges involved in delivering this information, and the potential of digital services to overcome these challenges. Two cases of recent disasters are used to illustrate these challenges, considerations and approaches. . It also considers how information systems and digital services have addressed the information needs of communities and individuals affected by disruptive events and empowered them in their individual and joint decision-making processes. From these cases key lessons learned, success factors and design principles that empower communities through digital services are extracted. The chapter also examines what challenges, risks, and other effects one encounters when developing and deploying these services.

THE AGE OF DIGITAL SERVICES

ICT advancements have contributed to a proliferation of digital services in many different fields, including governmental services ('government to citizen') (Fang & management, 2002), improved service to customer in the commercial sector ('business to consumer'), between consumers to exchange goods and services ('consumer to consumer'), and many other interactions. Meanwhile the acceptance of employing digital services in daily life by the society and individuals accelerates its adoption and subsequently reinforces the growth of the digital information era (Gantz & Reinsel, 2012).

The democratization of information technologies, including more accessible programming languages, cloud services, and public platforms have enabled more and more organizations to develop and offer services as an extension of their business model. Moreover, these developments are no longer exclusive to companies with large ICT-budgets or people with a highly specialized knowledge. Today, smaller organizations and even individuals have access to both the technologies and knowledge to develop digital services. In fact, thanks to digital services, this access to the building blocks has increased globally. For example, there are numerous websites that offer free tutorials on programming, cloud providers offer low-cost entry plans for providing infrastructure, and social media allow people to promote or exchange ideas. The push towards digital services is further spurred on by economic initiatives and the increased attention to STEM-education across the globe (National Research Council, 2011)

In addition to the increased accessibility of ICT from the supply side, what really boosts the popularity of digital services is the societal embrace. Many initiatives, such as Germany's industry 4.0 and China's 'internet plus' a few years ago, have been promoted substantively to encourage the integration of digitalization in traditional services. The convenience in daily work and life brought by ICT has been quickly recognized. The demand for faster, cheaper, and more innovative digital services soars due to the available new technologies and the inspiration triggered from existing services. When people are familiar with new technologies, they tend to start using it for new things or in new ways (Scarborough et al., 1991). For instance, mobile commerce has already infiltrated into the daily life of a large portion of the population (Lee et al., 2015) while mobile payment and digital governments are common services nowadays in many developed and developing countries.

DEFINING DIGITAL SERVICES DEVELOPMENT & DELIVERY

Before we look into the context of disaster management, let us get familiar with what a typical lifecycle of digital services look like and understand the unique characteristics of digital service provision. A digital service is usually delivered through a series of iterative activities across planning, design, development, and delivery. This sounds similar to the software development life cycle (SDLC) which includes stages of planning, analysis, design, implementation, and maintenance (Ragunath et al., 2010). If only looking from a product perspective, digital services include use of digital components which could follow the SDLC if new ICT development is planned. Nevertheless, the concept of a digital service is broader than software. It combines both ICT system and service management (Williams et al., 2008). Therefore, it is helpful to adopt a holistic view that integrates methods and practices from these two fields in digital service delivery. The following components highlight the key considerations of providing digital services.

Planning and Design

Service provision is in essence a value co-creation process that configures the interactions of people, technology, the shared information, and the value proposition concerning the internal and external service systems (Maglio & Spohrer, 2008). Such a co-creation process can be achieved via a user-centered (Tien et al., 2003) approach throughout the service lifecycle.

The first step in new digital service planning is often the requirement engineering (Wegmann et al., 2008). It is a user-centered process (Düchting et al., 2007) that looks into the challenges the end users face, collects their needs, assesses the feasibility, and selects available solution directions. In addition, the value delivered by the service envisioned should be configured in an environment that represents the same conditions where the users encounter. Understanding the essential problems and getting the right requirements are crucial to define a clear concept of the service. In service management, a good service concept is of importance to the service delivery (Goldstein et al., 2002). The design requirements focus on services perceived added value by customers. Thus, it is critical to ensure that the same vision is well communicated and understood by all involved stakeholders.

One unique characteristic of digital services is that it does not necessary requires the delivery of new ICT systems, as the keyword here is 'service'. Digital services come in different forms of digitization, ranging from pure computational services like automated data analysis, to manual/human service delivered through a digital platform such as a call center service. A clear vision of the service and its implementation environment allows providers to evaluate and design the most suitable solutions according to needs (Hevner et al., 2004).

Since the wide adoption of service-oriented architecture (SOA) (Bieberstein et al., 2005) in business system design, software component and service reuse has led to tremendous innovation in digital service design. It is popular to adopt a best-of-breed manner in composing existing web services in new services and consider including existing service platforms that are already adopted by end users in the new design.

Development

The development of digital services usually involves multiple roles for various tasks that are defined in the plan according to requirements. Depending on the development scale, there could be teams working on providing service contents, developing the digital components, defining the service workflow, or managing the implementation handover (Heikkinen et al., 2007). Here it discusses a few important development issues regarding service delivery.

Outsourcing is common in digital service development, especially those who build upon other existing (digital) service components. The clear benefits of doing development in an outsourcing environment is to make good use of the core capabilities and expertise. In an agile development approach (Dingsøyr et al., 2012), the service providers or contracted suppliers can better allocate their capacity in specialized tasks.

The specification and division of development tasks also brings to another important issue in service delivery, i.e. communication and coordination in the service networks. Stakeholders involved in the delivery process interact with one or more of others on the tasks they are assigned to. This means that not all of them would know about each other and the service components the others work on. For instance, the end users would care more about a user-friendly interface rather than the system backend design; the content providers focus on providing the right material. However they are not involved in developing the data processing algorithm.

An important practice to ensure continuous service delivery is the DevOps approach (Virmani, 2015), which can also be widely applied in digital service delivery. The digital service delivery is a series of collaborative activities that involve users and domain experts in various steps. Each step or task has its own scope and objectives due to the current culture of agile development. Handing over the service product crossing different teams could cause layers of information filtering and feedback gaps. It is important for different stakeholders to be aware of the divergent scope across the delivery lifecycle and actively create a feedback loop

Implementation and Governance

Once the development is completed, digital services are ready to go live. The implementation of digital services is not only centered around technology deployment. It also includes the acceptance and adoption of the service within the target user network. What really makes digital services active in a long term is how well the services fulfill the community demands and are able to evolve according to the potential change of demands over time.

One of the most convincing indicators about the service adoption is the number of active users. The social network of the active users disseminates the information about perceived service performance swiftly (Christakis & Fowler, 2013). One reason behind this word-of-mouth effect is the empathy among the users and their network, i.e. they may encounter the same challenges or face similar situations. To facilitate the social integration, one important task is to provide continuous service operation and maintenance. A good governance structure is thus needed to coordinate service components and capture users' feedback. It is equally important to reserve or build *governance capacity* in the post-implementation stage as getting development capacity.

POTENTIAL OF DIGITAL SERVICES IN DISASTER RESPONSE

Due to climate change, extreme weather events, and increased urbanization more and more people around the world are at risk be confronted to disasters (International Federation of Red Cross, 2006; Thomas & López, 2015). Dense population, complex infrastructure, and intense resource use have made communities and cities susceptible to disasters (Korset, 2013). Over the past decade efforts in disaster risk reduction, preparedness, and improvements in emergency response have reduced immediate loss of life. Nevertheless, the secondary effects and lasting impacts have increased (Thomas & López, 2015).

Figure 1. The number of natural disasters is increasing, and while the (direct) fatalities are declining, the number of people affected are increasing (EM-DAT, 2020)

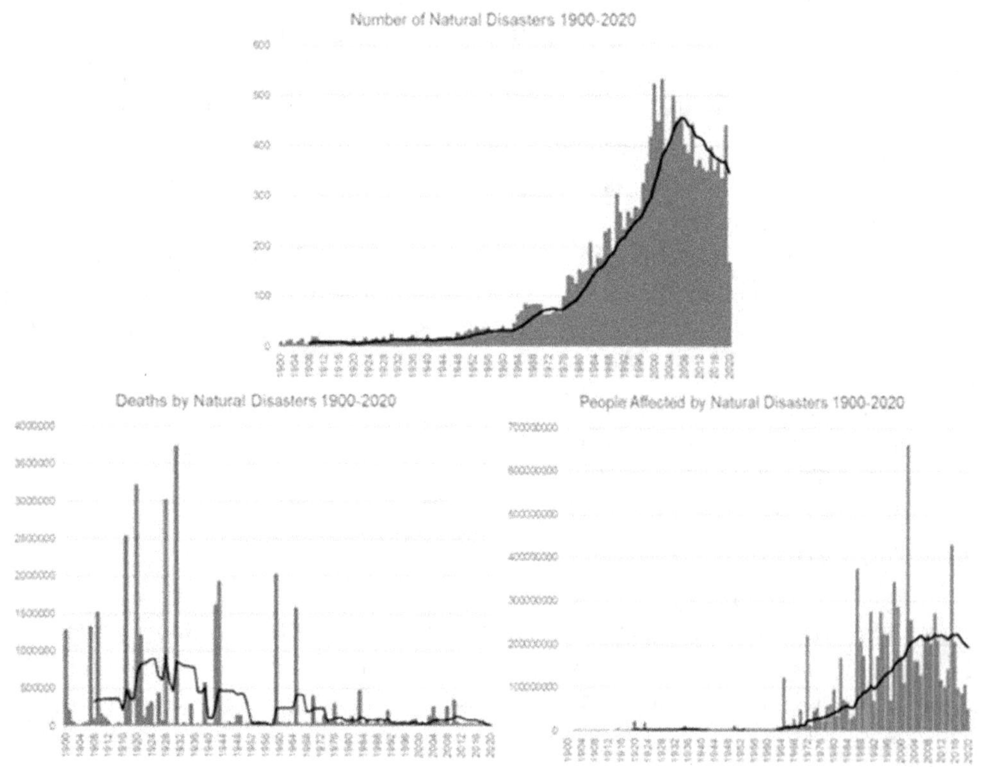

In the aftermath of these events, many responders, agencies, and other stakeholders are faced with making decisions in highly uncertain and volatile situations. Information is vital to reducing this uncertainty and enable decision makers to effectively allocate scarce resources, anticipate humanitarian needs, or secure additional capacity. Therefore, it is no surprise that the important role of information has long been recognized in the field of crisis response, disaster management and humanitarian aid (Comfort et al., 2004).

Information Resides in Dynamic Environment

When a disaster strikes, efficiently processing information of the affected communities and effectively communicating it with key stakeholders are important premises of making right decisions to respond to and recover from hazardous situations (Hong et al., 2018). During a large-scale emergency, the number of stakeholders can be numerous, each faced with a large number of decisions (Gralla et al., 2015).

There have been numerous studies examining these decisions, the related information needs, and potential for technologies to support the decisions making process (Lissenden et al., 2015; van den Homberg et al., 2014; Weijman & Meesters, 2020). Such studies often focus on governmental agencies or international organizations at an institutional level, and on humanitarian aid workers at an individual level. Often because these organizations and individuals have a formal mandate, or principal responsibility in mitigating the effects of a disaster. The decisions that these organizations face, are often also more clearly delineated, their decision-making process is more clearly defined, or more information is documented. For example, in standard operating procedures (SOPs) used by Search & Rescue teams, in missions' reviews in after-action reports, or information stored throughout the missions such as published documents and logbooks.

Often the communities that are affected play an important part in the data-collection process (i.e. needs assessment) (Guha-Sapir & Lechat, 1986; Hong et al., 2018). The information collected helps emergency responders and agencies to understand the effects of the disaster on the community and plan accordingly. However, this is in sharp contrast to the core humanitarian principles, which include "the participation of beneficiaries" and "use information activities to portray victims as dignified human beings, not hopeless objects" (Gostelow, 1999; Hilhorst & Schmiemann, 2002). More importantly the "The principle of humanity", states that humanitarian action must "ensure respect for the individual." (American Red Cross, 2015). Local affected communities are also key decision makers. They do not only have important local knowledge and resources but are also looking for information to inform their individual and community (joint) decision making process (Cardia et al., 2017; Raymond et al., 2015).

However, the information needs, and decision-making processes of communities are less clearly defined compared to those of emergency responders. Foremost, communities are heterogeneous groups. And while they often share similar values and belief, individual community members each face their own unique situations (Twigg, 2009). They live in different circumstances and may have been affected differently by the disaster. Therefore the information needs and decision-making processes of a community can be wide and varied between and within communities (Baharmand et al., 2016). In addition, the needs can quickly evolve and expand throughout a crisis (Meesters et al., 2019).

Information as Humanitarian Aid

Providing correct, timely and relevant information supports affected communities in their decision making. In similar fashion to formal responders and agencies, it enables communities to use their resources in the most effective manner. Information can also provide communities with a situational overview that allows them to make key decisions about their immediate response as well as their recovery process (Korset, 2013). Moreover, the ability to collect, process and share information transforms communities from a passive role to an active one and enables them to engage with other stakeholders and participate in joint decision-making processes with actors (Helsloot et al., 2004; Hong et al., 2018).

In short, information and digital services allows communities to transition from pure information provider to an active decision maker (Leong et al., 2015). Humanitarian aid provision and decision making relies on the availability of quality information. In many situations, the bottleneck of aid delivery, is no longer the shortage of tangible commodities. It is rather the limited access to critical information (Gralla et al., 2015). Therefore, information is a life-saving item for affected communities and providing quality information is a humanitarian aid activity. Today, technology allows us to develop digital services with the potential to provide this crucial and lifesaving information to a wide community. However, designing, developing, and deploying such systems in these situations generates specific design requirements and warrant special considerations.

Digital Services to Manage Information

Over the last decade, many advances in the field of computer science, ICT and data science have resulted in the development of new digital systems, services and platforms that enable crisis responders, government agencies, and humanitarian organizations to rapidly collect, process and share critical, life-saving information (Initiative, 2010; Meier, 2011; Meissner et al., 2002).

These developments contributed to myriad of digital services that have been used in crisis responses around the world. These digital services can span across broad range of applications; from collaborative mapping efforts to platforms for the exchange of data, and from using algorithms to improve the information quality to virtual agents providing contextual information (Imran et al., 2014; Meier, 2015; Tsai et al., 2019). And many other developments gloom over the horizon. Each of these developments provide a service through and with ICTs with the aim to collect, process and present the right information, at the right time to the right person. To have a closer look at digital services used for community response and recovery, two real-world cases of sudden onset disasters are presented in the following sections. It focuses on the approach of the service development and delivery, as well as the effectiveness and long-term impact on empowering local communities.

CASE 1: NEPAL EARTHQUAKE AND KATHMANDU LIVING LABS

In April 2015 Nepal, and specifically the Gorkha district, was struck by a 7.8 Mw earthquake. The shocks of the earthquake were felt in southwestern parts of China, Pakistan, and Bhutan, up to 1900 km away from the epicenter. The earthquake was followed by over 38 aftershocks in the days after, in strength up to 6.8 Mw. As a result of the earthquake and the aftershocks, nearly 9000 people lost their lives and over 22.000 people were injured. The earthquake left thousands of houses damaged, and even entire villages destroyed. The earthquake also directly and indirectly caused a major economic loss for a country with an already fragile economy.

Following the earthquake, a large-scale national and international crisis response and humanitarian aid effort was mobilized. Local communities, groups, private companies, and other countries support the immediate response efforts. Over 450 organizations responded to the earthquake within the first few days. For the international response over 100 international search and rescue and medical teams arrived in Nepal within 24 hours mainly through the Tribhuvan International Airport in Kathmandu.

As the earthquakes affected predominantly remote mountain villages, rescue and humanitarian operations took place in extremely challenging terrain. These challenging circumstances and the difficult access to the affected areas, meant that many Nepalese in the hard-hit areas had to largely fend for themselves in the first days. This also meant that information about the situation in these areas was limited in these first days. Nevertheless, gaining this situational awareness was key for the delivery of critical aid and planning of rescue operations. In lieu of physical access and the possibility to assess the situation in person, organizations had to rely on remote mapping options and digital services to exchange information.

Kathmandu Living Labs

Kathmandu Living Labs (KLL) is a non-profit organization that focuses on the civic use and development of technology. Specifically, KLL is focused on initiatives that involve communities in data collection, mapping and generating overviews. In the 2015 Nepal earthquake Kathmandu Living Labs played an important part in the crisis response. KLL mobilized their members and within a day launched the website 'QuakeMap.org'. The website employed a combination of crowdsourcing, mapping, and social media technologies to gather and map data coming from a variety of channels and reports (Poiani et al., 2016). Additional geographical information (based on satellite imagery) was provided by remote volunteers, in collaboration with the Humanitarian OpenStreetMap team.

The result of these efforts were a range of maps and information products that provided key insights into the disaster affected areas. On April 27, less than 48 hours after the initial earthquake, over 1500 joined the mapping efforts. Additional information was collected through social media channels and interactions as well as traditional text-messaging systems when internet connectivity was not available or limited. As a specific example, the Nepal Army used over 300 QuakeMap reports marked as 'critical for prioritizing and planning its emergency rescue operations (Kathmandu Living Labs, 2015).

However, the key value of KLL was not only delivered through the use of technology. However, most of all their knowledge of the local context, their team capacities, and their network consisting of local and international connections. The latter transformed KLL into a linking pin between their network of local actors and knowledge holders on the one hand, and the international community of responders on the other hand. This enabled KLL to provide an added value in the form of information to both the communities delivered through their digital services. For example, providing information to the international responders on the situation in the affected and hard to reach areas. Or informing affected communities about aid mobilization and relief or response efforts undertaken.

Figure 2. KLL working space in 2015 (Kathmandu Living Labs)

Post-Earthquake

In the wake of the earthquake, as with most crisis situations, the situation stabilized over time. While a few months later many Nepalese still suffered from the (secondary) effects of the earthquake, the chaos of the initial response to the situation stabilized. The humanitarian response entered the next stage and moved to recovery and reconstruction efforts rather than rescue and relief. While the programs continued to run and support the recovery efforts throughout the country, the operational presence was being reduced. Organizations ended their (initial) programs, for example due to the end of funding, or transferred their programs to local organizations and government agencies.

As the situation stabilized, so did the information flow. And as the country's digital infrastructure was being restored many rapidly deployed digital services were phased out. Either because they were no longer necessary, made redundant by native Nepalese digital services, or the organization providing the service was no longer active. Nevertheless, Kathmandu Living Labs, as a local organization, remains active in Nepal. Kathmandu Living Labs has used the momentum to further develop their digital services in support of both the Nepalese government agencies, specifically in the field of humanitarian aid and emergency response as well as the Nepalese communities.

The international attention and assistance that KLL has received following their contribution in the earlier stages of the emergency response was now leveraged to further establish the organization and expand their activities. Over time KLL has increasingly involved local communities in their projects and focused on developing digital services that support the information needs of local communities. Moreover, their projects increasingly include capacity building to empower communities (Kathmandu Living Labs, 2016). Through their projects and experiences KLL keeps exploring the possibilities, expanding technological know-how, and developing the network of people, organizations, and communities. They remain ready to scale up and use these building blocks to provide and support even more improved digital services in a next emergency.

Capacity Building

The case of Kathmandu Living Labs illustrates several key considerations for the development of the digital services. First, KLL employed several well-known and widely adopted systems and approaches for the delivery of the digital services, such as OpenStreetMap for mapping satellite imagery and crowdsourcing approaches for the collection of data. Rather than developing their own approaches and completely new tools, the use of existing 'building blocks' allowed them to not only quickly deploy services, it also ensures their compatibility with other initiatives. Second, they build on *local capacities and networks*. Through their contacts with the local communities KLL was able to mobilize local knowledge, and firsthand observations from the field. However, their networks were not only used for the provision and validation of the information. Moreover, it also to disseminate key information to communities. For example, updates on the international response efforts and incoming assistance. Working closely with news agencies and radio's stations, it created a reciprocal feedback loop that added value to all users of the services.

Finally, and most important, the local and permanent presence of KLL provided both the international humanitarian organizations as well as the local organizations with an 'anchoring' point. As the nature and scale of crisis and humanitarian organizations changed, KLL provided a strong base for the design, development, and delivery of digital services. Not only in terms of technical resources and capacities but through the knowledge of the local context, their networks, and their ability to build on existing socio-technical systems.

CASE 2: JAKARTA FLOODS AND HUMANITARIAN OpenStreetMap

Over the past years the city of Jakarta, Indonesia has reportedly sunk by 2.5 meters in the last 10 years and is continuing to sink at a rate of up to 15 cm per year in some areas. The city and wider metropolitan area face frequent flooding due to this sinking rate, heavy urban development, the increasing frequency of extreme weather events, and sea-level rise. The land subsidence coupled with the increased flood risk have exposed communities in the area to regular flooding events. In February 2020 Jakarta was hit again by rapid floods due to heavy rains. Overnight, the rivers burst the banks and water flooded residential and commercial areas up to five feet deep, forcing thousands to evacuate their homes. The floods also caused outages at the Cipto Mangunkusumo hospital, damaging medical machines and equipment (USA Today, 2020). One month earlier, heavy floods and landslides cost the lives of over 60 people and displaced hundreds of thousands.

The floods (both riverine and coastal) have become a frequent occurrence, and local communities have become accustomed to these risks (Padawangi & Douglass, 2015). When flood warnings are issued residents move their furniture to the upper floor, keeping them safe from water damage. Many houses in the flood-prone areas for example also have tiled floors and walls on the lower floor of their houses. This allows residents to quickly clean-up after the water subsides. In addition, communities have also become accustomed to sharing information about the floods such as early warning messages. Communities use popular instant messaging applications such as WhatsApp to inform each other about early warning signs, actions to be taken, and even coordinate internal relief efforts. Over time communities developed rudimentary digital services to share information, exchange knowledge, and develop action plans in case of flooding (Meesters et al., 2019).

One specific group that supports communities in their efforts to deal with the flood risks is the Humanitarian OpenStreetMap group. OpenStreetMap (OSM) is an online collaborative project in which maps are being produced with the input of thousands of volunteers across the globe (Haklay & Weber, 2008). OpenStreetMap is sometimes referred to as the 'Wikipedia of Maps' and is in fact inspired by the same principles as the online encyclopedia. Humanitarian OpenStreetMap (HOTOSM) is an international NGO that strongly relies on OpenStreetMap and related technologies to provide (mapping and GIS) support in humanitarian operations and crisis response. In Indonesia HOTOSM has an office working on a range of projects, including those related to flooding and flood risks in Indonesia.

MURIA Project

The MURIA (Muranda Urban Resilience in Action) program is a collaborative effort between Humanitarian OpenStreetMap, several local and international NGOs, and the Marunda community (Open Street Map, 2018). This community (administrative village, Kelurahan) is located in the north of Jakarta and directly adjacent to Jakarta bay. The area is downstream of several canals and streams. Because of the location the community faces several floods per year. And while most of the floods have a limited impact, floods with a more devastating impact can happen at any moment. These risks have prompted different NGOs to work together to implement a Disaster Risk Reduction (DRR) program: the MURIA project.

The MURIA project is centered around the collection of (flood) risk information on the area. The data collection includes for example the mapping of public facilities, roads, and housing. This data is collected through a crowdsourcing and mapping platform called 'Ushahidi', developed by a Kenyan organization and frequently used for crowdsourcing and crowd mapping projects (Okolloh & action, 2009). The collected data is further augmented by flood risk information, for example obtained from modeling using digital evaluation models, and software developed by the OpenStreetMap. The resulting information is combined into risk maps. These maps were subsequently used by communities to plan specific interventions to reduce the risks. The collection of data to create the maps and develop the website where information was accessible by the community, were done by local youth groups. Various members of these groups received training in technologies such as OpenStreetMap, Ushahidi and the use of mobile devices for mapping.

DRR and Mobile Technology

The MURIA project has not only contributed to the improved understanding of the risks that the community faces, but also improved the awareness of the risks throughout the community. Because the data collection and mapping activities of local youth groups, a dialog was encouraged in the community about flood risks. This dialog was then further supported by the resulting information products. Because the ownership resides with the community, and the project had a strong focus on local capacity building, the information products and maps that have resulted from this project have been updated over time. More importantly, the community has used the maps and risk information to plan evacuation routes for the village.

Figure 3. Youth Mapping group in the MURIA project (OpenStreetMap Indonesia)

It is interesting to note that, communities have also embraced the introduction of mobile technology and software within their own community management. WhatsApp for example, is frequently used to establish groups and share information between community members. The aforementioned youth groups that initially supported the project in data collection continued their efforts and actually expanded to their work to include other activities related to disaster management and emergency response. For example, the groups have formed 'disaster response' teams. When a warning is issued these teams would go around the village checking up on people who may need assistance, using the digital maps and website to identify high-risk areas.

Inclusive Approach

Humanitarian OpenStreetMap Indonesia, in a collaborative effort with local and community organizations. As the project mentions on their website: "With locals involved in making a map, we expected information collected will be better and valid because it locals know their territory, also is expected the map produced from this project will be a map derived from the community and to a community itself." This case highlights the importance of capacity building in local communities. More than the digital services provided themselves, the inclusion of the local community in the design, development and implementation of the digital services provides them with the necessary skills, capacities, and tools to maintain and expand these services.

Moreover, through the inclusion of the community the tools have also become embedded in the community (Meesters et al., 2019). For example, through frequent use for flood risk assessment, the social structures for information sharing, and fit with the local commonly used technologies. This inclusive approach also allows communities and the community members to effectively use these digital services during emergencies. Because of the high uncertainty and the cognitive load during a (sudden) emergency, users will default back to known and existing solutions. The familiarity with the services and tools across the community allows them to quickly update and utilize this information. Additionally, because the project builds on OpenStreetMap, the information that is collected and processed by the community is available to the wider emergency response community, including international agencies and digital volunteers.

REFLECTIONS

The cases introduced in this chapter are two examples of digital services used during an emergency to support local communities. Over the past decade, with the shift towards a more community-inclusive approach towards crisis response, an increasing number of digital services have been introduced to facilitate the communication, information exchange, and joint decision making with communities. The above cases for example show the use of crowdsourcing, social media, and other web 2.0 technologies to create digital services to provide information to communities in times of disaster. While these are just two (illustrative) examples of digital services provided, they show several important elements that relate back to overall concepts of digital services. The cases show that the circumstances during disruptive events warrant specific considerations in designing, developing, and delivering these services. More in-depth reflections are discussed below with respect to the core components of delivering digital services introduced earlier in this chapter.

Planning and Design of Digital Service in Disaster Response

Designing digital services that empower communities in the aftermath of a major disruptive event require special considerations. The context, challenges, and requirements outlined in the previous section illustrate several important considerations that are required in the design of successful, reliable, and useful digital services that empower communities. Many of these elements are part of the planning and design of the digital service. Many digital services in the field of humanitarian aid and crisis response are provided by altruistic motivations and do not rely on economic incentives for the users and other stakeholders. Rather, the value is created through the available information offered via the services. More specifically, information provided by the digital services should be *actionable* (Chatfield et al., 2013). In other words, the information provided should be related and relevant to key decisions that are made by the community or community members.

To support and empower communities affected by disasters, the information needs of the communities should be thoroughly understood. Furthermore, we also need to pay attention to the reciprocity during the information collection and digital service implementation process (Paulus et al., 2018). As described, communities are often viewed as an information source. However, to ensure that there is a continuous exchange of information, digital services should provide an added value to all users involved, especially to information providers. Value co-creation should be the underlying principle in the service design for the disaster affected communities. This also means that information should be delivered to the user in a manner that closely aligns with their decision-making process (Meesters & Van de Walle, 2014). Digital services allow the delivery of tailored information in a timely manner. Rather than presenting all information and risking information overload, presenting users with relevant information also increases the actionability and value of the service. It should also be considered that information can be delivered via 'proxy'-users. Especially in communities, members can relay information to each other via their own networks. For example, if a person has access to the digital service, they can act as proxy for family members who may not have direct access themselves.

Development of Digital Service in Disaster Response

The advancement of information and community technology also enables much wider and faster information access in remote locations. Instant Messaging applications like WhatsApp, social media platforms like Facebook, and a multitude of other applications and technologies allow people to connect, collect and share information (Bieberstein et al., 2005). Similar developments have occurred in the field of crisis response and disaster management. Specific data and information services have been developed and used by humanitarian aid organizations (e.g. HDX, HumanitarianResponse.info). Many organizations and volunteer groups are providing digital services (e.g. MapAction, Ushahidi, Humanitairan OpenSteetMap) (Meier, 2015).

These two worlds of digital services, commonly used services in daily life versus bespoke services used in the field of emergency response could be disconnected. Advances are being made to integrate these services, for example using chatbots to collect information from communities through Facebook Messenger (Piccolo et al., 2018). While currently these interactions are often one-way (broadcasting information, or data collection), the introduction of multi-service platforms offers possibilities for further integration in the existing ecosystem of digital services. Examples of such platforms are AliPay, or GO-JEK (Natadjaja & Setyawan, 2016). These platforms offer features such as payments, and -most importantly- are commonly used by communities.

Additionally, there is a strong network effect. Many digital services today use some mechanism of crowdsourcing, rely on a pool of users (for example to match offer/demand), or require information from others to create value. Often in an emergency there is (initially) limited data available to leverage this network effect (Nespeca et al., 2018). For example, when a digital service relies on delivering information, the data needed to create that information first has to be available. Therefore, services that build on existing data-sets, integrate the data collection into their services, or can rely on partners and networks to quickly obtain the needed information or build on a large existing user-base are key to create a digital services that delivers an added value for the community' (Meesters & Wang, 2020).

Implementation and Governance of Digital Service in Disaster Response

A key success factor is the adoption of the platform by the end-users (communities) and the integration in the community's (potential) response to disasters events. This not only requires a careful examination of the existing approaches, process, and procedures in a community to ensure digital services are well aligned and integrated, but also the involvement of the community in the development process. A clear implementation benefit from user-centered service design is that users are not completely new to the service concept and are willing to try it out. In addition, increasingly communities have younger generations that have ideas, skills, and know-how to develop (rudimentary) systems and services (National Research Council, 2011). For example, grass-roots development groups in Indonesia have made several apps that inform communities about imminent floods through social media and crowdsourcing.

A key concern for the effective use of systems is their limited application scope. Systems purely designed for (large-scale) emergencies are rarely used. This implies that communities will become less familiar with these systems, and the maintenance and ownerships will be low (Meesters et al., 2019). With the risk of digital services becoming oblivious, communities will not use them when unexpected situations occur. It is therefore key to ensure that digital services not only have an added value during an emergency but also have merit in more frequently occurring situations.

When planning, designing, and implementing digital services there is often a direct need to exchange certain information, especially in humanitarian situations. However, humanitarian situations, especially those followed by sudden or severe crisis, are often volatile in nature. Subsequently, the response organization, such as the involved agencies and their operations will also change. This provides another argument for a community-centric approach to the development of digital services. In contrast with -especially international- agencies who operate often on basis of a 'mission' with a fixed term, communities have a permanent presence.

Engaging and empowering communities has been observed as a key stable factor in the whole response. Not only the ownership of these systems should reside with the community for the purpose of continuity, additionally it is also a best practice to ensure that the systems are being maintained, updated, and improved over time. However, the success of this ownership depends on the available (or provided) capacities, capabilities, and resources. Communities can also pool their resources or work closely with local government agencies to adopt these services for example.

A key aspect is therefore to 'keep an eye' out for not only existing tools, but the (potential) organizations, developers and communities that could be brought in to develop these digital services. In more recent disasters, communities organized their own data-gathering efforts for example as illustrated in the cases. Enabling organization to build on existing initiatives rather than deploying new services during a crisis. A key element to leveraging this potential are partnerships. Including local organizations such as schools, local NGOs, and community initiatives (Streefkerk et al., 2014). These initiatives could also be connected to larger international communities of volunteer or organizations that can offer more long-term support.

Ethical Considerations of Digital Service in Disaster Response

Digital services used in a crisis response deal not only with a high uncertainty in a volatile situation. In contrast with digital services used in daily situations, digital services during critical events have different motivations, objectives, and incentives for use. As mentioned, information is a critical resource during an emergency as it supports key decisions. Especially during life threatening situations these decisions and thus the information provided can be critical. Therefore, careful considerations must be given to not only ensure that the information is correct, and the services operate accordingly, but also accountability and responsibility are carefully considered. Especially when digital services at critical moments fail, a responsibility has to be allocated, along with legal disclaimers and broader ethical considerations regarding accountability.

For example, digital services in crisis situations are not geared towards optimization of resources, rather than ensuring fair and equal access. These humanitarian principles are key in the design of digital services for these situations. These values should be integrated in all aspects of the digital services delivery (Raymond et al., 2015). Not only in the design of the services itself and for example algorithms that process information, but also in the (equal and inclusive) access to service, and (future) ownership of the service and data (Raymond et al., 2016). Even the mere fact of offering a digital service to provide access to information can create inequality. The digital divide can provide advantages to those who are able to access digital services over those who lack the skills or resources for example (Iwasaki, 2013).

Another common ethical consideration is regarding the privacy and protection of the data collected from the communities. This issue could be a paradox for digital service delivery in disastrous situations. On the one hand, no one should be left behind under critical conditions especially when it is life-threatening. Making needs for support known and being visible to the aid organizations is necessary for community members to be helped or survive. Community members who are affected by disruptive events sometimes are not given the choice to use or not to use certain digital services if such digital channel is the only way of collecting information. One could argue the priorities between basic needs and privacy for the community members in those situations. Nevertheless, this should not be a reason to ignore this ethical consideration for implementing digital services.

On the other hand, data associated with affected population and area is very sensitive due to the vulnerability of the population group. Respecting and protecting the privacy of personal data is part of the professional ethics and integrity requirement for the aid organizations, so is for the disaster management digital service providers. The technological possibility should be assessed together with the ethical and privacy guidance (The Center for Humanitarian Data, 2019, 2020) for humanitarian data management. This applies to the entire process of collecting, processing, analyzing, and disseminating the community data. The responsibility of data protection lies with all stakeholders involved in the digital service delivery lifecycle.

Integration into Humanitarian Programming

The use of digital services for communities in humanitarian programs and crisis response is often an addition, a nice-to-have. Despite the increased robustness and the increased use of technologies and resulting digital services, they remain a niche and are used ad-hoc. The use and integration of these services often depend on the availability of certain staff, funding, or other incentives. As a result, such services are often once-off pilots and not an integral part of the crisis response approaches and humanitarian programming. Digital services, however, can greatly strengthen and enhance the impact of crisis response and humanitarian programs (Derczynski et al., 2018; Piccolo et al., 2018). The direct information exchange between communities and the responders creates an added value for both users. Integrating digital service delivery in response and recovery programs enable a more deliberate approach compared to ad-hoc interventions. This enables organizations to also build up knowledge, tools, and capacity structurally to effectively deliver these services.

Moreover, the approach of designing, developing, and delivering of a digital service, can even support the objectives of humanitarian programs. The process of jointly creating digital services does not only improve the impact of the service through a better understanding of the requirements (Piccolo et al., 2017). It can in fact strengthen the local capacity of communities to design, develop and maintain their own digital services. The collaborative design efforts of these programs builds the social capital of communities, which will not only strengthen their recovery efforts in the current crisis inasmuch it also provides transformative resilience towards future crisis events (Bodin & Wiman, 2004; Coles & Buckle, 2004; Norris et al., 2008)

Figure 4. Workshop to design digital services in Nepal

CONCLUSION

Over the past years, digital services have become part of our daily personal and professional lives. Such services support us in the decision-making process and allow us to 'optimize' our decisions. For example, by saving time, reducing costs, and in general increase our decisiveness and confidence. At the same time, the possibilities for developing new and innovative digital services have increased. Moreover, they have also become accessible to a broad audience. In fact, it is these digital services that have also created more accessible Information and Communication Technologies, improved education and (global) knowledge exchange, and increased the availability of digital data and information. Digital services themselves have also further spurred on the development and uptake of digital services in a wide range of fields.

Technology Push vs. Community Driven Development

Following the 2010 Haiti Earthquake, a new era began in which an increasing number of technologies were introduced in the crisis response field (Comes et al., 2019). Such technologies leveraged novel methods for data collection, for example creating (remote) situational awareness through social-media analyses. Other innovations build on Web 2.0 advancements to mobilize large groups and used crowd-sourcing approaches. In the following years more digital services were developed using new tools, platforms, and innovations. Inspired by these possibilities, through organizational and international policies, and by increased funding opportunities, a strong 'technology-push' resulted (Lissenden et al., 2015).

Even with this drive for innovation and new possibilities, the structural use of these technologies to deliver services as part of mission, response efforts or humanitarian programs remains limited. Many of the project and innovations remain in the pilot stage, and/or the duration of a mission. However, digital services that did become more structurally adapted, rely strongly on partnerships, such as larger international (digital) communities or local capacities (Meesters et al., 2019). In fact, the local capacity to develop digital services has been increasing as the development of information and communication technologies become more accessible in terms of both hard- and software, as well as the knowledge needed to create digital services with these tools.

Empowerment and Transformative Resilience

Despite their importance in the decision-making process, information and data do not "do" anything by themselves. Rather they can empower and encourage the recipient to take decisive, informed, or justified action. Especially in circumstances with a high degree of uncertainty and an urgent need to act, delivering information becomes a crucial part of the response operations. Additionally, the delivery of humanitarian and emergency aid has shifted in the last decade from providing direct relief to community driven response (Twigg, 2004). In this approach organizations do not only provide direct relief items, additionally they work closely with communities to build local capacities to strengthen their resilience (Sapirstein, 2006). Often, relief and recovery programs include a transformative element to 'build back better' (Coles & Buckle, 2004).

A similar approach can be employed to digital services. It is not just creating a digital service *for* communities in disaster responses, inasmuch *with* communities. This inclusive, co-creation, approach not only ensures a better fit with the needs of the communities, or longevity of the services, it also the ability for communities to articulate, design and even develop their own services (Kera et al., 2013). Rather than providing digital services directly to affected communities, co-creating these digital services also encourages, strengthens, and extends local capacities. Not only does using digital services encourages transformative resilience, they also provide the tools, capacities, and knowledge to support the transformative capacity of the communities themselves (Soden et al., 2014).

Towards Open and Inclusive Development

Therefore, we need to ensure that we foster an 'open development' culture. While there are some bespoke applications and systems that can be brought in to support emergency operations in -especially a sudden-onset- natural disaster, it is important that as the humanitarian aid delivery collaborations are fostered that not only build on partnerships for physical aid delivery, but also in regards to information management and digital services.

There have been numerous developments that have enabled more collaborative efforts in a technical sense. For example, the United Nations Office for the Coordination of Humanitarian Affairs (UN OCHA) offers standards and framework to build upon such as the Humanitarian Exchange Language (HXL). Other organizations such as OpenStreetMap provide tools, training, and volunteers to support development efforts. These platforms, tools and standards provide key building blocks that allow organization to quickly develop digital services. In addition to these technical building blocks, partnerships and networks are key to the successful development of digital services. Partnerships with existing service providers for example allow for easier adoption, civic organization can support training, or international organizations can provide additional capacity.

More important that the technical developments and/or the partnerships, is the awareness of these possibilities of digital services by program managers, country directors, and indeed community leaders themselves. The development of digital services in their missions, projects, and programs. This combination of technology, partnerships, organizational commitment, and local community involvement allow the potential of digital services to be leveraged towards more empowered and resilient communities.

ACKNOWLEDGMENT

The authors thank all participants who have contributed to this research. Specifically, the authors appreciate the support of the local communities and initiatives in Jakarta, Indonesia, and Nepal for sharing their insights.

Part of this research has conducted as part of the 'COMRADES'-project, supported by the European Union's Horizon 2020 research and innovation programme under grant agreement No 687847.

REFERENCES

American Red Cross. (2015). *A Principled Approach to Innovation*. The Red Cross and the Red Crescent.

Apple. (2009). *iPhone 3g Commercial "There's An App For That"*. https://www.youtube.com/watch?v=szrsfeyLzyg

Baharmand, H., Boersma, K., Meesters, K., Mulder, F., & Wolbers, J. (2016). *A multidisciplinary perspective on supporting community disaster resilience in Nepal*. ISCRAM.

Bieberstein, N., Bose, S., Walker, L., & Lynch, A. J. I. j. (2005). *Impact of service-oriented architecture on enterprise systems, organizational structures, and individuals*. Academic Press.

Bodin, P., & Wiman, B. J. E. b. (2004). *Resilience and other stability concepts in ecology: Notes on their origin, validity, and usefulness*. Academic Press.

Cardia, I. V., Holzer, A., Xu, Y., Maitland, C., & Gillet, D. (2017). Towards a principled approach to humanitarian information and communication technology. *Proceedings of the Ninth International Conference on Information and Communication Technologies and Development.*

Chatfield, A. T., Scholl, H. J. J., & Brajawidagda, U. J. G. i. q. (2013). *Tsunami early warnings via Twitter in government: Net-savvy citizens' co-production of time-critical public information services*. doi:10.1145/3136560.3136588

Christakis, N. A., & Fowler, J. H. J. S. i. m. (2013). *Social contagion theory: examining dynamic social networks and human behavior*. Academic Press.

Coles, E., & Buckle, P. J. A. J. o. E. M. (2004). *Developing community resilience as a foundation for effective disaster recovery*. Academic Press.

Comes, T., Meesters, K., Torjesen, S. J. S., & Infrastructure, R. (2019). *Making sense of crises: the implications of information asymmetries for resilience and social justice in disaster-ridden communities*. Academic Press.

Comfort, L. K., Ko, K., & Zagorecki, A. J. A. B. S. (2004). *Coordination in rapidly evolving disaster response systems: the role of information.* Academic Press.

Derczynski, L., Meesters, K., Bontcheva, K., & Maynard, D. J. a. p. a. (2018). *Helping crisis responders find the informative needle in the tweet haystack.* Academic Press.

Düchting, M., Zimmermann, D., & Nebe, K. (2007). Incorporating user centered requirement engineering into agile software development. *International Conference on Human-Computer Interaction.*

EM-DAT. D. G.-S. (2020). *EM-DAT: The Emergency Events Database.* www.emdat.be

Fang, Z. J. I. j. o. t. C., the Internet, & Management. (2002). *E-government in digital era: concept, practice, and development.* Academic Press.

Gantz, J., & Reinsel, D. J. I. i. I. A. t. f. (2012). *The digital universe in 2020: Big data, bigger digital shadows, and biggest growth in the far east.* Academic Press.

Goldstein, S. M., Johnston, R., Duffy, J., & Rao, J. J. J. o. O. m. (2002). *The service concept: the missing link in service design research?* Academic Press.

Gostelow, L. J. D. (1999). *The Sphere Project: the implications of making humanitarian principles and codes work.* Academic Press.

Gralla, E., Goentzel, J., & Van de Walle, B. (2015). *Understanding the information needs of field-based decision-makers in humanitarian response to sudden onset disasters.* ISCRAM.

Guha-Sapir, D., & Lechat, M. F. J. D. (1986). *Information systems and needs assessment in natural disasters: an approach for better disaster relief management.* Academic Press.

Haklay, M., & Weber, P. J. I. P. C. (2008). *Openstreetmap: User-generated street maps.* Academic Press.

Heikkinen, M. T., Mainela, T., Still, J., & Tähtinen, J. J. I. M. M. (2007). *Roles for managing in mobile service development nets.* Academic Press.

Helsloot, I., Ruitenberg, A. J. J. o. c., & Management, c. (2004). *Citizen response to disasters: a survey of literature and some practical implications.* Academic Press.

Hevner, A. R., March, S. T., Park, J., & Ram, S. J. M. q. (2004). *Design science in information systems research.* Academic Press.

Hilhorst, D., & Schmiemann, N. J. D. i. P. (2002). *Humanitarian principles and organisational culture: Everyday practice in Meédecins Sans Frontié res-Holland.* Academic Press.

Hong, L., Fu, C., Wu, J., & Frias-Martinez, V. J. I. S. F. (2018). *Information needs and communication gaps between citizens and local governments online during natural disasters.* Academic Press.

Imran, M., Castillo, C., Lucas, J., Meier, P., & Vieweg, S. (2014). AIDR: Artificial intelligence for disaster response. *Proceedings of the 23rd International Conference on World Wide Web.*

Initiative, H. H. (2010). Disaster Relief 2.0: The future of information sharing in humanitarian emergencies. In Disaster Relief 2.0: The future of information sharing in humanitarian emergencies. HHI; United Nations Foundation; OCHA; The Vodafone Foundation.

Iwasaki, N. J. J. o. E.-G. (2013). *Usability of ICT applications for elderly people in disaster reduction*. Academic Press.

Kathmandu Living Labs. (2015). *QuakeMap*. http://www.kathmandulivinglabs.org/projects/quakemaporg

Kathmandu Living Labs. (2016). *Map for everyone*. http://www.kathmandulivinglabs.org/projects/map-for-everyone

Kera, D., Rod, J., Peterova, R. J. N. d. a. F. D. S., Political, & Issues, E. (2013). *Post-apocalyptic citizenship and humanitarian hardware*. Academic Press.

Korset, R. (2013). *World Disasters Report 2013*. Academic Press.

Lee, S.-Y. T., Phang, C. W. D. J. E. C. R., & Applications. (2015). *Leveraging social media for electronic commerce in Asia: Research areas and opportunities*. Academic Press.

Lelis, S., & Howes, A. (2011). Informing decisions: how people use online rating information to make choices. *Proceedings of the SIGCHI conference on human factors in computing systems.*

Leong, C. M. L., Pan, S. L., Ractham, P., & Kaewkitipong, L. J. J. o. t. A. f. I. S. (2015). *ICT-enabled community empowerment in crisis response: Social media in Thailand flooding 2011*. doi:10.1145/1978942.1979278

Lissenden, J., Maley, S., & Mehta, K. J. J. o. H. E. (2015). *An era of Appropriate Technology: Evolutions, oversights and opportunities.* Academic Press.

Maglio, P. P., & Spohrer, J. J. J. o. t. a. o. m. s. (2008). *Fundamentals of service science*. Academic Press.

Malhotra, N. K. J. J. o. c. r. (1982). *Information load and consumer decision making*. Academic Press.

Meesters, K., Nespeca, V., & Comes, T. (2019). *Designing Disaster Information Management Systems 2.0: Connecting communities and responders*. ISCRAM.

Meesters, K., & Van de Walle, B. (2014). Increasing efficiency of humanitarian organizations with volunteer driven information products. *2014 47th Hawaii International Conference on System Sciences*.

Meesters, K., & Wang, Y. (2020). *Information Management in Large-scale Disaster Exercises: An Integrated Perspective*. Academic Press.

Meier, P. (2015). *Digital humanitarians: how big data is changing the face of humanitarian response*. CRC Press. doi:10.1201/b18023

Meier, P. J. I. l. R. R. C. (2011). *New information technologies and their impact on the humanitarian sector*. Academic Press.

Meissner, A., Luckenbach, T., Risse, T., Kirste, T., & Kirchner, H. (2002). Design challenges for an integrated disaster management communication and information system. *The First IEEE Workshop on Disaster Recovery Networks (DIREN 2002).*

Natadjaja, L., & Setyawan, P. B. (2016). *Creating Community through Design: The Case of Go-Jek Online*. Petra Christian University.

National Research Council. (2011). *Successful K-12 STEM education: Identifying effective approaches in science, technology, engineering, and mathematics*. National Academies Press.

Nespeca, V., Meesters, K., & Comes, T. (2018). *Evaluating Platforms for Community Sense-making: Using the Case of the Kenyan Elections*. ISCRAM.

Norris, F. H., Stevens, S. P., Pfefferbaum, B., Wyche, K. F., & Pfefferbaum, R. L. J. A. p. (2008). *Community resilience as a metaphor, theory, set of capacities, and strategy for disaster readiness*. Academic Press.

Okolloh, O. J. P. l., & Action. (2009). *Ushahidi, or 'testimony': Web 2.0 tools for crowdsourcing crisis information*. Academic Press.

Open Street Map. (2018). *Marunda Urban Resilience in Action*. https://openstreetmap.id/en/project/marunda-urban-resilience-in-action-alliance-muria/

Padawangi, R., & Douglass, M. J. P. A. (2015). *Water, water everywhere: Toward participatory solutions to chronic urban flooding in Jakarta*. Academic Press.

Paulus, D., Meesters, K., & Van de Walle, B. (2018). *Turning data into action: supporting humanitarian field workers with open data*. ISCRAM.

Piccolo, L., Meesters, K., & Roberts, S. (2017). *Co-designing for Community Resilience Beyond the Local*. Academic Press.

Piccolo, L. S., Roberts, S., Iosif, A., & Alani, H. (2018). Designing chatbots for crises: a case study contrasting potential and reality. *Proceedings of the 32nd International BCS Human Computer Interaction Conference (HCI)*.

Poiani, T. H., dos Santos Rocha, R., Degrossi, L. C., & De Albuquerque, J. P. (2016). Potential of collaborative mapping for disaster relief: A case study of OpenStreetMap in the Nepal earthquake 2015. *2016 49th Hawaii International Conference on System Sciences (HICSS)*.

Ragunath, P., Velmourougan, S., Davachelvan, P., Kayalvizhi, S., Ravimohan, R. J. I. J. C. S., & Security, N. (2010). *Evolving a new model (SDLC Model-2010) for software development life cycle*. SDLC; doi:10.14236/ewic/HCI2018.56

Raymond, N., Al Achkar, Z., Verhulst, S., Berens, J., Barajas, L., Easton, M. J. O. P., & Series, S. (2016). *Building data responsibility into humanitarian action*. Academic Press.

Raymond, N. A., Card, B. L. J. S. P. o. H. S., & Technology, H. H. I. (2015). *Applying humanitarian principles to current uses of information communication technologies: Gaps in doctrine and challenges to practice*. Academic Press.

Sapirstein, G. J. J. J. o. D. R. S. D. o. v. (2006). *Social resilience: the forgotten dimension of disaster risk reduction*. Academic Press.

Scarborough, N. M., Zimmerer, T., & Naumes, W. (1991). *Effective small business management*. Merrill Pub. Co.

Soden, R., Budhathoki, N., & Palen, L. (2014). *Resilience-building and the crisis informatics agenda: Lessons learned from open cities Kathmandu*. ISCRAM.

Streefkerk, J. W., Neef, M., Meesters, K., Pieneman, R., & van Dongen, K. (2014). HCI challenges for community-based disaster recovery. *International Conference on Digital Human Modeling and Applications in Health, Safety, Ergonomics and Risk Management.*

The Center for Humanitarian Data. (2019). *OCHA Data Responsibility Guidelines working draft.* https://centre.humdata.org/wp-content/uploads/2019/03/OCHA-DR-Guidelines-working-draft-032019.pdf

The Center for Humanitarian Data. (2020). *Guidance Note: Humanitarian Data Ethics*. https://centre.humdata.org/guidance-note-humanitarian-data-ethics/

Thomas, V., & López, R. J. A. D. B. E. W. P. S. (2015). *Global increase in climate-related disasters*. Academic Press.

Tien, J. M., Berg, D. J. J. o. s. s., & Engineering, s. (2003). *A case for service systems engineering*. Academic Press.

Tsai, M.-H., Chen, J. Y., & Kang, S.-C. J. W. (2019). *Ask Diana: A Keyword-Based Chatbot System for Water-Related Disaster Management*. Academic Press.

Twigg, J. (2004). *Disaster risk reduction: mitigation and preparedness in development and emergency programming*. Academic Press.

Twigg, J. (2009). *Characteristics of a disaster-resilient community: a guidance note* (version 2). Academic Press.

USA Today. (2020). *Jakarta floods 2020: Flooding, rain submerges Indonesia capital*. https://eu.usatoday.com/story/news/world/2020/02/25/jakarta-floods-2020-indonesia-capital/4865913002/

van den Homberg, M., Meesters, K., & Van de Walle, B. (2014). *Coordination and information management in the Haiyan response: Observations from the field*. Academic Press.

Virmani, M. (2015). Understanding DevOps & bridging the gap from continuous integration to continuous delivery. *Fifth International Conference on the Innovative Computing Technology (INTECH 2015).*

Weber, M. J. E. j. o. o. r. (1987). *Decision making with incomplete information*. doi:10.1109/INTECH.2015.7173368

Wegmann, A., Regev, G., Garret, G.-A., & Maréchal, F. (2008). Specifying services for ITIL service management. *2008 International Workshop on Service-Oriented Computing: Consequences for Engineering Requirements*.

Weijman, S., & Meesters, K. (2020). Shifting Control and Trust: Exploring Implications of Introducing Delegated Decision Support Systems. *Proceedings of the 17th ISCRAM Conference*.

Williams, K., Chatterjee, S., & Rossi, M. J. E. j. o. i. s. (2008). *Design of emerging digital services: a taxonomy*. Academic Press.

KEY TERMS AND DEFINITIONS

Community Empowerment: A process to enable communities to increase control over their lives.

Crowdsourcing: A practice of engaging a group for a common goal, such as data collection, data analysis or dissemination information.

Governance Capacity: The resource to define common directions of an organization and translate the vision into policy and practices.

Information System: A socio-technical system designed to collect, process, store and disseminate information.

Mapping: A matching process where the points of one set are linked to points of another one.

Participatory Design: An approach that actively involves all stakeholders in the design process to ensure the products or services meet the needs of target users.

Service Delivery: A process of plan, design, development, implement and maintain of an application that provides added value to customers. The application could be digital, intangible or a mix of both formats.

User-Centered Design: A type of design approach that centers around users' problem and needs to define products or services.

Compilation of References

Aaker, J. (1997). Dimensions of Brand Image. *JMR, Journal of Marketing Research*, *34*(3), 347–356. doi:10.1177/002224379703400304

Abdullah, S., & Wu, X. (2011). An Epidemic Model for News Spreading on Twitter. *International Conference on Tools with Artificial Intelligence*, 163-169. 10.1109/ICTAI.2011.33

Acar, A., & Muraki, Y. (2011). Twitter for crisis communication: Lessons learned from Japan's tsunami disaster. *International Journal of Web Based Communities*, *7*(3), 392–402. doi:10.1504/IJWBC.2011.041206

Accenture Development Partnerships. (2012, April). *Global Alliance for Clean Cookstoves: Rwanda market assessment*. Global Alliance for Clean Cookstoves. doi:10.1163/9789004322714_cclc_2016-0006-001

Adam, B., Beck, U., & van Loon, J. (Eds.). (2000). *The Risk Society and Beyond: Critical Issues for Social Theory*. Sage. doi:10.4135/9781446219539

Adams, J. (2003). Risk and morality: three framing devices. In R. V. Ericson & A. Doyle (Eds.), *Risk and morality* (pp. 87–103). University of Toronto Press. doi:10.3138/9781442679382-006

Adebayo, O. (2017). *The application of Facebook to crisis communication management: a case study of Malaysia airlines* [Unpublished doctoral dissertation]. Salford Business School, University of Salford, UK.

Adibi, S. (Ed.). (2015). *Mobile Health: A Technology Road Map*. Springer. doi:10.1007/978-3-319-12817-7

ADiLife – Covid 19. (2020). *Covid-19*. https://www.adilife.net/covid-19/

Adlakha, K., & Sharma, S. (2019). Brand positioning using multidimensional scaling technique: An application to herbal healthcare brands in Indian market. *Vision (Basel)*, *1*(11). Advance online publication. doi:10.1177/0972262919850930

Adler, R. B., Rosenfield, L. B., & Proctor, R. F. (2017). *Interplay: The process of interpersonal communication* (14th ed.). Oxford University Press.

Agamben, G. (2020). *Biosicurezza e politica*. https://www.quodlibet.it/giorgio-agamben-biosicurezza

Agamben, G. (2003). *Stato di eccezione*. Bollati Boringhieri.

Ahmed, S. (2019). Credit Cities and limits of the Social Credit System. In N. D. Wright (Ed.), Artificial intelligence, China, Russia and the Global Order (pp. 55-61). Air University Press.

Ahmed, R., Vveinhardt, J., & Streimikiene, D. (2017). Interactive digital media and impact of customer attitude and technology on brand awareness: Evidence from the South Asian countries. *Journal of Business Economics and Management*, *18*(6), 1115–1134. doi:10.3846/16111699.2017.1400460

Ahmouda, A., Hochmair, H. H., & Cvetojevic, S. (2018). Analyzing the effect of earthquakes on OpenStreetMap contribution patterns and tweeting activities. *Geo-Spatial Information Science*, *21*(3), 195–212. doi:10.1080/10095020.2018.1498666

Aiello, A. E., Renson, A., & Zivich, P. N. (2020). Social media – and Internet-Based Disease Surveillance for Public Health. *Annual Review of Public Health*, *41*(1), 101–118. doi:10.1146/annurev-publhealth-040119-094402 PMID:31905322

Ajami, R. (2020). Globalization, the Challenge of COVID-19 and Oil Price Uncertainty. *Journal of Asia-Pacific Business*, *21*(02), 1–3. doi:10.1080/10599231.2020.1745046

Alexander, G. (2010). *Defining crisis communication*. www.sagepub.com/upm.37705_1.pdf

Ali, I. (2020). *The COVID-19 pandemic: Making sense of rumor and fear*. Academic Press.

Altheide, D. L., & Schneider, C. J. (2013). Qualitative Media Analysis. *Sage (Atlanta, Ga.)*.

Álvarez, F., Almon, L., Lieser, P., Meuser, T., Dylla, Y., Richerzhagen, B., & Steinmetz, R. (2018). Conducting a large-scale field test of a smartphone-based communication network for emergency response. *Proceedings of the 13th Workshop on Challenged Networks*, 3-10. 10.1145/3264844.3264845

Alves, C. (2007). *Comportamento Organizacional: a Gestão de Crise nas Organizações*. Escolar Editora.

Ambios. (2019). *Estudo de avaliação de risco à saúde humana em localidades atingidas pelo rompimento da barragem do Fundão – MG*. Retrieved from: https://apublica.org/wp-content/uploads/2019/11/ambios-arsh-mariana-e-barra-linga-final-20190417.pdf

American Academy of Social Work and Social Welfare. (2019). *The 12 challenges*. Grand Challenges of Social Work. https://grandchallengesforsocialwork.org/grand-challenges-initiative/12-challenges/

American Psychiatric Association. (2013). *Diagnostic and statistical manual of mental disorders (DSM-5®)*. American Psychiatric Association.

American Red Cross. (2015). *A Principled Approach to Innovation*. The Red Cross and the Red Crescent.

Amit, M., Kimhi, H., Bader, T., Chen, J., Glassberg, E., & Benov, A. (2020). *Mass-surveillance technologies to fight coronavirus spread: the case of Israel*. doi: 10.1038/s41591-020-0927-z

Anderson, C. (2014). *Makers: The New Industrial Revolution*. Currency.

Anderson, R. (2008). New MRC guidance on evaluating complex interventions. *BMJ 337*(1). *Article a, 1937–9*. Advance online publication. doi:10.1136/bmj.a1937

Andersson, G., & Titov, N. (2014). Advantages and limitations of internet-based interventions for common mental disorders. *World Psychiatry: Official Journal of the World Psychiatric Association*, *13*(1), 4–11. doi:10.1002/wps.20083

Andrade, J. (2008). *A sociedade da informação e as organizações em momentos de crise: uma investigação interpretativista do fluxo de comunicação* (Dissertação do Mestrado em Ciências da Comunicação). Lisboa: Universidade Católica Portuguesa.

Andrade, J. (2009). Gestão de crises organizacionais e a Web 2.0. In *6º Congresso SOPCOM - Sociedade dos Media: Comunicação, Política e Tecnologi*, (pp. 3948 - 3953). Associação Portuguesa de Ciências da Comunicação.

Andrade, J. (2015). Portugal no "mensalão" - um estudo de caso sobre o envolvimento da Portugal Telecom na crise brasileira de compra de votos parlamentares. *Comunicação & Marketing*, *6*(4), 54–71.

Annan-Diab, F., & Jensen, B. (2017). Customer Value Perception: Understanding Customer Response to Company's CSR Initiatives. In M. Aluchna & S. Idowu (Eds.), *CSR, Sustainability, Ethics & Governance* (pp. 53–70). Springer., doi:10.1007/978-3-319-39089-5

Anthonissen, P. F. (2008). No thrillers, but hard reality. Crisis communication, practical PR strategies for reputation management and company survival, 7 – 23.

Anthony, K. E., Sellnow, T. L., & Millner, A. G. (2013). Message convergence as a message-centered approach to analysing and improving risk communication. *Journal of Applied Communication Research, 41*(4), 346–364. doi:10.1080/00909882.2013.844346

Antunes, M. N., Oliveira, A. E., & Rebouças, E. (2018). Zika and press releases: reflections on the communication of risk and emergency under the perspective of culture and media industries. *Revista Brasileira de Pesquisa em Saúde, 20*, 110-120. http://periodicos.ufes.br/rbps/article/view/21236

Antunes, M. N., Silva, C. H., Guimarães, M. C. S., & Rabaço, M. H. L. (2014). Monitoramento de informação em mídias sociais: O e-Monitor Dengue. *Transinformação, 26*(1), 9–18. doi:10.1590/S0103-37862014000100002

Apple. (2009). *iPhone 3g Commercial "There's An App For That"*. https://www.youtube.com/watch?v=szrsfeyLzyg

Appleby-Arnold, S., Brockdorff, N., Fallou, L., & Bossu, R. (2019). Truth, trust, and civic duty: Cultural factors in citizens' perceptions of mobile phone apps and social media in disasters. *Journal of Contingencies and Crisis Management, 27*(4), 293–305. doi:10.1111/1468-5973.12282

Aquino, V., & Monteiro, N. (2020). *CORONAVÍRUS | Brasil confirma primeiro caso da doença.* Retrieved from https://www.saude.gov.br/noticias/agencia-saude/46435-brasil-confirma-primeiro-caso-de-novo-coronavirus

Arkin, E. (1989). Translation of Risk Information for the Public: Message Development. In V. Covello, D. McCallum, & M. Pavlova (Eds.), *Effective Risk Communication. Contemporary Issues in Risk Analysis* (pp. 127–135). Springer. doi:10.1007/978-1-4613-1569-8_19

Arli, D. (2017). Does social media matter? Investigating the effect of social media features on consumer attitudes. *Journal of Promotion Management, 23*(4), 521–539. doi:10.1080/10496491.2017.1297974

Arnberg, F. K., Linton, S. J., Hultcrantz, M., Heintz, E., & Jonsson, U. (2014). Internet-delivered psychological treatments for mood and anxiety disorders: A systematic review of their efficacy, safety, and cost-effectiveness. *PLoS One, 9*(5), e98118. Advance online publication. doi:10.1371/journal.pone.0098118

Athanasis, N., Themistocleous, M., Kalabokidis, K., Papakonstantinou, A., Soulakellis, N., & Palaiologou, P. (2018). The emergence of social media for natural disasters management: A big data perspective. *International Archives of the Photogrammetry, Remote Sensing and Spatial Information Sciences - ISPRS Archives, 42*(3W4), 75–82. doi:10.5194/isprs-archives-XLII-3-W4-75-2018

Auer, M. R. (2011). The policy sciences of social media. *Policy Studies Journal: the Journal of the Policy Studies Organization, 39*(4), 707–736. doi:10.1111/j.1541-0072.2011.00428.x

Auger, P., Burke, P., Devinney, T. M., & Louviere, J. (2003). What will consumers pay for social product features? *Journal of Business Ethics, 42*(3), 281–304. doi:10.1023/A:1022212816261

Austin, L. L., & Jin, Y. (Eds.). (2017). *Social Media and crisis communication*. Routledge. doi:10.4324/9781315749068

Avery, E. J. (2017). Public information officers' social media monitoring during the Zika virus crisis, a global health threat surrounded by public uncertainty. *Public Relations Review, 43*(3), 468–476. doi:10.1016/j.pubrev.2017.02.018

Aydın, H. (2019). Consumer perceptions and responsiveness toward csr activities: A sectoral outlook. In I. Altinbasak-Farina & S. Burnaz (Eds.), *Ethics, Social Responsibility and Sustainability in Marketing, Accounting, Finance, Sustainability, Governance & Fraud: Theory and Application* (pp. 45–62). Springer. doi:10.1007/978-981-13-7924-6_3

Babiak, K. (2010). CSR and environmental responsibility: Motives and pressures to adopt green management practices. *Corporate Social Responsibility and Environmental Management, 24*(March), 11–24. doi:10.1002/csr

Baharmand, H., Boersma, K., Meesters, K., Mulder, F., & Wolbers, J. (2016). *A multidisciplinary perspective on supporting community disaster resilience in Nepal.* ISCRAM.

Bakker, D., Kazantzis, N., Rickwood, D., & Rickard, N. (2016). Mental health smartphone apps: Review and evidence-based recommendations for future developments. *JMIR Mental Health, 3*(1), e7. Advance online publication. doi:10.2196/mental.4984

Balcázar, F. E., Keys, C. B., Kaplan, D. L., & Suárez-Balcázar, Y. (1998). Participatory Action Research and people with disabilities: Principles and challenges. *Canadian Journal of Rehabilitation, 12*, 105–112. https://www.researchgate.net/profile/Christopher_Keys/publication/239921190_Participatory_Action_Research_and_People_with_Disabilities_Principles_and_Challenges/links/0deec53260cdd4e251000000.pdf

Bardin, L. (1977). *Análise de Conteúdo.* Lisboa: Edições 70.

Baron, G. (2010). *Response suggestions and public participation: the new norm in response management.* http://www.emergencymgmt.com/emergency-blogs/crisis-comm/Response-suggestions-and-Public-111510.html

Barrenechea López, G. (2017). *Gender in Transitional Justice Contexts: Reproductive Rights in the Spanish Case of Stolen Babies* [Master Thesis]. Lund University, Lund, Scania Province, Sweden.

Barsky, A. E. (2017). Social work practice and technology: Ethical issues and policy responses. *Journal of Technology in Human Services, 35*(1), 1–12. doi:10.1080/15228835.2017.1277906

Barsky, A. E., & Reamer, F. G. (2018). *New practice standards on social work and technology: A situation-based discussion on best practices.* National Association of Social Workers, NASW Special Policy Services.

Bauman, Z. (1999). *La società dell'incertezza* [The society of uncertainty]. il Mulino.

Bauman, Z. (2001). *Modernidade Líquida.* Rio de Janeiro: Jorge Zahar Editor.

Baum, N., Rahav, G., & Sharon, M. (2014). Heightened susceptibility to secondary traumatization: A meta-analysis of gender differences. *The American Journal of Orthopsychiatry, 84*(2), 111–122. doi:10.1037/h0099383 PMID:24826927

Bazago, F., Guardia, M., & García, J. (2020). Environmental discourse in natural disaster scenarios. *Ekonomska Istrazivanja, 33*(1), 3093–3107. doi:10.1080/1331677X.2019.1694425

BBC News. (2020). *São Paulo registra primeira morte por coronavirus e investiga outras quatro.* Retrieved from https://www.bbc.com/portuguese/brasil-51931934

Bebes Robados Catalunya, S. O. S. (2017, June 16). *Te Estamos Buscando* [We are looking for you]. Created by Escudero, C. [Video]. YouTube. https://www.youtube.com/watch?v=MfOMKmqKgsM

Bebes Robados Catalunya, S. O. S. (2018, November 29). *Te Estamos Buscando* [We are looking for you]. Created by Escudero, C. [Video]. YouTube. https://www.youtube.com/watch?v=G0D6Txc08hU

Beck, U. (1992). Risk Society: Towards a New Modernity. *Sage (Atlanta, Ga.).*

Beck, U. (2000). *La società del rischio. Verso una seconda modernità.* Carocci. (Original work published 1986).

Beck, U. (2002). *La sociedad del Riesgo Global*. Siglo XXI Editores.

Bennett, P., Calman, K., Curtis, S., & Smith, D. F. (2010). *Risk Communication and Public Health*. Oxford University Press., doi:10.1093/acprof:oso/9780199562848.001.0001

Benoit, W. L. (1997). Image Repair Discourse and Crisis Communication. *Public Relations Review*, *23*(2), 177–186. doi:10.1016/S0363-8111(97)90023-0

Ben-Porat, A. (2015). Vicarious post-traumatic growth: Domestic violence therapists versus social service department therapists in Israel. *Journal of Family Violence*, *30*(7), 923–933. doi:10.100710896-015-9714-x

Bent-Goodley, T. B. (2018). Being intentional about self-care for social workers. *Social Work*, *63*(1), 5–6. doi:10.1093wwx058 PMID:29177487

Beristain, C. (2007). Reparations to victims in the inter-American human rights system. In *The role of the justice system in the face of massive violations of human rights*. Abuelas de Plaza de Mayo.

Berliner, L., Bisson, J., Cloitre, M., Forbes, D., Goldbeck, L., & Jensen, T. … Shapiro, F. (2019). *ISTSS PTSD prevention and treatment guidelines methodology and recommendations*. International Society for Traumatic Stress Studies. https://www.istss.org/treating-trauma/new-istss-prevention-and-treatment-guidelines

Berndt, H., & Herczeg, M. (2019). The Role of Mental Models and Situation Awareness for Computer System Support in Mass Casualty Incident Management. In *Proceedings of the 31st European Conference on Cognitive Ergonomics*. Association for Computing Machinery.

Berzin, S. C., Singer, J., & Chan, C. (2015). *Practice innovation through technology in the digital age: A grand challenge for social work* (Grand Challenges for Social Work Initiative Working Paper No. 12). American Academy of Social Work and Social Welfare.

Bhavaraju, S. K. T., Beyney, C., & Nicholson, C. (2019). Quantitative analysis of social media sensitivity to natural disasters. *International Journal of Disaster Risk Reduction*, *39*(July), 101251. doi:10.1016/j.ijdrr.2019.101251

Bieberstein, N., Bose, S., Walker, L., & Lynch, A. J. I. j. (2005). *Impact of service-oriented architecture on enterprise systems, organizational structures, and individuals*. Academic Press.

Bigo, D., & Tsoukala, A. (Eds.). (2008). *Terror, insecurity and liberty. Illiberal practices of liberal regimes after 9/11*. Routledge. doi:10.4324/9780203926765

Bigwood, G., Rehunathan, D., Bateman, M., Henderson, T., & Bhatti, S. (2008). Exploiting self-reported social networks for routing in ubiquitous computing environments. In *IEEE International Conference on Wireless and Mobile Computing, Networking and Communications* (pp. 484-489). IEEE. 10.1109/WiMob.2008.86

Blomgren, A. (2011). Is the CSR craze good for society? The welfare economic approach to corporate social responsibility. *Review of Social Economy*, *69*(4), 495–515. doi:10.1080/00346764.2011.592329

Bodin, P., & Wiman, B. J. E. b. (2004). *Resilience and other stability concepts in ecology: Notes on their origin, validity, and usefulness*. Academic Press.

Boin, A. (2019). The Transboundary Crisis:Why we are unprepared and the road ahead. *Contingenciesand Crisis Management*, 94–99. doi:10.1111/1468-5973.12241BOIN|99

Boin, A., 't Hart, P., Stern, E., & Sundelius, B. (2005). *The Politics of Crisis Management. Public 212 Leadership under Pressure*. New York: Cambridge University Press.

Boulianne, S., Minaker, J., & Haney, T. J. (2018). Does compassion go viral? Social media, caring, and the Fort McMurray wildfire. *Information Communication and Society*, *21*(5), 697–711. doi:10.1080/1369118X.2018.1428651

Bradbury, H., & Reason, P. (2001). Conclusion: Broadening the band with of validity: issues and choice-points for improving the quality of action research. In P. Reason & H. Bradbury (Eds.), *The SAGE Handbook of Action Research: Participatory Inquiry and Practice*. SAGE. https://www.researchgate.net/publication/281504513_Conclusion_Broadening_the_bandwidth_of_validity_Issues_and_choice-points_for_improving_the_quality_of_action_research

Branco, P., Firth, P., Encarnação, L. M., & Bonato, P. (2005, April). Faces of emotion in human-computer interaction. In *CHI'05 Extended Abstracts on Human factors in computing systems*. ACM. doi:10.1145/1056808.1056885

Branson, D. C. (2019). Vicarious trauma, themes in research, and terminology: A review of literature. *Traumatology*, *25*(1), 2–10. doi:10.1037/trm0000161

Branson, D. C., & Miller, K. A. (in press). Social work ethics: Harnessing technology for the good of the discipline. *Journal of Sociology and Social Work*.

Branson, D. C., Radu, M. B., & Loving, J. D. (2019). Adverse Childhood Experiences (ACE) scores: When social work students and trauma mix. *The Journal of Baccalaureate Social Work*, *24*(1), 339–360. doi:10.18084/1084-7219.24.1.339

Braziel, R., Straub, F., Watson, G., & Hoops, R. (2016). Bringing calm to chaos: A critical incident review of the San Bernardino public safety response to the December 2, 2015, terrorist shooting incident at the Inland Regional Center. U.S. Community Oriented Policing Services, Department of Justice.

Brengarth, L. B., & Mujkic, E. (2016). WEB 2.0: How social media applications leverage nonprofit responses during a wildfire crisis. *Computers in Human Behavior*, *54*, 589–596. doi:10.1016/j.chb.2015.07.010

Bride, B. E. (2007). Prevalence of secondary traumatic stress among social workers. *Social Work*, *52*(1), 63–70. doi:10.1093w/52.1.63 PMID:17388084

Brønn, P., & Vrioni, A. (2001). Corporate social responsibility and cause-related marketing: An overview. *International Journal of Advertising*, *20*(2), 207–222. doi:10.1080/02650487.2001.11104887

Brown, J. (Ed.). (2014). *The Future of Policing*. Routledge.

Brown, P. (2020). Studying COVID-19 in light of critical approaches to risk and uncertainty: Research pathways, conceptual tools, and some magic from Mary Douglas. *Health Risk & Society*, *22*(1), 1–14. doi:10.1080/13698575.2020.1745508

Brown, S. (2018). The impact of resiliency on nurse burnout: An integrative literature review. *Medsurg Nursing*, *27*(6), 349–378.

Brownstein, J. S., Freifeld, C. C., & Madoff, L. C. (2009). Digital disease detection—Harnessing the Web for public health surveillance. *The New England Journal of Medicine*, *360*(21), 2153–2157. doi:10.1056/NEJMp0900702 PMID:19423867

Brownstein, J. S., Freifeld, C. C., Reis, B. Y., & Mandl, K. D. (2008). Surveillance sans frontières: Internet-based emerging infectious disease intelligence and the HealthMap project. *PLoS Medicine*, *5*(7), 151. doi:10.1371/journal.pmed.0050151 PMID:18613747

Bruns, A., & Burgess, J. (2014). Crisis Communication in Natural Disasters: The Queensland Floods and Christchurch Earthquakes. In A. Bruns, M. Mahrt, K. Weller, J. Burgess, & C. Puschmann (Eds.), *Twitter and Society* (pp. 373–384). Peter Lang.

Bueno Morales, M. M., & González Besteiro, J. M. (Eds.). (2018). *Bebés robados en Andalucía desde 1936. Buscando la verdad, la justicia y la reparación* [Stolen babies in Andalusia since 1936. Looking for truth, justice and reparation]. Dirección General de Memoria Democrática. Junta de Andalucía.

Bueno, W. C. (2018). A cobertura jornalística de catástrofes ambientais: Entre a vigilância e aespetacularização da notícia. *Comunicação e Sociedade*, *39*(1), 21–41. doi:10.15603/2175-7755/cs.v39n1p21-41

Bundy, J., Pfarrer, M., Short, C., & Coombs, W. (2016). Crises and Crisis Management: Integration, Interpretation, and Research Development. *Journal of Management*, *43*(6), 1661–1692. doi:10.1177/0149206316680030

Bureau of Labor Statistics. (2020, July). *Occupational outlook handbook, social workers*. U. S. Department of Labor. https://www.bls.gov/ooh/community-and-social-service/social-workers.htm

Burke, J. A., Estrin, D., Hansen, M., Parker, A., Ramanathan, N., Reddy, S., & Srivastava, M. B. (2006). *Participatory sensing* [Paper presentation]. World Sensor Web Workshop, ACM SenSys 2006, Boulder, CO, United States. https://escholarship.org/uc/item/19h777qd

Butare, A., & Munyampundu, A. (n.d.). *Towards an improved cook stoves program: Market based solutions to eliminate energy poverty. Enquiry on the use of improved cooking stoves in Bugesera, Kirehe and Ngororero Districts.* Draft report. Africa Energy Services Group. (Unknown issue date) https://waterportal.rwfa.rw/sites/default/files/inline-files/SNV_cookstoves%20AESG%20final%20ICS%20draft%20report.pdf

Butler, L. D., Mercer, K. A., McClain-Meeder, K., Horne, D. M., & Dudley, M. (2019). Six domains of self-care: Attending to the whole person. *Journal of Human Behavior in the Social Environment*, *29*(1), 107–124. doi:10.1080/10911359.2018.1482483

Byambasuren, O., Sanders, S., Beller, E., & Glasziou, P. (2018). Prescribable mHealth apps identified from an overview of systematic reviews. *NPJ Digital Medicine, 1*(1), Article 12. https://www.nature.com/articles/s41746-018-0021-9?report=reader

Cabrero, S., García, R., Pañeda, X. G., & Melendi, D. (2015). Understanding opportunistic networking for emergency services: Analysis of one year of GPS traces. *Proceedings of the 10th ACM MobiCom Workshop on Challenged Networks*, 31-36. 10.1145/2799371.2799381

Caetano, J., Vasconcelos, M., & Vasconcelos, P. (2006). *Gestão de crise*. Editorial Presença.

Callister, E., & Plante, T. (2017). Compassion predictors in undergraduates: A Catholic college example. *Pastoral Psychology*, *66*(1), 1–11. doi:10.100711089-016-0729-x

Calvo, M. G., Gutiérrez-García, A., Avero, P., & Lundqvist, D. (2013). Attentional mechanisms in judging genuine and fake smiles: Eye-movement patterns. *Emotion (Washington, D.C.)*, *13*(4), 792–802. doi:10.1037/a0032317 PMID:23627721

Cameron, M. A., Power, R., Robinson, B., & Yin, J. (2012). Emergency situation awareness from twitter for crisis management. In *Proceedings of the 21st International Conference on World Wide Web*. ACM. 10.1145/2187980.2188183

Camps, S. (1999). *Periodismo sobre catástrofes*. Ediciones Paulinas.

Cao, D. (2012). Linguistic uncertainty and legal transparency: statutory interpretation in China and Australia. In V. K. Bhatia, C. A. Hafner, L. Miller, & A. Wagner (Eds.), *Transparency, power and control. Perspective on legal communication* (pp. 13–30). Ashgate.

Cardia, I. V., Holzer, A., Xu, Y., Maitland, C., & Gillet, D. (2017). Towards a principled approach to humanitarian information and communication technology. *Proceedings of the Ninth International Conference on Information and Communication Technologies and Development.*

Carmo, H., & Ferreira, M. (2008). *Metodologia da Investigação: guia para a auto-aprendizagem* (2nd ed.). Universidade Aberta.

Carmona, M. (2020, May 13). *What's the difference between telehealth and teletherapy?* The Recovery Village. https://www.therecoveryvillage.com/treatment-program/online-counseling/faq/difference-between-telehealth-teletherapy/

Carroll, A. (1999). Corporate social responsibility: Evolution of a definitional construct. *Business & Society, 38*(3), 268–295. doi:10.1177/000765039903800303

Cassandro, D. (2020). *Siamo in guerra! Il coronavirus e le sue metafore*. https://www.internazionale.it/opinione/daniele-cassandro/2020/03/22/coronavirus-metafore-guerra

Castells, M. (2011). A era da informação: Economia, Sociedade e Cultura. Volume I: A sociedade em rede (4th ed.). Lisboa: Calouste Gulbenkian.

Castillo, C. (2016). *Big crisis data - Social media in disasters and time-critical situations*. Cambridge University Press. doi:10.1017/CBO9781316476840

Cavalletti, A. (2005). *La città biopolitica. Mitologie della sicurezza*. Mondadori.

Center for Medicare and Medicaid Services. (2020, March 17). *Medicare telemedicine health care provider facts sheet*. https://www.cms.gov/newsroom/fact-sheets/medicare-telemedicine-health-care-provider-fact-sheet

Centers for Disease Control and Prevention. (2013). *CDC competition encourages use of social media to predict flu*. Press Release. 25. https://www.cdc.gov/flu/news/predict-flu-challenge.htm

Cernvall, M., Sveen, J., Johannesson, K. B., & Arnberg, F. (2018). A pilot study of user satisfaction and perceived helpfulness of the Swedish version of the mobile app PSTD Coach. *European Journal of Psychotraumatology, 9*(sup1). Advance online publication. doi:10.1080/20008198.2018.1472990

Chao, M., Chenji, H., Yang, C., Stoleru, R., Nikolova, E., & Altaweel, A. (2020). EAR: Energy-aware risk-averse routing for disaster response networks. *Ad Hoc Networks, 103*, 1–15. doi:10.1016/j.adhoc.2020.102167

Chatfield, A. T., Scholl, H. J. J., & Brajawidagda, U. J. G. i. q. (2013). *Tsunami early warnings via Twitter in government: Net-savvy citizens' co-production of time-critical public information services*. doi:10.1145/3136560.3136588

Cheng, Y. (2016). How social media is changing crisis communication strategies: Evidence from the updated literature. *Journal of Contingencies and Crisis Management*, 1 – 11. Doi:10.1111/1468-5973.12130

Cheung, M., Pires, G., & Rosenberger, P. (2019). Developing a conceptual model for examining social media marketing effects on brand awareness and brand image. In International Journal of Economics and Business Research (pp. 243–261). doi:10.1504/IJEBR.2019.098874

Chida, K., Morishima, Y., Inaba, Y., Taura, M., Ebata, A., Takeda, K., Shimura, H., & Zuguchi, M. (2012). Physician received scatter radiation with angiography systems used for interventional radiology: Comparison among many x-ray systems. *Radiation Protection Dosimetry, 149*(4), 410–416. doi:10.1093/rpd/ncr312

Cho, H., Ippolito, D., & Yu, Y. W. (2020). *Contact Tracing Mobile Apps for COVID-19: Privacy Considerations and Related Trade-offs*. https://arxiv.org/pdf/2003.11511.pdf

Choo, C.W. (2003) *A organização do conhecimento: como as organizações usam a informação para criar significado, construir conhecimento e tomar decisões*. tradution Eliana Rocha. São Paulo: Editora Senac São Paulo.

Choo, C. W., & Auster, E. (1993). Environmental scanning: Acquisition and use of information by managers. *Annual Review of Information Science & Technology, 28*, 279–281.

Cho, S. E., Jung, K., & Park, H. W. (2013). Social Media use during Japan's 2011 Earthquake: How Twitter transforms the locus of crisis communication. *Media International Australia*, *149*(1), 28–40. doi:10.1177/1329878X1314900105

Chowdhury, R. M. M. I. (2019). The moral foundations of consumer ethics. *Journal of Business Ethics*, *158*(3), 585–601. doi:10.100710551-017-3676-2

Christakis, N. A., & Fowler, J. H. J. S. i. m. (2013). *Social contagion theory: examining dynamic social networks and human behavior*. Academic Press.

Citarella, P. (2017). *Social media e P.A. La comunicazione istituzionale ai tempi di Facebook* [Social media and PAs. Institutional communication in the age of Facebook]. Angeli.

Civil Defense of Minhas Gerais. (2019). *Disaster Information Tailings Dam in Brumadinho.* Retrieved from: http://www.defesacivil.mg.gov.br/index.php/component/gmg/page/787-informacoes-do-desastre-barragem-de-rejeitos-em-brumadinho-28-12-19

Claeys, A., & Coombs, T. (2020). Organizational Crisis Communication: Suboptimal Crisis Response Selection Decisions and Behavioral Economics. *Communication Theory*, *30*(3), 290–309. doi:10.1093/ct/qtz002

Clément, G. (2020). *Strategia della paura*. Academic Press.

Clement, J. (2020). *Annual number of global mobile app downloads 2016*–2019. Statista. Retrieved 4 April 2020, from https://www.statista.com/statistics/271644/worldwide-free-and-paid-mobileapp-store-downloads/

Clement, J. (2020). *Global social networks ranked by number of users*. Retrieved from https://www.statista.com/statistics/272014/global-social-networks-ranked-by-number-of-users/

Climate change impacts and adaptation in Europe - JRC Science for Policy Report. (2020). doi:10.2760/171121

Cmeciu, C., Coman, I., & Coman, C. (2018). *Beyond The Ministry Of Health's Vaccination Campaign during the Measles Outbreak in Romania – Public Assessment and interaction*. Paper presented at the ICA conference, Prague.

Çolak, E., & Sunar, F. (2020). The importance of ground-truth and crowdsourcing data for the statistical and spatial analyses of the NASA FIRMS active fires in the Mediterranean Turkish forests. *Remote Sensing Applications: Society and Environment*, *19*(March), 100327. doi:10.1016/j.rsase.2020.100327

Coles, E., & Buckle, P. J. A. J. o. E. M. (2004). *Developing community resilience as a foundation for effective disaster recovery*. Academic Press.

Comes, T., Meesters, K., Torjesen, S. J. S., & Infrastructure, R. (2019). *Making sense of crises: the implications of information asymmetries for resilience and social justice in disaster-ridden communities*. Academic Press.

Comfort, L. K., Ko, K., & Zagorecki, A. J. A. B. S. (2004). *Coordination in rapidly evolving disaster response systems: the role of information*. Academic Press.

Comfort, L. K. (2007). Crisis management in hindsight: Cognition, communication, coordination, and control. *Public Administration Review*, *67*(s1), 189–197. doi:10.1111/j.1540-6210.2007.00827.x

Comitato Invisibile. (2019). *L'insurrezione che viene – Ai nostri amici – Adesso* (M. Tarì, Trans.). Nero. (Original work published 2007, 2014, 2017).

Comunello, F., & Mulargia, S. (2017). Tra risposte protocollate e «social sensing». L'uso dei social media per la comunicazione d'emergenza nelle istituzioni locali italiane [Between registered answers and social sensing. The use of social media in PAs crisis communication]. *Sociologia e ricerca sociale*, *112*, 111-137.

Comunello, F. (2014). *Social media e comunicazione d'emergenza* [Social media and crisis communication]. Guerini e Associati.

Cook, S., Conrad, C., Fowlkes, A. L., & Mohebbi, M. H. (2011). Assessing Google flu trends performance in the United States during the 2009 influenza virus A (H1N1) pandemic. *PLoS One*, *6*(8), 23610. doi:10.1371/journal.pone.0023610 PMID:21886802

Coombs, T. (2007). *Crisis Management and Communications.* Institute for Public Relations. http://www.instituteforpr.org/crisis-management-and-communications/

Coombs, W. T. (2010*)*. Parameters for crisis communication. In W. T. Coombs & S. J. Holladay (Eds.), *The handbook of crisis communication* (pp. 17 – 53). Wiley Blackwell.

Coombs, T. (2010). Parameters for Crisis Communication. In T. Coombs & S. Holladay (Eds.), *The Handbook of Crisis Communication, (pp. 17 - 53).* Wiley-Blackwell. doi:10.1002/9781444314885.ch1

Coombs, W. T. (1999). *Ongoing crisis communication: Planning, managing, and responding*. Sage.

Coombs, W. T. (2002). Assessing Online Issue Threats: Issue Contagions and Their Effect on Issue Prioritisation. *Journal of Public Affairs*, *2*(4), 215–229. doi:10.1002/pa.115

Coombs, W. T. (2002). Deep and surface threats: Conceptual and practical implications for "crisis" vs "problem.". *Public Relations Review*, *28*(4), 339–345. doi:10.1016/S0363-8111(02)00167-4

Coombs, W. T. (2004). Impact of Past Crises on Current Crisis Communications: Insights From Situational Crisis Communication Theory. *Journal of Business Communication*, *41*(3), 265–289. doi:10.1177/0021943604265607

Coombs, W. T. (2007). Protecting organisation reputations during a crisis: The development and application of situational crisis communication theory. *Corporate Reputation Review*, *10*(3), 163–176. doi:10.1057/palgrave.crr.1550049

Coombs, W. T. (2007b). Ongoing Crisis Communication – Planning, Managing and Responding. *Sage (Atlanta, Ga.).*

Coombs, W. T. (2009). Conceptualising crisis communication. In R. L. Heath & H. D. O'Hair (Eds.), *Handbook of crisis and risk communication*. Routledge.

Coombs, W. T. (2014). Applied Crisis Communication and Crisis Management: Cases and Exercises. *Sage (Atlanta, Ga.).* Advance online publication. doi:10.4135/9781544308531

Coombs, W. T., Claeys, A., & Holladay, S. (2017). Social Media's Value in a Crisis. Channel Effect or Stealing Thunder? In *Social Media and Crisis Communication* (pp. 159–167). Routledge. doi:10.4324/9781315749068-12

Coombs, W. T., Frandsen, F., Holladay, S. J., & Johansen, W. (2010). Why a concern for apologia and crisis communication? *Corporate Communications*, *15*(4), 337–349. doi:10.1108/13563281011085466

Coombs, W. T., & Holladay, S. J. (2014). How publics react to crisis communication efforts. *Journal of Communication Management (London)*, *18*(1), 40–57. doi:10.1108/JCOM-03-2013-0015

Coombs, W. T., & Holladay, S. J. (Eds.). (2010). *Handbook of Crisis Communication*. John Wiley & Sons. doi:10.1002/9781444314885

Coombs, W. T., & Laufer, D. (2018). Global Crisis Management – Current Research and Future Directions. *Journal of International Management*, *24*(3), 199–203. doi:10.1016/j.intman.2017.12.003

Cooper, M., & Lesser, J. G. (2014). *Clinical social work practice: An integrated approach* (5th ed.). Pearson Education, Inc.

Corporation, S. T. (2011, July 26). Seikatsusha no fuan wo kaisho- surutame shutodaigaku Tokyo to kyoudou kaihatsu kateiyou houshasen sokuteiki air-counter wo shinhatsubai [S.T. and Tokyo Metropolitan University launch Air Counter, a home radiation measuring instrument developed in collaboration to alleviate anxiety among consumers]. *S.T. Corporation*. https://www.st-c.co.jp/release/2011/20110726_000266.html

Costa, H., de Rigo, D., Libertà, G., Houston Durrant, T., & San-Miguel-Ayanz, J. (2020). *European wildfire danger and vulnerability in a changing climate : towards integrating risk dimensions*., doi:10.2760/46951

Coté, W., & Simpson, R. (2000). Covering violence. A guide to ethical reporting about victims and trauma. New Yok: Columbia University Press.

Covello, V. T. (2006). Risk communication and message mapping: A new tool for communicating effectively in public health emergencies and disasters. *Journal of Emergency Management (Weston, Mass.)*, *4*(3), 25–40. doi:10.5055/jem.2006.0030

Cowley, B., Filetti, M., Lukander, K., Torniainen, J., Henelius, A., Ahonen, L. ... Ravaja, N. (2016). The psychophysiology primer: A guide to methods and a broad review with a focus on human-computer interaction. *Foundations and Trends® in HumanComputer Interaction, 9*(34), 151–308.

Craciun, C., & Baban, A. (2012). "Who will take the blame?": Understanding the reasons why Romanian mothers decline HPV vaccination for their daughters. *Vaccine*, *30*(48), 6789–6793. doi:10.1016/j.vaccine.2012.09.016 PMID:23017603

Craglia, M., Ostermann, F., & Spinsanti, L. (2012). Digital Earth from vision to practice: Making sense of citizen-generated content. *International Journal of Digital Earth*, *5*(5), 398–416. doi:10.1080/17538947.2012.712273

Craun, S., Bourke, M., & Coulson, F. (2015). The Impact of Internet crimes against children work on relationships with families and friends: An exploratory study. *Journal of Family Violence*, *30*(3), 393–402. doi:10.100710896-015-9680-3

Creswell, J. (2014). *Research design: qualitative, quantitative, and mixed methods approaches* (4th ed.). SAGE.

Cross, N. (1982). Designerly ways of knowing. *Design Studies*, *3*(4), 221–227. doi:10.1016/0142-694X(82)90040-0

Cruz, M. M., & Reis, A. C. (2011). Monitoramento & Avaliação como uma das funções gestoras do Sistema Único de Saúde. In Qualificação e Gestores do SUS. Fiocruz.

Cuijpers, P., & Schuurmans, J. (2007). Self-help interventions for anxiety disorders: An overview. *Current Psychiatry Reports*, *9*(4), 284–290. doi:10.100711920-007-0034-6

Cummings, S. (2016). *Social work ethics and "everyday" technology*. National Association of Social Workers. NASW Press.

Cunha, M. (2006). Crises Organizacionais: Ameaças ou oportunidades? In A. Mendes & F. Pereira (Eds.), *Crises: de ameaças a oportunidades - gestão estratégica de comunicação de crises* (pp. 143–159). Edições Sílabo.

Dalvit, L. (2018). From Access to Proficiency: Reconceptualizing Digital Inclusion in a Rural Area in South Africa. In M. Ragnedda & B. Mutsvairo (Eds.), *Digital Inclusion: An International Comparative Analysis*. Lexington Books.

Davino, C., & Villani, L. (2020). La società del controllo. Lettura di spazi e fatti urbani attraverso il paradigma della trasparenza. In S. Casini, F. Di Blasio, & G. Perletti (Eds.), Trasparenze – Elephant&Castle, 22. https://cav.unibg.it/elephant_castle

de Albuquerque, J. P., Herfort, B., Brenning, A., & Zipf, A. (2015). A geographic approach for combining social media and authoritative data towards identifying useful information for disaster management. *International Journal of Geographical Information Science*, *29*(4), 667–689. doi:10.1080/13658816.2014.996567

De Dominicis, C., Mazzotti, D., Piccinelli, M., Rinaldi, S., Vezzoli, A., & Depari, A. (2012). Evaluation of Bluetooth Hands-Free profile for sensors applications in smartphone platforms. In *Sensors Applications Symposium Proceedings* (pp. 1-6). IEEE. 10.1109/SAS.2012.6166305

de Rigo, D., Libertà, G., Houston Durrant, T., Artés Vivancos, T., & San-Miguel-Ayanz, J. (2017). *Forest fire danger extremes in Europe under climate change: variability and uncertainty*. JRC Science Hub., doi:10.2760/13180

De Rosa, P. (2020). Da locale a virale. L'emergenza permanente dello "Sceriffo" Vincenzo De Luca. In V. De Luca & M. Spalletta (Eds.), *Pandemie mediali. Comunicazioni, narrazioni, socializzazioni e consumi del MediaVirus* [Media Pandemics. Communication, storytelling, socialization and consumption practices of the MediaVirus]. Aracne.

Dearnaley, G., & Northrop, D. C. (1966). *Semiconductor counters for nuclear radiations* (2nd ed.). Wiley.

Dedeoğlu, B., van Niekerk, M., Küçükergin, K., De Martino, M., & Okumuş, F. (2019). Effect of social media sharing on destination brand awareness and destination quality. *Journal of Vacation Marketing*, *24*(1), 33–56. doi:10.1177/1356766719858644

DelTosta, J. E., Ellis, M. V., & McNamara, M. L. (2019). Trainee vicarious traumatization: Examining supervisory working alliance and trainee empathy. *Training and Education in Professional Psychology*, *13*(4), 300–306. doi:10.1037/tep0000232

Denef, S., Bayerl, P., & Kaptein, N. (2013). *Social media and the Police - Tweeting Pratices of British Police Forces during the August 2011*. Fraunhofer Institute for Applied Information Technology. doi:10.1145/2470654.2466477

Denscombe, M. (2017). *The Good Research Guide: for small-scale social research projects* (6th ed.). McGraw-Hill Education.

Derczynski, L., Meesters, K., Bontcheva, K., & Maynard, D. J. a. p. a. (2018). *Helping crisis responders find the informative needle in the tweet haystack*. Academic Press.

Deuze, M., & Witschge, T. (2018). Beyond journalism: Theorizing the transformation of journalism. *Journalism: Theory. Practice & Criticism*, *19*(2), 165–181. doi:10.1177/1464884916688550 PMID:29417952

Dias, P., & Andrade, J. (2015) Desafios das RP na Era dos Media Sociais. Novos media e novos públicos.

Dias, A. O., Luz, G. S., Assunção, V. K., & Gonçalves, T. M. (2018). Mariana, o maior desastre ambiental do Brasil: uma análise do conflito socioambiental. In N. I. Ladwig & H. Schwalm (Eds.), *Planejamento e gestão territorial: a sustentabilidade dos ecossistemas urbanos* (pp. 455–476). Ediunesc. doi:10.18616/pgt20

Dias, P., & Andrade, J. (2017). O papel dos social media nas crises reputacionais. In *A Comunicação Organizacional e os desafios tecnológicos: estudos sobre a influência tecnológica nos processos de comunicação nas organizações* (pp. 55–69). CECS - Centro de Estudos de Comunicação e Sociedade Universidade do Minho.

DiStaso, M. W., Vafeiadis, M., & Amaral, C. (2015). Managing a health crisis on Facebook: How the response strategies of apology, sympathy, and information influence public relations. *Public Relations Review*, *41*(2), 222–231. doi:10.1016/j.pubrev.2014.11.014

Domino's. (2020). *Domino's 101: Fun Facts*. https://biz.dominos.com/web/public/about-dominos/fun-facts

Donker, T., Petrie, K., Proudfoot, J., Clarke, J., Birch, M. R., & Christensen, H. (2013). Smartphones for smarter delivery of mental health programs: A systematic review. *Journal of Medical Internet Research*, *15*(11), e247. Advance online publication. doi:10.2196/jmir.2791

Droppert, H., & Bennett, S. (2015). Corporate social responsibility in global health: An exploratory study of multinational pharmaceutical firms. *Globalization and Health, 11*(1), 1–8. doi:10.118612992-015-0100-5 PMID:25886175

Du Plessis, Ch. (2018). Social media crisis communication: Enhancing a discourse of renewal through dialogic content. *Public Relations Review, 44*(5), 829–838. doi:10.1016/j.pubrev.2018.10.003

Duarte, R. (2004). Entrevistas em pesquisas qualitativas. *Review of Education, 24*(24), 213–225. doi:10.1590/0104-4060.357

Ducci, G. (2017). *Relazionalità consapevole: la comunicazione pubblica nella società connessa* [Conscious relationality: public communication in the connected society]. Angeli.

Düchting, M., Zimmermann, D., & Nebe, K. (2007). Incorporating user centered requirement engineering into agile software development. *International Conference on Human-Computer Interaction.*

Dugdale, J., Negre, E., & Turoff, M. (2020). Introduction to the Minitrack on Information and Communication Technologies for Crisis and Emergency Management. *Proceedings of the 53rd Hawaii International Conference on System Sciences*, 573-575. 10.24251/HICSS.2020.070

Duncan, S., & Barrett, L. F. (2007). Affect is a form of cognition: A neurobiological analysis. *Cognition and Emotion, 21*(6), 1184–1211. doi:10.1080/02699930701437931 PMID:18509504

Dzarasov, R. (2016). The global crisis and its impact on the Eurasian Economic Union. *European Political Science, 17*(May), 23–34. doi:10.1080/23745118.2016.1171272

Eckert, S., Sopory, P., Day, A., Wilkins, L., Padgett, D., Novak, J., Allen, T., Alexander, N., Vanderford, M., & Gamhewage, G. (2018). Health-related disaster communication and social media: Mixed-method systematic review. *Health Communication, 33*(12), 1389–1400. doi:10.1080/10410236.2017.1351278 PMID:28825501

Edwards, C., Spence, P. R., Gentile, C. J., Edwards, A., & Edwards, A. (2013). How much Klout do you have… A test of system generated cues on source credibility. *Computers in Human Behavior, 29*(5), A12–A16. doi:10.1016/j.chb.2012.12.034

Egan, R., Maidment, J., & Connolly, M. (2017). Trust, power and safety in the social work supervisory relationship: Results from Australian research. *Journal of Social Work Practice, 31*(3), 307–321. doi:10.1080/02650533.2016.1261279

Egbunike, N. (2015). Framing the #Occupy Nigeria Protests in Newspapers and Social Media. *Open Access Library Journal, 2*(05), e1486. doi:10.4236/oalib.1101486

Ekman, P. (1994). Strong evidence for universals in facial expressions: A reply to Russell's mistaken critique. *Psychological Bulletin, 115*(2), 268–287. doi:10.1037/0033-2909.115.2.268 PMID:8165272

Elbarazi, I., Loney, T., Yousef, S., & Elias, A. (2017). Prevalence of and factors associated with burnout among health care professionals in Arab countries: A systematic review. *BMC Health Services Research, 17*(1), 1–10. doi:10.118612913-017-2319-8 PMID:28716142

EM-DAT. D. G.-S. (2020). *EM-DAT: The Emergency Events Database.* www.emdat.be

Endsley, M. R. (1995). Toward a theory of situation awareness in dynamic systems. *The Journal of the Human Factors and Ergonomics Society, 37*(1), 32–64. doi:10.1518/001872095779049543

Endsley, M. R., & Garland, D. J. (Eds.). (2000). *Situation Awareness Analysis and Measurement.* Lawrence Erlbaum Associates Publishers. doi:10.1201/b12461

Eriksson, M. (2018). Lessons for crisis communication on social media: A systematic review of what research tells the practice. *International Journal of Strategic Communication, 12*(5), 526–551. doi:10.1080/1553118X.2018.1510405

Eriksson, M., & Olsson, E. K. (2016). Facebook and Twitter in Crisis Communication: A Comparative Study of Crisis Communication Professionals and Citizens. *Journal of Contingencies and Crisis Management*, *24*(4), 198–208. doi:10.1111/1468-5973.12116

Escudero, C. (2020). Digital Mutation, a Result of Motivation and Resilience. Stolen Babies in Catalunya. *Review of Journalism and Mass Communication*, *1*(8), 1–8. doi:10.15640/rjmc.v8n1a1

Escudero, C. (2020). Giving voice to the traumatic event, Spanish mothers of stolen babies. Three strategies to silence mothers during and after the dictatorship. *International Journal of Humanities and Social Science*, *3*(10). doi:10.30845/ijhss.v10n3p1

Escudero, C. (2020). Stolen Babies in Spain: Mediated Stories for Recovery. Mothers' activism through online campaigns. *International Journal of Business and Social Science*, *11*(3). Advance online publication. doi:10.30845/ijbss.v11n3a3

Espindola, H. S., Nodari, E. S., & Santos, M. (2019). Rio Doce: Risks and Uncertainties of the Mariana Disaster (MG). *Revista Brasileira de História*, *39*(81), 141–162. doi:10.1590/1806-93472019v39n81-07

Esposito, R. (2002). *Immunitas. Protezione e negazione della vita*. Einaudi.

Estellés-Arolas, E., & González-Ladrón-de-Guevara, F. (2012). Towards an integrated crowdsourcing definition. *Journal of Information Science*, *38*(2), 189–200. doi:10.1177/0165551512437638

European Commission. (2018) Privacy Code of Conduct on mobile health (mHealth) apps. In *Shaping Europe's digital future*. https://ec.europa.eu/digital-single-market/en/privacy-code-conduct-mobile-health-apps

European Union – iBorderCtrl. (2019). *iBorderCtrl: Intelligence Portable Control System*. https://www.iborderctrl.eu/

Ewart, J., & McLean, H. (2019). Best practice approaches for reporting disasters. *Journalism*, *20*(12), 1573–1592. doi:10.1177/1464884918757130

Eysenbach, G. (2001). What is e-health? *Journal of Medical Internet Research*, *3*(2), e20. Advance online publication. doi:10.2196/jmir.3.2.e20

Eysenbach, G. (2006). Infodemiology: Tracking flu-related searches on the web for syndromic surveillance. *AMIA ... Annual Symposium Proceedings - AMIA Symposium. AMIA Symposium*, *2006*, 244–248. PMID:17238340

Faasse, K., Casey, J., Chatman, C. J., Leslie, R., & Martin, L. R. (2016). A comparison of anguage use in pro- and anti-vaccination comments in response to a high profile Facebook post. *Vaccine*, *34*(47), 5808–5814. doi:10.1016/j.vaccine.2016.09.029 PMID:27707558

Faccioli, F. (2006). *Comunicazione pubblica e cultura del servizio* [Public communication and culture of service]. Carocci.

Fahy, A. (2007). The unbearable fatigue of compassion: Notes from a substance abuse counselor who dreams of working at Starbuck's. *Clinical Social Work Journal*, *35*(3), 199–205. doi:10.100710615-007-0094-4

Fals Borda, O., Bonilla, V., & Castillo, G. (1972). *Causa popular, ciencia popular: una metodología del conocimiento científico a través de la acción* [Popular cause, popular science: a methodology of scientific knowledge through action]. Publicaciones de La Rosca de Investigación y Acción Social.

Fang, Z. J. I. j. o. t. C., the Internet, & Management. (2002). *E-government in digital era: concept, practice, and development*. Academic Press.

Fatma, M., Rahman, Z., & Khan, I. (2015). Building company reputation and brand equity through CSR: The mediating role of trust. *International Journal of Bank Marketing*, *33*(6), 840–856. doi:10.1108/IJBM-11-2014-0166

Faulkner, L. (2003). Beyond the five-user assumption: Benefits of increased sample sizes in usability testing. *Behavior Research Methods, Instruments, & Computers, 35*(3), 379–383. doi:10.3758/BF03195514

Fearn-Banks, K. (1996). *Crisis communication: A casebook approach*. Lawrence Erlbaum Associates.

Fearn-Banks, K. (2011). *Crisis Communication: a casebook approach* (Vol. 4). Routledge.

Felgueiras, S. (2015). Ação Policial face à ação coletiva: teoria para uma estratégia de policiamento de multidões. Lição Inaugural da Abertura Solene do ano letivo 2015/2016. Lisboa: ISCPSI.

Ferrigni, N., & Spalletta, M. (2018). La paura viene twittando. Social media, terrorismo e percezione della sicurezza [Fear comes with Twitter. Social media, terrorism and social perception of security]. *Sociológia, 1*, 193–204.

Fetters, M. D., Curry, L. A., & Creswell, J. W. (2013). Achieving integration in mixed methods designs: Principles and practices. *Health Services Research, 48*(6), 2134–2156. doi:10.1111/1475-6773.12117 PMID:24279835

Figley, C. R. (Ed.). (1995). *Compassion fatigue: Coping with secondary traumatic stress disorder in those who treat the traumatized*. Routledge.

Finklestein, M., Stein, E., Greene, T., Bronstein, I., & Solomon, Z. (2015). Posttraumatic stress disorder and vicarious trauma in mental health professionals. *Health & Social Work, 40*(2), 25–31. doi:10.1093/hsw/hlv026

Finley, B. A., & Sheppard, K. G. (2017). Compassion fatigue: Exploring early-career oncology nurses' experiences. *Clinical Journal of Oncology Nursing, 21*(3), 61–66. doi:10.1188/17.CJON.E61-E66 PMID:28524893

Fisher, M. (2018). *Realismo capitalista* (V. Mattioli, Trans.). Nero. (Original work published 2009).

Flu Trends Team. (2015). The next chapter for flu trends. *Google AI Blog*. https://ai.googleblog.com/2015/08/the-next-chapter-for-flu-trends.html

Folkes, V. S. (1988). The availability heuristic and perceived risk. *The Journal of Consumer Research, 15*(1), 13–23. doi:10.1086/209141

Forest fires. (2020). Retrieved from https://www.eea.europa.eu/data-and-maps/indicators/forest-fire-danger-3/assessment

Foucault, M. (2005). *Sicurezza, territorio, popolazione. Corso al Collège de France (1977-1978)* (P. Napoli, Trans.). Feltrinelli. (Original work published 2004).

Foucault, M. (1995). *Discipline and punish. The birth of the prison* (A. Sheridan, Trans.). Vintage Books. (Original work published 1975).

Fraser, J. (1981). Italy: Society in Crisis, Society in Transformation. Routledge & Kegan.

Freberg, K., Palenchar, M. J., & Veil, S. R. (2013). Managing and sharing H1N1 crisis information using social media bookmarking services. *Public Relations Review, 39*(3), 178–184. doi:10.1016/j.pubrev.2013.02.007

Freimuth, V. S., Hilyard, K. M., Barge, J. K., & Sokler, L. A. (2008). Action, not talk: A simulation of risk communication during the first hours of a pandemic. *Health Promotion Practice, 9*(4_suppl), 35S-44S.

Freimuth, V. S. (2006). Order out of chaos: The self-organization of communication following the Anthrax attacks. *Health Communication, 20*(2), 141–148. doi:10.120715327027hc2002_5 PMID:16965251

Freitas, F., & Mendes, J. M. (2019). Disaster, reconstruction, and data for social good: The case of wildfires in Portugal. *International Journal of Disaster Resilience in the Built Environment, 10*(4), 239–247. doi:10.1108/IJDRBE-06-2019-0035

Friedman, K. (2017). Counselor self-care and mindfulness. *Contemporary Buddhism, 18*(2), 321–330. doi:10.1080/14639947.2017.1373437

Fujibuchi, T., Inoue, A., Ishigaki, Y., & Matsumoto, Y. (2019a). Development of a wireless multisensor active personal dosimeter–tablet system. *Progress in Nuclear Science and Technology, 6*(0), 73–76. doi:10.15669/pnst.6.73

Fujibuchi, T., Nozaki, Y., Ishigaki, Y., & Matsumoto, Y. (2019b). Estimation of the characteristics of gamma-ray dose measurements with an experimental wireless dose monitoring system. *Progress in Nuclear Science and Technology, 6*(0), 77–80. doi:10.15669/pnst.6.77

Fundação Renova. (2020). *A Fundação*. Retrieved from https://www.fundacaorenova.org/

Fundación Telefónica. (2019). *Sociedad Digital en España 2018*. https://www.fundaciontelefonica.com/cultura-digital/publicaciones/sociedad-digital-en-espana-2018/655/

Gama, M. G. (2000). Quando o inferno desce à terra: A gestão de crises e a sua problemática. *Comunicação e Sociedade, 14*, 535–542. doi:10.17231/comsoc.2(2000).1419

Gantz, J., & Reinsel, D. J. I. i. I. A. t. f. (2012). *The digital universe in 2020: Big data, bigger digital shadows, and biggest growth in the far east*. Academic Press.

García Ramírez, M., Albar-Marín, M. J., Morano Báez, R., & Castro, V. P. (2017). Metodología de la intervención social: implicaciones para la investigación y la práctica de la psicología comunitaria [Methodology of social intervention: implications for research and practice in community psychology]. In I. Maya-Jariego, M. García Ramírez, F. J. Santolaya -Soriano (Eds.). Estrategias de Intervención Psicosocial [Strategies for Psychosocial Intervention]. Psicología Pirámide.

García-Ramírez, M., Balcázar, F., & Suárez-Balcázar, Y. (2003). Investigación acción participativa en la psicología comunitaria: Una herramienta para entender la diversidad humana [Participatory Action Research in community psychology: a tool for understanding human diversity]. *Apuntes de Psicología, 21*(3), 409–417.

Garnelo-Gomez, I., & Saraeva, A. (2019). Yes, we can! Encouraging responsible management through effective CSR communication. In F. Farache, G. Grigore, A. Stancu, & D. McQueen (Eds.), *Responsible People, Palgrave Studies in Governance, Leadership and Responsibility* (pp. 115–134). Springer. doi:10.1007/978-3-030-10740-6_6

Garnett, J., & Kouzmin, A. (2007). Communicating throughout Katrina: Competing and complementary conceptual lenses on crisis communication. *Public Administration Review, 67*(1), 171–188. doi:10.1111/j.1540-6210.2007.00826.x

Garnett, J., & Kouzmin, A. (2009). Crisis communication post-Katrina: What are we learning? *Public Organization Review, 9*(4), 385–398. doi:10.100711115-009-0096-5

Garret, J. J. (2003). *The Elements of User Experience*. American Institute of Graphic Arts.

Gerken, F., Van der Land, S. & van der Meer, T. (2016). Crisis in the air: an investigation of AirAsia's crisis-response effectiveness based on frame alignment. *Public Relations Review,* 1 – 14. doi:10.1016/j.pubrev.2016.09.002

Giddens, A. (1990). *The Consequences of Modernity*. Polity Press.

Gills, B., & Gray, K. (2012). People power in the era of global crisis: Rebellion, resistance, and liberation. *Third World Quarterly, 33*(2), 205–224. doi:10.1080/01436597.2012.664897

Giordano, A. L., Prosek, E. A., Stamman, J., Callahan, M. M., Loseu, S., Bevly, C. M., Cross, K., Woehler, E. S., Calzada, R. R., & Chadwell, K. (2016). Addressing trauma in substance abuse treatment. *Journal of Alcohol and Drug Education, 60*(2), 55–71.

Giunti, G., Mylonopoulou, V., & Romero, O. R. (2018). More stamina, a gamified mhealth solution for persons with multiple sclerosis: Research through design. *JMIR mHealth and uHealth*, *6*(3), e51. doi:10.2196/mhealth.9437

Global Justice. (2016). *Vale da Lama: Relatório de inspeção em Mariana após o rompimento da barragem de rejeitos do Fundão*. Retrieved from: http://www.global.org.br/wp-content/uploads/2016/03/Vale-de-Lama-Justi--a-Global.pdf

Godard, F. (2018). *Il consenso nell'epoca del terrorismo* (A. L. Carbone, Trans.). Eleuthera. (Original work published 2016).

GODIAC. (2013a). *Field Study Handbook: GODIAC - Good practice for dialogue and communication as strategic principles for policing political manifestations in Europe*. http://www.polisen.se

GODIAC. (2013b). *Recommendations for policing political manifestations in Europe. Good practice for dialogue and communication as strategic principles for policing political manifestations in Europe*. http://www.polisen.se

Goldberg, J. H., Stimson, M. J., Lewenstein, M., Scott, N., & Wichansky, A. M. (2002). Eye tracking in web search tasks: design implications. In *Proceedings of the 2002 symposium on Eye tracking research & applications*. ACM.

Goldingay, S., & Boddy, J. (2017). Preparing social work graduates for digital practice: Ethical pedagogies for effective learning. *Australian Social Work*, *70*(2), 209–220. doi:10.1080/0312407X.2016.1257036

Goldstein, S. M., Johnston, R., Duffy, J., & Rao, J. J. J. o. O. m. (2002). *The service concept: the missing link in service design research?* Academic Press.

Google. (n.d.). *Google Trends*. Retrieved May 25, 2020, from https://trends.google.pt/trends/?geo=PT

Gorringe, H., Stott, C., & Rosie, M. (2012). Dialogue Police, Decision Making, and the Management of Public Order During Protest Crowd Events. *Journal of Investigative Psychology and Offender Profiling*, *9*(2), 111–125. doi:10.1002/jip.1359

Gostelow, L. J. D. (1999). *The Sphere Project: the implications of making humanitarian principles and codes work*. Academic Press.

Gralla, E., Goentzel, J., & Van de Walle, B. (2015). *Understanding the information needs of field-based decision-makers in humanitarian response to sudden onset disasters*. ISCRAM.

Grandi, R. (2013). *La comunicazione pubblica* [Public communication]. Carocci.

Grant, J. M., & Mack, D. A. (2004). Preparing for the Battle: Healthy Leadership During Organizational Crisis. *Organizational Dynamics*, *33*(4), 409–425. doi:10.1016/j.orgdyn.2004.09.007

Gregor-Haack, J. (2018). Erstattung von Health-Apps durch die gesetzliche Krankenversicherung [Reimbursement of health apps by the German statutory health insurance]. *Bundesgesundheitsblatt, Gesundheitsforschung, Gesundheitsschutz*, *61*(3), 328–333. doi:10.100700103-018-2689-z

Greinacher, A., Derezza-Greeven, C., Herzog, W., & Nikendei, C. (2019). Secondary traumatization in first responders: A systematic review. *European Journal of Psychotraumatology*, *10*(1), 1–11. doi:10.1080/20008198.2018.1562840 PMID:30719236

Groshong, L., & Phillips, D. (2015). The impact of electronic communication on confidentiality in clinical social work practice. *Clinical Social Work Journal*, *43*(2), 142–150. doi:10.100710615-015-0527-4

Grosser, K., & Moon, J. (2005). The role of corporate social responsibility in gender mainstreaming. *International Feminist Journal of Politics*, *7*(4), 532–554. doi:10.1080/14616740500284524

Grosser, K., & Moon, J. (2019). CSR and feminist organization studies: Towards an integrated theorization for the analysis of gender issues. *Journal of Business Ethics*, *155*(2), 321–342. doi:10.100710551-017-3510-x

Grundlingh, H., Knight, L., Naker, D., & Devries, K. (2017). Secondary distress in violence researchers: A randomized trial of the effectiveness of group debriefings. *BMC Psychiatry*, *17*(1), 1–14. doi:10.118612888-017-1327-x PMID:28578682

Grundy, Q. H., Wang, Z., & Bero, L. A. (2016). Challenges in assessing mobile health app quality: A systematic review of prevalent and innovative methods. *American Journal of Preventive Medicine*, *51*(6), 1051–1059. doi:10.1016/j.amepre.2016.07.009

Gryszkiewicz, A., & Chen, F. (2012). Temporal aspects in crisis management and its implications on interface design for situation awareness. *Cognition Technology and Work*, *14*(2), 169–182. doi:10.100710111-011-0199-y PMID:32214898

Guha-Sapir, D., & Lechat, M. F. J. D. (1986). *Information systems and needs assessment in natural disasters: an approach for better disaster relief management*. Academic Press.

Guidry, J. P. D., Jinb, Y., Orr, C. A., Messner, M., & Meganck, S. (2017). Ebola on Instagram and Twitter: How health organizations address the health crisis in their social media engagement. *Public Relations Review*, *43*(3), 477–486. doi:10.1016/j.pubrev.2017.04.009

Guigoni, A., & Ferrari, R. (2020). *Pandemia 2020. La vita quotidiana in Italia con il Covid-19* [Pandemic 2020. The Italian everyday life with the Covid-19]. M & J Publishing House.

Guimarães, M. C. S., Silva, C. H., & Antunes, M. N. (2008). Monitoramento de informação como estratégia de e-health: Um estudo prospectivo. *Textos de la Cibersociedad*, *16*, 216.

Guo, L. (2014). Toward the Third Level of Agenda Setting Theory: A Networked Agenda Setting Model. In T. Johnson (Ed.), *Agenda Setting in a 2.0 World. New Agendas in Communication*. Routledge.

Guo, L., & McCombs, M. (2011). Networked Agenda Setting: A Third Level of Media Effects. *Annual Conference of the International Communication Association.*

Haklay, M., & Weber, P. J. I. P. C. (2008). *Openstreetmap: User-generated street maps*. Academic Press.

Han, B. C. (2020, April 7). La cura al virus è lo Stato di polizia? *Avvenire*.

Han, B. C. (2017). *Psychopolitics. Neoliberalism and new technologies of power* (E. Butler, Trans.). Verso. (Original work published 2014).

Hanitzsch, T., Van Dalen, A., & Steindl, N. (2018). Caught in the Nexus: A Comparative and Longitudinal Analysis of Public Trust in the Press. *The International Journal of Press/Politics*, *23*(1), 3–23. doi:10.1177/1940161217740695

Hansen, P. G., Skov, L. R., & Skov, K. L. (2016). Making healthy choices easier: Regulation versus nudging. *Annual Review of Public Health*, *37*(1), 237–251. doi:10.1146/annurev-publhealth-032315-021537

Harari, Y. N. (2020). *The world after coronavirus*. https://www.ft.com/content/19d90308-6858-11ea-a3c9-1fe6fedcca75

Haraway, D. J. (1991). *Simians, cyborgs and women: the reinvention of nature*. Routledge.

Hart, V., Siddarth, D., Cantrell, B., Tretikov, L., Eckersley, P., Langford, J., Leibrand, S., Kakade, S., Latta, S., Lewis, D., Tessaro, S., & Weyl, G. (2020). *Outpacing the Virus: Digital Response to Containing the Spread of COVID-19 while Mitigating Privacy Risks.* Academic Press.

Harte, R., Quinlan, L. R., Glynn, L., Rodríguez-Molinero, A., Baker, P. M., Scharf, T., & ÓLaighin, G. (2017). Human-Centered Design Study: Enhancing the usability of a mobile phone app in an integrated falls risk detection system for use by older adult users. *JMIR mHealth and uHealth*, *5*(5), e71. Advance online publication. doi:10.2196/mhealth.7046

Hartmann, P., Apaolaza-Ibáñez, V., & Forcada-Sainz, F. J. (2005). Green branding effects on attitude: Functional versus emotional positioning strategies. *Marketing Intelligence & Planning*, *23*(1), 9–29. doi:10.1108/02634500510577447

Hassan, S., & Craft, S. (2012). Examining world market segmentation and brand positioning strategies. *Journal of Consumer Marketing*, *29*(5), 344–356. doi:10.1108/07363761211247460

Hassenzahl, M. (2006). Hedonic, emotional and experiential perspectives on product quality. In C. Ghaoui (Ed.), *Encyclopedia of human computer interaction* (pp. 266–272). IGI Global. doi:10.4018/978-1-59140-562-7.ch042

Haus, M., Ding, A., & Ott, J. (2020). Multimodal Co-Presence Detection with Varying Spatio-Temporal Granularity. *IEEE International Conference on Pervasive Computing and Communications Workshops (PerCom Workshops)*, 1-7. 10.1109/PerComWorkshops48775.2020.9156105

Havelund, J., Ilum, J., Jensen, M., Nielsen, B., Rasmussen, K., & Stott, C. (2011). *Event Policing - Dialogue in the policing of mass events in Denmark.* European Police Science and research. *Bulletin*, *4*, 3–7.

Haworth, B. T. (2018). Implications of Volunteered Geographic Information for Disaster Management and GIScience: A More Complex World of Volunteered Geography. *Annals of the Association of American Geographers*, *108*(1), 226–240. doi:10.1080/24694452.2017.1321979

Haworth, B., Whittaker, J., & Bruce, E. (2016). Assessing the application and value of participatory mapping for community bushfire preparation. *Applied Geography (Sevenoaks, England)*, *76*, 115–127. doi:10.1016/j.apgeog.2016.09.019

Hazin, C. A. F., Gazineu, M. H. P., & de Farias, E. E. G. (2008). *Uranium and thorium in zircon sands processed in Northeastern Brazil* [Paper presentation]. IRPA 12: 12 International Congress of the International Radiation Protection Association (IRPA): Strengthening radiation protection worldwide, Buenos Aires, Argentina. 10.1148/86.2.354b

Hazra, K., Shah, V., Bilal, M., Silvestri, S., Das, S., Nandi, S., & Saha, S. (2019). A Novel Network Architecture for Resource-constrained Post-disaster Environments. In *11th International Conference on Communication Systems & Networks (COMSNETS)* (pp. 328-335). IEEE. 10.1109/COMSNETS.2019.8711166

Healey, J. A., & Picard, R. W. (2005). Detecting stress during real-world driving tasks using physiological sensors. *IEEE Transactions on Intelligent Transportation Systems*, *6*(2), 156–166. doi:10.1109/TITS.2005.848368

Heath, R. L., & Palenchar, M. J. (2016). Paradigms of risk and crisis communication in the 21st century. In A. Schwarz, M.W. Seeger, & C. Auer (Eds.), The handbook of international crisis communication research (pp. 437-446). Bognor Regis: John Wiley & Sons Ltd.

Heath, R. (2010). Crisis Communication: defining the Beast and De-marginalizing Key Publics. In T. Coombs & S. Holladay (Eds.), *The Handbook of Crisis Communication, 1 - 14.* Wiley-Blackwell. doi:10.4324/9780203891629

Heide, M., & Simonsson, C. (2019). *Internal Crisis Communication: Crisis Awareness, Leadership and Coworkership.* Routledge. doi:10.4324/9780429425042

Heikkinen, M. T., Mainela, T., Still, J., & Tähtinen, J. J. I. M. M. (2007). *Roles for managing in mobile service development nets.* Academic Press.

Heinberg, M., Ozkaya, H., & Taube, M. (2017). The influence of global and local iconic brand positioning on advertising persuasion in an emerging market setting. *Journal of International Business Studies*, *48*(8), 1009–1022. doi:10.105741267-017-0071-2

Helsloot, I., Ruitenberg, A. J. J. o. c., & Management, c. (2004). *Citizen response to disasters: a survey of literature and some practical implications*. Academic Press.

Hempel, M. (2014). *The use of social media in environmental health research and communication: an evidence review*. Environmental Public Health.

Hensel, J. M., Ruiz, C., Finney, C., & Dewa, C. S. (2015). Meta-analysis of risk factors for secondary traumatic stress in therapeutic work with trauma victims. *Journal of Traumatic Stress*, *28*(2), 83–91. doi:10.1002/jts.21998 PMID:25864503

Her Majesty's Chief Inspector of Constabulary (HMIC). (2008). *Adapting to protest– nurturing the British model of policing*. Author.

Herman, J. (1992). *Trauma and Recovery. The Aftermath of Violence - From domestic abuse to political terror*. Pandora Edition.

Herman, J. (1997). *Trauma and recovery: The aftermath of violence-from domestic abuse to political terror*. Basic Books.

Hermann, C. F. (1963). Some consequences of crisis which limit the viability of organisations. *Administrative Science Quarterly*, *8*(1), 61–82. doi:10.2307/2390887

Hevner, A. R., March, S. T., Park, J., & Ram, S. J. M. q. (2004). *Design science in information systems research*. Academic Press.

Heyman, B. (2010). The concept of risk. In B. Heyman, M. Shaw, A. Alaszewski, & M. Titterton (Eds.), *Risk, safety and clinical practice: Healthcare through the lens of risk* (pp. 15–36). Oxford University Press.

Hight, J., & Smyth, F. (2003). *Tragedias & Periodistas. Guía para una cobertura más eficaz*. Dart Center for Journalism and Trauma.

Hilding-Hamann, S. (2012). *The effects of social media on crisis communication: A case study of the response strategy performed by Research in Motion* [Unpublished B. A Thesis]. Aarhus University, Denmark.

Hilhorst, D., & Schmiemann, N. J. D. i. P. (2002). *Humanitarian principles and organisational culture: Everyday practice in Meédecins Sans Frontié res-Holland*. Academic Press.

Hiltz, S. R., Hughes, A. L., Imran, M., Plotnick, L., Power, R., & Turoff, M. (2020). Exploring the usefulness and feasibility of software requirements for social media use in emergency management. *International Journal of Disaster Risk Reduction*, *42*, 101367. doi:10.1016/j.ijdrr.2019.101367

Hoffman, J. E., Wald, L., Kuhn, E., Greene, C., Ruzek, J., & Weingardt, K. (2011). *PTSD Coach* [mobile application]. US Department of Veterans Affairs. https://itunes.apple.com/de/app/ptsd-coach/id430646302?mt=8

Holgersson, S. (2010). *Dialogue Police: experiences, observations and opportunities*. Swedish National Police Board.

Hong, L., Fu, C., Wu, J., & Frias-Martinez, V. J. I. S. F. (2018). *Information needs and communication gaps between citizens and local governments online during natural disasters*. Academic Press.

Hootsuite & We Are Social. (2020). *Digital 2020 Portugal. Global Digital Insights*. Author.

Hopwood, T. L., Schutte, N. S., & Loi, N. M. (2019). Stress responses to secondary trauma: Compassion fatigue and anticipatory traumatic reaction among youth workers. *The Social Science Journal*, *56*(3), 337–348. doi:10.1016/j.soscij.2018.08.008

Hornmoen, H., & Backholm, K. (Eds.). (2018). *Social Media Use in Crisis and Risk Communication. Emergency, Concern and Awareness*. Emerald. doi:10.1108/9781787562691

Hossain, M., & Ray, S. (2018). A smartphone-assisted device-to-device communication for post-disaster recovery. *Springer International Conference on Smart Grid Inspired Future Technologies*, 11-20. 10.1007/978-3-319-94965-9_2

Hosseinali-Mirza, V., Marcellis-Warin, N., & Warin, T. (2015). Crisis communication strategies and reputation risk in the online social media environment. *International Journal of Business and Social Science*, *6*(5), 7–21.

Houston, J. B., Hawthorne, J., Perreault, M. F., Park, E. H., Goldstein Hode, M., Halliwell, M. R., Turner McGowen, S. E., Davis, R., Vaid, S., McElderry, J. A., & Griffith, S. A. (2014). Social media and disasters: A functional framework for social media use in disaster planning, response, and research. *Disasters*, *39*(1), 1–22. doi:10.1111/disa.12092 PMID:25243593

Hovland, C. I., Janis, I. L., & Kelley, H. H. (1953). *Communication and persuasion: Psychological studies of opinion change*. Yale University Press.

Hricova, M., & Lovasova, S. (2018). Stress, secondary trauma and burnout: Risk characteristics in helping professions. *Ad Alta: Journal of Interdisciplinary Research*, *8*(2), 161–165.

Huang, R., & Saragollu, E. (2014). How Brand Awareness Relates to Market Outcome, Brand Equity, and the Marketing Mix. In T.-M. Choi (Ed.), *Fashion Branding and Consumer Behaviors* (pp. 113–132). Springer., doi:10.1007/978-1-4939-0277-4_8

Huang, Y. C., Wu, F., & Huang, Q. (2016). Does research on digital public relations indicate a paradigm shift? An analysis and critique of recent trends. *Telematics and Informatics*. Advance online publication. doi:10.1016/j.tele.2016.08.012

Hughes, A. L., & Shah, R. (2016). Designing an application for social media needs in emergency public information work. In *Proceedings of the 19th International Conference on Supporting Group Work*. ACM Press. 10.1145/2957276.2957307

Hur, W. M., Kim, H., & Woo, J. (2014). How CSR leads to corporate brand equity: Mediating mechanisms of corporate brand credibility and reputation. *Journal of Business Ethics*, *125*(1), 75–86. doi:10.100710551-013-1910-0

Hussain, M., Al-Haiqi, A., Zaidan, A. A., Zaidan, B. B., Kiah, M. L. M., Anuar, N. B., & Abdulnabi, M. (2015). The landscape of research on smartphone medical apps: Coherent taxonomy, motivations, open challenges and recommendations. *Computer Methods and Programs in Biomedicine*, *122*(3), 393–408. doi:10.1016/j.cmpb.2015.08.015

Hyvärinen, J., & Vos, M. (2016). Communication concerning disasters and pandemics: coproducing community resilience and crisis response. In A. Schwarz, M. W. Seeger, & C. Auer (Eds.), *The handbook of international crisis communication research* (pp. 96–108). John Wiley & Sons Ltd. doi:10.1002/9781118516812.ch10

Iannella, R., & Henricksen, K. (2007). Managing information in the disaster coordination centre: Lessons and opportunities. *4th International Conference on Information Systems for Crisis Response and Management.*

Ikeda, S., & Maeda, Y. (Eds.). (2013, March 11). *Emerging issues learned from the 3.11 disaster as multiple events of earthquake, tsunami, and Fukushima nuclear accident*. The Society for Risk Analysis. http://www.sra-japan.jp/cms/uploads/311Booklet.pdf

Ilharco, F. (2005). A Tecnologia como Contexto ou a Ordenação Informacional e Comunicacional do Mundo (Prisma. Com, Ed.). Porto: Faculdade de Letras da Universidade do Porto.

Ilharco, F. (2003). *Filosofia da Informação: uma introdução à informação como fundação da acção, da comunicação e da decisão*. Universidade Católica Editora.

Illich, I. (1975). *Medical nemesis. The exploration of health*. Pantheon Books.

Imran, M., Castillo, C., Lucas, J., Meier, P., & Vieweg, S. (2014). AIDR: Artificial intelligence for disaster response. *Proceedings of the 23rd International Conference on World Wide Web.*

INE - Instituto Nacional de Estatística (Ed.). (2019). *Estatísticas do Ambiente 2018. Instituto Nacional de Estatística.* INE - Instituto Nacional de Estatística.

Iniewski, K. (Ed.). (2010). *Semiconductor radiation detection systems.* CRC Press. doi:10.1201/9781315222608

Iniewski, K. (Ed.). (2011). *Electronics for radiation detection.* CRC Press.

Initiative, H. H. (2010). Disaster Relief 2.0: The future of information sharing in humanitarian emergencies. In Disaster Relief 2.0: The future of information sharing in humanitarian emergencies. HHI; United Nations Foundation; OCHA; The Vodafone Foundation.

Institute for Strategic Dialogue. (2014). *Case Study: Dialogue Police.* Author.

International Atomic Energy Agency. (2014). *IAEA report on international experts meeting radiation protection after the Fukushima Daiichi accident: Promoting confidence and understanding.* International Atomic Energy Agency. https://www-pub.iaea.org/MTCD/Publications/PDF/IEM-6_web.pdf

International Atomic Energy Agency. (2016). *IAEA Incident and Trafficking Database (ITDB): Incidents of nuclear and other radioactive material out of regulatory control – 2016 Fact Sheet.* International Atomic Energy Agency. doi:10.1787/9789264244047-24-en

International Commission on Radiological Protection. (2007). The 2007 recommendations of the International Commission on Radiological Protection (ICRP Publication 103). *Annals of the ICRP, 37*(2–4). Advance online publication. doi:10.1177/ANIB_37_2-4

Iqbal, A. (2015). The ethical considerations of counselling psychologists working with trauma: Is there a risk of vicarious traumatisation? *Counselling Psychology Review, 30*(1), 44–51.

Ishigaki, Y., Matsuno, Y., Bando, K., & Tanaka, K. (2017). Wisdom of crowds for reliable discussion and need finding: A case study of information sharing regarding radiation after the Fukushima Nuclear Disaster. *Proc. of the 50th Hawaii International Conference on System Sciences (HICSS)*, 323–331. https://www.researchgate.net/publication/317121029_Wisdom_of_Crowds_for_Reliable_Discussion_and_Need_Finding_A_Case_Study_of_Information_Sharing_Regarding_Radiation_after_the_Fukushima_Nuclear_Disaster

Ishigaki, Y., Pradana, H. A., Permana, S., & Tanaka, K. (in press). Dose-rate mapping using smartphone for risk awareness in local residents and workers at Zircon Sand Facility in Bangka Island in Indonesia (Extended abstract). *Proc. of 15th International Congress of the International Radiation Protection Association (IRPA15).*

Ishigaki, Y., Matsumoto, Y., Pradana, H. A., & Tanaka, K. (2017). *Citizen sensing for environmental risk communication action research on PM2.5 air quality monitoring in East Asia* [Paper presentation]. *The Second International Conference on Cyber-Technologies and Cyber-Systems*, Barcelona, Spain.

Ishigaki, Y., Matsuno, Y., & Tanaka, K. (2017). *Agile way of risk awareness by smartphone-connected environmental sensors* [Paper presentation]. *14th International Conference on Information Systems for Crisis Response and Management*, Albi, France.

Ishigaki, Y., & Tanaka, K. (2017). *Pollution mapping by smartphone becoming real possibility* [Paper presentation]. *International Conference on Civil, Disaster Management and Environmental Sciences*, Bali, Indonesia.

ISO 9241. (2018). *Ergonomics of human-system interaction—Part 210: Human-centered design for interactive systems.* Geneva: International Standard Organization.

Italian Army. (2020). *Operazione "Strade Sicure"*. http://www.esercito.difesa.it/operazioni/operazioni_nazionali/Pagine/Operazione-Strade-Sicure.aspx

Iwasaki, N. J. J. o. E.-G. (2013). *Usability of ICT applications for elderly people in disaster reduction*. Academic Press.

Jagannath, J., Furman, S., Jagannath, A., & Drozd, A. (2019). *Energy Efficient Ad Hoc Networking Devices for Off-the-Grid Public Safety Networks. In 16th IEEE Annual Consumer Communications \& Networking Conference (CCNC)*. IEEE.

Jain, P., Vyas, V., & Chalasani, D. P. (2016). Corporate social responsibility and financial performance in SMEs: A structural equation modelling approach. *Global Business Review*, *17*(3), 630–653. doi:10.1177/0972150916630827

Jalkala, A., & Keränen, J. (2014). Brand positioning strategies for industrial firms providing customer solutions. *Journal of Business and Industrial Marketing*, *29*(3), 253–264. doi:10.1108/JBIM-10-2011-0138

Jamali, D., & Mirshak, R. (2007). Corporate Social Responsibility (CSR): Theory and practice in a developing country context. *Journal of Business Ethics*, *72*(3), 243–262. doi:10.100710551-006-9168-4

Japan Atomic Energy Agency. (2020). *PHITS: Particle and Heavy Ion Transport Code System*. https://phits.jaea.go.jp/index.html

Jarrad, R. A., & Hammad, S. (2020). Oncology nurses' compassion fatigue, burnout and compassion satisfaction. *Annals of General Psychiatry*, *19*(1), 1–8. doi:10.118612991-020-00272-9 PMID:32265998

Jasanoff, S. (2004). The Idiom of co-production. In S. Jasanoff (Ed.), *States of knowledge: the co-production of science and social order* (pp. 1–12). International Library of Sociology. doi:10.4324/9780203413845

Jasmontaite, L., & Dimitrova, D. (2017). Online disaster management: Applicability of the European data protection framework and its key principles. *Journal of Contingencies and Crisis Management*, *25*(1), 23–30. doi:10.1111/1468-5973.12142

Jin, Y., Austin, L., Santosh Vijaykumar, S., Jun, H., & Nowak, G. (2019). Communicating about infectious disease threats: Insights from public health information officers. *Public Relations Review*, *45*(1), 167–177. doi:10.1016/j.pubrev.2018.12.003

Jin, Y., & Liu, B. F. (2010). The blog-mediated crisis communication model: Recommendations for responding to influential external blogs. *Journal of Public Relations Research*, *22*(4), 429–455. doi:10.1080/10627261003801420

Johansson, J., Dimofte, C., & Mazvancheryl, S. (2012). The performance of global brands in the 2008 financial crisis: A test of two brand value measures. *International Journal of Research in Marketing*, *29*(3), 235–245. doi:10.1016/j.ijresmar.2012.01.002

Johnson, H. A., Wagner, M. M., Hogan, W., Chapman, W., & Olszewski, R. T. (2004). Analysis of Web access logs for surveillance of influenza. *Studies in Health Technology and Informatics*, *107*, 1202–1206. PMID:15361003

Johnston, K. A., & Taylor, M. (2018). Engagement as communication pathways, possibilities, and future directions. In The handbook of communication engagement (pp. 1-15). John Wiley & Sons.

Joint Committee on Human Rights. (2009). *Demonstrating respect for rights? A human rights approach to policing protest*. House of Commons & House of Lords.

Jona, L. (2019). *La macchina della verità*. https://www.rai.it/programmi/report/inchieste/La-macchina-della-verita-27b730ca-6997-419d-a74c-deea8237e4e3.html

Jorge, N. (2010). *Reputação: um elemento diferenciador e protector face a crises organizacionais*. Escola Superior de Comunicação Social.

Josephson, S. (2005). Eye tracking methodology and the Internet. In K. L. Smith, S. Moriarty, G. Barbatsis, & K. Kenney (Eds.), *Handbook of visual communication: Theory, methods, and media* (pp. 63–80). Erlbaum.

Juang, P., Oki, H., Wang, Y., Martonosi, M., Peh, L. S., & Rubenstein, D. (2002). Energy-efficient computing for wildlife tracking: Design tradeoffs and early experiences with ZebraNet. *10th international conference on Architectural support for programming languages and operating systems*, 96-107.

Jurdak, R., Zhao, K., Liu, J., AbouJaoude, M., Cameron, M., & Newth, D. (2015). Understanding Human Mobility from Twitter. *PLoS One*, *10*(7), e0131469. Advance online publication. doi:10.1371/journal.pone.0131469 PMID:26154597

Jurgens, M., & Helsloot, I. (2018). The effect of social media on the dynamics of (self) resilience during disasters: A literature review. *Journal of Contingencies and Crisis Management*, *26*(1), 79–88. doi:10.1111/1468-5973.12212

Ju, Y. J., Park, E. C., Ju, H. J., Lee, S. A., Lee, J. E., Kim, W., Chun, S. Y., & Kim, T. H. (2018). The influence of family stress and conflict on depressive symptoms among working married women: A longitudinal study. *Health Care for Women International*, *39*(3), 275–288. doi:10.1080/07399332.2017.1397672 PMID:29095122

Kaplan, A. M., & Haenlein, M. (2010). Users of the world, unite! The challenges and opportunities of social media. *Business Horizons*, *53*(1), 59–68. doi:10.1016/j.bushor.2009.09.003

Kathmandu Living Labs. (2015). *QuakeMap*. http://www.kathmandulivinglabs.org/projects/quakemaporg

Kathmandu Living Labs. (2016). *Map for everyone*. http://www.kathmandulivinglabs.org/projects/map-for-everyone

Kauer, S. D., Mangan, C., & Sanci, L. (2014). Do online mental health services improve help-seeking for young people? A systematic review. *Journal of Medical Internet Research*, *16*(3), e66. doi:10.2196/jmir.3103 PMID:24594922

Kaufmann, M. (2015). Resilience 2.0: Social Media Use and (Self-)care During the 2011 Norway Attacks. *Media Culture & Society*, *37*(7), 972–987. doi:10.1177/0163443715584101 PMID:29708120

Kazlauskas, E., Javakhishvilli, J., Meewisse, M., Merecz-Kot, D., Şar, V., Schäfer, I., Schnyder, U., & Gersons, B. P. R. (2016). Trauma treatment across Europe: Where do we stand now from a perspective of seven countries. *European Journal of Psychotraumatology*, *7*(1), 29450. Advance online publication. doi:10.3402/ejpt.v7.29450

Keen, S. M., & Roberts, N. (2017). Preliminary evidence for the use and efficacy of mobile health applications in managing posttraumatic stress disorder symptoms. *Health Systems (Basingstoke, England)*, *6*(2), 122–129. doi:10.1057/hs.2016.2

Kemp, S. (2019, January 30). *Digital trends 2019: Every single stat you need to know about the Internet*. The Next Web. https://thenextweb.com/contributors/2019/01/30/digital-trends-2019-every-single-stat-you-need-to-know-about-the-internet/

Kera, D., Rod, J., Peterova, R. J. N. d. a. F. D. S., Political, & Issues, E. (2013). *Post-apocalyptic citizenship and humanitarian hardware*. Academic Press.

Kerst, A., Zielasek, J., & Gaebel, W. (2020). Smartphone applications for depression: A systematic literature review and a survey of health care professionals' attitudes towards their use in clinical practice. *European Archives of Psychiatry and Clinical Neuroscience*, *270*(2), 139–152. doi:10.100700406-018-0974-3 PMID:30607530

Kessler, R. C., Aguilar-Gaxiola, S., Alonso, J., Benjet, C., Bromet, E. J., Cardoso, G., ... Ferry, F. (2017). Trauma and PTSD in the WHO world mental health surveys. *European Journal of Psychotraumatology*, *8*(sup5). *Article*, *1353383*. Advance online publication. doi:10.1080/20008198.2017.1353383

Keyton, J. (2019). *Communication Research: Asking Questions, Finding Answers* (5th ed.). McGraw-Hill Education.

Kingshott, B. (2011). Effective police management of the media. Criminal Justice Studies: A Critical Journal of Crime. *Law and Society*, *24*(3), 241–253. doi:10.1080/1478601X.2011.592728

Kirchgässner, G. (2009). The Global Crisis and the Answer of Economics. *Schweizerische Zeitschrift für Volkswirtschaft und Statistik*, *145*(4), 381–385. doi:10.1007/BF03399285

Kitaguchi, H., Miyai, H., Izumi, S., & Kaihara, A. (1996). Silicon semiconductor detectors for various nuclear radiations. *IEEE Transactions on Nuclear Science*, *43*(3), 1846–1850. doi:10.1109/23.507234

Kloog, I., Ridgway, B., Koutrakis, P., Coull, B. A., & Schwartz, J. D. (2013). Long- and short-term exposure to PM2.5 and mortality: Using novel exposure models. *Epidemiology (Cambridge, Mass.)*, *24*(4), 555–561. doi:10.1097/EDE.0b013e318294beaa

Knight. (2010). Indirect trauma in the field practicum: Secondary traumatic stress, vicarious trauma, and compassion fatigue among social work students and their field instructors. *The Journal of Baccalaureate Social Work, 15*, 32-57.

Knoll, G. F. (2010). *Radiation detection and measurement* (4th ed.). Wiley.

Knowles, A. J., & Cooner, T. S. (2016). International collaborative learning using social media to learn about social work ethics and social media. *Social Work Education*, *35*(3), 260–270. doi:10.1080/02615479.2016.1154662

Koenen, K. C., Ratanatharathorn, A., Ng, L., McLaughlin, K. A., Bromet, E. J., Stein, D. J., Karam, E. G., Meron Ruscio, A., Benjet, C., Scott, K., Atwoli, L., Petukhova, M., Lim, C. C. W., Aguilar-Gaxiola, S., Al-Hamzawi, A., Alonso, J., Bunting, B., Ciutan, M., de Girolamo, G., ... Kessler, R. C. (2017). Posttraumatic stress disorder in the world mental health surveys. *Psychological Medicine*, *47*(13), 2260–2274. doi:10.1017/S0033291717000708

Kohnhauser, F., Stute, M., Baumgartner, L., Almon, L., Katzenbeisser, S., Hollick, M., & Freisleben, B. (2017). SEDCOS: A secure device-to-device communication system for disaster scenarios. In *42nd Conference on Local Computer Networks (LCN)* (pp. 195--198). IEEE. 10.1109/LCN.2017.47

Koppeschaar, C. E., Colizza, V., Guerrisi, C., Turbelin, C., Duggan, J., Edmunds, W. J., Kjelsø, C., Mexia, R., Moreno, Y., Meloni, S., Paolotti, D., Perrotta, D., van Straten, E., & Franco, A. O. (2017). Influenzanet: Citizens among 10 countries collaborating to monitor influenza in Europe. *JMIR Public Health and Surveillance*, *3*(3), e66. Advance online publication. doi:10.2196/publichealth.7429 PMID:28928112

Korset, R. (2013). *World Disasters Report 2013*. Academic Press.

Kotler, P., & Keller, K. (2012). *Marketing Management* (14th ed.). Prentice Hall., doi:10.1080/08911760903022556

Kozinets, R. (2013). *Netnography: Redefined. Netnography: Redefined.* doi:10.1002/9781118767771.wbiedcs067

Kozinets, R. (2016). Netnography. The Blackwell Encyclopedia of Sociology, 1–2. doi:10.1002/9781405165518.wbeos0782

Kozinets, R. (2002). The Field Behind the Screen: Using Netnography For Marketing Research in Online Communities. *JMR, Journal of Marketing Research*, *39*(1), 61–72. doi:10.1509/jmkr.39.1.61.18935

Kozinets, R. (2012). Marketing Netnography: Prom/ot(Ulgat)ing a New Research Method. *Methodological Innovations Online*, *7*(1), 37–45. doi:10.4256/mio.2012.004

Kozinets, R. (2019). *Netnography: The Essential Guide to Qualitative Social Media Research*. Sage.

Kreibig, S. D. (2010). Autonomic nervous system activity in emotion: A review. *Biological Psychology*, *84*(3), 394–421. doi:10.1016/j.biopsycho.2010.03.010 PMID:20371374

Kremen, C., Ullman, K. S., & Thorp, R. W. (2011). Evaluating the quality of citizen scientist data on pollinator communities. *Conservation Biology, 25*(3), 607–617. doi:10.1111/j.1523-1739.2011.01657.x

Krippendorff, K. (2004). Content Analysis: An Introduction to Its Methodology. *Sage (Atlanta, Ga.).*

Kucharska, W., & Kowalczyk, R. (2019). How to achieve sustainability? Employee's point of view on company's culture and CSR practice. *Corporate Social Responsibility and Environmental Management, 26*(2), 453–467. doi:10.1002/csr.1696

Kucuk, S. (2011). Push-based brand awareness: The role of product availability and in-store merchandising. *International Review of Retail, Distribution and Consumer Research, 21*(3), 201–213. doi:10.1080/09593969.2011.578793

Kuhn, E., Greene, C., Hoffman, J., Nguyen, T., Wald, L., Schmidt, J., Ramsey, K. M., & Ruzek, J. (2014). Preliminary evaluation of PTSD Coach, a smartphone app for post-traumatic stress symptoms. *Military Medicine, 179*(1), 12–18. doi:10.7205/MILMED-D-13-00271

Kuhn, E., Kanuri, N., Hoffman, J. E., Garvert, D. W., Ruzek, J. I., & Taylor, C. B. (2017). A randomized controlled trial of a smartphone app for posttraumatic stress disorder symptoms. *Journal of Consulting and Clinical Psychology, 85*(3), 267–273. doi:10.1037/ccp0000163

Kuhn, E., van der Meer, C., Owen, J. E., Hoffman, J. E., Cash, R., Carrese, P., Olff, M., Bakker, A., Schellong, J., Lorenz, P., Schopp, M., Rau, H., Weidner, K., Arnberg, F. K., Cernvall, M., & Iversen, T. (2018). PTSD Coach around the world. *mHealth, 4*(5). Advance online publication. doi:10.21037/mhealth.2018.05.01

Kuipers, T., Van't Wout, C., & Bader, F. (2011). iPhone als stralingsdetector [iPhone as radiation detector]. *Nederlands Tijdschrift voor Stralingsbescherming, 2*(2), 31–34.

Kukula, E. P., Sutton, M. J., & Elliott, S. J. (2010). The human–biometric-sensor interaction evaluation method: Biometric performance and usability measurements. *IEEE Transactions on Instrumentation and Measurement, 59*(4), 784–791. doi:10.1109/TIM.2009.2037878

Kula, I., Branaghan, R. J., Atkinson, R. K., & Roscoe, R. D. (2018). Assessing User Experience via Biometric Sensor Affect Detection. In R. Roscoe, S. Craig, & I. Douglas (Eds.), *End-User Considerations in Educational Technology Design* (pp. 123–139). IGI Global. doi:10.4018/978-1-5225-2639-1.ch006

Kulkarni, S., Bell, H., Hartman, J. L., & Herman-Smith, R. L. (2013). Exploring individual and organizational factors contributing to compassion satisfaction, secondary traumatic stress, and burnout in domestic violence service providers. *Journal of the Society for Social Work and Research, 4*(1), 114-130. https://doi. doi:10.5243/jsswr.2013.8

Kunst, M. J. J., Saan, M. C., Bollen, L. J. A., & Kuijpers, K. F. (2017). Secondary traumatic stress and secondary posttraumatic growth in a sample of Dutch police family liaison officers. *Stress and Health, 33*(5), 570–577. doi:10.1002mi.2741 PMID:28127898

Kurosaki, T. (2017). Household-Level Recovery after Floods in a Tribal and ConflictRidden Society. *World Development, 94*(Supplement C), 51–63. doi:10.1016/j.worlddev.2016.12.039

Kuziemsky, C. E., Borycki, E. M., Purkis, M. E., Black, F., Boyle, M., Cloutier-Fisher, D., ... Tschanz, C. (2009). An interdisciplinary team communication framework and its application to healthcare 'e-teams' systems design. *BMC Medical Informatics and Decision Making, 9*(1), 43. Advance online publication. doi:10.1186/1472-6947-9-43

Lam, A. (2012). Visualizing the terrorist risk in president Bush's War on Terror and Peter Jackson's "The two towers". In V. K. Bathia, C. A. Hafner, L. Miller, & A. Wagner (Eds.), *Transparency, power and control. Perspective on legal communication* (pp. 165–181). Ashgate.

Landers, A. L., Dimitropoulos, G., Mendenhall, T. J., Kennedy, A., & Zemanek, L. (2020). Backing the blue: Trauma in law enforcement spouses and couples. *Family Relations*, *69*(2), 308–319. doi:10.1111/fare.12393

Lawrence, A. K. (2016). A study of the media's portrayal of the measles outbreak. University Honors Program Theses. *Paper*, *175*, 1–38.

Le Bas, C. (2016). Frugal innovation, sustainable innovation, reverse innovation: Why do they look alike? Why are they different? *Journal of Innovation Economics & Management*, *3*(21), 9–26. doi:10.3917/jie.021.0009

Lee, S.-Y. T., Phang, C. W. D. J. E. C. R., & Applications. (2015). *Leveraging social media for electronic commerce in Asia: Research areas and opportunities*. Academic Press.

Lee, J. (2013). The netizen movement: a new wave in the social movements in Korea. In H. Y. Cho, L. Surendra, & H. J. Cho (Eds.), *Contemporary South Korean Society. A critical perspective* (pp. 346–397). Routledge.

Lee, J., Kim, Y., & Won, J. (2018). Sports brand positioning: Positioning congruence and consumer perceptions toward brands. *International Journal of Sports Marketing & Sponsorship*, *19*(4), 450–471. doi:10.1108/IJSMS-03-2017-0018

Lelis, S., & Howes, A. (2011). Informing decisions: how people use online rating information to make choices. *Proceedings of the SIGCHI conference on human factors in computing systems.*

Leong, C. M. L., Pan, S. L., Ractham, P., & Kaewkitipong, L. J. J. o. t. A. f. I. S. (2015). *ICT-enabled community empowerment in crisis response: Social media in Thailand flooding 2011*. doi:10.1145/1978942.1979278

Lerbinger, O. (2012). *The crisis manager. Facing disasters, conflicts and failures* (2nd ed.). Routledge. doi:10.4324/9780203222133

Lewandowsky, S., Ecker, U. K. H., Seifert, C. M., Schwarz, N., & Cook, J. (2012). Misinformation and its correction: Continued influence and successful debiasing. *Psychological Science in the Public Interest*, *13*(3), 106–131. doi:10.1177/1529100612451018 PMID:26173286

Lewin, K. (1946). Action research and minority problems. *Human Relations*, *1*(2), 34–46. doi:10.1111/j.1540-4560.1946.tb02295.x

Lin, K. H. E., Khan, S., Acosta, L. A., Alaniz, R., & Olanya, D. R. (2020). The dynamism of post disaster risk communication: A cross-country synthesis. *International Journal of Disaster Risk Reduction*, *48*, 101556. doi:10.1016/j.ijdrr.2020.101556

Lin, L., Savoia, E., Agboola, F., & Viswanath, K. (2014). What have we learned about communication inequalities during the H1N1 pandemic: A systematic review of the literature. *BMC Public Health*, *21*(1), 1–13. doi:10.1186/1471-2458-14-484 PMID:24884634

Lin, S. P., & Yang, H. Y. (2009). Exploring key factors in the choice of e-health using an asthma care mobile service model. *Telemedicine Journal and e-Health*, *15*(9), 884–890. doi:10.1089/tmj.2009.0047

Lin, X., Spence, P. R., Sellnow, T. L., & Lachlan, K. A. (2016). Crisis communication, learning and responding: Best practices in social media. *Computers in Human Behavior*, *65*, 601–605. doi:10.1016/j.chb.2016.05.080

Lissenden, J., Maley, S., & Mehta, K. J. J. o. H. E. (2015). *An era of Appropriate Technology: Evolutions, oversights and opportunities.* Academic Press.

Littlefield, R. S., & Sellnow, T. L. (Eds.). (2015). *Risk and Crisis Communication: Navigating the Tensions between Organizations and the Public*. Lexington Books.

Liu, B. F., Austin, L. L., & Jin, Y. (2011). How publics respond to crisis communication strategies: The interplay of information form and source. *Public Relations Review*, *37*(4), 345–353. doi:10.1016/j.pubrev.2011.08.004

Liu, B. F., Jin, Y., & Austin, L. L. (2013). The Tendency To Tell: Understanding Publics' Communicative Responses To Crisis Information Form and Source. *Journal of Public Relations Research*, *25*(1), 51–67. doi:10.1080/1062726X.2013.739101

Liu, B. F., & Kim, S. (2011). How organizations framed the 2009 H1N1 pandemic via social and traditional media: Implications for U.S. health communicators. *Public Relations Review*, *37*(3), 233–244. doi:10.1016/j.pubrev.2011.03.005

Liu, S. B. (2014). Crisis crowdsourcing framework: Designing strategic configurations of crowdsourcing for the emergency management domain. *Computer Supported Cooperative Work*, *23*(4–6), 389–443. doi:10.100710606-014-9204-3

Liu, Y., Zhu, J., Shao, X., Adusumilli, N. C., & Wang, F. (2020). Diffusion patterns in disaster-induced internet public opinion: Based on a Sina Weibo online discussion about the 'Liangshan fire' in China. *Environmental Hazards*, *0*(0), 1–25. doi:10.1080/17477891.2020.1758608

Loureiro, S., Serra, J., & Guerreiro, J. (2019). How Fashion Brands Engage on Social Media: A Netnography Approach. *Journal of Promotion Management*, *25*(3), 367–378. doi:10.1080/10496491.2019.1557815

Lourenço, N. (2012). Legitimidade e confiança nas polícias. *Revista do Ministério Público*, 181-198.

Lovari, A. (2013). *Networked citizens: comunicazione pubblica e amministrazioni digitali* [Networked citizens: public communication and digital administration]. Angeli.

Lugosi, P., & Quinton, S. (2018). More-than-human netnography. *Journal of Marketing Management*, *34*(3/4), 287–313. doi:10.1080/0267257X.2018.1431303

Luhmann, N. (1993). *Risk: a sociological theory*. Walter de Gruyter. (Original work published 1991).

Lundgren, R., & McMakin, A. (2004). *Risk Communication: A Handbook for Communicating Environmental, Safety, and Health Risks*. Battelle Press.

Lundgren, R., & McMakin, A. (2018). *Risk Communication A Handbook for Communicating Environmental, Safety, and Health Risks* (6th ed.). IEEE Press Wiley.

Lunn, D., & Harper, S. (2010). Using Galvanic Skin Response Measures to Identify Areas of Frustration for Older Web 2.0 Users. In *Proceedings of the 2010 International Cross Disciplinary Conference on Web Accessibility*. ACM. 10.1145/1805986.1806032

Lu, Z., Cao, G., & La Porta, T. (2017). Teamphone: Networking smartphones for disaster recovery. *IEEE Transactions on Mobile Computing*, *16*(12), 3554–3567. doi:10.1109/TMC.2017.2695452

Macias, W., Hilyard, K., & Freimuth, V. (2009). Blog functions as risk and crisis communication during Hurricane Katrina. *Journal of Computer-Mediated Communication*, *15*(1), 1–31. doi:10.1111/j.1083-6101.2009.01490.x

Mackert, M., Champlin, S. E., Pasch, K. E., & Weiss, B. D. (2013). Understanding health literacy measurement through eye tracking. *Journal of Health Communication*, *18*(1), 185–196. doi:10.1080/10810730.2013.825666 PMID:24093355

Madoff, L. C., & Woodall, J. P. (2005). The Internet and the global monitoring of emerging diseases: Lessons from the first 10 years of ProMED-mail. *Archives of Medical Research*, *36*(6), 724–730. doi:10.1016/j.arcmed.2005.06.005 PMID:16216654

Madsen, W. (2018). History in health: Health promotion's underexplored tool for change. *Public Health*, *154*, 118–122. doi:10.1016/j.puhe.2017.10.028

Maglio, P. P., & Spohrer, J. J. J. o. t. a. o. m. s. (2008). *Fundamentals of service science*. Academic Press.

Maireder, A., & Ausserhofer, J. (2014). Political Discourse on Twitter Networking. Topics, Objects and People. In A. Bruns, K. Weller, J. Burgess, C. Puschmann, & M. Mahrt (Eds.), *Twitter and Society* (pp. 305–318). Peter Lang.

Malhotra, N. K. J. J. o. c. r. (1982). *Information load and consumer decision making*. Academic Press.

Malhotra, S., Chakrabarti, S., & Shah, R. (2019). A model for digital mental healthcare: Its usefulness and potential for service delivery in low- and middle-income countries. *Indian Journal of Psychiatry*, *61*(1), 27–36. doi:10.4103/psychiatry.IndianJPsychiatry_350_18 PMID:30745651

Malone, T. W. (2004). *The future of work*. Harvard Business School Press., doi:10.1007/0-387-28918-6_2

Manfredi, G. (2019). *Infodemia: I meccanismi complessi della comunicazione nelle emergenze* [Infodemia. The complex mechanisms of emergency communication]. Guaraldi.

Mantineo, A. (2020). *Fakecrazia. L'informazione e le sfide del coronavirus* [Fakecracy. Coronavirus information and challenges]. Media Books.

Martinengo, L., Van Galen, L., Lum, E., Kowalski, M., Subramaniam, M., & Car, J. (2019). Suicide prevention and depression apps' suicide risk assessment and management: A systematic assessment of adherence to clinical guidelines. *BMC Medicine*, *17*(1), 1–12. doi:10.118612916-019-1461-z PMID:31852455

Masamura, T. (2013). Why was the trust for the science and the mass media lost? *Trends in the Sciences*, *18*(1), 42–45. doi:10.5363/tits.18.1_42

Matsuno, Y., Ishigaki, Y., Bando, K., & Kido, K. (2016). *Developing SNS tool for consensus building on environmental safety using assurance cases* [Paper presentation]. International Conference on Computer Safety, Reliability, and Security. https://doi:10.1007/978-3-319-45480-1_5

Maxwell, J. (2005). *Qualitative research design: an interactive approach*. Sage.

Mayo-Cubero, M. (2017). *Journalistic specialization as a strategic tool in disaster communication*. Complutense University of Madrid. https://eprints.ucm.es/42138/

Mayo-Cubero, M. (2019b). Assessment of journalistic coverage and public institutional communication of the most recent crises and emergencies in Spain. *University of Seville and Spanish Association of Investigation in Communication Conference*, 1–10. https://www.researchgate.net/publication/ 334479921_ Assessment_of_ journalistic _coverage _and_public_institutional_ communication_of_the _most_recent_ crises_and_emergencies _in_Spain_ Evaluacion_de_ la_cobertura_ periodistica_y_de_la_ comunicacion_institucional

Mayo-Cubero, M., Lavín, E., & Gallardo-Camacho, J. (2017). The news coverage of the Lorca Earthquake in Spain: the journalist's social responsibility. *Ámbitos. Revista Internacional de Comunicación*, *35*, 1–17. https://idus.us.es/handle/11441/66415

Mayo-Cubero, M. (2019a). Use of social media in news media coverage of the crisis, disaster, and emergencies in Spain. *Revista Española de Comunicación en Salud*, *1*, 43–54. doi:10.20318/recs.2019.4428

Mayo-Cubero, M. (2020). News sections, journalists and information sources in the journalistic coverage of crises and emergencies in Spain. *El Profesional de la Información*, *29*(2), 1–12. doi:10.3145/epi.2020.mar.11

Mazzoleni, G. (2012). La comunicazione politica [Mulino.]. *Political Communication*, il.

Mbembe, A. (2016). *Necropolitica* (R. Beneduce, Trans.). Ombre Corte. (Original work published 2003).

McAuliffe, D., & Nipperess, S. (2017). e-Professionalism and the ethical use of technology in social work. *Australian Social Work*, *70*(2), 131–134. doi:10.1080/0312407X.2016.1221790

McCann, I. L., & Pearlman, L. A. (1990). *Psychological trauma and the adult survivor: Theory, therapy, and transformation*. Brunner and Mazel.

McCombs, M. E. (2004). *Setting the Agenda: Mass Media and Public Opinion*. Polity Press.

McCombs, M. E., & Shaw, D. L. (1972). The Agenda-Setting Function of Mass Media. *Public Opinion Quarterly*, *36*(2), 176–187. doi:10.1086/267990

McDuff, D., El Kaliouby, R., Senechal, T., Amr, M., Cohn, J. F., & Picard, R. (2013). Affectiva-MIT Facial Expression Dataset (AM-FED): Naturalistic and Spontaneous Facial Expressions Collected" In-the-Wild". In *2013 IEEE Conference on Computer Vision and Pattern Recognition Workshops*. IEEE. 10.1109/CVPRW.2013.130

McLuhan, H. M. (1964). *Understanding media. The extension of man*. McGraw-Hill.

Meesters, K., & Van de Walle, B. (2014). Increasing efficiency of humanitarian organizations with volunteer driven information products. *2014 47th Hawaii International Conference on System Sciences*.

Meesters, K., Nespeca, V., & Comes, T. (2019). *Designing Disaster Information Management Systems 2.0: Connecting communities and responders*. ISCRAM.

Meesters, K., & Wang, Y. (2020). *Information Management in Large-scale Disaster Exercises: An Integrated Perspective*. Academic Press.

Meier, P. J. I. I. R. R. C. (2011). *New information technologies and their impact on the humanitarian sector*. Academic Press.

Meier, P. (2015). *Digital humanitarians: how big data is changing the face of humanitarian response*. CRC Press. doi:10.1201/b18023

Meissner, A., Luckenbach, T., Risse, T., Kirste, T., & Kirchner, H. (2002). Design challenges for an integrated disaster management communication and information system. *The First IEEE Workshop on Disaster Recovery Networks (DIREN 2002)*.

Mendes, A. M., & Pereira, F. C. (2006). *Crises: de ameaças a oportunidades - gestão estratégica de comunicação de crises*. Edições Sílabo.

Metzger, M. J., Flanagin, A. J., & Medders, R. B. (2010). Social and heuristic approaches to credibility evaluation online. *Journal of Communication*, *60*(3), 413–439. doi:10.1111/j.1460-2466.2010.01488.x

Meurisch, C., Nguyen, T., Wullkotte, S., Niemczyk, S., Kohnhausar, F., & Muhlhausar, M. (2017). NICER911: Ad-hoc Communication and Emergency Services Using Networking Smartphones and Wireless Home Routers. In *Proceedings of the 18th ACM International Symposium on Mobile Ad Hoc Networking and Computing* (pp. 1-2). ACM. 10.1145/3084041.3084075

MEXT. (2012, February 21). *Real-time sen-ryosokutei system no Fukushima-ken naino Kuukansenryo-ritsu no real-time sokuteikekka no kouhyou ni tsuite* [Release of the results of real-time air dose rate measurements in Fukushima prefecture using the real-time dosimetry system]. Ministry of Education, Culture, Sports, Science, and Technology.

Mezghani, F., Kortoc, P., Mitton, N., & Francesco, M. D. (2019). A Multi-tier Communication Scheme for Drone-assisted Disaster Recovery Scenarios. In *30th Annual International Symposium on Personal, Indoor and Mobile Radio Communications (PIMRC)* (pp. 1-7). IEEE. 10.1109/PIMRC.2019.8904140

Mezghani, F., & Mitton, N. (2017). Opportunistic alert diffusion in disaster scenario—Stay alive longer! *28th Annual International Symposium on Personal, Indoor, and Mobile Radio Communications (PIMRC)*, 1, 1-5. 10.1109/PIMRC.2017.8292230

Miner, A., Kuhn, E., Hoffman, J. E., Owen, J. E., Ruzek, J. I., & Taylor, C. B. (2016). Feasibility, acceptability, and potential efficacy of the PTSD coach app: A pilot randomized controlled trial with community trauma survivors. *Psychological Trauma: Theory, Research, Practice, and Policy*, *8*(3), 384–392. doi:10.1037/tra0000092

Ministerio del Interior. (2020). *AlertCops* [mobile application]. https://alertcops.ses.mir.es/mialertcops/

Ministério Público Federal. (2018). *Official document.* Retrieved from: http://www.mpf.mp.br/mg/sala-de-imprensa/docs/termo-de-compromisso-prescricao

Mojtabai, R., Olfson, M., Sampson, N. A., Jin, R., Druss, B., Wang, P. S., Wells, K. B., Pincus, H. A., & Kessler, R. C. (2011). Barriers to mental health treatment: Results from the National Comorbidity Survey Replication. *Psychological Medicine*, *41*(8), 1751–1761. doi:10.1017/S0033291710002291

Molyneux, L., Holton, A., & Lewis, S. C. (2018). How journalists engage in branding on Twitter: Individual, organizational, and institutional levels. *Information Communication and Society*, *21*(10), 1386–1401. doi:10.1080/1369118X.2017.1314532

Moon, J. (2020). *Discorso "Tre anni di Mandato" Presidente della Repubblica di Corea Moon Jae-in.* http://overseas.mofa.go.kr/it-it/brd/m_8792/view.do?seq=760669&srchFr=&srchTo=&srchWord=&srchTp=&multi_itm_seq=0&itm_seq_1=0&itm_seq_2=0&company_cd=&company_nm=&page=1

Moraes, I. H. S., & Vasconcellos, M. M. (2006). Gestão da informação e comunicação em saúde: Desenho e implementação de uma proposta de ensino-aprendizagem. In Cenários possíveis: experiências e desafios do mestrado profissional na saúde coletiva. Rio de Janeiro: Editora FIOCRUZ.

Moraes, I. H. S. (2014). Governança e modelo de gestão da informação e inovação em sistemas e serviços de atenção à saúde. In *Conhecimento, inovação e comunicação em serviços de saúde* (pp. 19–41). Editora FIOCRUZ.

Morcellini, M. (2020). *Antivirus. Una società senza sistemi immunitari alla sfida del Covid-19* [Antivirus. A society without immune systems to the challenge of Covid-19]. Lit Edizioni.

Morgan, M. G., Fischhoff, B., Bostrom, A., & Atman, C. J. (2002). *Risk communication: A mental models approach.* Cambridge University Press.

Municipality of Rome. (2020). *Sistema Unico di Segnalazione.* https://www.comune.roma.it/web/it/di-la-tua-segnala.page

Muralidharan, S., Rasmussen, L., Patterson, D., & Shin, J. H. (2011). Hope for Haiti: An analysis of Facebook and Twitter usage during the earthquake relief efforts. *Public Relations Review*, *37*(2), 175–177. doi:10.1016/j.pubrev.2011.01.010

Murdock, G., Petts, J., & Horlick, J. T. (2003). After amplification: rethinking the role of media in risk communication. In N. Pidgeon, R. Kasperson, & P. Slovic (Eds.), *The social amplification of risk.* Cambridge University Press. doi:10.1017/CBO9780511550461.008

Mythen, G. (2004). *Ulrich Beck: A Critical Introduction to the Risk Society.* Pluto Press.

Natadjaja, L., & Setyawan, P. B. (2016). *Creating Community through Design: The Case of Go-Jek Online.* Petra Christian University.

National Association of Social Workers. Council on Social Work Education, Association of Social Work Boards, & Clinical Social Work Association. (2017). Practice standards on social work and technology. NASW Press.

National Consumer Affairs Center of Japan. (2011a, September 8). Hikakuteki anka na houshasen sokuteiki no seinou [Performance of relatively inexpensive radiation measuring instruments]. *National Consumer Affairs Center of Japan.* http://www.kokusen.go.jp/news/data/n-20110908_1.html

National Consumer Affairs Center of Japan. (2011b, December 22). Hikakuteki anka na houshasen sokuteiki no seinou dai 2 dan [Performance of relatively inexpensive radiation detectors – Part 2. *National Consumer Affairs Center of Japan.* http://www.kokusen.go.jp/news/data/n-20111222_1.html

National Consumer Affairs Center of Japan. (2012, May 24). Digital shiki kojin senryo-kei no test kekka [Test results of digital personal dosimeter]. *National Consumer Affairs Center of Japan.* http://www.kokusen.go.jp/pdf/n-20120524_1.pdf

National Institute for Defense Studies. Japan. (2015). CBRN defense: Responding to growing threats. In East Asian Strategic Review 2015 (pp. 308–310). doi:10.17070/aeaas.2015.12.40

National Institute of Mental Health. (2019, September). *Technology and the future of mental health treatment.* https://www.nimh.nih.gov/health/topics/technology-and-the-future-of-mental-health-treatment/index.shtml

National Research Council. (2011). *Successful K-12 STEM education: Identifying effective approaches in science, technology, engineering, and mathematics.* National Academies Press.

Natural disaster 2018. (2019). Retrieved from https://www.emdat.be/publications

Nava, F., Gambino, G., Lucarelli, S., Di Benedetto Montaccini, V., Telese, L., & Revelli, M. (2020). *Epidemia colposa? Le verità nascoste sulla mancata zona rossa nella Val Seriana* [A negligent epidemic? The hidden truths about the missing red zone in Val Seriana]. The Post Internazionale.

NE - Instituto Nacional de Estatística (Ed.). (2020). *Anuário Estatístico de Portugal 2019 / Statistical Yearbook of Portugal 2019.* INE - Instituto Nacional de Estatística.

Nespeca, V., Meesters, K., & Comes, T. (2018). *Evaluating Platforms for Community Sense-making: Using the Case of the Kenyan Elections.* ISCRAM.

Netting, F. E., Kettner, P. M., McMurtr, S. L., & Thomas, M. L. (2017). *Social work macro practice.* Pearson Education, Inc.

Neuendorf, K. A. (2002). The Content Analysis Guidebook. *Sage (Atlanta, Ga.).*

Newsom, K., & Myers-Bowman, K. (2017). "I am not a victim. I am a survivor": Resilience as a journey for female survivors of child sexual abuse. *Journal of Child Sexual Abuse, 26*(8), 927–947. doi:10.1080/10538712.2017.1360425 PMID:28857725

Nguyen, M. X., Go, V. F., Bui, Q. X., Gaynes, B. N., & Pence, B. W. (2019). Perceived need, barriers to and facilitators of mental health care among HIV-infected PWID in Hanoi, Vietnam: A qualitative study. *Harm Reduction Journal, 16*(1), 1–9. doi:10.118612954-019-0349-8 PMID:31878934

Nielsen, R. K., Fletcher, R., Newman, N., Brennen, J. S., & Howard, P. N. (2020). *Navigating the "Infodemic": How People in Six Countries Access and Rate News and Information about Coronavirus.* https://reutersinstitute.politics.ox.ac.uk/infodemic-how-people-six-countries-access-and-rate-news-and-information-about-coronavirus

Nielsen, J. (1999). *Designing Web usability: The practice of simplicity.* New Riders Publishing.

Nielsen, J., & Tahir, M. (2001). *Homepage Usability: - 50 Websites Deconstructed.* New Riders Publishing.

Nihei, Y. (2014). Fukushima Dai-ichi Genshiryoku-hatsudensho Jikogo no Housyasei-busshitsu kakusan ni-yoru "Fuuhyou-higai" ha Shinriteki "Hanka-higai" Dearu: Shakai-kankei-shihon toshiteno Chiikikan Sougo Shinrai no Soushitsu heno Eikyou [Financial damage caused by radioactive fallout from the Fukushima No.1 Nuclear Power Plant is not the result of "damage from rumors", but of psychological generalization of fear: Changes in inter-regional trust as social capital in Japan]. *Hakuoh Journal of the Faculty of Education, 8*(1), 131–162, 2014–05. https://ci.nii.ac.jp/naid/110009814955

Nonaka, I., & Takeuch, I. H. (1997). *Criação do Conhecimento na Empresa: como as empresas geram a dinâmica da inovação*. Campus.

Norman, C. D., & Skinner, H. A. (2006). eHEALS: The eHealth Literacy Scale. *Journal of Medical Internet Research, 8*(4), e27. Advance online publication. doi:10.2196/jmir.8.4.e27

Norris, F. H., Stevens, S. P., Pfefferbaum, B., Wyche, K. F., & Pfefferbaum, R. L. J. A. p. (2008). *Community resilience as a metaphor, theory, set of capacities, and strategy for disaster readiness*. Academic Press.

Norton, M. I., Mochon, D., & Ariely, D. (2012). The IKEA effect: When labor leads to love. *Journal of Consumer Psychology, 22*(3), 453–460. doi:10.1016/j.jcps.2011.08.002

Nwogwugwu, D. I. (2018). Influence of crisis communication strategies on stakeholders' perception of organisational reputation: A review of research trends. *Journal of Communication and Media Research, 10*(2), 125–138.

Nyirimanzi, C. (2012). Rwanda's literacy rate rises. *National Institute of Statistics of Rwanda*. http://statistics.gov.rw/node/1086

O'Dea, S. (2020, August 20). *Number of smartphone users worldwide from 2016 to 2021*. Retrieved August 25, 2020, from Statista: https://www.statista.com/statistics/330695/number-of-smartphone-users-worldwide/

O'Malley, D. A., & Latimer-Cheung, A. E. (2013). Gaining perspective: The effects of message frame on viewer attention to and recall of osteoporosis prevention print advertisements. *Journal of Health Psychology, 18*(11), 1400–1410. doi:10.1177/1359105312456323 PMID:23188916

Office of Community Oriented Policing Services (COPS). (2015). After-Action Assessment of the Police Response to the August 2014 Demonstrations In Ferguson, Missouri. Washington, DC: Institute for Intergovernmental Research.

Ogundoyin, O., Olagunju, L., & Nwogwugwu, D. (2020, February). *Influence of Digital Technologies on Journalistic Practices and Quality of News Reporting in Nigeria*. Paper presented at the @frica Digital Media Conference, J. Valenti School of Communication, University of Houston.

Okolloh, O. J. P. l., & Action. (2009). *Ushahidi, or 'testimony': Web 2.0 tools for crowdsourcing crisis information*. Academic Press.

Okumura, N., Hayashi, K., Igarashi, K., & Tanaka, A. (2019). Japan's media fails its watchdog role: Lessons learned and unlearned from the 2011 earthquake and the Fukushima disaster. *Journalism*, (1), 2–6. doi:10.1177/1464884919891270

Olff, M. (2015). Mobile mental health: A challenging research agenda. *European Journal of Psychotraumatology, 6*(1), 27882. Advance online publication. doi:10.3402/ejpt.v6.27882

Oliveira, D., Machin, L., Deliza, R., Rosenthal, A., Walter, E. H., Gimenez, A., & Ares, G. (2016). Consumers' attention to functional food labels: Insights from eye-tracking and change detection in a case study with probiotic milk. *Lebensmittel-Wissenschaft + Technologie, 68*, 160–167. doi:10.1016/j.lwt.2015.11.066

Oliver, D. P., Washington, K., Wittenberg-Lyles, E., Gage, A., Mooney, M., & Demiris, G. (2015). Lessons learned from a secret Facebook support group. *Health & Social Work, 40*(2), 125–133. doi:10.1093/hsw/hlv007 PMID:26027421

Olson, D. R., Konty, K. J., Paladini, M., Viboud, C., & Simonsen, L. (2013). Reassessing Google Flu Trends data for detection of seasonal and pandemic influenza: A comparative epidemiological study at three geographic scales. *PLoS Computational Biology*, *9*(10), e1003256. Advance online publication. doi:10.1371/journal.pcbi.1003256 PMID:24146603

Open Street Map. (2018). *Marunda Urban Resilience in Action.* https://openstreetmap.id/en/project/marunda-urban-resilience-in-action-alliance-muria/

Orduña, O. (2004). *A comunicação em momentos de crise.* de Biblioteca on-line de Ciências da Comunicação (BOCC): http://www.bocc.ubi.pt/pag/orduna-octavio-comunicacao-em-momentos-de-crise.pdf

Orsini, A. (2019). *Viva gli immigrati! Gestire la politica migratoria per tornare protagonisti in Europa.* Rizzoli.

Osterling, O. (2013). Dialogue police work: a balancing act on a slack. In The anthology: GODIAC -, Good practice for dialogue and communication as strategic principles for policing political manifestations in Europe, (pp. 54-60). Swedish National Police Board and GODIAC.

Owen, J. E., Jaworski, B. K., Kuhn, E., Makin-Byrd, K. N., Ramsey, K. M., & Hoffman, J. E. (2015). mHealth in the wild: Using novel data to examine the reach, use, and impact of PTSD coach. *JMIR Mental Health*, *2*(1), e7. Advance online publication. doi:10.2196/mental.3935

Owen, J. E., Kuhn, E., Jaworski, B. K., McGee-Vincent, P., Juhasz, K., Hoffman, J. E., & Rosen, C. (2018). VA mobile apps for PTSD and related problems: Public health resources for veterans and those who care for them. *mHealth*, *4*(7), 28. doi:10.21037/mhealth.2018.05.07

Padawangi, R., & Douglass, M. J. P. A. (2015). *Water, water everywhere: Toward participatory solutions to chronic urban flooding in Jakarta.* Academic Press.

Pal, A., Raj, M., Kant, K., & Das, S. (2020). A Smartphone-based Network Architecture for Post-disaster Operations Using WiFi Tethering. *ACM Transactions on Internet Technology*, *20*(1), 1–27. doi:10.1145/3372145

Paldam, M. (2000). Social capital: One or many? Definition and measurement. *Journal of Economic Surveys*, *14*(5), 629–653. doi:10.1111/1467-6419.00127

Palen, L., & Anderson, K. M. (2016). Crisis informatics: New data for extraordinary times. *Science*, *353*(6296), 224–225. doi:10.1126cience.aag2579 PMID:27418492

Palen, L., Vieweg, S., Liu, S. B., & Hughes, A. L. (2009). Crisis in a networked world: Features of computer-mediated communication in the April 16, 2007, Virginia Tech event. *Social Science Computer Review*, *27*(4), 467–480. doi:10.1177/0894439309332302

Palen, L., Vieweg, S., Sutton, J., Liu, S. B., & Hughes, A. (2007). Crisis informatics: Studying crisis in a networked world. *Proceedings of the Third International Conference on E-Social Science.*

Palos-Sánchez, P., Saura, J. R., & Álvarez-García, J. (2019). Innovation and Creativity in the Mobile Applications Industry: A Case Study of Mobile Health Applications (e-Health Apps). In M. Peris-Ortiz, M. R. Cabrera-Flores, & A. Serrano-Santoyo (Eds.), *Cultural and Creative Industries. A path to Entrepreneurship and Innovation.* Springer. doi:10.1007/978-3-319-99590-8_7

Palttala, P., & Vos, M. (2012). Quality indicators for crisis communication to support emergency management by public authorities. *Journal of Contingencies and Crisis Management*, *20*(1), 39–51. doi:10.1111/j.1468-5973.2011.00654.x

Pariser, E. (2012). *Il filtro. Quello che Internet ci nasconde* (B. Tortorella, Trans.). Il Saggiatore. (Original work published 2003).

Park, J., Rajagopal, P., Dillon, W., Chaiy, S., & DeSarbo, W. (2017). A new bayesian spatial model for brand positioning. *Journal of Modelling in Management, 12*(3), 404–431. doi:10.1108/JM2-12-2015-0100

Park, S., Boatwright, B., & Avery, E. J. (2019). Information channel preference in health crisis: Exploring the roles of perceived risk, preparedness, knowledge, and intent to follow directives. *Public Relations Review, 45*(5), 101794. doi:10.1016/j.pubrev.2019.05.015

Paulus, D., Meesters, K., & Van de Walle, B. (2018). *Turning data into action: supporting humanitarian field workers with open data.* ISCRAM.

Pearlman, L. A., & Saakvitne, K. W. (1995). *Trauma and the therapist: Countertransference and vicarious traumatization in psychotherapy with incest survivors.* Norton.

Pearrow, M. (2000). *Web site usability handbook.* Charles River Media.

Pekar, V., Binner, J., Najafi, H., Hale, C., & Schmidt, V. (2020). Early detection of heterogeneous disaster events using social media. *Journal of the Association for Information Science and Technology, 71*(1), 43–54. doi:10.1002/asi.24208

Peñalva-Velez, A., Napal-Fraile, M., & Mendioroz-Lacambra, A. M. (2018). Competencia digital y alfabetización digital de los adultos (profesorado y familia) [Digital skills and adult literacy (faculty and family)]. *International Journal of Nursing Education, 1*(1). Advance online publication. doi:10.24310/IJNE1.1.2018.4892

Perry, D. C., Taylor, M., & Doerfel, M. L. (2003). Internet-based communication in crisis management. *Management Communication Quarterly, 17*(2), 206–232. doi:10.1177/0893318903256227

Perry, R. W. (2007). What Is a Crisis? In H. Rodriguez, E. L. Quarantelli, & R. R. Dynes (Eds.), *Handbook of Disaster Research* (pp. 1–15). Springer. doi:10.1007/978-0-387-32353-4_1

Petti, A. (2007). *Arcipelaghi e enclave.* Mondadori.

Picard, R. W. (2003). Affective computing: Challenges. *International Journal of Human-Computer Studies, 59*(1–2), 55–64. doi:10.1016/S1071-5819(03)00052-1

Piccolo, L. S., Roberts, S., Iosif, A., & Alani, H. (2018). Designing chatbots for crises: a case study contrasting potential and reality. *Proceedings of the 32nd International BCS Human Computer Interaction Conference (HCI).*

Piccolo, L., Meesters, K., & Roberts, S. (2017). *Co-designing for Community Resilience Beyond the Local.* Academic Press.

Pinho, C. (2015). *Repensar a Gestão de Crises em Redes Sociais: o caso do voo MH370.* Universidade do Minho.

Pinto-Coelho, R. M. (2015). Is there water governance in Brazil? The sudy case: The Fundão Dam Brech, Mariana (MG). In Arquivos do Museu de História Natural e Jardim Botânico. Universidade Federal de Minas Gerais.

Poels, K., & Dewitte, S. (2006). How to capture the heart? Reviewing 20 years of emotion measurement in advertising. *Journal of Advertising Research, 46*(1), 18–37. doi:10.2501/S0021849906060041

Pogorzelski, J. (2018). Perceptual Branding. In Managing Brands in 4D (p. 180). doi:10.1108/978-1-78756-102-120181003

Poiani, T. H., dos Santos Rocha, R., Degrossi, L. C., & De Albuquerque, J. P. (2016). Potential of collaborative mapping for disaster relief: A case study of OpenStreetMap in the Nepal earthquake 2015. *2016 49th Hawaii International Conference on System Sciences (HICSS).*

Police Executive Research Forum. (2016). *Recommendations on advancing Community Policing in the Pasco Police Department: Critical Response Initiative.* Office of Community Oriented Policing Services.

Pratt, C. B. (2012). Theoretical approaches to and sociocultural perspectives in crisis communication. In A. George & C. Pratt (Eds.), *Case studies in crisis communication. International perspectives on hits and misses* (pp. 3–27). Routledge.

Presidency of the Council of Ministers – Immuni Italia. (2020). *Aiuta te stesso, la tua famiglia, il tuo paese.* https://www.immuni.italia.it/?gclid=Cj0KCQjwoPL2BRDxARIsAEMm9y9YLjUzg32CTv5rH7GXusHnHmrS4O5uAEvh-KI2FBTtEIzKVA_1vv8aAs5JEALw_wcB

Presidency of the Council of Ministers. (2020). *Immuni's High-Level Description.* https://github.com/immuni-app/immuni-documentation

Price, M., Yuen, E. K., Goetter, E. M., Herbert, J. D., Forman, E. M., Acierno, R., & Ruggiero, K. J. (2014). mHealth: A mechanism to deliver more accessible, more effective mental health care. *Clinical Psychology & Psychotherapy, 21*(5), 427–436. doi:10.1002/cpp.1855

Price, R. H. (1989). Bearing witness. *American Journal of Community Psychology, 17*(2), 151–167. doi:10.1007/BF00931004

Prilleltensky, I., & Nelson, G. (2002). Doing Psychology Critically. Making a Difference in Diverse Settings. *Journal of Community & Applied Social Psychology, 13*(4). Advance online publication. doi:10.1002/casp.711

Primo, P. P. B., Antunes, M. N., Ramos, M. P., & Emmerich, A. (2018). Diante da dor dos outros: Desastres e a Violação de Direitos Humanos. In Questões sobre Direitos Humanos: Justiça, Saúde e Meio Ambiente, 3, 169 - 192. doi:10.30712/isbn9788565276474.169-192

Procter, R., Crump, J., Karstedt, S., Voss, A., & Cantijoch, M. (2013). Reading the riots: What were the police doing on Twitter? *Policing and Society, 23*(4), 413–436. doi:10.1080/10439463.2013.780223

Pryce, J. G., Shackelford, K. K., & Pryce, D. H. (2007). *Secondary traumatic stress and the child welfare professional.* Lyceum Books.

PubMed Central. Table 1. (2018). *Mhealth, 4*(28). http://mhealth.amegroups.com/article/viewFile/20524/20219/123518

PubMed Central. Table 2. (2018). *Mhealth, 4*(28). http://mhealth.amegroups.com/article/viewFile/20524/20219/123519

Pu, C., & Zhou, X. (2019). RescueMe: Smartphone-based self rescue system for disaster rescue. In *9th Annual Computing and Communication Workshop and Conference (CCWC)* (pp. 832-837). IEEE. 10.1109/CCWC.2019.8666565

Pulker, C., Trapp, G., Scott, J., & Pollard, C. (2018). Global supermarkets' corporate social responsibility commitments to public health: A content analysis. *Globalization and Health, 14*(1), 121. doi:10.118612992-018-0440-z PMID:30497500

Quarantelli, E. L. (2000). *Emergencies, Disaster and Catastrophes are different phenomena.* DRC Preliminary Paper.

Quarantelli, E. L. (2005). A Social Science Research Agenda for the Disasters of the 21st Century. In R. W. Perry & E. L. Quarantelli (Eds.), *What Is a Disaster? New Answers to Old Questions* (pp. 325–296). Xlibris.

Quitangon, G. (2019). Vicarious trauma in clinicians: Fostering resilience and preventing burnout. *The Psychiatric Times, 36*(7), 18–19.

Quivy, R., & Campenhoudt, L. (1998). *Manual de Investigação em Ciências Sociais* (2nd ed.). Gradiva.

Raffaetà, R. (2020). Another Day in Dystopia. Italy in the Time of COVID-19. *Medical Anthropology, 39*(5), 1–3. doi:10.1080/01459740.2020.1746300 PMID:32212942

Ragnedda, M. (2017). *The Third Digital Divide: A Weberian Approach to Digital Inequalities.* Routledge. doi:10.4324/9781315606002

Ragnedda, M., & Mutsvairo, B. (2018). Digital Inclusion: Empowering People through Information and Communication Technologies. In M. Ragnedda & B. Mutsvairo (Eds.), *Digital Inclusion: An International Comparative Analysis*. Lexington Books.

Ragunath, P., Velmourougan, S., Davachelvan, P., Kayalvizhi, S., Ravimohan, R. J. I. J. C. S., & Security, N. (2010). *Evolving a new model (SDLC Model-2010) for software development life cycle*. SDLC; doi:10.14236/ewic/HCI2018.56

Rahman, R. (2018). Building brand awareness: The role of celebrity endorsement in advertisements. *Journal of Global Scholars of Marketing Science*, *28*(4), 363–384. doi:10.1080/21639159.2018.1509366

Rajagopal. (2019). *Competitive Branding Strategies: Maning Performance in Emerging Markets*. Mexico City: Palgrave Macmillan. doi:10.1007/978-3-030-24933-5

Ranucci, S. (2020). *Wuhan città aperta?* https://www.rai.it/dl/doc/1586270571878_wuhan_citta_aperta_report.pdf

Rasmussen, J., & Ihlen, Ø. (2017). Risk, Crisis, and Social Media. A systematic review of seven years' research. *Nordicom Review*, *38*(2), 1–17. doi:10.1515/nor-2017-0393

Ratzan, S., & Moritsugu, K. (2014). Ebola crisis - Communication chaos we can avoid. *Journal of Health Communication*, *19*(11), 1213–1215. doi:10.1080/10810730.2014.977680 PMID:25356719

Raymond, N. A., Card, B. L. J. S. P. o. H. S., & Technology, H. H. I. (2015). *Applying humanitarian principles to current uses of information communication technologies: Gaps in doctrine and challenges to practice*. Academic Press.

Raymond, N., Al Achkar, Z., Verhulst, S., Berens, J., Barajas, L., Easton, M. J. O. P., & Series, S. (2016). *Building data responsibility into humanitarian action*. Academic Press.

Rea, B., Wang, Y., & Stoner, J. (2014). When a brand caught fire: The role of brand equity in product-harm crisis. *Journal of Product and Brand Management*, *23*(7), 532–542. doi:10.1108/JPBM-01-2014-0477

Reamer, F. G. (2014, July). Eye on ethics: Elastic boundaries in social work-proceed with caution. *Social Work Today*. https://www.socialworktoday.com/news/eoe_070214.shtml

Reamer, F. G. (2017, August). Eye on ethics: New NASW code of ethics standards for the digital age. *Social Work Today*. https://www.socialworktoday.com/archive/081617.shtml

Reamer, F. G. (2015). Clinical social work in a digital environment: Ethical and risk-management challenges. *Clinical Social Work Journal*, *43*(2), 120–132. doi:10.100710615-014-0495-0

Rees, C. S., & Maclaine, E. (2015). A systematic review of videoconference-delivered psychological treatment for anxiety disorders. *Australian Psychologist*, *50*(4), 259–264. doi:10.1111/ap.12122

Regester, M., & Larkin, J. (2008). *Risk issues and crisis management in public relations: a casebook of best practice* (4th ed.). Kogan Page.

Reicher, S., Stott, C., Drury, J., Adang, O., Cronin, P., & Livingstone, A. (2007). Knowledge-Based Public Order Policing: Principles and Practice. *Policing*, *1*(4), 403–415. doi:10.1093/police/pam067

Reid, E., & Duffy, K. (2018). A netnographic sensibility: Developing the netnographic/social listening boundaries. *Journal of Marketing Management*, *34*(3-4), 263–286. doi:10.1080/0267257X.2018.1450282

Repubblica Italiana. (2017). *Disposizioni integrative e correttive al decreto legislativo 26 agosto 2016, n. 179, concernente modifiche ed integrazioni al Codice dell'amministrazione digitale, di cui al decreto legislativo 7 marzo 2005, n. 82, ai sensi dell'articolo 1 della legge 7 agosto 2015, n. 124, in materia di riorganizzazione delle amministrazioni pubbliche* (GU n.9 del 12-1-2018). Gazzetta Ufficiale.

Repubblica Italiana. (2020). *Misure urgenti per la funzionalità dei sistemi di intercettazioni di conversazioni e comunicazioni, ulteriori misure urgenti in materia di ordinamento penitenziario, nonche' disposizioni integrative e di coordinamento in materia di giustizia civile, amministrativa e contabile e misure urgenti per l'introduzione del sistema di allerta Covid-19*. Gazzetta Ufficiale.

Reuter, C., Hughes, A. L., & Kaufhold, M. A. (2018). Social media in crisis management: An evaluation and analysis of crisis informatics research. *International Journal of Human-Computer Interaction, 34*(4), 280–294. doi:10.1080/10447318.2018.1427832

Reuter, C., & Mentler, T. (2018). Human–computer interaction and social media in safety-critical systems. *Journal of Contingencies and Crisis Management, 26*(1), 2–3. doi:10.1111/1468-5973.12192

Reuter, C., Stieglitz, S., & Imran, M. (2020). Social media in conflicts and crises. *Behaviour & Information Technology, 39*(3), 241–251. doi:10.1080/0144929X.2019.1629025

Reynolds, B., & Seeger, M. (2005). Crisis and Emergency Risk Communication as An Integrative Model. *Journal of Health Communication, 10*, 43–55. doi:10.1080/10810730590904571 PMID:15764443

Riccardi, M. T. (2016). The power of crowdsourcing in disaster response operations. *International Journal of Disaster Risk Reduction, 20*(November), 123–128. doi:10.1016/j.ijdrr.2016.11.001

Richards, L., Brew, N., & Smith, L. (2020, March 12). *Australian bushfires—frequently asked questions: a quick guide.* Retrieved August 25, 2020, from Parliament of Australia: https://www.aph.gov.au/About_Parliament/Parliamentary_Departments/Parliamentary_Library/pubs/rp/rp1920/Quick_Guides/AustralianBushfires

Ries, A., & Trout, J. (1981). *Positioning: The Battle for Your Mind: How to Be Seen and Heard in the Overcrowded Marketplac*. McGraw-Hill.

Ritchie, H., & Roser, M. (2014). Indoor air pollution. *Published online at OurWorldInData.org*. https://ourworldindata.org/indoor-air-pollution

Roberts, T., & Marchais, G. (2018). Assessing the role of social media and digital technology in violence reporting. *Contemporary Readings in Law and Social Justice, 10*(2), 9–42. doi:10.22381/CRLSJ10220181

Rocha, E. M., Moraes, L. G. M., Almeida, L. V., Dalvi, L. R., Andreato, L. C., Bergamaschi, L. K., Bernardina, L. S. D., Pereira, W. B., Gimenez, V. G., Neto, O. C., & Almeida, H. S. (2016). Impacto do Rompimento da barragem em Mariana –MG na saúde da população ribeirinha da cidade de Colatina – ES. *Tempus., 10*(3), 31–45. doi:10.18569/tempus.v10i3.1902

Roden-Foreman, J. W., Bennett, M. M., Rainey, E. E., Garrett, J. S., Powers, M. B., & Warren, A. M. (2017). Secondary traumatic stress in emergency medicine clinicians. *Cognitive Behaviour Therapy, 46*(6), 522–532. doi:10.1080/16506073.2017.1315612 PMID:28452256

Rodríguez-Martínez, M., & Garzón-Alfonso, C. C. (2018). Twitter Health Surveillance (THS) system. *Proc. IEEE Int. Conf. Big Data 2018*, 1647–54.

Rodríguez, P., & Odriozola-Farré, B. (2012). Catástrofes y periodismo: El relato, los escenarios, las interacciones y las necesidades prácticas y psicológicas de todos los implicados. *Estudios sobre el Mensaje Periodístico, 18*(2), 577–594. doi:10.5209/rev_ESMP.2012.v18.n2.41033

Rodriguez-Paras, C., Tippey, K., Brown, E., Sasangohar, F., Creech, S., Kum, H. C., Lawley, M., & Benzer, J. K. (2017). Posttraumatic Stress Disorder and Mobile Health: App Investigation and Scoping Literature Review. *JMIR mHealth and uHealth, 5*(10), e156. Advance online publication. doi:10.2196/mhealth.7318

Roma Capitale [@RomaCapitaleOfficialPage]. (2020, March 26). *Hai notato un assembramento nella tua zona?* [Status update]. Facebook.

Romaniuk, J., Wight, S., & Faulkner, M. (2017). Brand awareness: Revisiting an old metric for a new world. *Journal of Product and Brand Management*, *26*(5), 469–476. doi:10.1108/JPBM-06-2016-1242

Rosas, E., Garay, F., & Hidalgo, N. (2020). Context-aware self-adaptive routing for delay tolerant network in disaster scenarios. *Ad Hoc Networks*, *102*, 1–17. doi:10.1016/j.adhoc.2020.102095

Rosyidi, M., Puspita, R., Kashihara, S., Fall, D., & Ikeda, K. (2018). A Design of IoT-Based Searching System for Displaying Victim's Presence Area. In *42nd Annual Computer Software and Applications Conference (COMPSAC)* (pp. 8--13). IEEE. 10.1109/COMPSAC.2018.10195

Roth, S. P., Tuch, A. N., Mekler, E. D., Bargas-Avila, J. A., & Opwis, K. (2013). Location matters, especially for non-salient features–An eye-tracking study on the effects of web object placement on different types of websites. *International Journal of Human-Computer Studies*, *71*(3), 228–235. doi:10.1016/j.ijhcs.2012.09.001

Rousseau, J. J. (1997). *Il contratto sociale* (M. Garin, Trans.). Laterza. (Original work published 1762).

Russell, M., & Cowan, J. (2018). The making of compassion stress injury: A review of historical and etiological models toward a de-stigmatizing neurobehavioral conceptualization. *Challenges*, *9*(1), 7. doi:10.3390/challe9010007

Sachdeva, S., McCaffrey, S., & Locke, D. (2017). Social media approaches to modeling wildfire smoke dispersion: Spatiotemporal and social scientific investigations. *Information Communication and Society*, *20*(8), 1146–1161. doi:10.1080/1369118X.2016.1218528

Salathé, M. (2018). Digital epidemiology: What is it, and where is it going? *Life Sciences, Society and Policy*, *14*(1), 1. doi:10.118640504-017-0065-7 PMID:29302758

Sander, L. B., Schorndanner, J., Terhorst, Y., Spanel, K., Pryss, R., Baumeister, H., & Messner, E. M. (2020). 'Help for trauma from the app stores?' A systematic review and standardized rating of apps for Post-Traumatic Stress Disorder (PTSD). *European Journal of Psychotraumatology*, *11*(1), 1701788. Advance online publication. doi:10.1080/20008198.2019.1701788

San-Miguel-Ayanz, J., Durrant, T., Boca, R., Libertà, G., Branco, A., Rigo, D., & de, … Leray, T. (2019). Forest Fires in Europe. *Middle East and North Africa*, *2018*. Advance online publication. doi:10.2760/1128

Sansbury, B. S., Graves, K., & Scott, W. (2015). Managing traumatic stress responses among clinicians: Individual and organizational tools for self-care. *Trauma*, *17*(2), 114–122. doi:10.1177/1460408614551978

Santos, R. (2015). *Dialogue Policing: uma nova abordagem à gestão de multidões* (Dissertação do Mestrado integrado em Ciências Policiais). Lisboa: ISCPSI.

Santos, R. (2006). Crises no e-world. In A. Mendes & F. Pereira (Eds.), *Crises: de ameaças a oportunidades - gestão estratégica de comunicação de crises, (pp. 191 - 244)*. Edições Sílabo.

Sapirstein, G. J. J. J. o. D. R. S. D. o. v. (2006). *Social resilience: the forgotten dimension of disaster risk reduction*. Academic Press.

Sardana, D., Gupta, N., Kumar, V., & Terziovski, M. (2020). CSR 'sustainability' practices and firm performance in an emerging economy. *Journal of Cleaner Production*, *258*, 120766. doi:10.1016/j.jclepro.2020.120766

Sareen, J. (2014). Posttraumatic stress disorder in adults: Impact, comorbidity, risk factors, and treatment. *Canadian Journal of Psychiatry*, *59*(9), 460–467. doi:10.1177/070674371405900902

Scaglioni, M., & Sala, M. (Eds.). (2020). *L'altro virus. Comunicazione e disinformazione al tempo del Covid-19* [The other virus. Communication and disinformation in the age of the Covid-19]. Vita & Pensiero.

Scarborough, N. M., Zimmerer, T., & Naumes, W. (1991). *Effective small business management*. Merrill Pub. Co.

Schaefer, R. T. (2017). *Sociology: A brief introduction* (12th ed.). McGraw Hill Education.

Schaefer, S., Terlutter, R., & Diehl, S. (2020). Talking about CSR matters: Employees' perception of and reaction to their company's CSR communication in four different CSR domains. *International Journal of Advertising*, *38*(2), 191–212. doi:10.1080/02650487.2019.1593736

Schellong, J., Lorenz, P., & Weidner, K. (2019). Proposing a standardized, step-by-step model for creating post-traumatic stress disorder (PTSD) related mobile mental health apps in a framework based on technical and medical norms. *European Journal of Psychotraumatology*, *10*(1), 1611090. doi:10.1080/20008198.2019.1611090

Schmeltz, L. (2012). Consumer-oriented CSR communication: Focusing on ability or morality? *Corporate Communications*, *17*(1), 29–49. doi:10.1108/13563281211196344

Schmitt, C. (1984). *Le categorie del "politico": saggi di teoria politica* (G. Miglio & P. Schiera, Eds., SchieraP., Trans.). Il Mulino. (Original work published 1934).

Schmuck, G., San-Miguel-Ayanz, J., Durrant, T., Boca, R., Libertà, G., Petroliagkis, T., ... Schulte, E. (2018). *Forest Fires in Europe, Middle East and North Africa 2017. Scientific and Technical Research series*. doi:10.2788/1082

Sciullo, L., Trotta, A., & Di Felice, M. (2020). Design and performance evaluation of a LoRa-based mobile emergency management system (LOCATE). *Ad Hoc Networks*, *96*, 1–17. doi:10.1016/j.adhoc.2019.101993

Sebastião, S. (2009). *Comunicação Estratégica: as Relações Públicas*. Instituto Superior de Ciências Sociais e Políticas.

Seeger, M. W., Sellnow, T. L., & Ulmer, R. R. (1998). Communication, organisation, and crisis. *Annals of the International Communication Association*, *21*(1), 231–276. doi:10.1080/23808985.1998.11678952

Seeger, M. W., Sellnow, T. L., & Ulmer, R. R. (2003). *Communication and organizational crisis*. Praeger.

Segal, E. A., Gerdes, K. E., & Steiner, S. (2019). *An introduction to the profession of social work: Becoming a change agent* (6th ed.). Cengage.

Segault, A., Tajariol, F., Ishigaki, Y., & Roxin, I. (2016). Sharing radiation measurements through social media: A methodological user-oriented proposal set of guidelines. *International Journal of Information Systems for Crisis Response and Management*, *8*(2), 17–30. doi:10.4018/IJISCRAM.2016040102

Seife, C. (2013, November 27). 23andMe Is Terrifying, but not for the Reasons the FDA Thinks. *Scientific American*. https://www.scientificamerican.com/article/23andme-is-terrifying-but-not-for-the-reasons-the-fda-thinks/

Sellnow-Richmond, D. D., George, A. M., & Sellnow, D. D. (2018). An IDEA model analysis of instructional risk communication in the time of Ebola. *Journal of International Crisis and Risk Communication Research*, *1*(1), 135–166. doi:10.30658/jicrcr.1.1.7

Sellnow, T. L., & Seeger, M. W. (2013). *Theorising crisis communication*. Wiley.

Sellnow, T. L., Ulmer, R. R., Seeger, M. W., & Littlefield, R. S. (2009). *Effective risk communication. A message-centered approach*. Springer. doi:10.1007/978-0-387-79727-4

Sellnow, T., & Seeger, M. (2013). *Theorizing Crisis Communication*. Wiley-Blackwell.

Seo, S., & Jang, S. (2013). The roles of brand equity and branding strategy: A study of restaurant food crises. *International Journal of Hospitality Management*, *34*(1), 192–201. doi:10.1016/j.ijhm.2013.02.014

Shan, L., Regan, A., Brun, A., Barnett, J., van der Sanden, M., Wall, P., & McConnon, A. (2013). Food crisis coverage by social and traditional media: A case study of the 2008 Irish dioxin crisis. *Public Understanding of Science (Bristol, England)*, *23*(8), 911–928. doi:10.1177/0963662512472315 PMID:23825281

Shannon, P. J., Simmelink-McCleary, J., Im, H., Becher, E., & Crook-Lyon, R. E. (2014). Experiences of stress in a trauma treatment course. *Journal of Social Work Education*, *50*(4), 678–693. doi:10.1080/10437797.2014.947901

Sharma, T., & Basir, M. (2020). *Use of apps in the COVID-19 response and the loss of privacy protection.* https://www.nature.com/articles/s41591-020-0928-y.pdf

Sharma, R., Ahuja, V., & Alavi, S. (2018). The Future Scope of Netnography and Social Network Analysis in the Field of Marketing. *Journal of Internet Commerce*, *17*(1), 26–45. doi:10.1080/15332861.2017.1423533

Sherratt, C. S., & Schlabach, M. L. (1990). The application of concept mapping in reference and information services. *RQ*, *30*, 60-69. Retrieved from www.jstor.org/stable/25828679

Shu, K., Mahudeswaran, D., Wang, S., & Liu, H. (2019). *Hierarchical propagation networks for fake news detection: Investigation and exploitation*. Retrieved from https://www.semanticscholar.org

Silva, C. H., Guimarães, M. C. S., Esteves, M. A., Rabaço, M. H. L., Antunes, M. N., Abdala, R., & Marcondes, C. H. (2007). Monitoramento da informação na sociedade de risco: O caso da pandemia de gripe aviária. *Informacao e Sociedade-Estudos*, *17*, 1–10.

Silver, C., & Lewins, A. (2014). *Using software in qualitative research: A step-by-step guide*. SAGE. doi:10.4135/9781473906907

Smith, J. (2008). Interpretive Inquiry. In L. Given (Ed.), *The SAGE Encyclopedia of qualitative research methods* (Vol. I & II, pp. 459–461). SAGE.

Smolinski, M. S., Crawley, A. W., Baltrusaitis, K., Chunara, R., Olsen, J. M., Wójcik, O., Santillana, M., Nguyen, A., & Brownstein, J. S. (2015). Flu Near You: Crowdsourced symptom reporting spanning 2 influenza seasons. *American Journal of Public Health*, *105*(10), 2124–2130. doi:10.2105/AJPH.2015.302696 PMID:26270299

Snoijers, E. M., Poels, K., & Nicolay, C. (2014). #universitycrisis: The impact of social media type, source, and information on student responses towards a University Crisis. *Social Science Computer Review*, *32*(5), 647–661. doi:10.1177/0894439314525025

Sodeke-Gregson, E. A., Holttum, S., & Billings, J. (2013). Compassion satisfaction, burnout, and secondary traumatic stress in UK therapists who work with adult trauma clients. *European Journal of Psychotraumatology*, *4*(1), 1–10. doi:10.3402/ejpt.v4i0.21869 PMID:24386550

Soden, R., Budhathoki, N., & Palen, L. (2014). *Resilience-building and the crisis informatics agenda: Lessons learned from open cities Kathmandu*. ISCRAM.

Sofia, R., Firdose, S., Lopes, L. A., Moreira, W., & Mendes, P. (2016). NSense: A people-centric, non-intrusive opportunistic sensing tool for contextualizing nearness. In *18th International Conference on e-Health Networking, Applications and Services (Healthcom)* (pp. 1-6). IEEE. 10.1109/HealthCom.2016.7749490

Sontag, S. (1979). *Malattia come metafora* (E. Capriolo, Trans.). Einaudi. (Original work published 1978).

Spence, P. R., Lachlan, K. A., Westerman, D., & Spates, S. A. (2013). Where the gates matter less: Ethnicity and perceived source credibility in social media health messages. *The Howard Journal of Communications*, *24*(1), 1–16. doi:10.1080/10646175.2013.748593

Spieler, H. (2005). *Semiconductor detector systems*. Oxford University Press. doi:10.1093/acprof:oso/9780198527848.001.0001

Spies, T. A., White, E. M., Kline, J. D., Paige Fischer, A., Ager, A., Bailey, J., ... Hammer, R. (2014). Examining fire-prone forest landscapes as coupled human and natural systems. *Ecology and Society*, *19*(3), art9. Advance online publication. doi:10.5751/ES-06584-190309

Spinsanti, L., & Ostermann, F. (2011). Retrieve volunteered geographic information for forest fire. *CEUR Workshop Proceedings*, 704.

Spinsanti, L., & Ostermann, F. (2013). Automated geographic context analysis for volunteered information. *Applied Geography (Sevenoaks, England)*, *43*, 36–44. doi:10.1016/j.apgeog.2013.05.005

Spyropoulos, T., Psounis, K., & Raghavendra, C. (2005). *Spray and wait: an efficient routing scheme for intermittently connected mobile networks. In ACM SIGCOMM workshop on Delay-tolerant networking*. ACM. doi:10.1145/1080139.1080143

State Council of the People's Republic of China. (2014). *Planning Outline for the Construction of a Social Credit System (2014-2020).* https://chinacopyrightandmedia.wordpress.com/2014/06/14/planning-outline-for-the-construction-of-a-social-credit-system-2014-2020/

Stecula, D. A., Kuru, O., & Hall Jamieson, K. (2020). How Trust in Experts and Media Use Affect Acceptance of Common Anti-Vaccination Claims. *The Harvard Kennedy School (HKS). Misinformation Review*, *1*(1), 1–11. doi:10.37016/mr-2020-007

Stephens, D., Hill, J., & Greenberg, S. (2011). *Strategic Communication Practices: A Toolkit for Police Executives*. The Office of Community Oriented Policing Services (COPS).

Stephens, K. K., Malone, P. C., & Bailey, C. M. (2005). Communicating with stakeholders during a Crisis. Evaluating message strategies. *Journal of Business Communication*, *42*(4), 390–419. doi:10.1177/0021943605279057

Stern, R. M., Ray, W. J., & Quigley, K. S. (2001). *Psychophysiological Recording*. Oxford University Press.

Stieglitz, S., Mirbabaie, M., & Potthoff, T. (2018). Crisis Communication on Twitter during a Global Crisis of Volkswagen – The Case of "Dieselgate." In *Proceedings of the 51st Hawaii International Conference on System Sciences* (pp. 513–522). 10.24251/HICSS.2018.066

Stiglitz, J. (2009). The global crisis, social protection and jobs. *International Labour Review*, *148*(1–2), 1–13. doi:10.1111/j.1564-913X.2009.00046.x

Stoecker, R. (1999). Are Academics Irrelevant?: Roles for Scholars in Participatory Research. *The American Behavioral Scientist*, *42*(4), 840–854. doi:10.1177/00027649921954561

Stoklas, J. (2017). *Sicherheit im Schengen-Raum: Eine unendliche Datensammelei?* https://www.repo.uni-hannover.de/bitstream/handle/123456789/4587/Datensammelei.pdf?sequence=1&isAllowed=y

Stott, C. (2009). *Crowd Psychology & Public Order Policing: An Overview of Scientific Theory and Evidence*. School of Psychology - University of Liverpool.

Stott, C., Havelund, J., Lundberg, F., Khan, S., Joern, L., Hoggett, J., & Vestergren, S. (2016). *Policing football in Sweden: enabling an evidence-based approach.* ENABLE.

Strand, R., & Oughton, D. (2009). *Risk and Uncertainty as a Research Ethics Challenge*. National Committees for Research Ethics in Norway.

Streefkerk, J. W., Neef, M., Meesters, K., Pieneman, R., & van Dongen, K. (2014). HCI challenges for community-based disaster recovery. *International Conference on Digital Human Modeling and Applications in Health, Safety, Ergonomics and Risk Management.*

Sturges, D. L. (1994). Communicating through crisis: A strategy for organisational survival. *Management Communication Quarterly, 7*(3), 297–316. doi:10.1177/0893318994007003004

Substance Abuse and Mental Health Services Administration. (2014). Trauma-informed care in behavioral health services. Treatment Improvement Protocol (TIP) Series 57. Author.

Su, K.-C., Wu, H.-M., Chang, W.-L., & Chou, Y.-H. (2012). Vehicle-to-vehicle communication system through wi-fi network using android smartphone. *International conference on connected vehicles and expo (ICCVE)*, 191-196. 10.1109/ICCVE.2012.42

Svedin, L. (2012). *Accountability in Crises and Public Trust in Governing Institutions*. Routledge. doi:10.4324/9780203120149

Sweetser, K. D., & Metzgar, E. (2007). Communicating during crisis: Use of blogs as a relationship management tool. *Public Relations Review, 33*(3), 340–342. doi:10.1016/j.pubrev.2007.05.016

Takayama, T., Katoh, K., Imamura, F., Kawata, Y., Murata, S., & Takahashi, S. (2018). Chapter 4: Prevention and Mitigation of Tsunami Disasters. In Tsunami: To Survive from Tsunami 2nd edition, Advanced Series on Ocean Engineering Book 46, English Edition (p. 341). World Scientific Publishing Company.

Tanaka, K., & Itoh, M. (2003). Saigaiji ni tekikaku na kikenkaihikoudou wo michibikutameno jouhou communication [Communication and information inducing suitable danger-avoidance actions from disaster]. *Journal of Disaster Information Studies, 1*, 61–69.

Tandoc, E. C. Jr, & Takahashi, B. (2018). Journalists are humans, too: A phenomenology of covering the strongest storm on earth. *Journalism, 19*(7), 917–933. doi:10.1177/1464884916657518

Tavra, M., Racetin, I., & Peroš, J. (2019). Combining social media and authoritative data for crisis mapping: A case study of a wildfire reaching Croatian city of split. *International Archives of the Photogrammetry, Remote Sensing and Spatial Information Sciences - ISPRS Archives, 42*(3/W8), 415–420. doi:10.5194/isprs-archives-XLII-3-W8-415-2019

Tehrani, N. (2016). Extraversion, neuroticism and secondary trauma in Internet child abuse investigators. *Occupational Medicine, 66*(5), 403–407. doi:10.1093/occmed/kqw004 PMID:26928859

Terasaki, K., Fujibuchi, T., Murazaki, H., Kuramoto, T., Umezu, Y., Ishigaki, Y., & Matsumoto, Y. (2017). Evaluation of basic characteristics of a semiconductor detector for personal radiation dose monitoring. *Radiological Physics and Technology, 10*(2), 189–194. doi:10.100712194-016-0384-z

Tetrault Sirsly, C. A., & Lvina, E. (2019). From doing good to looking even better: The dynamics of CSR and reputation. *Business & Society, 58*(6), 1234–1266. doi:10.1177/0007650315627996

Thaichon, P., & Quach, T. N. (2015). From Marketing Communications to Brand Management: Factors Influencing Relationship Quality and Customer Retention. *Journal of Relationship Marketing, 14*(3), 197–219. doi:10.1080/15332667.2015.1069523

The Center for Humanitarian Data. (2019). *OCHA Data Responsibility Guidelines working draft*. https://centre.humdata.org/wp-content/uploads/2019/03/OCHA-DR-Guidelines-working-draft-032019.pdf

The Center for Humanitarian Data. (2020). *Guidance Note: Humanitarian Data Ethics.* https://centre.humdata.org/guidance-note-humanitarian-data-ethics/

Thelen, P. D., & Robinson, K. L. (2019). Crisis Communication in Institutions of Higher Education: Richard Spencer at the University of Florida. *Communication Quarterly*, *67*(4), 444–476. doi:10.1080/01463373.2019.1616586

Thieleman, K., & Cacciatore, J. (2014). Witness to suffering: Mindfulness and compassion fatigue among traumatic bereavement volunteers and professionals. *Social Work*, *59*(1), 34–41. doi:10.1093wwt044 PMID:24640229

Thomas, V., & López, R. J. A. D. B. E. W. P. S. (2015). *Global increase in climate-related disasters.* Academic Press.

Thompson, B., Mazer, J. P., Payne, H. J., Jerome, A. M., Kirby, E. G., & Pfohl, W. (2016). Social Media and Active Shooter Events: A School Crisis Communication Challenge. *Qualitative Research Reports in Communication.* Advance online publication. doi:10.1080/17459435.2016.1247111

Thompson, I. A., Amatea, E. S., & Thompson, E. S. (2014). Personal and contextual predictors of mental health counselors' compassion fatigue and burnout. *Journal of Mental Health Counseling*, *36*(1), 58–77. doi:10.17744/mehc.36.1.p61m73373m4617r3

Tien, J. M., Berg, D. J. J. o. s. s., & Engineering, s. (2003). *A case for service systems engineering.* Academic Press.

Tirkkonen, P., & Luoma-aho, V. (2011). Online authority communication during an epidemic: A Finnish example. *Public Relations Review*, *37*(2), 172–174. doi:10.1016/j.pubrev.2011.01.004

Torniainen-Holm, M., Pankakoski, M., Lehto, T., Saarelma, O., Mustonen, P., Joutsenniemi, K., & Suviaari, J. (2016). The effectiveness of email-based exercises in promoting psychological wellbeing and healthy lifestyle: A two-year follow-up study. *BMC Psychology*, *4*(21), 1–12. https://doi.org. doi:10.118640359-016-0125-4 PMID:27184251

Torous, J., Jän Myrick, K., Rauseo-Ricupero, N., & Firth, J. (2020). Digital mental health and COVID-19: Using technology today to accelerate the curve on access and quality tomorrow. *JMIR Mental Health*, *7*(3), e18848. doi:10.2196/18848 PMID:32213476

Trump calls WHO's global death rate from coronavirus 'a false number.' (2020, March 5). *The Guardian.* Retrieved from https://bit.ly/2yNALff

Tsai, M.-H., Chen, J. Y., & Kang, S.-C. J. W. (2019). *Ask Diana: A Keyword-Based Chatbot System for Water-Related Disaster Management.* Academic Press.

Tsuchiya, K., Moritake, T., Ishigaki, Y., & Kosukegawa, N. (2018a). *Development of a radiation-detection simulator with smartphones and beacons for first responders against radiological threat* [Paper presentation]. 4th International CBRNe Workshop, Rome, Italy.

Tsuchiya, K., Moritake, T., Ishigaki, Y., Kosukegawa, N., Uehara, K., Matsumoto, Y., Tanabe, K., Kurosawa, K., Akiba, N., Kakuda, H., Imoto, D., Hirabayashi, M., Hawai, Y., & Kuroki, K. (2018b). *Development of effective training methods for first responders against radiological threat* [Paper presentation]. 8th European Academy of Forensic Science Conference, Lyon, France. 10.3408/jafst.759

Turco, M., Bedia, J., Di Liberto, F., Fiorucci, P., Von Hardenberg, J., Koutsias, N., Llasat, M.-C., Xystrakis, F., & Provenzale, A. (2016). Decreasing fires in mediterranean Europe. *PLoS One*, *11*(3), e0150663. Advance online publication. doi:10.1371/journal.pone.0150663 PMID:26982584

Turgoose, D., Ashwick, R., & Murphy, D. (2018). Systematic review of lessons learned from delivering teletherapy to veterans with post-traumatic stress disorder. *Journal of Telemedicine and Telecare*, *24*(9), 575–585. doi:10.1177/1357633X17730443 PMID:28958211

Turoff, M., van de Walle, B., & Hiltz, S. R. (Eds.). (2010). Emergency response information systems: Past, present, and future. In M. Turoff, B. van de Walle, & S. R. Hiltz (Eds.), Information systems for emergency management (pp. 369–388). Routledge. doi:10.4018/978-1-60960-609-1.ch002

Tursunbayeva, A., Franco, M., & Pagliari, C. (2017). Use of social media for e-Government in the public health sector: A systematic review of published studies. *Government Information Quarterly*, *34*(2), 270–282. doi:10.1016/j.giq.2017.04.001

Twigg, J. (2004). *Disaster risk reduction: mitigation and preparedness in development and emergency programming*. Academic Press.

Twigg, J. (2009). *Characteristics of a disaster-resilient community: a guidance note* (version 2). Academic Press.

Tyler, T., & Fagan, J. (2008). Legitimacy and Cooperation: Why Do People Help the Police Fight Crime in Their Communities? *Journal of Criminal Law*, *6*, 231–275.

United State Bureau of Labor Statistics. (2019). *Labor force statics from the current population survey*. https://www.bls.gov/cps/cpsaat11.htm

United States Environmental Protection Agency (US-EPA). (1992). *Framework for ecological risk assessment* (Report no. EPA/630/R-92/001). Risk Assessment Forum, United States Environmental Protection Agency. doi:10.2458/azu_geo_arizona_epasuperfund_1997

USA Today. (2020). *Jakarta floods 2020: Flooding, rain submerges Indonesia capital*. https://eu.usatoday.com/story/news/world/2020/02/25/jakarta-floods-2020-indonesia-capital/4865913002/

Utz, S., Schultz, F., & Glocka, S. (2013). Crisis communication online: How medium, crisis type and emotions affected public reactions in the Fukushima Daiichi nuclear disaster. *Public Relations Review*, *39*(1), 40–46. doi:10.1016/j.pubrev.2012.09.010

Vafeiadis, M., Bortree, D., Buckley, C., Diddi, P., & Xiao, A. (2018, December). (2019). Refuting fake news on social media: Nonprofits, crisis response strategies and issue involvement. *Journal of Product and Brand Management*. Advance online publication. doi:10.1108/jpbm-12-2018-2146

Vahdat, A., & Becker, D. (2000). Epidemic routing for partially connected ad hoc networks. *Technical Report CS-200006*, 1-14.

Vala, J. (1986). A análise de conteúdo. In *Metodologia das Ciências Sociais, 101-128*. Edições Afrontamento.

Valente, M. M. (2014). *Teoria Geral do Direito Policial* (4th ed.). Almedina.

Van de Weijer, B. (2013, July 13). Een onderzoeker in ieders broekzak: Smartphones in het regenwoud [A researcher in everyone's pocket: Smartphones in the rainforest]. *de Volkskrant*, 32–33.

van den Homberg, M., Meesters, K., & Van de Walle, B. (2014). *Coordination and information management in the Haiyan response: Observations from the field*. Academic Press.

van der Meer, T. G., & Jin, Y. (2020). Seeking Formula for Misinformation Treatment in Public Health Crises: The Effects of Corrective Information Type and Source. *Health Communication*, *35*(5), 560–575. doi:10.1080/10410236.2019.1573295 PMID:30761917

Van Dijck, J., Poell, T., & de Waal, M. (2018). Healthcare and health research. In *The platform society. Public values in a connective world*. Oxford University Press.

van Laere, J., & Lindblom, J. (2019). Cultivating a longitudinal learning process through recurring crisis management training exercises in twelve Swedish municipalities. *Contingencies and Crisis Management*, 38–49, doi:10.1111/1468-5973.12230

Van Zoonen, W., & van der Meer, T. (2015). The importance of source and credibility perception in times of crisis: Crisis communication in a socially mediated era. *Journal of Public Relations Research, 27*(5), 371–388. doi:10.1080/1062726X.2015.1062382

Vannini, A. (2009). Interpretive Theory. In Encyclopedia of Communication Theory, (pp. 557-562). SAGE. doi:10.4135/9781412959384.n209

Veil, S. R., Buehner, T., & Palenchar, M. J. (2011). A work-in-process literature review: Incorporating social media in risk and crisis communication. *Journal of Contingencies and Crisis Management, 19*(2), 110–122. doi:10.1111/j.1468-5973.2011.00639.x

Veil, S. R., Sellnow, T. L., & Petrun, E. L. (2012). Hoaxes and the Paradoxical Challenges of Restoring Legitimacy Dominos' Response to Its YouTube Crisis. *Management Communication Quarterly, 26*(2), 322–345. doi:10.1177/0893318911426685

Velde, B., Meijer, A., & Homburg, V. (2015). Police message diffusion on Twitter: Analysing the reach of *social media* communications. *Behaviour & Information Technology, 34*(1), 4–16. doi:10.1080/0144929X.2014.942754

Viegas, X., Almeida, M. F., Ribeiro, M., Almeida, M. A., Raposo, J., Viegas, M. T., … Viegas, C. X. (2019). *Analysis of the Forest Fires Ocurred in October 15, 260.* Retrieved from https://www.portugal.gov.pt/download-ficheiros/ficheiro.aspx?v=c2da3d7e-dcdb-41cb-b6ae-f72123a1c47d

Vieweg, S., Hughes, A. L., Starbird, K., & Palen, L. (2010). Microblogging during two natural hazards events: what twitter may contribute to situational awareness. In *Proceedings of the SIGCHI conference on human factors in computing systems*. ACM. 10.1145/1753326.1753486

Virmani, M. (2015). Understanding DevOps & bridging the gap from continuous integration to continuous delivery. *Fifth International Conference on the Innovative Computing Technology (INTECH 2015).*

Vitell, S., King, R., Howie, K., Toti, J., Albert, L., Hidalgo, E., & Yacout, O. (2016). Spirituality, moral identity, and consumer ethics: A multi-cultural study. *Journal of Business Ethics, 139*(1), 147–160. doi:10.100710551-015-2626-0

Vormittag, E. M. P. A. A., Oliveira, M. P., & Gleriano, J. S. (2018). Avaliação de Saúde da População de Barra Longa afetada pelo Desastre de Mariana, Brasil. *Ambiente & Sociedade, 21*, 1–22. doi:10.1590/1809-4422asoc0122r2vu18l1ao

Voss Horrell, S. C., Holohan, D. R., Didion, L. M., & Vance, G. T. (2011). Treating traumatized OEF/OIF veterans: How does trauma treatment affect the clinician? *Professional Psychology, Research and Practice, 42*(1), 79–86. doi:10.1037/a0022297

Waegemakers-Schiff, J., & Lane, A. M. (2019). PTSD symptoms, vicarious traumatization, and burnout in front line workers in the homeless sector. *Community Mental Health Journal, 55*(3), 454–462. doi:10.100710597-018-00364-7 PMID:30684127

Wagner, T., Lutz, R., & Weitz, B. (2009). Corporate hypocrisy: Overcoming the threat of inconsistent corporate. *Journal of Marketing, 73*(November), 77–91. doi:10.1509/jmkg.73.6.77

Wakefield, B. J., Turvey, C. L., Nazi, K. M., Holman, J. E., Hogan, T. P., Shimada, S. L., & Kennedy, D. R. (2017). Psychometric Properties of Patient-Facing eHealth Evaluation Measures: Systematic Review and Analysis. *Journal of Medical Internet Research, 19*(10), e346. Advance online publication. doi:10.2196/jmir.7638

Wallace, R., Costello, L., & Devine, A. (2018). Netnographic Slog: Creative Elicitation Strategies to Encourage Participation in an Online Community of Practice for Early Education and Care. *International Journal of Qualitative Methods*, *17*(1), 1–13. doi:10.1177/1609406918797796

Wallach, D., & Scholz, S. C. (2012). User-centered design: Why and how to put users first in software development. In A. Maedche, A. Botzenhardt, & L. Neer (Eds.), *Software for people* (pp. 11–38). Springer. doi:10.1007/978-3-642-31371-4_2

Wang, H. (2015). A new approach to network analysis for brand positioning. *International Journal of Market Research*, *57*(5), 727–742. doi:10.2501/IJMR-2015-060

Wang, Y. (2016). Brand crisis communication through social media: A dialogue between brand competitors on Sina Weibo. *Corporate Communications*, *21*(1), 56–72. doi:10.1108/CCIJ-10-2014-0065

Wang, Y. (2019a). User experiences in live video streaming: A netnography analysis. *Internet Research*, *29*(4), 638–658. doi:10.1108/IntR-01-2018-0029

Wang, Y. (2019b). Virtual cohabitation in online dating sites: A netnography analysis. *Online Information Review*, *43*(4), 513–530. doi:10.1108/OIR-11-2016-0338

Watson, I., & Jeong, S. (2020). *Coronavirus mobile apps are surging in popularity in South Korea.* https://edition.cnn.com/2020/02/28/tech/korea-coronavirus-tracking-apps/index.html

Weber, M. J. E. j. o. o. r. (1987). *Decision making with incomplete information.* doi:10.1109/INTECH.2015.7173368

Wegmann, A., Regev, G., Garret, G.-A., & Maréchal, F. (2008). Specifying services for ITIL service management. *2008 International Workshop on Service-Oriented Computing: Consequences for Engineering Requirements.*

Weick, K. E. (1987). Organizational culture as a source of high reliability. *California Management Review*, *29*(2), 112–127. doi:10.2307/41165243

Weijman, S., & Meesters, K. (2020). Shifting Control and Trust: Exploring Implications of Introducing Delegated Decision Support Systems. *Proceedings of the 17th ISCRAM Conference.*

Wenzel, M., Stanske, S., & Lieberman, M. B. (2020). Strategic responses to crisis. *Strategic Management Journal*, *41*, V7–V18.

White, C., & Plotnick, L. (2010). A framework to identify best practices: Social media and web 2.0 technologies in the emergency domain. *International Journal of Information Systems for Crisis Response and Management*, *2*(3), 25–35. doi:10.4018/jiscrm.2010070102

WHO chief warns that countries are not taking coronavirus seriously: 'This is not a drill.' (2020, March 6). *France24*. Retrieved from https://bit.ly/2S98uXH

Wiegand, S., & Middleton, S. E. (2016). Veracity and velocity of social media content during breaking news: Analysis of November 2015 Paris shootings. In *Proceedings of the 25th international conference companion on world wide web.* International World Wide Web Conferences Steering Committee 10.1145/2872518.2890095

Wietrzyk, B., Radenkovic, M., & Kostadinov, I. (2008). Practical MANETs for pervasive cattle monitoring. *Seventh International Conference on Networking*, 14-23. 10.1109/ICN.2008.78

Williams, K., Chatterjee, S., & Rossi, M. J. E. j. o. i. s. (2008). *Design of emerging digital services: a taxonomy.* Academic Press.

Williams, K. (2003). Peel's Principles and their acceptance by American Police: Ending 175 years of reinvention. *The Police Journal*, *76*(2), 97–120. doi:10.1177/0032258X0307600202

Willis, N. G., & Molina, V. (2019). Self-care and the social worker: Taking our place in the code. *Social Work, 64*(1), 83–86. doi:10.1093wwy049 PMID:30365017

Win-Shwe, T. T., Thein, Z. L., Aunf, W. Y., Yi, E.E.P.N, Maung, C., Nway, N. C., Thant, Z., Suzuki, T., Mar, O., Ishigaki, Y., & Nakajima, D. (in press). Improvement of GPS-attached Pocket PM2.5 measuring device for personal exposure assessment. *Journal of University of Occupational and Environmental Health.*

World Health Organization. (2006). *Fuel for life: Household energy and health.* World Health Organization. https://apps.who.int/iris/handle/10665/43421

World Health Organization. (2009). *Pandemic (H1N1) 2009 - update 76.* Retrieved from: www.who.int/csr/don/2009_11_27a/en/index.html

World Health Organization. (2009). *Public Heath Surveillance.* Retrieved from: https://www.who.int/immunization/monitoring_surveillance/burden/vpd/en/

World Health Organization. (2018a, May 2). *9 out of 10 people worldwide breathe polluted air, but more countries are taking action.* doi:10.1163/2210-7975_hrd-9841-20180002

World Health Organization. (2018b). *WHO Global Ambient Air Quality Database (update 2018).* World Health Organization. https://www.who.int/airpollution/data/cities/en/

World Information Service On Energy. (2020). *Chronology of major tailings dam failures.* Retrieved from: https://www.wise-uranium.org/mdaf.html

Wray, L. O., Pikoff, E., King, P. R., Hutchison, D., Beehler, G. P., & Maisto, S. A. (2016). Veterans' mental health beliefs: Facilitators and barriers to primary care-mental health use. *Families, Systems & Health, 34*(4), 404–413. doi:10.1037/fsh0000231 PMID:27819439

Wright, D. K., & Hinson, M. D. (2009). An updated look at the impact of social media on public relations practice. *The Public Relations Journal, 3*(2), 1–27.

Wukich, C. (2016). Government social media messages across disaster phases. *Journal of Contingencies and Crisis Management, 24*(4), 230–243. doi:10.1111/1468-5973.12119

Xharavina, N., Kapoulas, A., & Miaoulis, G. (2019). Netnography as a marketing research tool in the fashion industry in Southeast Europe. *International Journal of Market Research, 147078531985921.* Advance online publication. doi:10.1177/1470785319859210

Xing, B., Seada, K., & Venkatasubramanian, N. (2009). *An experimental study on wi-fi ad-hoc mode for mobile device-to-device video delivery. IEEE INFOCOM Workshops.* doi:10.1109/INFCOMW.2009.5072111

Xu, D., Zhuang, L., Deng, X., Qing, C., & Yong, Z. (2020). Media exposure, disaster experience, and risk perception of rural households in earthquake-stricken areas: Evidence from rural China. *International Journal of Environmental Research and Public Health, 17*(9), 3246. doi:10.3390/ijerph17093246 PMID:32384741

Xu, J. (2020). Does the medium matter? A meta-analysis on using social media vs. traditional media in crisis communication. *Public Relations Review, 46*(4), 10947. doi:10.1016/j.pubrev.2020.101947

Yaakop, M. B., Abd Malik, I. A., bin Suboh, Z., Ramli, A. F., & Abu, M. A. (2017). Bluetooth 5.0 throughput comparison for internet of thing usability a survey. In *International Conference on Engineering Technology and Technopreneurship (ICE2T)* (pp. 1-6). IEEE. 10.1109/ICE2T.2017.8215995

Yang, C., & Tian, W. (2017). Social media geo-sensing services for EO missions under sensor web environment: Users sensing information about the Ya'an earthquake from Sina Weibo. In *2017 6th International Conference on Agro-Geoinformatics* (pp. 1–6). IEEE. 10.1109/Agro-Geoinformatics.2017.8047032

Yatbaz, H., Cinar, B., Gokdemir, A., Ever, E., Al-Turjman, F., Nguyen, H., & Yazici, A. (2018). Hybrid approach for disaster recovery using P2P communications in android. *43rd Conference on Local Computer Networks Workshops (LCN Workshops)*, 46-52.

Yeager, C. M., & Benight, C. (2018). If we build it, will they come? Issues of engagement with digital health interventions for trauma recovery. *mHealth*, *4*, 37. doi:10.21037/mhealth.2018.08.04

Yi, E. E. P. N., Nway, N. C., Aung, W. Y., Thant, Z., Wai, T. H., Hlaing, K. K., Maung, C., Yagishita, M., Ishigaki, Y., Win-Shwe, T. T., Nakajima, D., & Mar, O. (2018). Preliminary monitoring of concentration of particulate matter (PM2.5) in seven townships of Yangon City, Myanmar. *Environmental Health and Preventive Medicine*, *23*(1), 53. doi:10.118612199-018-0741-0

Yin, R. K. (2009). *Case study research: Design and methods.* Sage Publications.

Yoda, T., Elisephane, I., Ishigaki, Y., Kim, M., Matsukawa, M., Matsumoto, Y., Munyegera, G., Yamaguchi, K., & Kondo, H. (2019). *Seven months observations of indoor air pollution in Kigali Rwanda* [Paper presentation]. *15th International Conference on Atmospheric Sciences and Applications to Air Quality*, Kuala Lumpur, Malaysia.

Yoo, S., Shin, Y., Kim, S., & Choi, S. (2014). Toward realistic WiFi simulation with smartphone "Physics". In *Proceeding of IEEE International Symposium on a World of Wireless, Mobile and Multimedia Networks* (pp. 1-6). IEEE.

Yoo, W., Choi, D.-H., & Park, K. (2016). The effects of SNS communication: How expressing and receiving information predict MERS-preventive behavioral intentions in South Korea. *Computers in Human Behavior*, *62*, 34–43. doi:10.1016/j.chb.2016.03.058 PMID:32288174

Young, C. L., & Flowers, A. (2012). Fight viral with viral: A case study of Domino's Pizza's crisis communication strategies. *Case Studies in Strategic Communication*, *1*, 93–106.

Yue, Y., Dong, K., Zhao, X., & Ye, X. (2019). Assessing wild fire risk in the United States using social media data. *Journal of Risk Research*, *0*(0), 1–15. doi:10.1080/13669877.2019.1569098

Zhang, C., Fan, C., Yao, W., Hu, X., & Mostafavi, A. (2019). Social media for intelligent public information and warning in disasters: An interdisciplinary review. *International Journal of Information Management*, *49*, 190–207. doi:10.1016/j.ijinfomgt.2019.04.004

Zhang, S., Pal, A., Kant, K., & Vucetic, S. (2018). Enhancing disaster situational awareness via automated summary dissemination of social media content. In *IEEE Global Communications Conference (GLOBECOM)* (pp. 1-7). IEEE. 10.1109/GLOCOM.2018.8647340

Zhong, X., Duckham, M., Chong, D., & Tolhurst, K. (2016). Real-time estimation of wildfire perimeters from curated crowdsourcing. *Scientific Reports*, *6*(1), 1–10. doi:10.1038rep24206 PMID:27063569

Zhouri, A., Oliveira, R., Zucarelli, M., & Vasconcelos, M. (2018). O desastre do Rio Doce: Entre as políticas de reparação e a gestão das afetações. In *Mineração,violências e resistências: um campo aberto à produção do conhecimento no Brasil*. Editorial Iguana.

Zhou, X., Snoswell, L. E., Harding, L. E., Bambling, M., Edrippulige, S., Bai, X., & Smith, A. C. (2020). The role of telehealth in reducing the mental health burden from COVID-19. *Telemedicine Journal and e-Health*, *26*(4), 1–3. doi:10.1089/tmj.2020.0068 PMID:32202977

Ziek, P. (2009). Making sense of CSR communication. *Corporate Social Responsibility and Environmental Management, 16*(3), 137–145. doi:10.1002/csr.183

Zimmerman, M. A., & Rappaport, J. (1988). Citizen participation, perceived control, and psychological empowerment. *American Journal of Community Psychology, 16*(5), 725–750. doi:10.1007/BF00930023

Zuboff, S. (2019). *Il capitalismo della sorveglianza. Il futuro dell'umanità nell'era dei nuovi poteri* (P. Bassotti, Trans.). Luiss. (Original work published 2018).

Zunino, G. (2020). *Coronavirus, app e sistemi per tracciare i positivi: come funzionano (nel mondo, in Italia).* Academic Press.

Zylberman, P. (2013). *Tempêtes microbiennes. Essai sur la politique de sécurité sanitaire dans le monde transatlantique.* Gallimard.

About the Contributors

Lídia Oliveira has a degree in Philosophy from the University of Coimbra (1990), a Master's degree in Educational Technology from the University of Aveiro, in partnership with the University of Valenciennes (France) and Mons (Belgium) (1995), and a Ph.D. in Sciences and Technologies of Communication from the University of Aveiro (2002). She has been a professor in the Department of Communication and Art (http://www.ua.pt) at the University of Aveiro since 1995 until the present. Her main scientific interest is in cyberculture studies and social network analysis. Her areas of interest are: Science and Technologies of Communication, Cultural Heritage and New Media, ICT and Education, Multimedia, Social Network Analysis, Cyberculture Studies, Sociology of Communication, Information Systems, Digital Libraries, Time Studies, Science Communication, Communication in the Scientific Community (theoretical and empirical studies). She has published both nationally and internationally in peer review journals, books, book chapters and conferences proceedings.

* * *

José Gabriel Andrade is Assistant Professor at the University of Minho. "European Doctor in Communication Sciences" by the Catholic University of Portugal. He received the title of master in Communication Sciences: Communication, Organization and New Technologies by the Catholic University of Portugal.

Michele Antunes is a journalist, Master in Public Health Epidemiology, PhD in Collective Health and Post-Doctorate in Postgraduate Program in Collective Health by the Federal University of Espírito Santo, Brazil, Profix / Fapes scholarship holder.

Klas Backholm (PhD) is a University Lecturer in Mass Communication, Political Science with Mass Communication, Åbo Akademi University (Finland). His research areas include: (a) the psychological wellbeing of journalists after work-related crisis exposure or harassment; and (b) best practices for crisis journalism and communication in the digital media landscape.

Belem Barbosa is Adjunct Professor at the University of Aveiro and member of GOVCOPP, the research unit on Governance, Competitiveness and Public Policy. She received her PhD in Business and Management Studies – specialisation in Marketing and Strategy from the University of Porto, Portugal. Her research interests lie primarily in the area of internet marketing and consumer behaviour, including word-of-mouth communication, sustainability marketing, tourism and events marketing.

Dana Branson, PhD, LCSW, is an assistant professor at Southeast Missouri State University, and she teaches in the Department of Criminal Justice, Social Work, and Sociology. Before joining the university in 2016, Branson was a practicing licensed clinical social worker for 21 years, with her clinical focus being co-occurring substance use disorders, mental health, and trauma in women and adolescent populations. Branson's academic focus is trauma and its aftermath, both in primary and secondary settings. Branson currently serves on the editorial board of the American Psychological Association Journal Traumatology.

Cássia Carvalho graduated in Social Communication - Journalism with post-graduation in Marketing Strategies and MBA in Organizational Communication from the University of Caxias do Sul, Brazil. Master's in marketing from the University of Aveiro, Portugal.

Camelia Cmeciu is a professor at the Faculty of Journalism and Communication Studies, University of Bucharest, Romania and the editor-in-chief of Styles of Communication, indexed in ERIH PLUS, EBSCO, DOAJ etc. Her research interests cover international organizational and political communication, crisis and risk communication, social movements, semiotics. She is the author of Strategii persuasive în discursul politic (Persuasive Strategies in Political Discourse), Semiotici textuale (Textual Semiotics) or Tendințe actuale în campanile de relații publice (Present Tendencies in Public Relations Campaigns). She was the project manager of the "European citizens - (De)legitimators of a networking Europe" (e-Eurociti) project http://eeurociti.fjsc.unibuc.ro/. Her work on international public relations, organizational discourse or social movements appears in academic journals like Semiotica, Public Relations Review, Comunicar, European Legacy, Communication & Society, or Communication Today.

Chiara Davino (1994) is researcher and architect. She studied at Universidade Autónoma de Lisbona and graduated with honors at Iuav - Architecture University of Venice where she is currently collaborating with Lorenza Villani in a research grant on the use of digital platforms in political Italian context. She partecipated at 15. Architecture Biennale ("Poliorceticon", Venice Pavilion). With Lorenza Villani: "A contradictory reading of facts and urban spaces: the rhetoric of progress" in "Cartha Magazine" (2019); "In-security refuge. New configurations in the age of global emergencies" in "OFFICINA" (2020); "The society of control. Readings of spaces and urban facts through the paradigm of transparency" in "Elephant&Castle" (2020); "Net-self. Digital persona in times of emergency" in "Trans Magazine" (2020); "State of emergency. A reading of social space in Europe" at XII Seminário Internacional em Urbanismo (2020). Their investigation field ranges from the militarization of urban space, a consequence of normalization of state of emergency and exception, to the narrative and ideological construction that supports today's visual policy.

Paola De Rosa is PhD Student in "New Technologies and Frontiers of Law, Economy and Society" at Link Campus University of Rome where she teaches "Crisis Communication". At the same University, she has been Research Fellow in the framework of the national interest research project (PRIN) "Media and terrorisms. How communication and digital media affect the perception of safety and security". She attended a postgraduate specialization course in Corporate Communication and Public Affairs at "Il Sole 24 Ore" Business School. In 2011, she held her Master's Degree in Government Science and Public Communication at LUISS University of Rome discussing a dissertation titled "Publish responsibly. Ethics as management tool for the publishing sector". In the same University, she received her Bachelor's Degree in Political Science in 2009. From 2013 to 2016 she has been alternate Member of "Media and Minors Committee" at Ministry of Economic Development, dealing with analysis concerning the relationship between young people and new technologies.

Carolina Escudero joined the University of Missouri, USA, in 2006 bringing expertise in international journalism, gender perspective and health communication, particularly related to gender and human rights in the international context. Dr. Escudero has lived and taught in Argentina, pursued research and communication work as a Leonardo Da Vinci Fellow in Belgium, studied and worked in France and Spain; she also was a student/correspondent in Serbia. This broad international experience has made her especially sensitive to and interested in the ways that the media portrays victims/ survivors and how the media can influence the recovery process, an awareness that she brings to her research and teaching. Her preferred approach is action research (quantitative and enriched by qualitative insights). Her dissertation research assessed how media can induce altruistic behaviors. She earned her doctorate degree at the Universidad Kennedy, Argentina; her Master's degree in Gender Studies at the University of Barcelona, Spain, and her Bachelor's degree at the University Robert Schuman, Strasbourg, France. In the academic and formative field, she has obtained grants including the Leonardo Da Vinci, European Union (2001); Global Programs Research support for the TEB campaign USA (2016-2017); Women & Gender Studies Research & Creativity scholarship USA (2018); Solidarity, Communication and Resistance Award, Spain (2017). Publications, documentaries and media campaigns about Dr. Escudero's research have been published and disseminated in Argentina, Germany, Belgium, Bolivia, China, Spain and the United States.

Dario Fanara is a PhD in Communication, Research, Innovation and Expert in the Theory and Techniques of Television and Culture and Television Industries. Communication officer for a social security institution, over time he has participated in projects for the conception and implementation of online communication strategies for public administrations, companies and in the political sphere. At the moment, he is also engaged in training courses on self-promotion strategies through the web for freelancers. Author of several publications, other research interests concern the relationship between media and terrorism and the analysis of gender issues in a comparative perspective between Western and Arab media.

Nuno Ferreira is Police officer since 2007. He received the title of master in police sciences and public security by the Higher Institute of Police Sciences and Internal Security – Portugal.

Joachim Högväg is a researcher at the Experience Lab at Åbo Akademi University. He holds a master's degree in cognitive science from the University of Umeå, and works mainly with research and development projects in the field of user centered design, user experience and usability.

Yo Ishigaki received the Master of Fine Arts degree from Tama Art University Tokyo, Japan, in 2010 and the Ph.D. degree in engineering from National University of Electro-Communications, Tokyo, Japan, in 2014. From 2002 to 2013, he was a researcher with the Intelligent Systems Laboratory of SECOM Corporation, Tokyo, Japan. His research interest includes the participatory monitoring by smart environmental sensors, development of medical devices inspired from media arts and entertainment technologies. His awards and honors include the Japan International Prize Heisei Memorial Research

Grant (The Japan Prize Foundation), Minister of Education, Culture, Sports, Science and Technology Award (Japan Institute of Invention and Innovation), RedDot Design Award (Germany) and Good Design Award (Japan).

Shahriar Kaisar is currently working as a Lecturer in the Department of Information Systems and Business Analytics of RMIT University, Australia. He received his PhD degree in Information Technology from Monash University, Australia in 2018. Earlier, he completed his M.Sc in Computer Science from the University of Saskatchewan in 2012. His research interests are in mobile ad-hoc networks, smartphone usage, data analytics, and cybersecurity.

Jenny Lindholm (PhD) is a University Lecturer and Researcher in Political Science with Mass Communication, Åbo Akademi University. Her main research interests are how different types of crises and disasters affect society, as well as laboratory and experimental research.

Marcos Mayo-Cubero (PhD) is an Associate Professor and researcher at Nebrija University (Spain) and Director of the MA in Television Journalism - Antena 3. PhD in Communication Sciences and MPhil from Complutense University (Spain). BA in Journalism from Carlos III University. He has collaborated as a lecturer at King Juan Carlos University, Complutense University, and Carlos III University. As an early-career researcher, he has published articles in JCR scientific journals, indexed in WOS and Scopus. He has also presented papers at international and national academic conferences. His research strands are journalistic practice, crisis communication, political information, and misinformation in social media. He received his PhD 'summa cum laude' with the doctoral thesis: 'Journalistic specialization as a strategic tool in disaster communication.' He worked as a TV journalist for over twelve years at one of Spain's most relevant news outlets, laSexta Noticias (2006-18). News coverage for the newscasts and informative programs (Al Rojo Vivo, Más Vale Tarde, laSexta Columna). Specialization in political and economic news. Investigative journalism in corruption cases, among them, in 'Panama Papers.' He coordinated as deputy head of the international news desk of laSexta for three years. He has also worked as a journalist for Spanish public TVE, radio Cadena SER, and the economic magazine Capital.

Kenny Meesters is a researcher in humanitarian information management at Delft University of Technology. In his research, he specializes in information management during crisis response and disaster management. He examines how information can be used to empower both responders and (affected) communities. In his research, he uses a combination of methods including serious gaming and field research.

José Manuel Mendes holds a PhD in Sociology by the Faculty of Economics of the University of Coimbra, where he is a tenured Associate Professor with Habilitation. He is also senior researcher at the Centre for Social Studies, where he has been working in the fields of risk and social vulnerability, planning, public policies and citizenship. He is coordinator of the Risk Observatory (OSIRIS) of the Centre for Social Studies and is the Editor-in-Chief of Revista Crítica de Ciências Sociais.

Daniel Ikesinachi Nwogwugwu is a lecturer in Communication Arts Programme, College of Computing and Communication Studies, Bowen University, Iwo, Nigeria. He is also a doctoral student in the Department of Communication and Language Arts, University of Ibadan, Nigeria. His area of specialization is in Applied Communication.

Adauto Oliveira is a PhD in Public Health - ENSP / Fiocruz, Full Professor at the Federal University of Espírito Santo, Brazil, and the Postgraduate Program in Public Health - Brazil. Researcher at PRINT - Institutional Program for Internationalization. CAPES / PRINT - Notice no. 41/2017.

Paola Primo graduated in Social Communication, Master in Public Management and PhD student in the Postgraduate Program in Public Health at the Federal University of Espírito Santo, Brazil.

Mariela Ramos graduated in Speech Therapy from Universidade Vila Velha. Master in Public Health from the Federal University of Espírito Santo. Specialist in Audiology and Health of the Worker. Public servant at the State Health Secretariat of Espírito Santo currently working at the Capixaba Institute for Teaching, Research and Innovation in Health.

Carlos Eduardo Siqueira graduated in Medicine at the Federal University of Rio de Janeiro, Master in Public Health at Johns Hopkins Bloomberg School of Public Health and Doctorate in Work Environment Policy at the University of Massachusetts Lowell. He is currently an Associate Professor at the School for the Environment at the University of Massachusetts Boston. Dr. Siqueira has had a long career in Public Health, with an emphasis on occupational and environmental health policy. He has conducted research in the following areas: occupational and environmental health policy, occupational and environmental health, health policy, international health, and Brazilian immigrant health in the United States.

Marica Spalletta (PhD) is Associate Professor in Media Sociology at Link Campus University (Rome), where she teaches "Media, Politics and Public Opinion" and "Public Communication". In the same University, she is the scientific manager of Link LAB (Social Research Centre) and member of the Academic Board of the Ph.D. Course in "New technologies and frontiers of law, economy and society". Her researches focus on mediatization of cultural processes and hybrid journalism; the relationship between political system and media system and its effects on political communication; the credibility issue and its effects on development of public opinion. Over the last twenty years, she has participated and directed several researches focuses on these topics, in partnership with Universities, associations, organizations and Institutions. She was the scientific director of the LCU research unit in the framework of the national interest research project focused on "Media and terrorism. The social effect of communication and digital network on the perceived insecurity". She is also participating in the LCU research unit in the framework of the H2020 research project "DETECt – Detecting Transcultural Identity in European Popular Crime Narratives".

Kenji Tanaka (M'08) received the B.Sc. degree in Science from Kyoto University, Kyoto, Japan, the M.Sc. and D.Sc. degrees in system science from the Tokyo Institute of Technology, Tokyo, Japan. He is a Professor with the Graduate School of Informatics and Engineering, University of Electro-Communications, Tokyo. His current research interests include system safety, risk management, medical safety, reliability and trust. Prof. Tanaka was a Vice-Chair of the IEEE-SMC Japan Chapter from 2003 to 2004.

Lorenza Villani (1993) is a researcher and an architect. She studied at the American University of Beirut and she graduated with honors at Iuav - Architecture University of Venice where she is currently collaborating with Chiara Davino in a research grant on the use of digital platforms in political Italian context. In 2016, she was co-curator of La città In-forme (Altamura, Italy). With Chiara Davino: "A contradictory reading of facts and urban spaces: the rhetoric of progress" in "Cartha Magazine" (2019); "In-security refuge. New configurations in the age of global emergencies" in "OFFICINA" (2020); "The society of control. Readings of spaces and urban facts through the paradigm of transparency" in "Elephant&Castle" (2020); "Net-self. Digital persona in times of emergency" in "Trans Magazine" (2020); "State of emergency. A reading of social space in Europe" at XII Seminário Internacional em Urbanismo (2020). Their investigation field ranges from the militarization of urban space, a consequence of normalization of state of emergency and exception, to the narrative and ideological construction that supports today's visual policy.

Yan Wang is an information management researcher and data steward at Delft University of Technology with an extensive business process management background in international settings. She is specialized in information / data management & coordination, customer-centric requirement engineering, serious gaming, simulation exercise design, and digital service management.

Index

E

F

G

H

I

J

K

L

M

N

O

P

Q

R

S

www.ingramcontent.com/pod-product-compliance
Lightning Source LLC
LaVergne TN
LVHW080009210926
848481LV00011B/2978
* 9 7 8 1 7 9 9 8 6 7 0 6 7 *